Mixed Effects Models for Complex Data

MONOGRAPHS ON STATISTICS AND APPLIED PROBABILITY

General Editors

V. Isham, N. Keiding, T. Louis, R. L. Smith, and H. Tong

Monographs on Statistics and Applied Probability 113

Mixed Effects Models for Complex Data

Lang Wu
University of British Columbia
Vancouver, Canada

CRC Press
Taylor & Francis Group
Boca Raton London New York

CRC Press is an imprint of the
Taylor & Francis Group, an **informa** business
A CHAPMAN & HALL BOOK

CRC Press
Taylor & Francis Group
6000 Broken Sound Parkway NW, Suite 300
Boca Raton, FL 33487-2742

First issued in paperback 2019

© 2010 by Taylor & Francis Group, LLC
CRC Press is an imprint of Taylor & Francis Group, an Informa business

No claim to original U.S. Government works

ISBN-13: 978-1-4200-7402-4 (hbk)
ISBN-13: 978-0-367-38491-3 (pbk)

Library of Congress Cataloging-in-Publication Data

Wu, Lang, 1963-
   Mixed effects models for complex data / Lang Wu.
      p. cm. -- (Monographs on statistics and applied probability ; 113)
   Includes bibliographical references and index.
   ISBN 978-1-4200-7402-4 (hardcover : alk. paper)
   1. Multilevel models (Statistics) 2. Mathematical statistics--Longitudinal methods. I.
Title. II. Series.

QA278.W85 2009
519.5'4--dc22   2009038084

Visit the Taylor & Francis Web site at
http://www.taylorandfrancis.com

and the CRC Press Web site at
http://www.crcpress.com

To Xiaowei and Mengmeng

Contents

Preface

Mixed effects models or random effects models have been widely used in the analysis of longitudinal data, clustered data, and other correlated data. In practice, especially in longitudinal studies, data are often complex or incomplete in the sense that there may be dropouts, missing data, measurement errors, censoring, and outliers. Ignoring any of these problems or the use of naive methods in data analysis may lead to biased or misleading results. In other words, standard mixed effects models may not be directly applicable for complex data. In this book, we first provide an overview of commonly used mixed effects models, including linear mixed effects (LME) models, generalized linear mixed models (GLMMs), nonlinear mixed effects (NLME) models, frailty models, and semiparametric and nonparametric mixed effects models. Then we discuss appropriate approaches to address missing data, measurement errors, censoring, and outliers in mixed effects models.

A mixed effects model may be viewed as an extension of the corresponding classical regression model for cross-sectional data by introducing random effects in the model. These random effects may be viewed as individual effects or cluster effects, which facilitate *individual-specific* inference, and they also incorporate *correlation* between the repeated measurements within each individual/cluster since each individual or cluster shares the same random effects. In this book, before introducing each class of mixed effect models we first review the corresponding class of regression models for cross-sectional data to make the link between them clear.

Likelihood methods are standard inferential tools for mixed effects models, and they are also convenient and useful tools for various incomplete data problems. Therefore, in this book we mainly focus on likelihood methods, although we discuss other approaches as well. For likelihood methods, a major challenge here is the associated computation since the likelihood for a mixed effects model typically involves an intractable multi-dimensional integral. Thus, we also discuss various computational strategies for likelihood estimations of mixed effects models.

Generalized estimating equation (GEE) methods are alternative and popular approaches in the analysis of longitudinal or clustered data. An advantage of

GEE methods is that they require no distributional assumptions, but GEE estimates may be less efficient than likelihood estimates if distributional assumptions hold. Since distributional assumptions are often made in mixed effects models, in this book we only provide relatively brief descriptions of GEE methods and refer interested readers to other books for more detailed discussions. Similarly, although Bayesian methods are useful and popular, we only provide relatively brief descriptions of Bayesian mixed effects models in this book, due to space consideration, and refer interested readers to other books for more detailed discussions.

The book begins with an introduction, some real data examples, and an outline of general approaches to the analysis of longitudinal or clustered data and incomplete data. In Chapter 2, we provide an overview of commonly used mixed effects models with some details. General approaches to the analysis of complex data with missing values, measurement errors, censoring, and outliers are reviewed in Chapter 3. The first three chapters provide an overview of general models and methods, with motivating examples and illustrations. From Chapter 4 to Chapter 11, each chapter focuses on a specific topic, and readers may skip some of these chapters since each chapter is relatively self-contained. Chapter 12 is an Appendix which provides some background materials such as likelihood methods and the Gibbs sampler.

The book should be accessible to graduate students or senior undergraduate students in statistics, practitioners, researchers in statistics and biostatistics, and researchers in other fields with particular interest in mixed effects models. The reader is assumed to have a basic knowledge of statistical methods and some familiarity with regression models. The mathematics level is mostly intermediate, at the level of first-year college calculus and linear algebra.

I am grateful to a number of people who have provided help and encouragements. I would like to thank all colleagues, as well as staff members and graduate students, in the Department of Statistics at the University of British Columbia (Vancouver) for providing a wonderful environment for research and many helps. In particular, Jim Zidek initially suggested that I write this book and has provided encouragement and useful suggestions and comments, and Paul Gustafson provided valuable suggestions and help, especially in the early stage. I would also like to thank Michael Perlman, my Ph.D. supervisor at the University of Washington, for his continuous advice and encouragement, and Hulin Wu and Victor DeGruttola for introducing me to HIV/AIDS research which motivated many of the materials in the book. I also thank Rob Calver, Sarah Morris, and Amber Donley at Chapman and Hall/CRC for their help. Finally, I thank the Natural Sciences and Engineering Research Council of Canada for research grant support.

CHAPTER 1

Introduction

1.1 Introduction

Longitudinal data are very common in practice, either in observational studies or in experimental studies. In a longitudinal study, individuals in the study are followed over a period of time and, for each individual, data are collected at multiple time points. For example, each individual's blood pressure may be measured repeatedly over time, or each student takes multiple quizzes throughout a semester. See Section 1.2 for a more detailed discussion of longitudinal data and Section 1.3 for some real-data examples. Thus, the defining feature of a *longitudinal study* is that multiple or repeated measurements of the same variables (e.g., blood pressure, quiz score) are made for each individual in the study over a period of time.

In economics and sociology, longitudinal studies are often called *panel studies*. A key characteristic of longitudinal data or panel data is that the multiple measurements of a variable on the same individuals are *correlated*, although data from different individuals are usually assumed to be independent. A major advantage of a longitudinal study is that it allows us to study *changes of variables over time*.

Longitudinal studies are closely related to *repeated measures studies*, in which repeated or multiple measurements of one or more variables are made on each individual in the study but these repeated measurements are not necessarily made over time. For example, air pollution may be measured at different locations of a city, so multiple measurements of air pollution are made over space at different cities.

More generally, longitudinal data or panel data or repeated measurement data are all examples of *clustered data* in which data fall into different clusters. For clustered data, data within the same cluster may be correlated, but data between different clusters are often assumed to be independent. For example, in a multi-center study, each center may be viewed as a cluster and data collected within the same center may be correlated, but data from different centers are usually assumed to be independent. *In a longitudinal study, each individual may be viewed as a cluster.* Longitudinal data, repeated measurement data, and clustered data are all in the class of *correlated data.* In the analysis of correlated data, the correlation should be incorporated in the analysis in order to avoid potential bias and loss of efficiency.

Longitudinal data are also related to *time series* in which a single long series of measurements are observed over time. In fact, when there is only one individual in a longitudinal study (i.e., when sample size is 1), longitudinal data reduce to a single time series. Compared with time series, a main advantage of longitudinal studies is that information can be borrowed across different individuals in statistical inference. Techniques for analyzing longitudinal data combine elements from multivariate methods and time series methods. Compared with classical multivariate analysis, methods for longitudinal data allow study of change over time and allow unbalanced data.

If the time dimension in a longitudinal setting is replaced by one or more spatial dimensions, we have *spatial data.* For example, air pollution may be measured over different regions of a city, and each region may be viewed as a cluster. Many of the techniques used to analyze longitudinal data may also be used for analyzing spatial data with minor modification.

Longitudinal studies differ from *cross-sectional studies.* In a cross-sectional study, only one measurement is made for a variable of interest for each individual in the study. At a *fixed* time point, data in a longitudinal study may be viewed as cross-sectional data. Statistical methods for the analysis of cross-sectional data are well developed, such as classical linear regression models and generalized linear models. An important assumption for cross-sectional data is that observations are *independent* of each other. Therefore, statistical methods for analyzing cross-sectional data cannot directly be used for analyzing longitudinal data or correlated data.

Statistical methods for the analysis of longitudinal data are similar to those for the analysis of clustered data or repeated measurement data. A common characteristic of these data is that observations within the same cluster may be correlated but observations between clusters are usually independent, which motivates most of the statistical methods for analyzing these correlated data. However, there are often complications in longitudinal data. For example, missing data or dropouts are very common in longitudinal studies, since one can hardly expect to collect all data at each time point.

Specifically, the following problems or complications arise frequently in practice:

- missing data or dropouts,

- measurement errors,
- censoring,
- outliers.

Extensive research has demonstrated that ignoring any one or more of the above problems in statistical analyses may lead to severely biased or misleading results. In other words, it is very important to address the above problems in data analyses in order to obtain reliable results. Although there are many books available for the analysis of longitudinal or clustered data, this book offers additional contributions to provide a relatively comprehensive descriptions of commonly used statistical methods to address these common problems in practice.

In this book we mainly focus on longitudinal data, but most methods can also be applied to analyses of other correlated data such as repeated measures data and clustered data. In Section 1.2, we discuss longitudinal data and clustered data in more details. In Section 1.3, we present several real-data examples that are used to motivated the models and methods throughout the book. Section 1.4 describes some commonly used regression models for longitudinal data or clustered data, including mixed effects models. A detailed description of mixed effects models is given in Section 1.5. In Section 1.6, we discuss various incomplete data problems which arise frequently in practice. Section 1.7 discusses some common statistical software. The notation used throughout the book is illustrated in Section 1.8. Section 1.8 also gives an outline of the book.

1.2 Longitudinal Data and Clustered Data

We first consider a simple example to illustrate characteristics of longitudinal data. Suppose that a researcher wishes to evaluate a treatment for reducing high blood pressure. The researcher would measure blood pressures of each subject in the study before and after the treatment. The researcher may also be interested in how blood pressures of the subjects *change over time* after the treatment, so he/she would measure blood pressures of each subject repeatedly after the treatment over a period of time, say once a month for 5 months. Thus, there are 6 repeated measurements of blood pressure on each subject, including the measurement before the treatment. For each subject, these 6 multiple measurements are likely to be correlated since they are taken on the same subject. For example, a subject with a very high initial blood pressure is likely to still have a high blood pressures in the follow-up measurements, compared with other subjects in the study.

In such a longitudinal study, the observed data are usually *incomplete*. For example, some subjects in the study may drop out early for various reasons such as treatment side effects. It is also known that blood pressures are usually *measured with errors*, i.e., the observed values may differ from the (unobserved) true values. To address the measurement errors or to increase measurement precision, blood pressures are usually taken repeatedly on each subject at each visit – these repeated measurements may be viewed as *clustered data* since they are taken on the same subjects at roughly the same time. Therefore, longitudinal data may also be *clustered*. Moreover, the treatment may be more effective for some subjects than others, so there may be substantial *variations* between the subjects. Finally, there may be *outliers*, such as subjects with unusually high or low blood pressures or data recorded with errors. These features are common in many longitudinal studies. In Section 1.3, we will provide more detailed examples in which these features are quite clear.

In other words, a longitudinal study typically has the following characteristics:

- there are more than one measurements on each individual, and the numbers of measurements and measurement times may vary across individuals, i.e., the observed data are often *unbalanced*;
- the within-individual repeated measurements may be *correlated*, although the measurements across individuals are usually assumed to be independent;
- there may be substantial *variation* among repeated measurements within each individual *and* among measurements between different individuals, i.e., there may be substantial variations among within-individual measurements and between-individual measurements;
- the observed data are often *complex* or *incomplete* in the sense that there may be missing data, dropouts, measurement errors, censoring, and outliers.

Because of the above special characteristics of longitudinal data, statistical models and methods for analyzing longitudinal data can be challenging.

One of the main objectives of statistical analyses is to address variations in the data. As noted earlier, for longitudinal data there are two sources of variations:

- *within-individual variation*, i.e., the variation in the repeated measurements within each individual;
- *between-individual variation*, i.e., the variation in the data between different individuals.

In longitudinal studies, modeling within-individual variation allows one to study change over time, while modeling between-individual variation allows one to understand the differences between individuals.

In many longitudinal studies, understanding the systematic variation between individuals receives great attention. Much of this variation may be explained by covariates such as age and gender. Therefore, regression models are often used to approximate the relationship between the longitudinal response and covariates. In a regression model for longitudinal data, the response contains longitudinal measurements, while the *covariates* or *explanatory variables* can either be longitudinal measurements, called *time-dependent covariates*, or variables which do not change over time, called *time-independent covariates* (e.g., gender and other *baseline covariates*). These covariates are used to partially explain the variations in the longitudinal responses. The responses are usually chosen based on study objectives.

The types of regression models are usually determined by the types of the response variables. For example, if the response is continuous and roughly normal, one may consider a normal linear regression model, and if the response is binary, one may consider a logistic regression model. See Section 1.4 for a more detailed discussion on regression models. In this book, we focus on a class of widely used regression models for longitudinal data or clustered data, called *mixed effects models* or *random effects models*. We present a detailed discussion of mixed effects models in Section 1.5. In the following, we give a brief overview of regression models for longitudinal data or clustered data.

The following three classes of models are commonly used for analyses of longitudinal data or clustered data:

- *mixed effects models* (or *random effects models*): in these models random effects are introduced to incorporate the between-individual variation and within-individual correlation in the data;

- *marginal models* (or *generalized estimating equations (GEE) models*): in these model the mean structure and the correlation (covariance) structure are modeled separately without distributional assumptions for the data;

- *transitional models*, in these models the within-individual correlation is modeled via Markov structures.

Diggle, Heagerty, Liang, and Zeger (2002) provided a comprehensive overview of various models for longitudinal data. Mixed effects models and GEE models are perhaps among the most popular. See Section 1.4.3 for a more detailed description of these models. This book focuses on mixed effects models since they are widely used in practice and they are also natural extensions of classic regression models for cross-sectional data. See Section 1.5 for a more detailed discussion of mixed effects models. A detailed discussion of GEE models is provided in Chapter 10.

There has been extensive research in statistical methods for longitudinal data or clustered data. Recent developments are reviewed in Diggle et al. (2002), Hedeker and Gibbons (2006), and Fitzmaurice et al. (2008), among others. Books exclusively focusing on mixed effects models include Davidian and

Giltinan (1995), Vonesh and Chinchilli (1997), Verbeke and Molenberghs (2001), Demidenko (2004), and McCulloch, Searle, and Neuhaus (2008).

1.3 Some Examples

In this section, we present several examples of longitudinal data or clustered data to show typical features of these data. These examples also motivate the models and methods presented in the rest of the book. Specifically, these examples show the following features of longitudinal data or clustered data: (i) longitudinal data may be *unbalanced*, i.e., the numbers and times of measurements may vary across individuals; (ii) there may be *correlation* between the within-individual or within-cluster measurements; (iii) there may be substantial between-individual and within-individual *variations*; (iv) *missing data* and *dropouts* are very common; (v) observed data may have *measurement errors*; (vi) some data may be *censored*; and (vii) there may be *outliers*.

In the following real-data examples, we will briefly discuss the importance of addressing missing data, measurement errors, censoring, and outliers in statistical analyses. Failing to address these issues appropriately may lead to severely biased or misleading results. More thorough discussions of each of these issues will be presented in later chapters, in which these examples will be used again to illustrate how biases may arise when naive methods are used and which valid methods can be used.

1.3.1 A Study on Mental Distress

This study investigates changes in subjects' mental distress over time in a treatment group and a control group. Mental distress in 239 subjects were measured at baseline, 4, 12, 24, and 60 months, based on their answers to questionnaires. All subjects are randomly assigned into two groups: a treatment group and a control group. A more detailed description of the dataset can be found in Murphy et al. (2003). The Global Severity Index (GSI), which is one of the most sensitive indicators of mental distress, is used to measure subjects' distress levels, with a higher GSI score indicating a higher level of mental distress. Other variables are also used to measure subjects' distress or depression levels and other characteristics of the subjects.

The main objectives of the study are to compare the treatment effect on reducing mental distress compared with the control group, to study how subjects' mental distress changes over time, and to investigate whether the changes in mental distress over time can be partially explained by the treatment and other covariates. Covariates in the study include subjects' education, annual income,

Table 1.1 *Summary statistics for some variables at baseline*

Variable	Mean	Standard Deviation
GSI score of subjects (0 – 10)	1.13	0.72
Education of subjects (in years)	13.70	2.36
Income of subjects (in $10,000)	4.68	1.90
Depression score of subjects (0 – 10)	1.55	0.99
Anxiety score of subjects (0 – 10)	1.23	0.92

Table 1.2 *Missing data rates for some variables over time*

Variable	baseline	3 months	6 months	18 months	60 months
GSI	0.04	0.14	0.19	0.36	0.39
Depression	0.03	0.13	0.19	0.36	0.39
Anxiety	0.03	0.13	0.19	0.36	0.39

depression, anxiety, etc. Summary statistics for some variables are given in Table 1.1. Table 1.2 shows missing data rates for some time-dependent variables. We see that missing data increase in late period of the study. At the end of the study (60 months or 5 years from the beginning of the study), nearly 40% data are missing, mostly due to dropouts. For a longitudinal study lasting several years, dropouts or missing data are almost inevitable. There may be various reasons for dropouts, such as subjects moving elsewhere and subjects finding the treatment ineffective.

Figure 1.1 shows the GSI scores for all subjects and for 10 randomly selected subjects respectively. We see that the GSI trajectories vary substantially across subjects. That is, there is a large between-individual variation, so a regression model with *individual-specific regression parameters* may be useful to model these data. Such a model is called a mixed-effects model or a random effects model in which random effects are included in the model to represent individual effects – see Section 1.5 for a detailed discussion.

The repeated measurements on each parent, or the within-individual measurements, may be correlated since they are measurements on the same individuals. For example, a subject with an higher than average initial GSI score is likely to have higher than average GSI scores in later measurements as well. Note that some variables such as depression scores and anxiety scores may be subject to measurement errors, since these measurements may be subjective and may be greatly influenced by subjects' emotional status at the times of measurements.

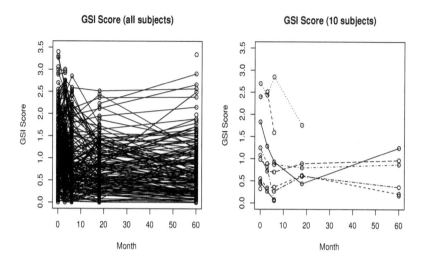

Figure 1.1 *GSI scores over time. Left figure: GSI scores for all subjects. Right figure: GSI scores for 10 randomly selected subjects. The open dots are the observed values.*

The repeated measurements of these variables allow us to at least partially address measurement errors (see Chapter 5). There seem also some outlying observations which do not follow the general patterns, so robust analysis may be needed.

Figure 1.2 displays boxplots of GSI scores from all subjects for the two groups at each measurement time. At each measurement time, the boxplot shows the *cross-sectional* summaries of GSI scores for the two groups, which allows us to compare the two groups at that particular time point using standard methods for cross-sectional data such as a t-test. On the other hand, all the four boxplots show the *longitudinal* trend of GSI scores, which allow us to study the changes of treatment effects *over time*. A formal method for longitudinal data analysis, such as a mixed effect model, allows us to evaluate the treatment effect and study its change over time *simultaneously*, so it is more desirable than separate analyses at each time point.

In this study, the measurement schedules are strictly *fixed* in advance for all subjects. However, there are many dropouts which lead to unbalanced observed data. In addition to dropouts, there are also missing data in some baseline variables such as income. As noted earlier, there may also be measurement errors and outliers. In later chapters, this example will be used again to show the consequences of ignoring any of these issues and how the results based on naive methods may be misleading. For example, the treatment effects may be mis-

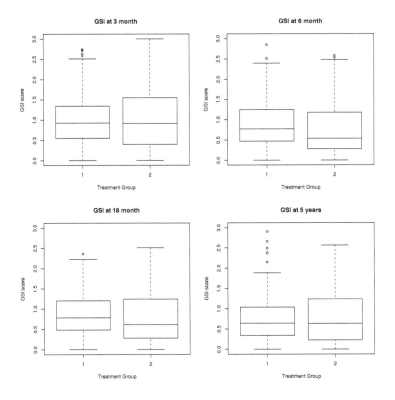

Figure 1.2 *Boxplots of GSI scores for all subjects at 3 months, 6 months, 18 months, and 5 years.*

leading if any of these issues are ignored or are handled inappropriately. Thus, it is important to address dropouts, missing data, measurement errors, and outliers simultaneously in order to avoid biased or misleading results.

1.3.2 An AIDS Study

In an AIDS study designed to evaluate an anti-HIV treatment, 53 HIV infected patients were treated with an antiviral regimen. *Viral load* (i.e., plasma HIV RNA) was repeatedly quantified on days 0, 2, 7, 10, 14, 21, and 28, and weeks 8, 12, 24, and 48 after initiation of the treatment. Immunologic markers known as CD4 and CD8 cell counts were also measured along with viral load, as well as some other variables. A more detailed description of this dataset can be found in Wu and Ding (1999). In this study, viral load has a lower detection limit of 100, i.e., viral loads below 100 are not quantifiable. Table 1.3 sum-

Table 1.3 *Summary statistics for viral load (RNA), CD4, and CD8 at five selected measurement times*

Variable	Day 2		Day 7		Day 14		Day 28		Day 56	
	Mean	S.D.[b]	Mean	S.D.	Mean	S.D.	Mean	S.D.	Mean	S.D.
RNA[a]	5.00	0.59	4.06	0.81	3.23	0.64	3.02	0.61	2.52	0.74
CD4	203	74	231	89	274	108	284	89	300	94
CD8	961	506	1026	643	1037	545	1086	627	1033	329

Note: a) RNA (viral load) is in $\log_{10}$ scale;   b) S.D.: standard deviation

Table 1.4 *Missing data rates for some variables at baseline*

Covariate	Definition	Missing Rate
AGE	age of the patient	0
WEIGHT	weight of the patient	0
LU20	NK activity	37.5%
TNF	plasma tumor necrosis factor	16.7%
APOP	% of cells that are apoptotic	0
CH50	complement CH50	18.75%
BIGG	gp120-binding IgG levels	22.92%
BIGC3	C3 binding to HIV-infected cells	27.08%

marizes the data for viral load, CD4, and CD8 measured at five selected time points.

Figure 1.3 shows viral load trajectories of all patients and six randomly selected patients respectively. Figure 1.4 shows CD4 and CD8 trajectories of six randomly selected patients. These figures indicate a number of features of this dataset: (a) different measurement times across patients; (b) different numbers of within-individual measurements across patients; (c) large variation between patients; (d) large variation in the data within each patient; (e) some patients dropping out of the study; (f) some viral loads being censored (i.e., below the limit of detection); (g) substantial measurement errors in the data; (h) complex long-term trajectories. (i) data being missing at measurement times. These features are common in many longitudinal studies, so statistical methods analyzing these data should address these features. Table 1.4 shows the missing data rates of some variables at baseline.

HIV viral dynamic models, which model viral load trajectories during an anti-HIV treatment, have received great attention in recent years (e.g., Ho et al. 1995; Perelson et al. 1996, 1997; Wu and Ding 1999; Wu 2005). These models

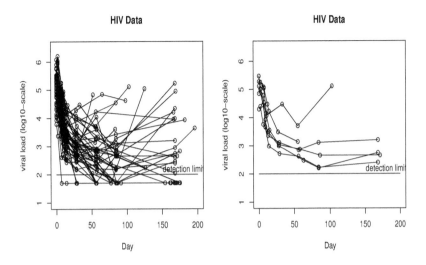

Figure 1.3 *Viral loads trajectories (in log_{10} scale). The open circles are observed values. The viral load detection limit in this study is $log_{10}(100) = 2$. Viral loads below the detection limit are substituted by half the limit.*

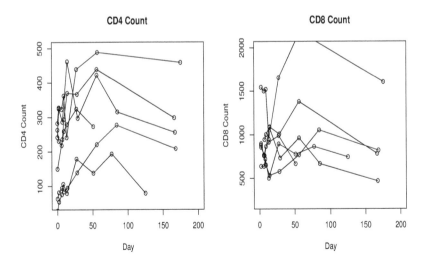

Figure 1.4 *CD4 Counts (left) and CD8 counts (right) of six randomly selected patients.*

describe the virus elimination and production processes during antiviral treatments. In an HIV viral dynamic model, the relationship between viral load and viral dynamic parameters is often nonlinear and the viral dynamic parameters often vary substantially across patients. Thus, nonlinear regression models with individual-specific regression parameters, i.e., nonlinear mixed effects models (see Section 1.5), have been widely used for modeling HIV viral dynamics.

AIDS researchers have also shown great interest in the relationship between viral loads and CD4 counts over time (Henry, Tebas, and Lane 2006; Rodriguez et al. 2006). Due to large between-individual variation in many AIDS datasets, regression models with individual-specific regression parameters (e.g., mixed effects models) are useful for modeling these types of data. See Section 1.5 for some specific examples.

A major challenge in statistical analyses of AIDS data is that these datasets typically contain missing values, censored values, measurement errors, and outliers. For example, CD4 counts are known to be measured with substantial errors, and patients often drop out because of drug side effects or other problems. In later chapters, we will show that it is important to address these issues appropriately in order to obtain reliable and valid results. For instance, if dropout patients are excluded in data analysis, a new anti-HIV treatment may be claimed as more effective than a standard treatment while in fact this may not be true. Thus, developments of valid statistical methods for missing data, censored data, measurement errors, and outliers, are very important in AIDS research.

1.3.3 A Study on Students' Performance

This dataset contains quiz scores of a statistics class with 53 students in a semester. Five quizzes were given in the class throughout the semester, with roughly equal time intervals between the quizzes. The objective is to study how students' performances change over time. Figure 1.5 shows the quiz scores for all 53 students in the class and the quiz scores for six randomly selected students. We see that students' performances appear to improve over time in general, but there seems to be substantial variation between students, so a model with student-specific parameters may be useful, such as a mixed effects model (see Section 1.5).

There are missing values in this dataset since some students failed to take some quizzes. Possible reasons for these missing data include i) some students were sick or had personal problems at the time of a quiz; ii) some students dropped out from the class since they found the course too challenging; iii) some students did not hand in their quizzes because they did very poorly. These reasons for the missing data lead to different *missing data mechanisms*, which suggest

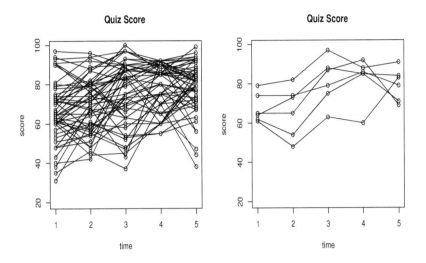

Figure 1.5 *Students' quiz scores throughout a semester. Left figure: all 53 students in the class. Right figure: 6 randomly selected students.*

different statistical methods for the missing data (see Chapter 3 for a detailed discussion). Moreover, there may be a few outliers in the data, so robust methods would also be useful.

1.3.4 A Study on Children's Growth

Goldstein (1987) described a well-known dataset which contains the heights of 26 boys from Oxford, England. It consists of the heights of these boys at different ages, each measured on nine occasions with roughly equal time intervals. The data are displayed in Figure 1.6.

A unique feature of this dataset is that, although children's initial heights are different, the rates of growth (i.e., the rates of increase in height) are similar across children and are roughly constant during the measurement period. For each child, the longitudinal trajectory of heights appears roughly linear, so the repeated measurement data on each child may be fitted by a straight line. That is, we may fit the data using a simple linear regression model with individual-specific intercepts but a constant slope. This is an extension of a standard simple regression model for cross-sectional data by introducing a random effect (or an individual effect) in the intercept but not in the slope.

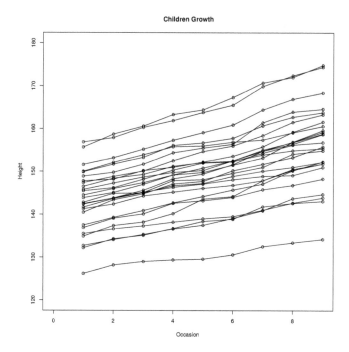

Figure 1.6 *Boys' heights (in cm) at nine measurement occasions. The open circles are observed values.*

1.4 Regression Models

1.4.1 General Concepts and Approaches

Regression models are perhaps the most widely used statistical methods. In either a cross-sectional study or a longitudinal study, data are usually collected on more than one variable. Depending on the objectives of the study, one variable may be treated as a *response* (or *dependent variable*) and some other variables may be treated as possible *covariates*, which are also called *explanatory variables* or *independent variables* or *predictors*. In this section, we provide a brief overview of the essential ideas of regression models. Chapter 2 provides a more detailed description of various regression models.

In a regression model, covariates are used to partially explain the *systematic variation* in the response. The remaining unexplained variation in the response is treated as random and is often assumed to follow a probability distribution – it reflects the uncertainty of the response aside from its systematic parts. This probability distribution is usually chosen based on the type of the response variable. In the systematic part of a regression model, the mean response is often assumed to link the covariates through a linear combination of the covariates, called a *linear predictor*. The resulting model is called a generalized linear model. The use of a linear predictor in a regression model is usually based on its simplicity and easy interpretation. In a nonlinear regression model or in a nonparametric or semiparametric regression model, however, the mean response may be linked to covariates in any nonlinear form. In a survival regression model, we may link the covariates to the hazard rather than the mean response.

The type of a regression model is usually determined by the type of the response variable. For example, if the response is a continuous variable, we may consider a *linear regression model* with normally distributed random errors. If the response is a binary variable, we may consider a *logistic regression model* with the random error following a binomial distribution. If the response is a count, we may consider a *Poisson regression model* with a Poisson random error. If the response is the time to an event of interest (e.g., time to death), we may consider a *survival regression model* with the random error following a Weibull distribution. In summary, the following types of regression models are commonly used in practice:

- linear models,
- generalized linear models,
- nonlinear models,
- survival regression models.

All these regression models can be extended to the analysis of longitudinal or clustered data.

In regression models, a main objective is to understand the dependence of the response on the covariates. Basic ideas and approaches of regression techniques apply to both cross-sectional data and longitudinal data. However, special considerations are needed for longitudinal or clustered data in order to incorporate the correlations within clusters.

In practice the *true* or exact relationship between a response and covariates may be very complicated. In a regression model, we often attempt to *approximate* this relationship using a simple and easy-to-interpret model, such as linear regression models or some generalized linear models. These simple models usually do not represent the true relationship between the responses and the covariates, even if the models fit the observed data well, and they are sometimes called empirical models. Prediction outside the observed-data range based on empirical models is often dangerous. These empirical models have been popular mainly because of their simplicity, which is especially important before modern computers become available.

In some cases, based on subject-area knowledge we may have a good understanding of the underlying mechanisms which generate the data. In these cases we may be able to derive the (approximate) true relationship between a response and covariates. The resulting models are often nonlinear, leading to *nonlinear regression models*. Therefore, nonlinear regression models are typically *mechanistic*, and they may represent the true (or approximately true) relationship between the response and the covariates. For this reason, nonlinear regression models often provide more reliable predictions than linear regression models. With the availability of modern computers, which greatly reduce computational burden, nonlinear models should be preferred if available.

In a broader sense, there are two general approaches for analyzing multivariate data. One approach is called *supervised learning*, where we treat one variable as a response and other variables as possible covariates. Regression models are examples of supervised learning. If more than one variables are treated as responses, we have multivariate regression models. The other approach is called *unsupervised learning*, where we treat all variables equally or symmetrically, i.e., no variables are treated as responses and no variables are treated as covariates. In unsupervised learning, the goal is to understand the underlying structures in all the variables. Examples of unsupervised learning include *principal components analysis, cluster analysis, factor analysis, contingency tables*, and *graphical models*. In the analysis of longitudinal data, regression models receive more attention and are thus the focus of this book.

In the following subsections we provide a brief review of commonly used regression models for cross-sectional data and for longitudinal data. A more detailed description of these models is given in Chapter 2.

1.4.2 Regression Models for Cross-Sectional Data

Regression models for cross-sectional data have been well developed. The most widely used regression models are probably *linear regression models*, where the relationship between the mean response and covariates is assumed to be linear and the random error is usually assumed to be normal. A comprehensive discussion of linear regression models can be found in Draper and Smith (1998) and Weisberg (2005), among others. Faraway (2004) discussed linear models using statistical software R, while Littell et al. (2002) illustrated statistical software SAS for linear models. In this section, we provide a brief overview of common regression models for cross-sectional data, with a focus on general concepts and approaches without technical details. In Chapter 2, we will return to this topic with some technical details. We begin with this review because mixed effects models are natural extensions of the corresponding regression models for cross-sectional data by introducing random effects in the models. Moreover, many of the ideas and approaches used in regression models for cross-sectional data can be extended to regression models for longitudinal data or clustered data.

Suppose that there are n individuals in the sample, and the data on individual i are $(y_i, x_{i1}, x_{i2}, \cdots, x_{ip})$, where y_i is the response and x_{ij}'s are covariates. A general (multiple) *linear regression model* can be written as

$$y_i = \beta_0 + \beta_1 x_{i1} + \cdots + \beta_p x_{ip} + \epsilon_i, \qquad i = 1, 2, \cdots, n, \quad (1.1)$$

$$\epsilon_i \quad \text{i.i.d.} \quad \sim N(0, \sigma^2),$$

where β_j's are unknown regression parameters linking the covariates to the response, and ϵ_i is a random error representing residual variation in the data. The standard assumptions for linear model (1.1) are:

- the random errors $\{\epsilon_1, \epsilon_2, \cdots, \epsilon_n\}$ or the response $\{y_1, y_2, \cdots, y_n\}$ are independent;
- the random errors ϵ_i's have a zero mean and a constant variance;
- the random errors ϵ_i's are normally distributed.

The *independence* of the observations $\{y_1, y_2, \cdots, y_n\}$ is an important assumption for classical linear regression models.

Unknown parameters in model (1.1) can be estimated by the *least-square method* or by the *maximum likelihood method*. Once the unknown parameters are estimated, model checking or model diagnostics should be performed to check the reasonability of the model and the assumptions, which can informally be based on residual plots and other graphical techniques. Variable transformations may be used to improve model fitting. Outliers and influential observations should also be checked since they may greatly affect the resulting estimates and may lead to misleading inference. Model selection can be based on standard statistical methods such as the stepwise method as well as on scientific considerations. In general, parsimonious models are preferred since they may avoid potential collinearity in the predictors and may improve precision of the main parameter estimates. In linear model (1.1), when all the covariates are *categorical* or *discrete*, the model is equivalent to an *analysis of variance (ANOVA)* model, which allows a specific decomposition of total variation into systematic part and random part.

Linear regression models have been widely used due to their simplicity, which is important in the pre-computer era since closed-form or analytic expressions of parameter estimates can be derived. However, linear models require strong assumptions such as linearity, and they may not be appropriate when the response is (say) categorical. Moreover, unlike nonlinear models, linear models usually do not describe the data-generating mechanisms and they often do not provide reliable prediction outside the observed data range. Therefore, extensions of linear models have received great attention in the last few decades, due partially to the developments of modern computers and computational tools. Linear models may be extended in two directions:

- *non-normal distributions* for the random errors;
- *nonlinear* relationships between the response and covariates.

These two extensions are briefly described below.

The first extension is to allow non-normal distributions for the responses or random errors. This is necessary for some responses whose distributions are clearly non-normal (even after transformations), such as binary responses taking only two possible values (say 0 or 1). A natural family of candidate distributions is the *exponential family*, which includes normal distributions, binomial distributions, Poisson distributions, and other distributions. For example, if the response of interest is an indicator of whether an individual has cancer or not, the response is a binary variable with only two possible values (say, 0 or 1). In this case, linear model (1.1) cannot be used since the covariates can take any real values and the response will not follow a normal distribution. However, we may assume that the response y_i follows a binomial (Bernoulli) distribution and consider the following nonlinear regression model, called a *logistic regression model*

$$\log\left\{\frac{P(y_i = 1)}{1 - P(y_i = 1)}\right\} = \beta_0 + \beta_1 x_{i1} + \cdots + \beta_p x_{ip}, \quad i = 1, 2, \cdots, n, \quad (1.2)$$

in which the mean response $E(y_i) = P(y_i = 1)$ and the covariates are linked through a monotone function $h(y) = \log(y/(1 - y))$, called a *logit link* function, and the response y_i is assumed to follow a binomial (Bernoulli) distribution.

More generally, we may assume that the response follows a distribution in the exponential family and then we link the mean response to the covariates via a linear predictor. The resulting model is called a *generalized linear model (GLM)*. Specifically, a GLM can be written as follows

$$h(E(y_i)) = \eta_i \equiv \beta_0 + \beta_1 x_{i1} + \cdots + \beta_p x_{ip}, \quad i = 1, 2, \cdots, n, \quad (1.3)$$

where $h(\cdot)$ is a known monotone link function and η_i is a linear predictor. Note that a GLM is a special nonlinear model. A more detailed discussion of GLMs is given in Chapter 2. When $h(y) = y$ (the identity link) and y follows a normal distribution, GLM (1.3) reduces to the standard normal linear model (1.1). When $h(y) = \log(y/(1 - y))$ (the logit link) and y follows a binomial distribution, GLM (1.3) reduces to a logistic regression model. When $h(y) = \log(y)$ and y follows a Poisson distribution, GLM (1.3) reduces to a Poisson regression model for count response. For comprehensive discussions of GLMs, see McCullagh and Nelder (1989), Fahrmeir et al. (2001), and McCulloch et al. (2008). Faraway (2005) describes GLM using software R.

Another extension of linear regression models is to allow the response to link the covariates in any nonlinear forms, leading to *nonlinear regression models*. We focus on the common class of nonlinear models in which the response and covariates may be linked in a nonlinear forms but the response or random error are assumed to be normal. A nonlinear regression model is often *mechanistic* in the sense that it usually describes or approximately describes the data-generating mechanism, i.e., the underlying mechanism which generates the observed data. Thus, nonlinear regression models often provide better predictions than linear regression models, and the parameters in nonlinear models often have natural physical interpretations.

A nonlinear regression model can be written as

$$y_i = g(x_{i1}, \cdots, x_{ip}, \boldsymbol{\beta}) + \epsilon_i, \qquad i = 1, 2, \cdots, n,$$

where $g(\cdot)$ is a nonlinear function and ϵ_i follows a normal distribution. Statistical inference for nonlinear regression models is more complex than that of linear models because closed-form expressions of parameter estimates are typically unavailable. Moreover, good choices of starting values are needed and are important for nonlinear models since the likelihoods may have multiple modes. See Chapter 2 for a more detailed discussion. For comprehensive discussions of nonlinear models, see Seber and Wild (2003) and Bates and Watts (2007).

A standard approach for statistical inference of regression models is the *maximum likelihood method*. The likelihood method is widely used because it is generally applicable to a wide variety of models and it offers nice asymptotic properties – the maximum likelihood estimates are consistent, asymptotically normal, and asymptotically most efficient under some regularity conditions. In Chapter 12 (Appendix), we will provide an overview of the likelihood method. A drawback of the likelihood method is that it requires distributional assumptions. So likelihood inference may be sensitive to departures from the assumed distributions and sensitive to outliers. A more robust approach is the so-called *quasi-likelihood method* and the closely related *generalized estimating equations (GEE)* method in which one only needs to specify the first two moments without distributional assumptions. However, the GEE estimates are less efficient than the likelihood estimates if the distributional assumptions hold. See Chapter 10 for a detailed discussion of the quasi-likelihood methods and GEE methods.

1.4.3 Regression Models for Longitudinal Data

The regression models for cross-sectional data presented in Section 1.4.2 can be extended to longitudinal data or clustered data. A key difference between cross-sectional data and longitudinal data is that the repeated measurements within an individual in a longitudinal study are typically correlated. Thus, regression models for longitudinal data must incorporate this correlation for valid inference. In other words, a regression model for longitudinal data must address *both* the relationship between the response and covariates *and* the correlation in the repeated measurements. As noted in Section 1.2, there are three common approaches to incorporate the correlation in longitudinal data or clustered data: a mixed effects modeling approach, a marginal GEE modeling approach, and a transitional modeling approach. In the following, we provide an overview for each of the three approaches.

Mixed Effects Models

For mixed effects models, we introduce random effects for each individual or cluster to incorporate the correlation between the repeated measurements within the individual or cluster. Since each individual or cluster shares the *same* random effects, the measurements within the individual or cluster are correlated. Moreover, the random effects facilitate individual-specific inference. Note that there are two sources of variations in longitudinal or clustered data: the between-individual variation and the within-individual variations. A mixed effects model specifically incorporates both sources of variations: it uses random effects or individual effects to represent deviations of individual longitudinal trajectories from the population average. Thus, a mixed effect model allows *subject-specific* inference, in addition to standard population-average inference.

A mixed effects model can be obtained from a standard regression model for cross-sectional data by introducing random effects to appropriate parameters, allowing these parameters to be *individual-specific*. For example, for the children growth data in Example 1.3.4 (page 13) in Section 1.3, since the intercepts vary greatly across children but the slopes remain roughly constant, we may consider a simple linear regression model with random intercepts but fixed slope:

$$\begin{aligned} y_{ij} &= \beta_0 + b_i + \beta_1 x_{ij} + e_{ij} \\ &= \beta_{0i} + \beta_1 x_{ij} + e_{ij}, \quad i = 1, 2, \cdots, n, \ j = 1, 2, \cdots, n_i, \quad (1.4) \end{aligned}$$

where y_{ij} and x_{ij} are the height and occasion of child i at measurement time j respectively, e_{ij} is the random error for the repeated measurements within child i, b_i is a random effect for child i, $\beta_{0i} = \beta_0 + b_i$ is a individual-specific parameter, and parameters β_0 and β_1 are called *fixed effects* or *population parameters*. We assume e_{ij} i.i.d. $\sim N(0, \sigma^2)$, $b_i \sim N(0, d^2)$, and e_{ij} and b_i are independent. Model (1.4) is a simple linear mixed-effects model. More examples of mixed effects models are given in Section 1.5, and the general theory will be presented in Chapter 2.

Marginal GEE Models

In a marginal model, we first model the mean structure of the response, i.e., the dependence of the mean response on covariates, and then we *separately* model the covariance structure of the response, i.e., the variances and correlations of the response measurements. The idea is similar to a quasi-likelihood approach which only requires specifications of the first moment (mean structure) and the second moment (covariance structure) of the responses without assuming any distributions for the data. So marginal models are robust to distributional assumptions. Moreover, if the mean structure is correctly specified, parameter estimates are consistent and asymptotically normal, even if the covariance structure is mis-specified. GEE models are usually more useful for non-normal data.

The parameters in a marginal model can be estimated by solving a set of estimating equations, called *generalized estimating equations (GEEs)*. The GEEs are analogous to the score equations in likelihood inference. Specifically, let $\mathbf{y}_i = (y_{i1}, y_{i2}, \cdots, y_{in_i})^T$ and $\mathbf{x}_i$ be the response measurements and covariates for individual i. A marginal GEE model for longitudinal or clustered data can be specified as follows:

- the marginal mean structure of the response $\mathbf{y}_i$:

$$\boldsymbol{\mu}_i(\boldsymbol{\beta}) \equiv E(\mathbf{y}_i|\mathbf{x}_i, \boldsymbol{\beta}) = g(\mathbf{x}_i, \boldsymbol{\beta}),$$

where $g(\cdot)$ is a known link function and $\boldsymbol{\beta}$ contains unknown regression parameters.
- a *working* covariance matrix $\Sigma_i(\boldsymbol{\beta}, \boldsymbol{\alpha})$ for the covariance structure of $\mathbf{y}_i$, which is often assumed to have a simple form such as an independence correlation structure or an auto-correlation structure, where $\boldsymbol{\alpha}$ contains unknown parameters.

The GEE for estimating $\boldsymbol{\beta}$ is given by

$$\sum_{i=1}^{n} D_i(\boldsymbol{\beta})\Sigma_i(\boldsymbol{\beta}, \boldsymbol{\alpha})(\mathbf{y}_i - \boldsymbol{\mu}_i(\boldsymbol{\beta})) = 0,$$

where $D_i(\boldsymbol{\beta}) = \partial g(\mathbf{x}_i, \boldsymbol{\beta})/\partial \boldsymbol{\beta}$. Note that, in the special case of independent observations with constant variances, the working covariance matrix can be chosen as $\Sigma_i = \sigma^2 I_{n_i}$, and the above GEE reduces to the usual form of the score function for cross-sectional data. A detailed discussion of marginal models and GEE methods is given in Chapter 10.

Transitional Models

A transitional model assumes a Markov structure for the longitudinal process to incorporate the correlation between the repeated measurements. The idea is to assume that the current response value depends on previous response values given observed data, which suggests a Markov process. Such a Markov structure may be reasonable in some applications. A transitional model can be specified by including previous response values as additional "covariates", so it is similar to a classical regression model for cross-sectional data. For example, if a first-order Markov assumption holds for a longitudinal process, a transitional model may be written as

$$E(y_{ij}) = g(y_{i,j-1}, \mathbf{x}_i, \boldsymbol{\beta}), \qquad i = 1, 2, \cdots,$$

where g is a known link function. Random effects may also be introduced in a transitional model to incorporate between individual variations, and such a mixed effects transitional model may be useful for modeling clustered longitudinal data (Cook, Yi, Lee, and Gladman 2004). We will not discuss details of transitional models. Interested readers are referred to Fahrmeir and Tutz (2001) and Diggle et al. (2002).

Comparison of the Models

Each of the three classes of regression models for longitudinal data has its own advantages and disadvantages. Mixed effects models are widely used in practice because (i) they allow for *individual-specific* (or *subject-specific*) inference; (ii) they are natural extensions of the corresponding regression models for cross-sectional data; and (iii) likelihood inference is conceptually straightforward and maximum likelihood estimates have very attractive asymptotic properties. Disadvantages of mixed effects models include the distributional assumptions, which sometimes may be restrictive, and computational challenges.

Advantages of marginal GEE models are: (i) they only require specifications of the first two moments, with no distributional assumptions; (ii) GEE estimates are consistent even if the covariance structure is misspecified, as long as the mean structure is correctly specified; and (iii) GEE estimates are asymptotically normal. Disadvantages of marginal models are that GEE estimates are not fully efficient if the covariance structure is mis-specified and marginal models do not allow for subject-specific inference.

Transitional models are similar to regression models for cross-sectional data if we view previous responses as additional "covariates", so transitional models share some characteristics with classical regression models. Note that, in mixed effects models or GEE models, we can use previous responses as additional "covariates", so mixed effects models or GEE models can be combined with transitional models.

In the presence of missing data or measurement errors, which are very common in longitudinal studies, mixed effects models are particularly attractive because it is conceptually straightforward to incorporate missing data or measurement errors in likelihood inference for mixed effects models, even when the missing data mechanism is non-ignorable (see Chapter 3). This is less straightforward for marginal models. Therefore, in the presence of incomplete data, mixed effects models offer another advantage over other models.

Other Models for Longitudinal Data

Other common models for longitudinal data include *nonparametric* models and *semiparametric* models. Since in practice longitudinal data can be very complex, parametric models may not be flexible enough to capture the longitudinal trajectories. Nonparametric or semiparametric models offer more flexible approaches for modeling complex longitudinal processes. Note that many nonparametric or semiparametric models may be approximated by parametric mixed effects models (see Chapter 3). This is yet another advantage of mixed effects models. We review some nonparametric or semiparametric models in Chapter 3.

Bayesian methods offer the advantage of incorporating prior information or information from similar studies. The advance of Markov chain Monte Carlo (MCMC) methods has led to rapid developments of Bayesian methods in the last few decades. In fact, mixed effects models have a Bayesian connection if the random effects are viewed as random parameters. Bayesian methods are closely connected to likelihood methods, when non-informative priors are considered. We discuss Bayesian methods in details in Chapter 11.

1.4.4 Regression Models for Survival Data

In a longitudinal study, often we are also interested in the times to an event of interest. For example, we may be interested in times to dropout, times to death, times to an accident, etc. Such data are called *survival data* or *event-time data*. Statistical analysis of survival data or event-time data is called *survival analysis*. In the following we give a brief review of survival data, and we will provide a more detailed discussion in Chapter 7.

There are some special features for survival data:

- survival data are often *censored*, i.e., the event of interest may not be observed for some subjects throughout the study period so the event times are censored;

- survival data are often *skewed*, i.e., they are usually not symmetric so a symmetric distribution such as a normal distribution should not be assumed for survival data (but such a distribution may be assumed if the survival data are transformed).

Moreover, survival data often have unequal follow-up times. Due to these features of survival data, special techniques are needed for analyzing survival data. In particular, nonparametric and semiparametric methods are widely used in survival analysis. In regression analysis of survival data, we are interested in the dependence of event times on covariates. The most commonly used regression models for survival data are probably the *Cox proportional hazards models*, which is a *semiparametric* regression model.

Parametric survival models are also available. These parametric models assume distributions for the survival data and may give more efficient estimates than nonparametric or semiparametric models *if* the distributional assumptions hold. A popular parametric distribution for survival data is the *Weibull distribution*, which includes the exponential distribution as a special case. The Weibull distribution plays a similar role as the normal distribution in linear regression models. Another class of models for survival data is *accelerated failure time models*. Accelerated failure time models offer attractive interpretations and are good choices in some applications.

Survival data may also be *clustered*. For example, survival data from different centers or hospitals may be clustered since data from the same centers or hospitals may be similar and thus may be correlated. For clustered survival data, random effects may be used to represent cluster effects, which also incorporate correlation within clusters. Survival models with random effects are called *frailty models*. A more detailed description of survival models and frailty models is given in Chapter 7.

Since survival data may arise in longitudinal studies, in which we may be interested in both the time to an event of interest and the longitudinal process, we may consider *joint models* for longitudinal data and survival data. In other words, we may simultaneously model survival data and longitudinal data and conduct joint inference, since the survival model and the longitudinal model in a study are usually linked in some ways so joint inference is desirable. Joint models have received great attention in recent years. We discuss joint models in details in Chapter 8.

1.5 Mixed Effects Models

1.5.1 Motivating Examples

In Section 1.4.3, we briefly discussed mixed effects models for longitudinal data and presented a simple example. Since mixed effects models are the main focus of this book, in this section we provide a more detailed description, including simple approaches for building mixed effects models in practice and an overview of commonly used mixed effects models. Technical details and more general descriptions of these mixed effects models will be presented in Chapter 2.

A mixed effects model may be viewed as an extension of the corresponding regression model for cross-sectional data by introducing random effects in the model to account for variations between individuals and correlations within individuals. Thus, the commonly used regression models described in Section 1.4.1 can be extended to the following commonly used mixed effects models:

- linear mixed effects (LME) models,
- generalized linear mixed models (GLMMs),
- nonlinear mixed effects (NLME) models,
- frailty models.

In regression models for cross-sectional data and in marginal GEE models for longitudinal data, all model parameters are *fixed*, i.e., the parameters are the same for all individuals, so the mean parameters are called *fixed effects* or *population parameters*. In some longitudinal studies, if the data vary greatly across individuals, we may allow each individual to have *individual-specific parameters*. These individual-specific parameters can be obtained by adding *random effects* or *individual effects* to the population parameters, so the random effects represent individual deviations from the population averages (plus random errors). Therefore, we can choose random effects informally based on the *heterogeneous* feature of the data. Formally, we can conduct hypothesis testing (e.g., the likelihood ratio test) to test the need of certain random effects, or we can use formal model selection methods, such as AIC or BIC criteria, to choose between models with and without random effects. We will return to this topic in Chapter 2.

The random effects in a mixed effects model not only incorporate heterogeneity in the data but also incorporate correlation between the multiple measurements within each individual or cluster. This is because the data within the same individuals or the same clusters share the same random effects or similar characteristics, which leads to correlation in the data. In the following, we present several examples to illustrate the selection of random effects.

We return to Example 1.3.4 (page 13) in Section 1.3. As evident in Figure 1.6 (page 14) in Section 1.3, there appear large variations in intercepts but relatively homogeneous slopes in the growth trajectories, so intuitively it may be reasonable to assume a model with fixed slope but random intercept, i.e.,

$$y_{ij} \quad = \quad \beta_{0i} + \beta_1 x_{ij} + e_{ij}, \tag{1.5}$$

$$\beta_{0i} = \beta_0 + b_i, \qquad i = 1, 2, \cdots, n; \quad j = 1, 2, \cdots, n_i, \qquad (1.6)$$

where y_{ij} is the height of child i at measurement time x_{ij}, e_{ij} is a random error, b_i is a random effect for the intercept of child i, and β_0 and β_1 are fixed effects. We assume that the random effect b_i and the error e_{ij} are independent and both follow normal distributions with zero means, i.e.,

$$b_i \sim N(0, d^2), \qquad e_{ij} \sim N(0, \sigma^2).$$

We also assume that $e_{i1}, e_{i2}, \cdots, e_{in_i}$ are conditional independent given b_i. Models (1.5) and (1.6) can be combined as follows:

$$y_{ij} = (\beta_0 + \beta_1 x_{ij}) + b_i + e_{ij}, \qquad (1.7)$$

where the fixed effects and the random effects are separated. Model (1.7) is an example of a *linear mixed-effects (LME) model*.

The LME model (1.5) and (1.6) is also called a *two-stage model*: in the first stage we model the within-individual variation by model (1.5), and in the second stage we model the between-individual variation by model (1.6). Such a two-stage approach can also be used in other mixed effects models.

For the AIDS data in Example 1.3.2 (page 9) in Section 1.3, a simple linear regression model may provide a reasonable fit to the viral load trajectories in the *first two weeks* after initiation of the anti-HIV treatment. For this initial period, Figure 1.3 (page 11) suggests that intuitively we may consider a linear model with random intercept and random slope, i.e.,

$$y_{ij} = \beta_{0i} + \beta_{1i} t_{ij} + e_{ij}, \quad i = 1, 2, \cdots, n; \; j = 1, 2, \cdots, n_i, \;\; (1.8)$$
$$\beta_{0i} = \beta_0 + b_{0i}, \quad \beta_{1i} = \beta_1 + b_{1i}, \qquad (1.9)$$

where y_{ij} is the $\log_{10}$-transformation of viral load (RNA) for individual i at time t_{ij}, β_{0i} and β_{1i} are individual-specific parameters, β_0 and β_1 are fixed effects, b_{0i} and b_{1i} are random effects, and e_{ij}'s are random errors. The $\log_{10}$-transformation of viral load is used to stabilize the variance of viral load and to make the viral load data more normally distributed. We assume that

$$(b_{1i}, b_{2i})^T \sim N(0, D), \qquad e_{ij} \text{ i.i.d. } \sim N(0, \sigma^2),$$

where D is a 2×2 covariance matrix. Models (1.8) and (1.9) can be combined as a single LME model

$$y_{ij} = (\beta_0 + \beta_1 t_{ij}) + (b_{0i} + b_{1i} t_{ij}) + e_{ij}, \qquad (1.10)$$

which separates the fixed effects and the random effects.

In models (1.8) and (1.9), the variation in the viral decay rates β_{1i} may be partially explained by variation in baseline CD4 values (denoted by z_i). In this case, we may replace the second-stage model (1.9) by the following model

$$\beta_{0i} = \beta_0 + b_{0i}, \quad \beta_{1i} = \beta_1 + \beta_2 z_i + b_{1i}. \qquad (1.11)$$

Then, we can combine models (1.8) and (1.9) and obtain the following single
LME model

$$y_{ij} = (\beta_0 + \beta_1 t_{ij} + \beta_2 z_i t_{ij}) + (b_{0i} + b_{1i} t_{ij}) + e_{ij}. \qquad (1.12)$$

In the above examples, the model selection approaches are quite informal and
are only used for illustration. More formally, one can use AIC or BIC criteria
or the likelihood ratio test to select good models or to confirm the informally
or empirically selected models. Pinheiro and Bates (2000) provided some detailed examples on model selection for mixed effects models using Splus.

1.5.2 LME Models

In this section we present LME models in general forms. Let $\mathbf{y}_i =
(y_{i1}, y_{i2}, \cdots, y_{in_i})^T$ be the n_i repeated measurements of the response vari-
able y on individual i, $i = 1, 2, \cdots, n$. A general *linear mixed-effects (LME)*
model can be written as

$$\mathbf{y}_i = X_i\boldsymbol{\beta} + Z_i\mathbf{b}_i + \mathbf{e}_i, \qquad i = 1, 2, \cdots, n, \qquad (1.13)$$
$$\mathbf{b}_i \sim N(0, D), \qquad \mathbf{e}_i| \sim N(0, R_i), \qquad (1.14)$$

where $\boldsymbol{\beta} = (\beta_1, \cdots, \beta_p)^T$ is a $p \times 1$ vector of fixed effects, $\mathbf{b}_i = (b_{i1}, \cdots, b_{iq})^T$
is a $q \times 1$ vector of random effects, the $n_i \times p$ matrix X_i and the $n_i \times q$
matrix Z_i are known design matrices which may contain covariates, $\mathbf{e}_i =
(e_{i1}, e_{i2}, \cdots, e_{in_i})^T$ represents random errors of the repeated measurements
within individual i, D is a $q \times q$ covariance matrix of the random effects, and
R_i is a $n_i \times n_i$ covariance matrix of the within-individual errors.

We often assume that $R_i = \sigma^2 I_{n_i}$ for simplicity, where I_{n_i} is the $n_i \times n_i$ identity matrix, i.e., the within-individual measurements are assumed to be inde-
pendent with constant variance. This assumption may be reasonable when the
within-individual measurements are relatively far apart (so they are approximately independent) and when the repeated measurements within individuals
roughly have a constant variance. The value of σ^2 represents the magnitude
of the within-individual variation, and the values of the diagonal elements of
D represent the magnitude of the between-individual variation. The simpli-
fied with-individual covariance structure R_i greatly reduces number of parameters and may avoid some identifiability problems. Wang and Heckman (2009)
showed that LME model (1.13) and (1.14) is always identifiable if $R_i = \sigma^2 I_{n_i}$.
This is an important advantage in many problems. See Section 2.2.2 in Chapter 2 for a more detailed discussion.

LME model (1.13) and (1.14) is an extension of the corresponding linear regression model by adding the random effects $\mathbf{b}_i$ in the model. In other words,
if the term with random effects $\mathbf{b}_i$ is omitted, LME model (1.13) and (1.14)

reduces to a standard linear regression model. A key characteristic of a LME model is that it is *linear* in both the mean parameters β and the random effects $\mathbf{b}_i$. Therefore, many analytic or closed-form expressions of parameter estimates can be obtained for LME models, which greatly reduces computational burden. This important advantage is unavailable for models which are nonlinear in either the mean parameters or the random effects or both.

In LME model (1.13) and (1.14), the fixed effects β are *population-level* parameters and are the same for all individuals, as in a classical linear regression model for cross-sectional data, while the random effects $\mathbf{b}_i$ are *individual-level* "parameters" representing individual variations from population-level parameters. The random effects $\mathbf{b}_i$ measure between-individual variation, and the random errors $\mathbf{e}_i$ measure within-individual variation. Since each individual shares the *same* random effects, the multiple measurements within each individual or cluster are *correlated*. Properties and statistical inference for LME models are presented in Chapter 2.

In a mixed effects model, the repeated measurements $\{y_{i1}, y_{i2}, \cdots, y_{in_i}\}$ of the response within each individual can be taken at *different* time points for different individuals, and the number of measurements n_i may also vary across individuals. In other words, a LME model allows *unbalanced data* in the response. This is an advantage of mixed effects models.

In LME model (1.13) and (1.14), the design matrix Z_i is often a submatrix of the design matrix X_i. For example, in model (1.7) (page 26) we have

$$X_i = \begin{pmatrix} 1 & x_{i1} \\ 1 & x_{i2} \\ \vdots & \vdots \\ 1 & x_{in_i} \end{pmatrix}, \quad Z_i = \begin{pmatrix} 1 \\ 1 \\ \vdots \\ 1 \end{pmatrix}, \quad \mathbf{b}_i = b_i, \qquad (1.15)$$

in model (1.10) (page 26) we have

$$X_i = Z_i = \begin{pmatrix} 1 & x_{i1} \\ 1 & x_{i2} \\ \vdots & \vdots \\ 1 & x_{in_i} \end{pmatrix}, \quad \mathbf{b}_i = \begin{pmatrix} b_{i0} \\ b_{i1} \end{pmatrix}, \qquad (1.16)$$

and in model (1.12) (page 27) we have

$$X_i = \begin{pmatrix} 1 & t_{i1} & z_i t_{i1} \\ 1 & t_{i2} & z_i t_{i2} \\ \vdots & \vdots & \vdots \\ 1 & t_{in_i} & z_i t_{in_i} \end{pmatrix}, \quad Z_i = \begin{pmatrix} 1 & t_{i1} \\ 1 & t_{i2} \\ \vdots & \vdots \\ 1 & t_{in_i} \end{pmatrix}, \quad \mathbf{b}_i = \begin{pmatrix} b_{i0} \\ b_{i1} \end{pmatrix}.$$

1.5.3 GLMM, NLME, and Frailty Models

In the previous section, we see that a LME model for longitudinal data can be obtained from the corresponding linear regression model for cross-sectional data by introducing random effects in the model to account for between-individual variation and within-individual correlation. This idea can be extended to other types of regression models. For example, if the longitudinal response is a binary variable, we can extend a logistic regression model for cross-sectional data to a longitudinal regression model by introducing random effects in the logistic model. The resulting longitudinal regression model is an example of generalized linear mixed models (GLMMs). We briefly illustrate the approach as follows.

Consider the study on mental distress in Section 1.3.1 (Example 1.3.1 on page 6). We may wish to study if a subject's mental distress at each measurement time throughout the study is above or below his/her baseline value. Let $y_{ij} = 1$ if the mental distress of subject i at measurement time j is above his/her baseline score and $y_{ij} = 0$ otherwise. Then, the data $\{y_{ij}, i = 1, \cdots, n; j = 1, \cdots, n_i\}$ are longitudinal binary data. Suppose that we are also interested in if the value of y_{ij} is related to the gender (x_i) of subject i. Then, the following *generalized linear mixed model (GLMM)* for longitudinal binary response may be considered:

$$\log \left(\frac{P(y_{ij} = 1)}{1 - P(y_{ij} = 1)} \right) = \beta_{i0} + \beta_1 x_i, \qquad (1.17)$$

$$\beta_{i0} = \beta_0 + b_i, \qquad b_i \sim N(0, d^2), \quad (1.18)$$

where b_i is a random effect used to incorporate the between-subject variation and within-subject correlation in the longitudinal data, $i = 1, \cdots, n; j = 1, \cdots, n_i$. We may also introduce a random effect for parameter β_1 if necessary. Here the responses y_{ij} are assumed to be conditionally independent and follow binomial distributions given the random effects.

More generally, a general GLMM may be written as

$$h(E(\mathbf{y}_i)) = X_i \boldsymbol{\beta} + Z_i \mathbf{b}_i, \qquad i = 1, \cdots, n, \qquad (1.19)$$

$$\mathbf{b}_i \sim N(0, D), \qquad (1.20)$$

where $\mathbf{y}_i = (y_{i1}, \cdots, y_{in_i})^T$ are the repeated measurements within individual or cluster i, $h(\cdot)$ is a known monotone link function, X_i and Z_i are known design matrices, $\boldsymbol{\beta}$ contains fixed effects, $\mathbf{b}_i$ contains random effects, and D is a covariance matrix. It is typically assumed that the responses y_{ij} are conditionally independent and follow distributions in the exponential family, given the random effects. Chapter 2 provides a more detailed discussion of GLMMs.

Next, we consider an example of nonlinear regression models. Consider the

AIDS study in Section 1.3.2 (page 9). The viral load trajectories in the first three months may be modeled by the following *nonlinear mixed effects (NLME) model* (Wu and Ding 1999)

$$y_{ij} = \log_{10}\left(\beta_{1i}e^{-\beta_{2i}t_{ij}} + \beta_{3i}e^{-\beta_{4i}t_{ij}}\right) + e_{ij}, \tag{1.21}$$

$$\boldsymbol{\beta}_i = \boldsymbol{\beta} + \mathbf{b}_i, \qquad \mathbf{b}_i \sim N(0, D), \qquad e_{ij} \sim N(0, \sigma^2), \tag{1.22}$$

where y_{ij} is the $\log_{10}$-transformation of viral load (RNA) for individual i at time t_{ij}, $\boldsymbol{\beta}_i = (\beta_{1i}, \beta_{2i}, \beta_{3i}, \beta_{4i})^T$ are individual-specific parameters, $\boldsymbol{\beta} = (\beta_1, \beta_2, \beta_3, \beta_4)^T$ contains fixed effects, $\mathbf{b}_i = (b_{i1}, b_{i2}, b_{i3}, b_{i4})^T$ contains random effects, and e_{ij} is a within-individual measurement error, $i = 1, \cdots, n$; $j = 1, \cdots, n_i$. A general NLME model can be written as

$$\mathbf{y}_i = g(\mathbf{x}_i, \boldsymbol{\beta}, \mathbf{b}_i) + \mathbf{e}_i, \qquad i = 1, \cdots, n, \tag{1.23}$$

$$\mathbf{b}_i \sim N(0, D), \qquad \mathbf{e}_i|\mathbf{b}_i \sim N(0, R_i), \tag{1.24}$$

where $g(\cdot)$ is a known nonlinear function, $\mathbf{x}_i$ contains covariates, $\mathbf{e}_i = (e_{i1}, \cdots, e_{in_i})^T$ are random errors, and R_i is a covariance matrix. More details of NLME models can be found in Chapter 2.

Similarly, if there are clusters in survival data, we can introduce random effects in a standard survival model to account for the variation between clusters and the correlation within clusters. The resulting survival model is called a *frailty model*. We will provide a detailed discussion of survival models and frailty models in Chapter 7.

The above approaches can be used to extend other regression models for cross-sectional data to longitudinal data or clustered data. That is, we can introduce random effects to a regression model for cross-sectional data and obtain a mixed effects model for longitudinal or clustered data. The distributions of the random effects are not limited to normal distributions – they can be other parametric distributions such as t-distributions and gamma distributions or even nonparametric distributions.

Statistical inference for mixed effects models is typically based on the likelihood methods. For mixed effects models, the observed-data likelihoods involve integrations with respect to the unobservable random effects, which can be intractable except for LME models. Therefore, computation can be intensive and challenging. See Chapter 2 for detailed discussions.

1.6 Complex or Incomplete Data

In longitudinal studies, the observed data are often complex or incomplete. For example, in a mixed effects model, there may be missing data in the response or covariates, dropouts, and censoring, so the observed data are incomplete.

The reason is that in practice it is almost impossible to obtain *all* data at *each* time point in a longitudinal study lasting a long time. This can be seen from the examples and figures in Section 1.3. In this book, we define *incomplete data* in a broad sense to include data with measurement errors and outliers, although most people may view incomplete data simply as missing data or dropouts. Data with measurement errors may be viewed as incomplete data since for these data their true values are not observed, i.e., the observed are incomplete. In a similar sense, outliers may also be viewed as incomplete data because their true values are not observed (here we assume that outliers are not true values, although in some cases outliers are in fact true values).

In this section we briefly discuss common incomplete data problems in longitudinal studies. Incomplete data include

- *completely missing data*: in this case there is no information available for the missing data. This is what *missing data* mean in the usual sense.

- *partially missing data*: in this case the true values are not observed but some closely related values are observed. Examples include data with measurement errors, and censored data.

In the following, we discuss these incomplete data problems in the context of regression models.

1.6.1 Missing Data

When data are completely missing, we do not have any information about the missing values. To address the missing data in statistical inference, a standard approach is to assume a probability distribution for the variables with missing data, and then "impute" the missing data based on the predictive distribution of the missing data given the observed data, and in the meantime take the missing data uncertainty into account. In regression models, missing data may arise in the response or covariates or both, as described below.

In a longitudinal study, measurements are taken repeatedly over time, often on a pre-specified measurement schedule. However, subjects in the study may fail to show up at a scheduled time for various reasons. This leads to missing data in the response and in time-dependent covariates. Note that there are two types of covariates in a longitudinal regression models: *time-independent covariates*, i.e., covariates whose values do not change over time such as gender and race, and *time-dependent covariates*, i.e., covariates whose values may change over time are repeatedly measured throughout the study. Missing data may occur for both types of covariates but are particularly common in time-dependent covariates. For a time-dependent covariate, missing data may arise in the following situations: i) the covariate is measured at time schedules different from the response measurement schedule, so covariate values are missing at the response measurement times; ii) subjects may fail to show up at scheduled measurement times; and iii) subjects may drop out.

Dropouts are especially common in longitudinal studies lasting a long time. Subjects may drop out before the end of the studies for various reasons: subjects may move to elsewhere, subjects may have drug side effects, subjects may be too busy, and so on. If a subject drops out, all data on that subject from the last visit are missing. If dropout subjects never return to the study, the resulting missing data have a *monotone* missing pattern (after perhaps a re-arrangement of the data). If dropout subjects return to the study at later times, the missing data are *intermittent missing* and have a *non-monotone* missing pattern.

In the presence of missing data, statistical analyses ignoring missing data or based on simple imputation methods may lead to biased results. Appropriate statistical methods for missing data must take the *missing data mechanisms* into account. In Chapter 3, we give an overview of common missing data methods. In Chapter 4, we focus on missing data problems in mixed effects models.

1.6.2 Censoring, Measurement Error, and Outliers

In practice some data may be viewed as partially missing or partially observed, such as censored data and mis-measured data. When data are partially missing, the true values of the variables are not observed but some versions of the true values are observed, i.e., partial information is available for the unobserved true values, which is different from completely missing data.

In some studies, some variables may be censored in the sense that too large or too small values are unobservable. For example, in Example 1.3.2 (page 9), some viral loads are left censored due to a low detection limit, i.e., a certain threshold below which viral loads are not quantifiable. When data are censored, their true values are missing but are known to be smaller (or larger) than known values, so partial information is available for the missing values. Mixed effects models with censored data will be discussed in details in Chapter 6.

In practice, some variables may be measured with errors, i.e., the observed data are not the true values but are contaminated ones. For example, blood pressure and air pollution are often measured with errors. Ignoring measurement errors may lead to misleading statistical inference. In regression models, we mainly focus on measurement errors in covariates. Statistical methods for covariate measurement errors will be studied in Chapter 5.

An *outlier* can be defined as an observation which appears to be inconsistent with the remainder of the data, i.e., an outlier is far away or is distinctly different from the rest of the data. If we assume that a statistical model approximates the true data-generating mechanism, we can treat outliers as observations that deviate from the true model (of course, if there are too many outliers, the assumed model may be questionable). Thus, in many cases outliers may be interpreted as observations that deviate from their true values or interpreted as incompletely observed values or mis-measured values. Outliers may arise from measurement errors, recording errors, or special individuals who are not representative of the population. Although some outliers may be true values, here we view outliers as incomplete data. However, our view does not affect the methods used to address outliers. It is known that maximum likelihood methods are sensitive to outliers, so robust methods are useful for mixed effects models. We discuss robust methods for regression models in Chapter 9.

1.6.3 Simple Methods

In the presence of missing data, a widely used method (often the default method in standard software) is the so-called *complete-case method*: it simply deletes any individuals (cases) which contain missing data. This simple method can lead to biased results if the missing data are informative (see Section 3.2.1, page 99, in Chapter 3 for a more detailed discussion and examples). Even if the missing data are not informative, the complete-case method may be inefficient because substantial observed data may be discarded in many cases. Note that a mixed effects model allows unbalanced response data, i.e., individuals or cases will not be discarded in standard software even if there are missing data in the response. However, if the missing responses are informative, the missing data mechanism must be incorporated for valid analysis (see Chapter 4). Moreover, a mixed effects model does not allow any missing data in covariates, so any cases with missing values may be discarded in a standard software.

Another commonly used simple method for missing data in longitudinal studies is the so-called *last-value-carried-forward (LVCF)* method: it imputes a missing value by the last observed value from the individual. This method may also lead to misleading results, since the validity of this method requires very strong assumptions (Cook, Zeng, and Yi 2004). Moreover, it fails to incorporate the missing data uncertainty.

There are other simple imputation methods. For example, we may impute a missing value by a mean value of the observed data or by a predicted value from a regression model. Little and Rubin (2002) provided a detailed discussion. A main drawback of these *single imputation methods* is that they fail to take the *missing data uncertainty* into account, so the standard errors of the main parameter estimates are likely to be under-estimated. More appropriate methods should either adjust the standard errors or impute several values for each missing value. In Chapter 3, we will provide a comprehensive overview of various missing data methods and literature in this area.

In regression analyses, covariate measurement errors are often ignored. However, if a covariate is measured with substantial error, ignoring the measurement error may lead to severely misleading inference. For example, a highly significant covariate may be claimed as not significant. We discuss details in Chapter 5.

A simple method for outliers is to remove them. However, outliers may contain valuable information. Moreover, for models with multiple variables, it may be difficult to detect outliers, due to the high dimensionality (so graphical tools may not be easily used). Therefore, robust methods which accommodate outliers or downweight outliers are very useful. These robust methods will be discussed in Chapter 9.

1.7 Software

Standard mixed effects models, including LME, GLMM, and NLME models, have been implemented in many commonly used statistical software, such as R or Splus, SAS, Stata, and SPSS. We use R or Splus for most of the examples and figures throughout the book. In this section we give a brief overview of common software for mixed effects models and incomplete data. More details are discussed in later chapters as appropriate.

Splus is a statistical software based on the S programming language developed at Bell Labs (now Lucent Technologies). R may be considered as a different implementation of S, and much of the code for Splus and R is the same. R is free and open source, and it can be downloaded from the R home page: *http://cran.r-project.org/*. The book by Venables and Ripley (2003) "Modern Applied Statistics with S-PLUS" (the *MASS* package) is a good reference for most common statistical analyses. Readers can go to the R webpage for information on R packages. To use a package, one needs to install it first (just once), and then in each R session one needs to load in the package using the **library()** function.

SAS is a widely used statistical software that is particularly popular in biomedical research (*www.sas.com*). Stata is a statistical software popular in social science and survey research (*www.stata.com*). It is becoming increasingly popular, and it offers some simple-to-implement methods such as methods for robust standard errors across a wide variety of models and cluster resampled bootstrap methods. SPSS is a statistical software that is popular in psychology and social science (*www.spss.com*). Other software, some are for specific statistical methods, includes WinBugs, NONMEM, HLM, MLWin, and more.

For fitting mixed effects models (LME, NLME, GLMM models), the R packages include **lme4, nlme, glmm**, and **glmmPQL**. Pinheiro and Bates (2002) is a good reference. The SAS procedures include **mixed, nlmixed**, and **glimmix**. The STATA command is **xtmixed**, and the SPSS command is **mixed**. See Section 2.7.1 in Chapter 2 for more details.

Software for missing data includes the Splus missing data library **library(missing)** in Splus version 7.0, Joseph Schafer's free software in Splus, R and Splus package MICE, SAS procedure **proc mi**, SPSS missing data library in SPSS version 12.0, and Bayesian package WinBugs. STATA offers software for measurement error in generalized linear models (*http://www.stata.com/merror/*). Robust methods are also widely implemented in many software, such as SAS procedure **proc robustreg** and Splus function **lmRobMM()** and many R packages. See Section 3.8 in Chapter 3 for more details.

Optimization procedures are also widely implemented. For example, R offers function **nlm()**, which carries out a minimization using a Newton-type algorithm, and function **optim()**, which offers general-purpose optimizations based on Nelder-Mead, quasi-Newton and conjugate-gradient algorithms and includes an option for box-constrained optimization and simulated annealing. Splus offers functions **ms()**, **nlmin()**, and **nlminb()**. SAS/OR software offers several procedures for optimization, including **proc lp** (for linear programming problems), **proc netflow** (for network optimization problems), **proc nlp** (for nonlinear programming problems), and more.

Available software can mostly be used to analyze standard models and problems. For many of the more advanced and more specific models and methods, available software may not be directly applicable. Users either need to modify existing packages or program their own functions.

1.8 Outline and Notation

Outline of the Book

In this book, we mainly focus on mixed effects models or random effects models for longitudinal or clustered data when the observed data may be complex or incomplete. These mixed effects models include LME models, GLMMs, NLME models, frailty models, and semiparametric or nonparametric mixed effects models. We first provide a comprehensive overview of these mixed effects models, and then we discuss various complex or incomplete data problems, including missing data, measurement error, censoring, and outliers.

Throughout the book, we mostly consider maximum likelihood methods for inference since likelihood methods are standard inferential tools for mixed effects models. However, for the analysis of longitudinal or clustered data, GEE methods and Bayesian methods are equally popular and in some cases may even be more desirable than likelihood methods. For example, for some problems software may be available for GEE methods and Bayesian methods, which makes these methods easy to implement, but software may not be available for likelihood methods since likelihood methods may require custom programming. There are many books available which focus on GEE methods and Bayesian methods. Thus, due to space limitation, in this book GEE methods and Bayesian methods will only be briefly discussed.

The book is organized as follows. In Chapter 2, we provide a comprehensive overview of commonly used mixed effects models, their properties, and inferential methods. In Chapter 3, we discuss missing data mechanisms and review general missing data methods, including the EM algorithm and multiple imputation methods. Chapter 4 focuses on missing data problems in mixed effects models. Covariate measurement errors in mixed effects models are discussed in Chapter 5. Mixed effects models with censored responses are discussed in Chapter 6. In Chapter 7, we consider survival models and frailty models with incomplete data. In Chapter 8, we discuss joint models of survival data and longitudinal data, in the presence of incomplete data. Robust methods for mixed effects models with incomplete data are discussed in Chapter 9. In Chapter 10, we consider marginal GEE models for longitudinal data or clustered data, in the presence of incomplete data. In Chapter 11, we discuss Bayesian methods for mixed effects models with incomplete data. Chapter 12 is an Appendix which provides some background materials.

Readers are recommended to read Chapters 1 – 3 to get an overview of various mixed effects models and general methods for incomplete data (Chapter 1 may be skipped if readers already have some background in mixed effects models), and then jump to certain chapters as needed without necessarily following the order of the chapters. Since some readers may be just interested in materials in certain chapters, we try to make each chapter somewhat self-contained if possible, so some essential materials may be briefly repeated in some chapters. Some background materials are provided in Chapter 12, including likelihood theory, the Gibbs sampler, rejection and importance sampling methods, numerical integration methods, optimization methods, bootstrap, and some matrix algebra.

Notation Used in the Book

Throughout the book, in general we will use capital letters to represent *matrices*, bold-face small letters to represent *vectors*, and small letters (not in bold type) to present *scalars*. Vectors will be in *column* formats. Transpose of a matrix or a vector will be denoted by T. Thus, for example, A represents a matrix,

A^T denotes the transpose of A, $\mathbf{y} = (y_1, y_2, \cdots, y_n)^T$ represents a $n \times 1$ vector with components y_j's, and y_j represents a scalar. To simplify notation, a sample point y_j may either represent a random variable or an observed value of this random variable (we rely on the context to distinguish the two).

In general we will use $f(\cdot)$ to denote a generic *probability density function*. Thus, for example, $f(x)$ is a density function of a random variable x and $f(y)$ is a density function of a random variable y, relying on the context to distinguish the two. The conditional distribution of y given x is denoted by $[y|x]$, and the corresponding conditional density function is denoted by $f(y|x)$. In general, such notation should not cause confusion, but in cases where confusion may arise, we will use $f_x(x)$ to denote the density function of x.

For longitudinal or clustered data, $\mathbf{y}_i = (y_{i1}, y_{i2}, \cdots, y_{in_i})^T$ represents the n_i repeated measurements on individual or cluster i ($i = 1, 2, \cdots, n$), so the natural unit is $\mathbf{y}_i$, not y_{ij}. The sample size is n, while n_i is the number of repeated measurements within individual or cluster i. For example, in the children growth data in Section 1.3.4 (page 13), $n = 26$ and $n_i = 9$ for all i. We will use the terms *individuals, subjects, clusters, patients, children*, etc., interchangeably for the experiment units based on context.

Although we try to make the notation consistent throughout the book, we also try to follow standard notation in the literature so that readers may find it easier to follow. Thus, notation in some chapters or sections are self-contained, which may differ slightly from that in other chapters or sections, so readers should not assume that the same notation always means the same variable throughout the book. In these cases, the notation will be re-defined in the particular chapters or sections. This approach should not cause much confusion.

CHAPTER 2

Mixed Effects Models

2.1 Introduction

In Sections 1.4 and 1.5 of Chapter 1, we reviewed regression models for cross-sectional data and longitudinal or clustered data. In this chapter, we provide a more detailed discussion of mixed effects models for longitudinal or clustered data, including their properties and methods for statistical inference.

A mixed effects model for longitudinal or clustered data can be obtained from the corresponding model for cross-sectional data by introducing random effects. Specifically, we have

- *linear mixed effects (LME) models*, which can be obtained from linear regression models by introducing random effects;
- *nonlinear mixed effects (NLME) models*, which can be obtained from nonlinear regression models by introducing random effects;
- *generalized linear mixed models (GLMMs)*, which can be obtained from GLMs by introducing random effects;
- *frailty models*, which can be obtained from survival models by introducing random effects.

For these mixed effects models, the random effects in the models represent the influence of each individual (cluster) on the repeated observations that is not captured by the observed covariates.

A mixed effects model may also be viewed as a multi-level or *hierarchical model* in which the level-1 observations are *nested* within the higher level-2 observations. For example, in longitudinal studies repeated observations from a subject are nested within this subject (cluster), and in multi-center studies observations from a center are nested within this center (cluster). Random effects are used to accommodate the heterogeneity in the data, which may arise from subject or clustering effects or from spatial correlation. The magnitude of the random effects measures the variability across individuals or measures the *between individual variations*. Mixed effects models allow *individual-specific* or *subject-specific* inference and are often called *subject-specific (SS) models*. This is to be contrasted with the *population averaged (PA) models* or *marginal models*.

With a mixed effects model, we can make inference or prediction for a specific individual or cluster, which is useful in some applications. For example, in some medical studies we may be able to provide individualized treatments. In mixed effects models, individuals are assumed to be drawn from a population that share common features, so one can *borrow strength* (information) from similar individuals or combine results from similar studies.

A LME model may be viewed as a special case of either a GLMM or a NLME model. A key difference between a LME model and a GLMM or a NLME or a frailty model is that the random effects are *linear* in the LME model, while the random effects are *nonlinear* in the other models. This difference leads to major computational challenges in likelihood estimation for models nonlinear in the random effects, since the likelihoods involve integrations with respect to these unobservable random effects. Moreover, the distinction between SS models and PA models may not be important for LME models but it is critical under nonlinearity (Heagerty 1999; Davidian and Giltinan 2003).

There are large parallel literatures of GLMMs and NLME models. Many of the estimation methods are similar, but their finite-sample performances may be different due to different types of the responses such as a continuous response and a binary response (see Section 2.6 for discussion). Moreover, NLME models are often "scientific" (or mechanistic) models while GLMMs (like LME models) are often empirical ones, so interpretations may be different. Survival and frailty models will be treated separately in Chapter 7 since survival data are often censored so special techniques are needed.

LME, GLMM, and NLME models are all *parametric* models in which the functional forms linking covariates to the responses are assumed to be known, and the objective is to estimate the unknown parameters in the models. For complex longitudinal data, however, parametric models may be too restrictive due to the complexity of longitudinal trajectories. In *nonparametric* models, the functional forms of the longitudinal trajectories are left to be unspecified, which can be any smooth functions without parametric forms, and a main objective is to estimate the smooth functions. Therefore, nonparametric models are more flexible in modeling longitudinal trajectories. *Semiparametric* models combine elements of parametric models and nonparametric models and sometimes may be more appealing. For longitudinal data, we can also introduce random effects or random processes in nonparametric or semiparametric models to incorporate between individual variation and within individual correlation. The resulting models are called nonparametric mixed effects models or semiparametric mixed effects models.

In this chapter, we discuss LME, GLMM, and NLME models in details. Frailty models will be discussed separately in Chapter 7. Since mixed effects models may be viewed as extensions of their corresponding models for cross-sectional data, in each section we first review the essential ideas of the corresponding models for cross-sectional data, and then we extend the models to mixed-effects models by introducing random effects. This approach also allows us to compare the models, estimates, and inferences for cross-sectional data and longitudinal or clustered data. Finally we describe commonly used statistical software to fit these mixed effects models.

2.2 Linear Mixed Effects (LME) Models

2.2.1 Linear Regression Models

In Section 1.4.2 of Chapter 1, we briefly discussed classical linear regression models for cross-sectional data. In this section, we provide a more detailed discussion.

A general linear regression model for cross-sectional data can be written as

$$y_i = \beta_0 + \beta_1 x_{i1} + \cdots + \beta_p x_{ip} + \epsilon_i = \mathbf{x}_i^T \boldsymbol{\beta} + \epsilon_i, \quad i = 1, 2, \cdots, n, \quad (2.1)$$

where y_i is the response for individual i, β_j's are unknown parameters, x_{ij} is the j-th covariate for individual i, $\mathbf{x}_i = (x_{i1}, \cdots, x_{ip})^T$, $\boldsymbol{\beta} = (\beta_0, \cdots, \beta_p)^T$, and ϵ_i's are random errors. Let

$$\mathbf{y} = \begin{pmatrix} y_1 \\ y_2 \\ \vdots \\ y_n \end{pmatrix}, \quad X = \begin{pmatrix} 1 & x_{11} & \cdots & x_{1p} \\ 1 & x_{21} & \cdots & x_{2p} \\ \cdot & \cdot & \cdots & \cdot \\ 1 & x_{n1} & \cdots & x_{np} \end{pmatrix}, \quad \boldsymbol{\epsilon} = \begin{pmatrix} \epsilon_1 \\ \epsilon_2 \\ \vdots \\ \epsilon_n \end{pmatrix}.$$

Model (2.1) can be written in a matrix form as follows

$$\mathbf{y} = X\boldsymbol{\beta} + \boldsymbol{\epsilon}. \quad (2.2)$$

The standard assumptions for models (2.1) or (2.2) are (i) the errors ϵ_i's are *independent*, (ii) the errors ϵ_i's have mean zero, i.e., $E(\boldsymbol{\epsilon}) = \mathbf{0}$, and *constant variance* σ^2, i.e., $Var(\boldsymbol{\epsilon}) = \sigma^2 I_n$, where I_n is the $n \times n$ identity matrix, and (iii) the errors ϵ_i's are normally distributed, i.e., $\boldsymbol{\epsilon} \sim N(\mathbf{0}, \sigma^2 I_n)$. The (marginal) distribution of $\mathbf{y}$ is given by

$$\mathbf{y} \sim N(X\beta, \sigma^2 I_n), \quad (2.3)$$

under the model assumptions.

The *least squares method* for estimating $\boldsymbol{\beta}$ is to minimize $Q(\boldsymbol{\beta}) = (\mathbf{y} -$

$X\beta)^T(\mathbf{y} - X\beta)$. The parameter estimates are

$$\hat{\beta} = (X^TX)^{-1}X^T\mathbf{y}, \qquad \hat{\sigma}^2 = \frac{RSS}{n-p-1} = \frac{\mathbf{r}^T\mathbf{r}}{n-p-1}, \qquad (2.4)$$

where $\mathbf{r} = \mathbf{y} - X\hat{\beta}$ is a vector of *residuals*, and $RSS = \mathbf{r}^T\mathbf{r}$ is the residual sum of squares. Residuals represent the differences between the fitted values and the observed values of the response, so they can be used to check if the model fits the data well. For linear model (2.2), the least square estimates are identical to the *maximum likelihood estimates (MLEs)*. It can be shown that

$$\hat{\beta} \sim N(\beta, \ \sigma^2(X^TX)^{-1}). \qquad (2.5)$$

This result can be used to construct confidence intervals and perform hypothesis testing for β.

Note that the *closed-form* or *analytic* expressions of the results in (2.3) – (2.5) depend on the *linearity* of the model in the parameters and other model assumptions. For nonlinear or generalized linear models, which are nonlinear in the parameters, such closed-form or analytic expressions are unavailable.

For model diagnostics, one should check i) whether the model assumptions are valid, i.e., goodness of fit, constant variance, and normality; ii) whether there are any outliers; and iii) whether there are any influential observations, which may have big impacts on parameter estimates but are not necessary outliers. Model diagnostics are often based on graphical tools such as residual plots, although some formal methods are also available.

References for linear regression models are extensive, including Draper and Smith (1998) and Weisberg (2005), among others. Interested readers can find more detailed discussions of linear models in these references.

2.2.2 LME Models

For longitudinal data or clustered data, the classical linear regression model (2.2) is inappropriate because the observations within each individual or cluster may be correlated (so the independence assumption for model model (2.2) does not hold). To incorporate the correlation within individuals or clusters and the variation between individuals or clusters, we can extend model (2.2) by introducing random effects in the model and thus obtain a LME model, as demonstrated by the examples in Section 1.5 of Chapter 1. In this section, we discuss general LME models and their properties.

The Model

Let y_{ij} be the response value for individual (or cluster) i at time t_{ij}, $i =$

$1, 2, \cdots, n$, $j = 1, 2, \cdots, n_i$, and let $\mathbf{y}_i = (y_{i1}, y_{i2}, \cdots, y_{in_i})^T$ be the n_i repeated observations within individual (or cluster) i. A general LME model can be written as (Laird and Ware 1982)

$$\mathbf{y}_i = X_i\boldsymbol{\beta} + Z_i\mathbf{b}_i + \mathbf{e}_i, \qquad i = 1, 2, \cdots, n, \tag{2.6}$$

$$\mathbf{b}_i \sim N(0, D), \qquad \mathbf{e}_i \sim N(0, R_i), \tag{2.7}$$

where $\boldsymbol{\beta} = (\beta_0, \beta_1, \cdots, \beta_p)^T$ are population parameters (or *fixed effects*), $\mathbf{b}_i = (b_{i1}, \cdots, b_{iq})^T$ are random effects, X_i is a $n_i \times (p + 1)$ design matrix containing covariates of individual i, Z_i is a $n_i \times q$ design matrix (Z_i is often a submatrix of X_i), $\mathbf{e}_i = (e_{i1}, e_{i2}, \cdots, e_{in_i})^T$ are random errors of within-individual measurements, R_i is a $n_i \times n_i$ variance-covariance matrix of within-individual measurements, and D is the variance-covariance matrix of the random effects.

A distributional assumption is made for the random effects in (2.7) since the sampled subjects are thought to represent a population of subjects. The matrix D is usually unstructured, but it can be structured such as a diagonal matrix (Jennrich and Schluchter 1986). The variances of the random effects $\mathbf{b}_i$, or the diagonal elements of D, are sometimes called *variance components*, which measure the variability of the longitudinal trajectories *between* individuals that are unexplained by covariates. The variances of e_{ij}'s, or the diagonal elements of R_i, measure the variability of the repeated measurements *within* each individual. We assume that $\mathbf{b}_i$ and $\mathbf{e}_i$ are independent.

The covariance matrix R_i is often assumed to depend on i only through their dimensions. For example, it is often assumed that $R_i = \sigma^2 I_{n_i}$. This suggests that the within individual measurements are often assumed to be conditionally independent given the random effects. This assumption may be reasonable when the within individual measurements are far apart so that within-individual autocorrelation is practically negligible, or that the between-individual variation is dominant. In many cases an accurate characterization of R_i is less critical. In fact, the conditional independence assumption is standard for GLMMs. Davidian and Giltinan (2003) provided a detailed discussion on the specification of matrix R_i.

LME model (2.6) and (2.7) specifically incorporates two sources of variability: the within-individual variation and the between-individual variation. Thus, it can be interpreted as a *hierarchical two-stage model*: stage 1 specifies the within-individual variation, which is given by (2.6), and stage 2 specifies the between-individual variation, which is given by (2.7).

Note that LME model (2.6) and (2.7) differs from the classical linear regression model (2.2) only by the term $Z_i\mathbf{b}_i$, which links the random effects to the response.

The marginal distribution of the response $\mathbf{y}_i$ is given by

$$\mathbf{y}_i \sim N(X_i\boldsymbol{\beta}, \ Z_iDZ_i^T + R_i). \tag{2.8}$$

Thus, the variance-covariance structure of the repeated observations within individual i is given by

$$Cov(\mathbf{y}_i) = V_i = Z_iDZ_i^T + R_i. \tag{2.9}$$

The marginal mean $E(\mathbf{y}_i) = X_i\boldsymbol{\beta}$ can be interpreted as an average over all random effects, so it does not reflect individual longitudinal trajectories. Instead, individual-specific inference is performed by *conditioning* on the random effects $\mathbf{b}_i$. Inference for the population parameters $\boldsymbol{\beta}$ is based on the marginal distribution (2.8).

Note that the ability to derive the marginal distribution of $\mathbf{y}_i$ in *closed form* in (2.8) depends on the assumptions that the random effects $\mathbf{b}_i$ and random errors $\mathbf{e}_i$ are *linear* in LME model (2.6) and that $\mathbf{b}_i$ and $\mathbf{e}_i$ are independent and normally distributed. In GLMM or NLME models, however, closed form expressions of the marginal distributions are unavailable, as will be seen in Sections 2.3.2 and 2.4.2.

Parameter identifiability (or model identifiability) is an important problem in mixed effects models. Parameters or models are called *non-identifiable* if two sets of different parameters lead to the same probability distribution. Demidenko (2004) and Wang and Heckman (2009) discussed parameter identifiability for LME models. In particular, Wang and Heckman (2009) showed that LME model (2.6) and (2.7) is always identifiable when $R_i = \sigma^2 I$.

Inference Based on Likelihood Methods

Statistical inference for a LME model is typically based on the maximum likelihood method or the restricted maximum likelihood method (Laird and Ware 1982; Lindstrom and Bates 1988).

Let $\boldsymbol{\eta}$ denote the vector of all distinct parameters in the variance-covariance matrices D and R_i, and let $\boldsymbol{\theta} = (\boldsymbol{\beta}, \boldsymbol{\eta})$ denote all parameters in LME model (2.6) and (2.7). The likelihood for the observed data $\mathbf{y} = \{\mathbf{y}_1, \cdots, \mathbf{y}_n\}$ is given by

$$
\begin{aligned}
L(\boldsymbol{\theta}|\mathbf{y}) &= \prod_{i=1}^{n} f(\mathbf{y}_i|\boldsymbol{\beta}, \boldsymbol{\eta}) \tag{2.10} \\
&= \prod_{i=1}^{n} \int f(\mathbf{y}_i|\mathbf{b}_i, \boldsymbol{\beta}, R_i) f(\mathbf{b}_i|D) \, d\mathbf{b}_i, \tag{2.11}
\end{aligned}
$$

where

$$f(\mathbf{y}_i|\mathbf{b}_i, \boldsymbol{\beta}, R_i) = (2\pi)^{-n_i/2} |R_i|^{-1/2} \exp\Big[-(\mathbf{y}_i - X_i\boldsymbol{\beta} - Z_i\mathbf{b}_i)^T R_i^{-1}$$

$$\times (\mathbf{y}_i - X_i \boldsymbol{\beta} - Z_i \mathbf{b}_i) \Big],$$

$$f(\mathbf{b}_i|D) \;=\; (2\pi)^{-q/2}|D|^{-1/2}\exp\left(-\mathbf{b}_i^T D^{-1}\mathbf{b}_i\right).$$

In the case $R_i = \sigma^2 I$, given the variance-covariance parameters $\boldsymbol{\eta}$ (or $V_i = V_i(\boldsymbol{\eta})$), the values of $\boldsymbol{\beta}$ and σ^2 that maximize the likelihood are

$$\hat{\boldsymbol{\beta}} \;=\; \left(\sum_{i=1}^{n} X_i^T V_i^{-1} X_i\right)^{-1} \sum_{i=1}^{n} X_i^T V_i^{-1} \mathbf{y}_i,$$

$$\hat{\sigma}^2 \;=\; \frac{1}{n}\sum_{i=1}^{n}(\mathbf{y}_i - X_i\hat{\boldsymbol{\beta}})^T V_i^{-1}(\mathbf{y}_i - X_i\hat{\boldsymbol{\beta}}),$$

which may be used to derive the profile likelihood.

Maximum likelihood estimates (MLEs) of the unknown parameters in LME model (2.6) and (2.7) can be obtained using an iterative algorithm such as an *expectation-maximization (EM) algorithm* or a *Newton-Raphson method* (Laird and Ware 1982; Lindstrom and Bates 1988; Verbeke and Molenberghs 2001). Pinheiro and Bates (2002) provided computational details. In Chapter 3 (Section 3.4) we will provide a detailed description of the EM algorithm in its general form.

The EM algorithm is a popular iterative algorithm typically used to compute MLEs in the presence of missing data or unobservables (Dempster, Laird, and Rubin 1977). It is general, simple, and stable. It can be used for likelihood inference in mixed effects models in which the random effects are viewed as unobservables or "missing data".

The EM algorithm iterates between an *E-step*, which computes the conditional expectation of the "complete-data" log-likelihood given the observed data and current parameter estimates, and an *M-step*, which maximizes the conditional expectation in the E-step to obtain updated parameter estimates. Given starting values of the unknown parameters, one iterates between the E-step and the M-step until convergence. At convergence, the final parameter estimates are (possibly local) maximizers of the observed-data likelihood. Standard errors of the MLEs are not directly produced in the EM algorithm, but they can be obtained separately.

For LME model (2.6) and (2.7), if we treat the random effects as "missing data", so we have "complete data" $(\mathbf{y}, \mathbf{b}) = \{(\mathbf{y}_i, \mathbf{b}_i), i = 1, 2, \cdots, n\}$. Let k denote the iteration number, $k = 0, 1, 2, \cdots$. The E-step of the EM algorithm at iteration k computes

$$Q(\boldsymbol{\theta}|\boldsymbol{\theta}^{(k)}) \;=\; E\left[\log L(\boldsymbol{\theta}|\mathbf{y}, \mathbf{b}) \mid \mathbf{y}, \boldsymbol{\theta}^{(k)}\right] \qquad (2.12)$$

$$= E\left[\sum_{i=1}^{n}\left\{(\log f(\mathbf{y}_i|\mathbf{b}_i,\boldsymbol{\beta},R_i) + \log f(\mathbf{b}_i|D))\mid \mathbf{y}_i,\boldsymbol{\theta}^{(k)}\right\}\right],$$

where the conditional expectation is taken with respect to the conditional distribution $f(\mathbf{b}_i|\mathbf{y}_i,\boldsymbol{\theta}^{(k)})$. For LME models, the E-step simplifies to the computation of the following sufficient statistics of the variance-covariance parameters $\boldsymbol{\eta}$ in the covariance matrices D and R_i:

$$\sum_{i=1}^{n} E\left(\mathbf{e}_i^T\mathbf{e}_i|\mathbf{y}_i,\hat{\boldsymbol{\theta}}^{(k)}\right) = \sum_{i=1}^{n}\left[\hat{\mathbf{e}}_i^{(k)T}\hat{\mathbf{e}}_i^{(k)} + \text{tr}\left(\text{Cov}\left(\mathbf{e}_i|\mathbf{y}_i,\hat{\boldsymbol{\theta}}^{(k)}\right)\right)\right],$$

$$\sum_{i=1}^{n} E\left(\mathbf{b}_i\mathbf{b}_i^T|\mathbf{y}_i,\hat{\boldsymbol{\theta}}^{(k)}\right) = \sum_{i=1}^{n}\left[\hat{\mathbf{b}}_i^{(k)}\hat{\mathbf{b}}_i^{(k)T} + \text{Cov}\left(\mathbf{b}_i|\mathbf{y}_i,\hat{\boldsymbol{\theta}}^{(k)}\right)\right],$$

where

$$\hat{\mathbf{e}}_i^{(k)} = \mathbf{y}_i - X_i\hat{\boldsymbol{\beta}}^{(k)} - Z_i\hat{\mathbf{b}}_i^{(k)},$$

$$\hat{\mathbf{b}}_i^{(k)} = D(\hat{\boldsymbol{\eta}}^{(k)})Z_i^T V_i^{-1}(\hat{\boldsymbol{\eta}}^{(k)})\left(\mathbf{y}_i - X_i\hat{\boldsymbol{\beta}}^{(k)}\right),$$

$$V_i(\hat{\boldsymbol{\eta}}^{(k)}) = Z_i D(\hat{\boldsymbol{\eta}}^{(k)})Z_i^T + R_i(\hat{\boldsymbol{\eta}}^{(k)}),$$

and the posterior covariance matrices can be obtained from the joint normal density of $(\mathbf{y}_i, \mathbf{b}_i, \mathbf{e}_i)$.

The M-step finds an updated parameter estimate $\boldsymbol{\theta}^{(k+1)}$ which maximizes $Q(\boldsymbol{\theta}|\boldsymbol{\theta}^{(k)})$, i.e.,

$$\hat{\boldsymbol{\beta}}^{(k+1)} = \left[\sum_{i=1}^{n} X_i^T \hat{V}_i^{-1}(\hat{\boldsymbol{\eta}}^{(k)})X_i\right]^{-1}\sum_{i=1}^{n} X_i^T \hat{V}_i^{-1}(\hat{\boldsymbol{\eta}}^{(k)})\mathbf{y}_i,$$

$$\hat{\sigma}^{(k+1)2} = \sum_{i=1}^{n} E\left(\mathbf{e}_i^T\mathbf{e}_i|\mathbf{y}_i,\hat{\boldsymbol{\theta}}^{(k)}\right)\Big/\sum_{i=1}^{n} n_i,$$

$$\hat{D}^{(k+1)} = \sum_{i=1}^{n} E\left(\mathbf{b}_i\mathbf{b}_i^T|\mathbf{y}_i,\hat{\boldsymbol{\theta}}^{(k)}\right)\Big/n, \qquad k = 0, 1, 2, \cdots.$$

Iterating the above procedure until convergence, we obtain the MLE of $\boldsymbol{\theta}$.

Note that, for LME models, the unobservable random effects can be integrated out in the E-step of the EM algorithm, so *analytic* (or *closed-form*) expressions of the E-step and M-step can be obtained, which is not possible for GLMM and NLME models. This greatly reduces computation effort since computing $Q(\boldsymbol{\theta}|\boldsymbol{\theta}^{(k)})$ in the E-step for GLMM and NLME models can be challenging, as will be seen in Section 2.6.

Confidence intervals and hypothesis tests for the parameters in a LME model are often based on asymptotic results (Miller 1977), since the exact finite-sample distribution of the MLEs cannot be explicitly derived. That is, we can use the asymptotic normal distribution of the MLE to obtain approximate confidence intervals and hypothesis tests for the parameters. See Verbeke and Molenberghs (2001) for a detailed discussion. Alternatively, we can use bootstrap methods (Efron and Tibshirani 1993), which generally gives more accurate approximation but is computationally more intensive.

Given a dataset, the random effects can be estimated by the following *empirical Bayesian estimators*

$$\hat{\mathbf{b}}_i = E(\mathbf{b}_i|\mathbf{y}_i, \hat{\boldsymbol{\theta}}) = D(\hat{\boldsymbol{\eta}})Z_i^T V_i^{-1}(\hat{\boldsymbol{\eta}})(\mathbf{y}_i - X_i\hat{\boldsymbol{\beta}}). \tag{2.13}$$

These random effect estimates can then be used for individual-specific inference. For example, a predicted response trajectory for individual j is given by

$$\hat{\mathbf{y}}_j = X_j\hat{\boldsymbol{\beta}} + Z_j\hat{\mathbf{b}}_j.$$

Standard errors and confidence intervals for $\hat{\mathbf{b}}_i$ and $\hat{\mathbf{y}}_j$ can also be obtained (see, e.g., Verbeke and Molenberghs 2001).

Restricted Maximum Likelihood (REML) Estimates

Note that MLEs of the variance components $\boldsymbol{\eta}$ are biased downward since the degrees of freedom lost in the estimation of the mean parameters $\boldsymbol{\beta}$ are not incorporated in the estimation of $\boldsymbol{\eta}$. This can be more easily seen from the following simple example. Suppose that $z_1, \cdots, z_n$ i.i.d. $\sim N(\mu, \sigma^2)$. The MLE of σ^2 is given by $\hat{\sigma}^2 = \sum_{i=1}^n (z_i - \bar{z})^2/n$, which is biased because $E(\hat{\sigma}^2) = (n-1)\sigma^2/n$. An unbiased estimate of σ^2 is given by $\tilde{\sigma}^2 = \sum_{i=1}^n (z_i - \bar{z})^2/(n-1)$, which is the REML estimate of σ^2 described as follows.

For a LME model, less biased estimates of the variance components $\boldsymbol{\eta}$ can be obtained using the *restricted (or residual) maximum likelihood (REML) estimates* (Harville 1977). REML estimates of $\boldsymbol{\eta}$ can be obtained by maximizing the following restricted likelihood

$$L_R(\boldsymbol{\eta}|\mathbf{y}) = \int L(\boldsymbol{\beta}, \boldsymbol{\eta}|\mathbf{y}) \, d\boldsymbol{\beta},$$

which may be interpreted as integrating the mean parameters $\boldsymbol{\beta}$ out of the likelihood $L(\boldsymbol{\beta}, \boldsymbol{\eta}|\mathbf{y})$ in a Bayesian framework with uniform prior for $\boldsymbol{\beta}$. So the restricted likelihood $L_R(\boldsymbol{\eta}|\mathbf{y})$ is a likelihood only involving $\boldsymbol{\eta}$. Laird and Ware (1982) gave expressions of an EM algorithm for REML estimation. Note that the bias of the MLE of $\boldsymbol{\eta}$ depends on the dimension of the mean parameters $\boldsymbol{\beta}$. The REML estimator is less biased than the MLE for estimating $\boldsymbol{\eta}$, but REML estimate may have larger mean square errors (Corbeil and Searle 1976).

More extensive discussions of LME models can be found in Verbeke and Molenberghs (2001), Demidenko (2004), Jiang (2007), and McCulloch, Searle, and Neuhaus (2008). Pinheiro and Bates (2002) provided nice illustration using Splus/R, and Littell et al. (2006) illustrated SAS implementation.

Example 2.1

We consider the following simple LME model to illustrate the correlation introduced by the random effect in the model:

$$
\begin{aligned}
y_{ij} &= \beta + b_i + e_{ij}, && i = 1, 2, \cdots, n; \ j = 1, 2, \cdots, m, \\
b_i &\sim N(0, \sigma_b^2), && e_{ij} \ \text{i.i.d.} \ \sim N(0, \sigma_e^2),
\end{aligned}
$$

where b_i is the random effect and e_{ij} is the random error. It can be shown that the correlation coefficient between the repeated measurements $\{y_{i1}, y_{i2}, \cdots, y_{im}\}$ within subject i is given by

$$
r = corr(y_{ij}, y_{ik}) = \frac{\sigma_b^2}{\sigma_b^2 + \sigma_e^2}, \qquad j \neq k, \ j, k = 1, 2, \cdots, m.
$$

Thus, the random effect b_i introduces correlation r between the repeated measurements. If there is no random effect (i.e., $\sigma_b^2 = 0$), there is no correlation (i.e., $r = 0$). If the between-individual variation is much larger than the within-individual variation (i.e., $\sigma_b^2 \gg \sigma_e^2$), the correlation r will be very high (close to 1).

Example 2.2 *Mental distress data*

Consider the mental distress data in Section 1.3.1 of Chapter 1 (page 6). We fit a simple linear regression model with GSI score (y_{ij}) as the response and time (t_{ij}, in month) as a covariate. If we ignore the correlation between the repeated measurements within each individual, we may simply fit the following linear regression model

$$
y_{ij} = \beta_1 + \beta_2 t_{ij} + e_{ij}, \qquad i = 1, \cdots, n; \ j = 1, \cdots, n_i, \tag{2.14}
$$

where e_{ij} i.i.d. $\sim N(0, \sigma^2)$. However, the repeated measurements within each individual are likely to be correlated, and there is also a large variation between individuals. Thus, it may be more appropriate to consider the following LME model

$$
\begin{aligned}
y_{ij} &= \beta_{1i} + \beta_{2i} t_{ij} + e_{ij}, && \tag{2.15} \\
\beta_{1i} &= \beta_1 + b_{1i}, && \beta_{2i} = \beta_2 + b_{2i}, && i = 1, \cdots, n; \ j = 1, \cdots, n_i,
\end{aligned}
$$

where e_{ij} i.i.d. $\sim N(0, \sigma^2)$, and $(b_{1i}, b_{2i})^T \sim N(0, D)$ are random effects.

There are missing data in this dataset. As an illustration, here we only use complete-cases, i.e., individuals with missing data are removed. Table 2.1 shows the estimates based on models (2.14) and (2.15). We see that there is a strong downward trend of GSI scores over time based on either model (significant negative estimates of β_2 in both models), suggesting that the depression levels decrease over time. However, the two models may produce different estimates. In particular, the two models produced quite different estimates of the

Table 2.1 *Estimates based on linear model (2.14) and LME model (2.15) for the mental distress data*

Model	Par.	Est.	S. E.	Par.	Est.	S. E.	Par.	Est.
Linear Model (2.14)	β_1	1.015	0.026	β_2	−0.276	0.062	σ	0.67
LME Model (2.15)	β_1	1.024	0.042	β_2	−0.222	0.038	σ	0.36

S.E.: standard error

standard errors. Estimates based on the LME model (2.15) should be more reliable. Moreover, by separating the within-individual and between-individual variations, the LME model (2.15) produced much smaller estimate of the residual standard error σ than the linear model (2.14).

2.3 Nonlinear Mixed Effects (NLME) Models

2.3.1 Nonlinear Regression Models

Classical linear regression models have been widely used because of their simplicity, which is an important advantage before modern computers become available. However, linear models usually only provide description of observed data, rather than trying to understand data, since they are usually chosen based on simplicity and reasonable fit of the observed data without necessarily understanding of the data-generation mechanisms. In other words, linear models usually provide little understanding of the data-generation mechanism, but only provide a local *approximation* to the true relationship between the response and covariates, if such true relationship exists. With the availability of modern computing power, there may be little reason to constrain one's choice to linear models.

Nonlinear regression models, on the other hand, attempt to *understand* the mechanics of data generation, so they are often called *mechanistic models* (or "scientific" models). The advantages of nonlinear models include:

- nonlinear models may provide better predictions outside the range of observed data than that of linear models, since nonlinear models are usually based on data-generating mechanisms;

- parameters in nonlinear models often have natural physical interpretations;

- nonlinear models may require few parameters than the corresponding linear models that fit the data equally well.

Note that, however, in many cases we do not know the data-generating mechanisms. In these cases, linear models would be good choices.

Unlike linear models, for nonlinear models there are typically *no analytic* or *closed-form* expressions for parameter estimates, so an iterative algorithm is generally required and computation sometimes may be intensive. Moreover, in fitting nonlinear models it is important to choose good *starting values* for the iterative algorithms since the likelihood may have multiple modes. In this section we briefly review the essential ideas and methods for nonlinear regression models for cross-sectional data.

Let y_i and $\mathbf{x}_i = (x_{i1}, \cdots, x_{ip})^T$ be the response and covariates for individual i respectively, $i = 1, 2, \cdots, n$. A general *nonlinear regression model* for cross-sectional data can be written as

$$y_i = g(\mathbf{x}_i, \boldsymbol{\beta}) + \epsilon_i, \qquad i = 1, 2, \cdots, n, \qquad (2.16)$$

where g is a known nonlinear function, $\boldsymbol{\beta}$ is a vector of regression parameters, and ϵ_i is the random error. Assumptions for a standard nonlinear regression model are the same to those for a standard linear model, i.e., (i) the errors ϵ_i's are independent, (ii) the errors ϵ_i's have mean zero and constant variance σ^2, and (iii) the errors ϵ_i's are normally distributed. Some of these assumptions may be relaxed for more general models.

Statistical inference for a nonlinear regression model can be based on the least squares method or the likelihood method. The ordinary least-squares estimator for $\boldsymbol{\beta}$ is to minimize the sum of squares $\sum_{i=1}^n (y_i - g(\mathbf{x}_i, \boldsymbol{\beta}))^2$. This can be achieved by solving the following estimating equation

$$\sum_{i=1}^n \frac{\partial g(\mathbf{x}_i, \boldsymbol{\beta})}{\partial \boldsymbol{\beta}} [y_i - g(\mathbf{x}_i, \boldsymbol{\beta})] = 0. \qquad (2.17)$$

An iterative algorithm such as the Newton-Raphson method is often needed to solve the above equation.

Alternatively, under the normality assumption for the errors, i.e., ϵ_i i.i.d. $\sim N(0, \sigma^2)$, the MLE of $\boldsymbol{\beta}$ can be obtained by maximizing the likelihood function

$$L(\boldsymbol{\beta}, \sigma^2 | \mathbf{y}) = \prod_{i=1}^n \frac{1}{\sqrt{2\pi\sigma^2}} \exp\left[-\frac{(y_i - g(\mathbf{x}_i, \boldsymbol{\beta}))^2}{2\sigma^2}\right].$$

So the MLE of $\boldsymbol{\beta}$ satisfies the following likelihood equation

$$\frac{\partial \log L(\boldsymbol{\beta}, \sigma^2 | \mathbf{y})}{\partial \boldsymbol{\beta}} = 0,$$

which is identical to the least-squares equation (2.17). Therefore, the ordinary least-squares estimator of $\boldsymbol{\beta}$ is the same as the MLE of $\boldsymbol{\beta}$, and estimation for a nonlinear regression model is analogous to that for a linear regression model.

For nonlinear regression models, analytic or closed-form expressions for parameter estimates are unavailable. However, statistical inference can still be

carried out based on the standard asymptotic results of likelihood methods under the usual regularity conditions (see Chapter 12). That is, under some regularity conditions, MLEs of the model parameters are consistent, asymptotically
normal, and asymptotically most efficient. Confidence intervals and hypothesis testing can be based on the asymptotic normality of the MLEs. Therefore,
with the availability of modern computers and software, statistical inference
for nonlinear models does not offer much more difficulties than that for linear
models.

In many cases, nonlinear models can be derived from a set of differential equations based on the understanding of the underlying data-generation mechanisms. Sometimes closed-form expressions of the models can be obtained by
solving the differential equations under plausible assumptions (see, e.g., Wu
and Ding 1999; Lindsey 2001). In some cases, however, the differential equations cannot be solved explicitly, so the nonlinear models are implicit (see, e.g.,
Ramsay et al. 2007; Huang and Lu 2009).

The developments of nonlinear models require close collaboration between
statisticians and subject-area scientists, but such models may not always available since the true data-generation mechanisms can be highly complex. Note
that, in principle, any smooth nonlinear functions can be approximated by a
high-order polynomial based on Taylor series expansions, if the functions are
sufficiently smooth. However, high order polynomials are often unstable in
replications of the data so are generally not recommended.

Nonlinear models have been widely used in practice, such as HIV viral dynamics, pharmacokinetics, pharmacodynamics, molecular genetics, and growth or
decay. More detailed discussions of nonlinear models can be found in Bates
and Watts (1988), Seber and Wild (1989), and Lindsey (2001).

Example 2.3   *Growth curve models*

In the analysis of *growth curves*, nonlinear models are usually necessary. There
are various growth curve models (see, e.g., Lindsey 2001). Here we consider
a simple monomolecular growth function. Let $y(t)$ be the size at time t (e.g.,
size of an animal), and let $\mu(t) = E(y(t))$. Suppose that the growth rate is
proportional to the remaining size. Then $\mu(t)$ satisfies the following differential
equation:
$$\frac{d\mu(t)}{dt} = \beta_1(\beta_0 - \mu(t)), \qquad \beta_1 > 0,$$
which can be solved analytically, with solution
$$\mu(t) = \beta_0 + \beta_2 e^{-\beta_1 t}.$$

Thus, given an observed sample, we can consider the following nonlinear regression model for estimating the parameters
$$y_{ij} = \beta_0 + \beta_2 e^{-\beta_1 t_{ij}} + e_{ij}, \qquad i = 1, \cdots, n, \ j = 1, \cdots, n_i, \qquad (2.18)$$

where y_{ij} is the size for individual i at measurement time t_{ij} and e_{ij} is the corresponding measurement error. If the parameters β_j's vary greatly across individuals, we can introduce random effects to account for the between-individual variation. Note that, when $\beta_0 + \beta_2 = 0$ (i.e., when the initial size is 0), the above model is called the *von Bertalanffy growth curve*, which is often used in ecology to describe animal growth.

Example 2.4 *Pharmacokinetics*

Studies of *pharmacokinetics* are important in drug developments (Gibaldi and Perrier 1982; Gutfreund 1995; Lindsey 2001). Pharmacokinetics studies the course of absorption, distribution, metabolism, and elimination of some substance in the body over time, given drug dose, i.e., how the drug moves through the body. Suppose that a substance enters the body via ingestion. Let $y(t)$ be the concentration of the substance in the body at time t (usually measured in the blood), and let $\mu(t) = E(y(t))$. Let $\mu_0(t)$ be the amount at the absorption site (e.g., stomach). A commonly used one-compartment model is based on the following differential equations

$$\frac{d\mu(t)}{dt} = \beta_1 \mu_0(t) - \beta_2 \mu(t),$$

$$\frac{d\mu_0(t)}{dt} = -\beta_1 \mu_0(t),$$

where β_1 is the absorption rate and β_2 is the elimination rate. The above differential equations have an analytic solution given by

$$\mu(t) = \frac{\beta_1 x}{(\beta_1 - \beta_2)\beta_3} \left(e^{-\beta_2 t} - e^{-\beta_1 t} \right),$$

where x is the dose of the substance and β_3 is the volume of distribution. Therefore, given an observed sample, we can consider the following nonlinear regression model for estimating the parameters

$$y_{ij} = \frac{\beta_1 x_i}{(\beta_1 - \beta_2)\beta_3} (e^{-\beta_2 t_{ij}} - e^{-\beta_1 t_{ij}}) + e_{ij}, \qquad (2.19)$$

$$i = 1, \cdots, n, \quad j = 1, \cdots, n_i,$$

where y_{ij} is the concentration for individual i at time t_{ij} and e_{ij} is the corresponding random error. This nonlinear model is widely used.

2.3.2 NLME Models

Nonlinear models for cross-sectional data can be extended to modeling longitudinal or clustered data by introducing random effects in the models to account for *correlation* among the repeated observations within each individual or cluster and the variation between individuals. The resulting models are called *nonlinear mixed effects (NLME) models*. NLME models are useful in many longitudinal studies, such as studies of growth and decay, HIV viral dynamics, and pharmacokinetics analysis.

The Model

We can write a NLME model as a two-stage *hierarchical nonlinear model* to specifically model intra-individual variation and inter-individual variation. In the first stage we specify the mean and covariance structure for a given individual or cluster (see equation (2.20) below), and in the second stage we model the between-individual variation through random effects (see equation (2.21) below). Specifically, let $\mathbf{y}_i = (y_{i1}, \cdots, y_{in_i})^T$ be the n_i repeated observations of the response for individual or cluster i. A general NLME model can be written as follows

$$
\begin{aligned}
y_{ij} &= g(t_{ij}, \boldsymbol{\beta}_i) + e_{ij}, & &\text{(2.20)} \\
\boldsymbol{\beta}_i &= h(\mathbf{x}_i, \boldsymbol{\beta}, \mathbf{b}_i), & i = 1, 2, \cdots, n, \; j = 1, 2, \cdots, n_i, & \text{(2.21)} \\
\mathbf{b}_i &\sim N(0, D), & \mathbf{e}_i \sim N(0, R_i), & \text{(2.22)}
\end{aligned}
$$

where $g(\cdot)$ is a known nonlinear function, $h(\cdot)$ is often chosen as a linear function, $\boldsymbol{\beta}_i$ and $\boldsymbol{\beta}$ are individual-specific parameters and fixed-effects parameters respectively, $\mathbf{x}_i$ contains covariates for individual i, R_i is a covariance matrix for the repeated observations within individual i, D is a covariance matrix for the random effects, $\mathbf{e}_i = (e_{i1}, e_{i2}, \cdots, e_{in_i})^T$ are random errors for observations within individual i, and $\mathbf{b}_i$'s are random effects. We assume that $\mathbf{e}_i$ and $\mathbf{b}_i$ are independent.

Davidian and Giltinan (1995, 2003) provided a detailed discussion about the choice of the covariance matrix R_i. The within individual error $\mathbf{e}_i$ may be partitioned into two sources: deviation of the observed trajectory for individual i from the assumed model and possible measurement error. The choice of R_i should be guided by practical considerations. In many cases, we can simply choose $R_i = \sigma^2 I_{n_i}$, which may be reasonable if the observation times t_{ij}'s are far apart for fixed i (so correlation between y_{ij} and y_{ik} is negligible) or if one emphasizes measurement errors. The advantage of this choice is that it greatly reduces the number of parameters and it may avoid parameter identifiability problems. Note that the accurate specification of R_i may be less critical when the between individual variation is dominant. This conditional independence assumption is standard for GLMMs. The covariance matrix D measures the between individual variation that is not explained by covariates $\mathbf{x}_i$. The function $h(\cdot)$ is often chosen to be simple (e.g., linear) and parsimonious.

A NLME model is a *subject-specific (SS)* model. The parameter $\boldsymbol{\beta}$ has an interpretation as the typical value of the individual-specific parameter $\boldsymbol{\beta}_i$, rather than the typical average response profile as in a population averaged (PA) model. In other words, the interpretations of β in SS models and PA models are not the same for NLME models. This distinction is important for NLME models, but not for LME models (Davidian and Giltinan 2003; Heagerty 1999). Thus, one should take this into account when choosing a NLME model in practice.

Statistical Inference

If the within-individual measurements are rich, i.e., if the n_i's are large, inference can be based on the following *two-step method*: in step 1 individual parameters β_i are estimated by fitting a nonlinear regression model to the repeated observations within each individual using standard estimation methods for nonlinear models such as the least square method, then in step 2 the individual estimates $\hat{\beta}_i$ are used to estimate the fixed parameters β and perform inference based on large-sample asymptotic results. Davidian and Giltinan (1995, 2003) described this two-step method in details. This two-step method is simple and requires no distributional assumptions, but it requires large n_i's. Moreover, no software is currently available for this two-step method. Therefore, in the following we focus on the likelihood method.

Let $\theta = (\beta, \eta, D)$ denote all parameters, where η is the collection of distinct parameters in the covariance matrix R_i. The marginal distribution of the response y_i is given by

$$f(\boldsymbol{y}_i|\boldsymbol{\theta}) = \int f(\mathbf{y}_i|\boldsymbol{x}_i, \boldsymbol{\beta}, \boldsymbol{\eta}, \boldsymbol{b}_i) f(\boldsymbol{b}_i|D) \, d\boldsymbol{b}_i, \qquad (2.23)$$

so the likelihood is

$$L(\boldsymbol{\theta}|\boldsymbol{y}) = \prod_{i=1}^{n} \int f(\mathbf{y}_i|\boldsymbol{x}_i, \boldsymbol{\beta}, \boldsymbol{\eta}, \boldsymbol{b}_i) f(\boldsymbol{b}_i|D) d\boldsymbol{b}_i. \qquad (2.24)$$

Unlike LME models, the marginal distribution (2.23) and the likelihood (2.24) typically do *not* have *analytic* or *closed form* expressions (except in special cases), since the NLME model is *nonlinear* in the random effects $\mathbf{b}_i$. This is a key difference between a LME model and a NLME model, and it leads to many computational problems which do not exist for LME models. This difference also leads to the distinction between a SS model and a PA model.

A major difficulty of likelihood inference for a NLME model is the evaluation of the intractable integral in the likelihood (2.24). Commonly used methods include numerical or Monte Carlo integration methods, EM algorithms, and approximate methods. Specifically, the following approaches are often used:

- *numerical or Monte Carlo methods.* These methods use Gauss-Hermite quadrature or Monte Carlo integration techniques to approximate the integral in the likelihood (2.24).

- *EM algorithms.* These methods use EM algorithms to indirectly maximize the likelihood. Typically, a Monte Carlo or stochastic or numerical method is needed in the E-step of the EM algorithm.

- *approximate methods.* These methods use Taylor expansions to *linearize* the NLME model and then iteratively solve the resulting LME models, or use a *Laplace approximation* to directly approximate the likelihood (2.24).

The Gauss-Hermite quadrature method works well when the random effects have a low dimension and follow a normal distribution, but the computational effort increases exponentially with the dimension of the random effects or the dimension of the integral (see, e.g., Evans and Swartz 2000). The EM algorithm is very general and stable and works for random effects with any dimensions and any distributions, but it can be computationally intensive. The approximate methods are computationally very efficient since they avoid the intractable integration, so they are implemented in many standard software such as Splus/R and SAS, but the approximations cannot be made arbitrarily accurate. Since these methods are also used in GLMMs, their detailed descriptions will be presented in Section 2.6, after GLMMs are discussed.

For a comprehensive discussion of NLME models, see Davidian and Giltinan (1995, 2003), Vonesh and Chinchilli (1997), and Demidenko (2004), among others.

Example 2.5 *Mixed effects growth and pharmacokinetics models*

The growth curve model and pharmacokinetics model, as examples of nonlinear models for cross-sectional data, presented in Section 2.3.1 can be extended to longitudinal data by introducing random effects to appropriate parameters, leading to NLME models.

Specifically, the corresponding NLME model for the growth curve model (2.18) (page 51) can be written as

$$
\begin{align}
y_{ij} &= \beta_{i0} + \beta_{i2} e^{-\beta_{i1} t_{ij}} + e_{ij}, \tag{2.25} \\
(\beta_{i0}, \beta_{i1}, \beta_{i2})^T &= (\beta_0, \beta_1, \beta_2)^T + (b_{i0}, b_{i1}, b_{i2})^T, \tag{2.26} \\
e_{ij} \;\; \text{i.i.d.} \;\; &\sim \;\; N(0, \sigma^2), \quad (b_{i0}, b_{i1}, b_{i2})^T \sim N(0, D). \tag{2.27}
\end{align}
$$

Note that, in the above second-stage model (2.26), we can incorporate covariates to partially explain the variation in the individual-specific parameters $(\beta_{i0}, \beta_{i1}, \beta_{i2})$. Moreover, some parameters may not vary substantially across individuals so the corresponding random effects may be unnecessary. See Section 2.7.1 for strategies of selecting random effects and covariates.

Similarly, the corresponding NLME model for the pharmacokinetics model

(2.19) (page 52) can be written as

$$y_{ij} = \frac{\beta_{i1} x_i}{(\beta_{i1} - \beta_{i2})\beta_{i3}} (e^{-\beta_{i2} t_{ij}} - e^{-\beta_{i1} t_{ij}}) + e_{ij}, \quad (2.28)$$

$$(\beta_{i1}, \beta_{i2}, \beta_{i3})^T = (\beta_1, \beta_2, \beta_3)^T + (b_{i1}, b_{i2}, b_{i3})^T, \quad (2.29)$$

$$e_{ij} \text{ i.i.d.} \sim N(0, \sigma^2), \quad (b_{i1}, b_{i2}, b_{i3})^T \sim N(0, D), \quad (2.30)$$

which is widely used in pharmacokinetics. Again, we can introduce covariates and select random effects for the second-stage model (2.29).

Example 2.6 *HIV viral dynamic models*

Consider the AIDS study in Section 1.3.2 of Chapter 1 (page 9). Viral dynamic models have been used to describe the virus elimination and production process during anti-HIV treatments. These HIV viral dynamic models have received great attention in recent years.

Wu and Ding (1999) proposed the following mathematical model for HIV dynamics

$$\begin{aligned}
dT_p/dt &= k^* T V_I - \delta_p T_p, \\
dV_I/dt &= (1 - \eta)P - cV_I, \quad (2.31) \\
dV_{NI}/dt &= \eta P + P^* + N\delta_p T_p - cV_{NI},
\end{aligned}$$

where T_p, V_I, V_{NI} are the concentrations of the productively infected cells, infectious virus, and non-infectious virus respectively, δ_p is the rate of infected cells T_p killed by HIV, η is the proportion of non-infectious virus produced by infected cells, P^* accounts for virus production from the ignored compartments, k^* is a constant, and N is the number of virions per cell.

Under some reasonable assumptions and approximations, model (2.31) leads to the following bi-exponential (two-compartments) dynamic model, which is appropriate for the early period after initiation of the treatments (say, first three months)

$$V(t) = P_1 e^{\lambda_1 t} + P_2 e^{\lambda_2 t}, \quad (2.32)$$

where $V(t)$ is the total virus at time t, λ_1 and λ_2 are the turnover rates of productively infected cells and long-lived and/or latently infected cells respectively, and P_1 and P_2 are baseline values.

Since the repeated measurements on each individual may be correlated and there are large variations in viral load across individuals and within individuals (see Figure 1.3, page 11), we introduce random effects and random error in the nonlinear model (2.32) and obtain the following NLME model

$$\begin{aligned}
y_{ij} &= \log_{10}(P_{1i} e^{-\lambda_{1i} t_{ij}} + P_{2i} e^{-\lambda_{2i} t_{ij}}) + e_{ij}, \quad (2.33) \\
P_{1i} &= P_1 + b_{1i}, \qquad \lambda_{1i} = \lambda_1 + b_{2i}, \\
P_{2i} &= P_2 + b_{3i}, \qquad \lambda_{2i} = \lambda_2 + b_{4i},
\end{aligned}$$

Table 2.2 *Estimates of viral dynamic parameters*

Model	P_1		λ_1		P_2		λ_2		σ
	Est.	S.E.	Est.	S.E.	Est.	S.E.	Est.	S.E.	Est.
Nonlin. Model (2.32)	12.20	0.32	0.40	0.05	7.70	0.33	0.02	0.005	0.66
NLME Model (2.33)	12.35	0.21	0.42	0.02	7.58	0.26	0.02	0.005	0.29
NLME Model (2.34)	12.38	0.19	0.46	0.06	7.45	0.27	0.02	0.005	0.27

S.E.: standard error

where $y_{ij} = \log_{10}(V(t_{ij}))$ and the $\log_{10}$-transformation is used to stabilize the variance and make the data more normally distributed, $\mathbf{b}_i = (b_{1i}, b_{2i}, b_{3i}, b_{4i})^T$ are random effects, and e_{ij} is a random error. We assume that e_{ij} i.i.d. $\sim N(0, \sigma^2)$ and $\mathbf{b}_i \sim N(0, D)$.

In model (2.33), the individual-specific parameters are used to incorporate the variation of these parameters across individuals, which are decomposed into fixed effects and random effects. Part of this variation may also be explained by covariates. For example, the variation in the first-phase viral decay rate λ_{1i} may be partially explained by variation in CD4 counts. Moreover, we may allow the individual-specific parameters to change over time. Therefore, an alternative NLME model is

$$
\begin{aligned}
y_{ij} &= \log_{10}(P_{1i}e^{-\lambda_{1ij}t_{ij}} + P_{2i}e^{-\lambda_{2i}t_{ij}}) + e_{ij}, & (2.34) \\
P_{1i} &= P_1 + b_{1i}, & \lambda_{1ij} = \lambda_1 + \beta\,CD4_{ij} + b_{2i}, \\
P_{2i} &= P_2 + b_{3i}, & \lambda_{2i} = \lambda_2 + b_{4i}.
\end{aligned}
$$

Other covariates may also be introduced in the model. One may select a good model based on AIC or BIC criteria as well as scientific considerations.

We fit the nonlinear regression model (2.32) using the nonlinear least-square method and fit the NLME models (2.33) and (2.34) using the approximate method of Lindstrom and Bates (1990) to the data in the first three months. In fitting the nonlinear model (2.32), we assume that the errors e_{ij}'s are i.i.d., so we ignore the correlation in the clusters. Table 2.2 shows some parameter estimates. Other parameter estimates are: $\hat{\beta} = -0.00013$ (with standard error 0.0002) and $diag(\hat{D})^{1/2} = (1.04, 0.05, 1.50, 0.03)$. We see that ignoring correlation in the data may lead to possibly biased estimates. Moreover, by splitting the variation in the data into within-individual measurement error e_{ij} and between-individual random effects $\mathbf{b}_i$, one can estimate the within-individual measurement error more accurately (note that the estimate of σ based on model (2.32) is more than twice as large as that based on NLME models (2.33) or (2.34). Figure 2.1 shows the fitted values and observed values based on NLME

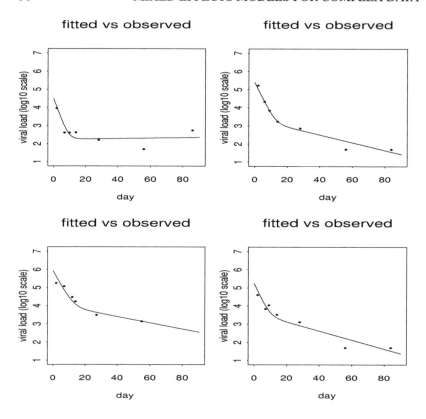

Figure 2.1 *Fitted curves based on NLME model (2.33) for four randomly selected subjects.*

model (2.33) for four randomly selected individuals. We see that NLME model (2.33) fits the observed data well.

2.4 Generalized Linear Mixed Models (GLMMs)

2.4.1 Generalized Linear Models (GLMs)

The nonlinear models in Section 2.3.1 extend classical linear models by allowing arbitrary nonlinear relationships between the mean response and covariates, but the response is assumed to be normally distributed, as in linear models. In practice, however, there are various types of responses, and many of them are unlikely to follow normal distributions even after transformations. For example, if the response is a binary variable taking only two possible values (say, 0 or 1), such a response cannot follow a normal distribution no matter what transformation is used. Generalized linear models (GLMs) extend classical linear models by allowing responses to follow distributions in the exponential family, as well as allowing nonlinear relationships between the response and covariates. The exponential family includes a wide range of commonly used distributions, such as normal, binomial, and Poisson distributions. Therefore, GLMs greatly extend the applicability and popularity of regression models. In this section, we briefly review GLMs for cross-sectional data.

The Model

Let $\{y_1, y_2, \cdots, y_n\}$ be a sample of i.i.d. observations from a distribution in the *exponential family*. Then the general probability density function of y_i can be expressed in the form

$$f(y|\theta, \phi) = \exp\left\{\frac{y\theta - b(\theta)}{a(\phi)} + c(y, \phi)\right\}, \qquad (2.35)$$

where $a(\cdot)$, $b(\cdot)$ and $c(\cdot)$ are known functions, θ is called the *canonical parameter* representing the *location*, and ϕ is called the *dispersion parameter* representing the *scale*. It can be shown that

$$E(y_i) = \mu = \partial b(\theta)/\partial\theta, \qquad Var(y_i) = a(\phi)\partial^2 b(\theta)/\partial\theta^2.$$

The following commonly used distributions belong to the exponential family: normal distribution, binomial distribution, Poisson distribution, gamma distribution, and inverse Gaussian distribution.

For a *normal distribution* $N(\mu, \sigma^2)$ with mean μ and variance σ^2, we have

$$\theta = \mu, \qquad \phi = \sigma^2,$$
$$a(\phi) = \phi, \qquad b(\theta) = \theta^2/2, \qquad c(y, \phi) = -(y^2/\phi + \log(2\pi\phi))/2.$$

For a *binomial distribution* with probability $P(y = k) = \begin{pmatrix} n \\ k \end{pmatrix} \mu^k (1 - \mu)^{n-k}$, $k = 0, 1, \cdots, n$, where $0 < \mu < 1$, we have

$$\theta = \log(\mu/(1 - \mu)), \qquad \phi = 1,$$
$$b(\theta) = -n\log(1 - \mu), \qquad c(y, \phi) = \log\begin{pmatrix} n \\ y \end{pmatrix},$$
$$E(y) = n\mu, \qquad Var(y) = n\mu(1 - \mu).$$

The binomial distribution reduces to the *Bernoulli distribution* when $n = 1$, i.e.,

$$P(y = k) = \mu^k (1 - \mu)^{1-k}, \qquad k = 0, 1,$$

with

$$E(y) = \mu = P(y = 1), \qquad Var(y) = \mu(1 - \mu).$$

For a *Poisson distribution* with probability $P(Y = k) = (k!)^{-1}e^{-\mu}\mu^k$, $k = 0, 1, 2, \cdots$, where $\mu > 0$, we have

$$\theta = \log(\mu), \qquad \phi = 1,$$
$$a(\phi) = 1, \qquad b(\theta) = e^\theta, \qquad c(y, \phi) = -\log(y!),$$

with $E(y) = Var(y) = \mu$. These three distributions are most commonly used in GLMs.

Let $\mu_i = E(y_i)$ be the mean response and $\boldsymbol{x}_i$ be a vector of covariates, $i = 1, \cdots, n$. Let

$$\eta_i = \boldsymbol{x}_i^T \boldsymbol{\beta} = x_{i1}\beta_1 + x_{i2}\beta_2 + \cdots + x_{ip}\beta_p$$

be the *linear predictor*, which combines the predictors (covariates) in a linear form, where $\boldsymbol{\beta} = (\beta_1, \cdots, \beta_p)^T$ is a $p \times 1$ vector of unknown parameters. A *generalized linear model (GLM)* for cross-sectional data can be written as

$$g(\mu_i) = \boldsymbol{x}_i^T \boldsymbol{\beta}, \qquad i = 1, 2, \cdots, n, \qquad (2.36)$$

where $g(\cdot)$ is a monotone and differentiable function, called the *link function*. Thus, a GLM has two components:

- the response follows a distribution in the exponential family;
- the link function describes how the mean response is related to a linear combination of predictors.

GLMs are special nonlinear regression models. They include classical linear models as a special case when the response distribution is normal and the link function is the identity function. Note that GLMs are still restrictive in that they involve essentially linear regression models and only cover distributions from the exponential family. The two most widely used GLMs for non-normal data are logistic regression models and Poisson regression models (see Example 2.7 and Example 2.8 below).

Statistical Inference

Statistical inference for GLMs can be based on the likelihood method. For a general GLM, the log-likelihood function is given by

$$l(\boldsymbol{\beta}, \phi) = \sum_{i=1}^{n} \left\{ \frac{y_i\theta_i - b(\theta_i)}{a(\phi)} + c(y_i, \phi) \right\}.$$

Note that the regression parameters $\boldsymbol{\beta}$ is implicit in the loglikelihood function $l(\boldsymbol{\beta}, \phi)$ since $g(E(y_i)) = \mathbf{x}_i^T\boldsymbol{\beta}$ and $E(y_i) = \partial b(\theta_i)/\partial\theta_i$. Since the loglikelihood $l(\boldsymbol{\beta}, \phi)$ is nonlinear in the parameters $\boldsymbol{\beta}$ and ϕ, MLEs are obtained using an iterative algorithm such as the Newton-Raphson method, which is equivalent to the *iteratively reweighted least squares* method described in McCullagh and Nelder (1989). Inference is often based on the *deviance*, which can be defined as the difference between the log-likelihoods for the full model and for the fitted model. The full model (or saturated model) is the most complex model where the data is explained exactly (i.e., it represents the data as being entirely systematic), while the null model is the smallest model where there is no relationship between the predictors and the response (i.e., it represents the data as being entirely random). When comparing two nested models, under the null hypothesis of no difference between the two models, the difference in the deviances asymptotically follows a χ^2 distribution with degrees of freedom being the difference of the number of parameters in the two models being compared.

Note that for some most common distributions in the exponential family, such as the binomial distribution and the Poisson distribution, the variance is completely determined by the mean or the variance has a strong relationship with the mean, unlike normal regression models where the mean and the variance parameters are separate and can vary freely. This is very restrictive in practice since the variation in the observed data may *not* agree with the theoretical variance assumed in the model. If the variation in the data is larger or smaller than the theoretical variance determined by the assumed distribution, the problem is called an *over-dispersion* or a *under-dispersion* problem, which must be addressed for correct inference (McCullagh and Nelder 1989; Dean 1998; McCulloch, Searle, and Neuhaus 2008). Overdispersion problems can arise in longitudinal or clustered data if the correlation within clusters are not addressed. One way to address these problems is to specify the mean and variance functions *separately* without a distributional assumption. This approach is called the *quasi-likelihood* method, which is closely related to the generalized estimating equation (GEE) method (see Chapter 10 for a detailed discussion).

There are extensive literature on GLMs. Interested readers are referred to Mc-Cullagh and Nelder (1989), Fahrmeir et al. (2001), Diggle et al. (2002), and McCulloch et al. (2008).

Example 2.7 *Logistic regression models*

Logistic regression models are perhaps the most widely used models in the GLM family. A logistic regression model is usually used when the response y is a binary variable taking only two possible values (say, 0 or 1). In this case, it may be reasonable to assume that y follows a binomial or Bernoulli distribution. There are several ways to link the mean response to covariates. The most popular choice of the link function is the following *logit link*

$$g(\mu) = \log\left(\frac{\mu}{1-\mu}\right),$$

where $\mu = E(y) = P(y = 1)$, which has an attractive interpretation as the odds in log-scale. With the logit link, the resulting GLM is the following *logistic regression model*

$$\log\left(\frac{\mu_i}{1 - \mu_i}\right) = x_{i1}\beta_1 + x_{i2}\beta_2 + \cdots + x_{ip}\beta_p, \qquad i = 1, \cdots, n, \quad (2.37)$$

where $\mu_i = E(y_i) = P(y_i = 1)$. Other common link functions for binary responses include probit link and complementary log-log link, but they do not have the attractive interpretation as the logit link.

Example 2.8 *Poisson regression models*

If the response y is a count, it may be reasonable to assume that y follows a Poisson distribution. Then we may choose the following log link

$$g(\mu) = \log(\mu),$$

where $\mu = E(y)$. The resulting GLM is the following Poisson GLM

$$\log(\mu_i) = x_{i1}\beta_1 + x_{i2}\beta_2 + \cdots + x_{ip}\beta_p, \qquad (2.38)$$

where $\mu_i = E(y_i)$. Over-dispersion arises when the observed variance in the response data is greater than the theoretical variance $Var(y_i) = \mu_i$, which must be addressed for valid inference.

2.4.2 GLMMs

GLMs for cross-sectional data can be extended to longitudinal data or clustered data by introducing random effects in the models in a way similar to LME and NLME models. The random effects incorporate correlation between the repeated observations within each individual or cluster and variation between individuals or clusters. The resulting models are called *generalized linear mixed models (GLMMs)*. In a GLMM, it is assumed that correlation arises among repeated observations within a given individual or cluster because of the shared random effects, but these repeated observations are assumed to be conditionally independent given the random effects.

The Model

Let $\mathbf{y}_i = (y_{i1}, y_{i2}, \cdots, y_{in_i})^T$ be the n_i repeated observations of the response within individual or cluster i, $i = 1, 2, \cdots, n$. We assume that, conditioning on the random effects $\mathbf{b}_i$, the repeated measurements $y_{i1}, y_{i2}, \cdots, y_{in_i}$ are independent and each follows a distribution in the exponential family, i.e., a general GLMM can be written as

$$\begin{aligned} g(\mu_{ij}) &= \mathbf{x}_{ij}^T\boldsymbol{\beta} + \mathbf{z}_{ij}^T\mathbf{b}_i, & (2.39) \\ \mathbf{b}_i &\sim N(0, D), & j = 1, 2, \cdots, n_i, \quad i = 1, 2, \cdots, n, \quad (2.40) \end{aligned}$$

where $\mu_{ij} = E(y_{ij}|\boldsymbol{\beta}, \boldsymbol{b}_i)$ is the conditional mean, $\boldsymbol{x}_{ij}$ and $\boldsymbol{z}_{ij}$ are vectors containing covariates ($\boldsymbol{z}_{ij}$ is often a sub-vector of $\boldsymbol{x}_{ij}$), $\boldsymbol{\beta}$ is a vector of fixed effects, and D is a covariance matrix. Since the model is specified based on the conditional mean, GLMMs are sometimes called *conditional models* or *subject-specific models*, in contrast to the marginal GEE models.

One can see that the difference between a GLM and a GLMM is the random effect term $\boldsymbol{z}_{ij}^T \boldsymbol{b}_i$ in (2.40). However, this term will cause much of the computational problems since the random effects $\boldsymbol{b}_i$ are unobservable and are nonlinear in the model, as in a NLME model. The interpretation of GLMMs is similar to NLME models (see Section 2.3.2).

Statistical Inference

Statistical inference for a GLMM is typically based on the likelihood method. In GLMM (2.39) and (2.40), the marginal distribution for $\boldsymbol{y}_i$ is

$$f(\boldsymbol{y}_i|\boldsymbol{\beta}, D) = \int \prod_{j=1}^{n_i} [f(y_{ij}|\boldsymbol{x}_{ij}, \boldsymbol{z}_{ij}, \boldsymbol{\beta}, \phi, \boldsymbol{b}_i)f(\boldsymbol{b}_i|D)]\, d\boldsymbol{b}_i, \qquad (2.41)$$

which usually does not have an analytic or closed-form expression since the model is nonlinear in the random effects $\boldsymbol{b}_i$. The likelihood for all observed data is given by

$$L(\boldsymbol{\beta}, D|\boldsymbol{y}) = \prod_{i=1}^{n} \left\{ \int \prod_{j=1}^{n_i} [f(y_{ij}|\boldsymbol{x}_{ij}, \boldsymbol{z}_{ij}, \boldsymbol{\beta}, \phi, \boldsymbol{b}_i)f(\boldsymbol{b}_i|D)]\, d\boldsymbol{b}_i \right\}. \quad (2.42)$$

Note that the likelihood (2.42) of a GLMM is similar to the likelihood (2.24) of a NLME model (see page 54) in the sense that both involve an intractable multi-dimensional integral with respect to the random effects. Therefore, likelihood inference for a GLMM is similar to that for a NLME model. In fact, most estimation methods for these two classes of models are conceptually the same, although there have been parallel developments in the literature. However, an important difference between these two classes of models is that the responses in GLMMs may have different types, such as binary or count, while the responses in NLME models are continuous and assumed to be normal. This difference may cause different performances of the same methods for the two classes of models. For example, for finite samples, the approximate methods based on Taylor approximations or Laplace approximations perform well for NLME models but may lead to biases for GLMMs. These issues will be discussed further in Section 2.6.

As noted above, the commonly used inference methods for GLMMs are similar to that for NLME models (see Section 2.3.2). These methods include "exact" methods based on Gauss-Hermite quadrature or Monte Carlo integration techniques, EM algorithms, and approximate methods based on Taylor approximations or Laplace approximations (Breslow and Clayton 1993; McCulloch 1994, 1997; McCulloch, Searle, and Neuhaus 2008; Lee, Nelder, and Pawitan 2006). In particular, the approximate methods based on first- or second-order Taylor expansions or Laplace approximations are called *penalized or predictive quasi-likelihood (PQL)* (Breslow and Clayton 1993). As in NLME models, numerical or Monte Carlo methods can be computationally intensive, especially when the dimension of the random effects is not small, and approximate methods which avoid the integrations are computationally much more efficient. Dean and Nielsen (2007) provided a recent review of these methods for GLMMs. We will discuss these methods in details in Section 2.6.

Unlike NLME models, however, approximate methods for GLMMs based on Taylor or Laplace approximations such as the PQL method may lead to non-negligible biases (towards zero) for non-normal responses, especially binary responses. Breslow and Lin (1995) and Lin and Breslow (1996) proposed methods for bias correction. Joe (2008) demonstrated that the performance of these approximate methods may depend on how discrete the response is – the more discrete the response is the worse the methods perform. The performance of these approximate methods may be improved by using higher order Taylor or Laplace approximations and the so-called *hierarchical likelihood method* (*h-likelihood* method) (Lee, Nelder, and Pawitan 2006).

Over-dispersion problems may also arise in GLMMs when the observed data variation is inconsistent with the theoretical variance based on the assumed model and distribution. In this case, a more robust method, which only assumes the first two moments and does not require distributional assumptions, is the *quasi-likelihood* method and the *GEE* method. See Chapter 10 for details.

There has been extensive literature in GLMMs, especially in recent years. More comprehensive discussions of GLMMs can be found in recent books (e.g., Jiang 2007; McCulloch, Searle, and Neuhaus 2008).

Example 2.9 *Logistic regression model with random effects*

Consider a longitudinal binary response y_{ij} taking only two possible values (say, 0 or 1), $i = 1, \cdots, n; j = 1, \cdots, n_i$. A simple logistic regression model with random intercept can be written as

$$\log \left(\frac{\mu_{ij}}{1 - \mu_{ij}} \right) = \beta_{0i} + \beta_1 t_{ij} = \beta_0 + b_i + \beta_1 t_{ij},$$

$$b_i \sim N(0, d^2),$$

where $\mu_{ij} = E(y_{ij}) = P(y_{ij} = 1)$ and $\beta_{0i} = \beta_0 + b_i$. A more general GLMM for binary longitudinal or clustered responses may be written as

$$\log \left(\frac{\mu_{ij}}{1 - \mu_{ij}} \right) = \boldsymbol{x}_{ij}^T \boldsymbol{\beta} + \boldsymbol{z}_{ij}^T \boldsymbol{b}_i,$$

Table 2.3 *Estimates for the distress data based on GLM and GLMM models*

Model	Parameter β_0		Parameter β_1		Parameter β_2		Parameter β_3	
	Est.	S.E.	Est.	S.E.	Est.	S.E.	Est.	S.E.
GLM (2.43)	–2.64	0.68	–0.73	0.21	0.04	0.29	1.39	0.31
GLMM (2.44)	–1.98	0.33	–0.63	0.20	0.02	0.13	1.04	0.15

S.E.: standard error

$$\mathbf{b}_i \quad \sim \quad N(0, D),$$

which can be used to model longitudinal or clustered binary data.

Example 2.10 *Poisson regression model with random effects*

For longitudinal or clustered count responses y_{ij}, we may consider the following Poisson regression models with random effects

$$\log(\mu_{ij}) = \boldsymbol{x}_{ij}^T\boldsymbol{\beta} + \boldsymbol{z}_{ij}^T\boldsymbol{b}_i,$$
$$\mathbf{b}_i \sim N(0, D),$$

where $\mu_{ij} = E(y_{ij})$. This is another example of a GLMM.

Example 2.11 *Mental distress data*

We return to the mental distress data in Chapter 1 (page 6). An alternative approach for analyzing this dataset is to check if a subject's mental distress is above or below average over time and if the changes can be partially explained by covariates such as gender and treatment group. We thus define $y_{ij} = 1$ if the GSI score of subject i at time t_{ij} is above average and $y_{ij} = 0$ otherwise. There are missing data in this dataset. Here we only use the complete-cases for illustration.

If we ignore the correlation between the repeated measurements on each individual, we may consider the following GLM:

$$\text{logit}(P(y_{ij} = 1)) = \beta_0 + \beta_1 t_{ij} + \beta_2 x_{i1} + \beta_3 x_{i2}, \qquad (2.43)$$

where x_{i1} is group and x_{i2} is gender for individual i, time t_{ij} is in month, and the y_{ij}'s are assumed to be independent. To incorporate possible correlation between the repeated measurements on each individual, we consider the following GLMM:

$$\text{logit}(P(y_{ij} = 1)) = \beta_0 + b_i + \beta_1 t_{ij} + \beta_2 x_{i1} + \beta_3 x_{i2}, \qquad (2.44)$$

where $b_i \sim N(0, d^2)$ is a random effect.

Table 2.3 presents the estimates based on the above two models. We see that the two models lead to somewhat different estimates of the parameters, so ignoring correlation in the repeated measurements may lead to possibly biased estimates. We also see that subjects' mental distress changes significantly over time, with more subjects becoming less depressed over time (since the estimate $\hat{\beta}_1$ is negative and significant at 5% level). Moreover, subject's mental distress seems to be significantly correlated with gender, with female subjects more likely to be distressed (since estimate $\hat{\beta}_3$ is positive and significant). However, there seems to be no significant difference between the two groups (the estimate $\hat{\beta}_2$ is not significant).

2.5 Nonparametric and Semiparametric Mixed Effects Models

2.5.1 Nonparametric and Semiparametric Regression Models

So far we have focused only on *parametric* regression models. In parametric regression models, the functions determining the relationship between the response and covariates are assumed to be *known*, and one only needs to estimate the unknown parameters in the models. In practice, many longitudinal processes are quite complex. In these cases, it may be difficult to specify a known functional form for the processes, either linear or nonlinear. One may consider polynomials with high orders, but they are known to be unstable.

Nonparametric regression models are very flexible for modeling complex time series or longitudinal processes since they leave the functional form completely unspecified, except the requirement of some smoothness. A semiparametric regression model, which contains both parametric and nonparametric parts, offers a good compromise between parametric and nonparametric regression models. In this section we briefly review the main ideas and approaches for nonparametric and semiparametric regression models for cross-sectional or i.i.d. data. The approaches are then extended to longitudinal data in Section 2.5.2.

Let y_i be a response and x_i be a covariate for individual i. We focus on the most common case where $x_i = t_i$ is the time, although x_i can be other covariate, and we first focus on one covariate. A general *nonparametric regression model* can be written as

$$y_i = g(t_i) + e_i, \qquad i = 1, 2, \cdots, n, \qquad (2.45)$$

where y_i is the response at time t_i, $g(\cdot)$ is a *unknown smooth function*, $t_1 < t_2 < \cdots < t_n$, and e_i is random noise with mean zero. We assume that e_i i.i.d. $\sim N(0, \sigma^2)$. *The objective is to estimate the function $g(\cdot)$ directly*. That is, we let the data to determine the suitable form of the smooth function $g(\cdot)$. This is different from parametric regression models where the function $g(\cdot)$ is known (specified in advance) and the objective is to estimate unknown parameters.

There are many approaches for estimating the smooth function $g(\cdot)$ in (2.45). The commonly used approaches are:

- regression splines,
- kernel estimation,
- local-polynomial regression,
- smoothing splines.

In the following, we briefly review the main ideas of each approach.

Regression Splines

Splines are piecewise polynomials within any two neighboring breakpoints. The idea is based on the fact that any smooth function can be approximated by a low-order polynomial within a small range, based on Taylor expansions. Specifically, let $s_1 < \cdots < s_m$ be m breakpoints or *knots* chosen from the entire time range $[t_1, t_n]$. One can approximate $g(t)$ by a polynomial piecewise in the intervals $[s_j, s_{j+1})$, $j = 1, \cdots, m-1$, which join smoothly at the knots. For *cubic splines*, the polynomials have continuous first and second derivatives at the knots. The *natural cubic spline* has zero second and third derivatives at the boundaries. A polynomial B-spline consists of polynomial pieces between the knots.

A regression spline can be constructed using a basis-based approach. That is, the smooth function $g(t)$ can be approximated by $\hat{g}_r(t)$ given by

$$\hat{g}_r(t) = \sum_{j=0}^{r} \gamma_j \psi_j(t), \qquad (2.46)$$

where $\{\psi_0(t), \psi_1(t), \cdots\}$ are known basis functions and γ_j are coefficients to be estimated. The number r of basis functions controls the smoothness of the estimated function $\hat{g}_r(t)$. There are many basis functions available for curve fitting, including global bases such as Legendre polynomials and Fourier series and local bases such as regression splines, B-splines, natural splines, and wavelet bases (Eubank 1988; de Boor 1978; Green and Silverman 1994, Ramsay and Silverman 2005). For example, the *power basis* functions are $\psi_j(t) = t^j$ (leading to polynomial models), the *Gaussian basis* functions are

$$\psi_j(t) = \exp(-|t - \mu_j|^2/(2h_j^2)),$$

and the *Fourier basis* functions are

$$\psi_0(t) = 1, \quad \psi_{2j}(t) = \sqrt{2}cos(2\pi jt), \quad \psi_{2j+1}(t) = \sqrt{2}sin(2\pi jt)$$

(useful for modeling seasonal effects).

The location and the number of knots s_j's are very important for splines since they strongly determine the degree of smoothing. One can place the knots uniformly (i.e., equally-spaced knot placing), or use all distinct time points as knots. The *percentile-based knot-placing rule* puts more (fewer) knots in the areas where more (fewer) data are available (Eubank 1988), i.e., we use the sample percentiles of the design time points as knots for constructing splines. This approach is useful for smoothing time series or longitudinal data which are not uniformly scattered.

Kernel Smoothing and Local Polynomial Kernel Smoothing

A *kernel estimation* procedure uses a weighted average as follows:

$$\hat{g}_h(t) = \frac{\sum_{i=1}^n w_i(t, h) y_i}{\sum_{i=1}^n w_i(t, h)} \tag{2.47}$$

where

$$w_i(t, h) = k\left(\frac{t - t_i}{h}\right)$$

is a weight function, $k(\cdot)$ is a kernel function, and h is the *bandwidth* (or smoothing parameter) of the kernel. The bandwidth is used to specify the size of the local neighborhood, e.g., a window with size $2h$ may be the interval $I(h, t) = [t - h, t + h]$. A *kernel* function $k(\cdot)$ is a continuous symmetric function with $\int k(t)dt = 1$. It is used to specify how the observations contribute to the fit at each time t, while the bandwidth h specifies the size of the local neighborhood. Typically, the weights $w_i(t, h)$ will decrease as the distance $|t - t_i|$ increases. The bandwidth (or window width) h determines how fast the weights decrease. For example, the *Gaussian kernel* is given by

$$k(t) = \exp(-t^2/2),$$

where the contribution of the data y_i's for fitting $\hat{g}_h(t)$ at time t is determined by the distance of t_i from t. The *uniform kernel* on interval $[-1, 1]$ (say) is given by $k(t) = 0.5$ if $-1 \leq t \leq 1$ and $k(t) = 0$ if $|t| > 1$, where all the data y_i's corresponding to the t_i's in the interval $[-h, h]$ contribute equally while observations outside this interval make no contribution.

The *local polynomial kernel smoothing* method locally approximates the function $g(\cdot)$ by a low-order polynomial, since any smooth function can be locally approximated by a polynomial based on Taylor expansion. Specifically, one minimizes the weighted least squares:

$$\hat{g}_h(t) = \sum_{i=1}^n w_i(t, h) \left\{ y_i - \left[\beta_0 + \beta_1(t_i - t) + \cdots + \beta_k(t_i - t)^k \right] \right\}^2, \tag{2.48}$$

at each fixed time t. The weights $w_i(t, h)$ in (2.48) can be chosen as the one discussed above. Local constant ($k = 0$), linear ($k = 1$), and quadratic ($k = 2$)

estimates are most commonly used. The local constant smoother corresponds to the weighted average (2.47), and the local linear and quadratic estimates are sometimes called *loess* or *lowess* estimators. See Fan and Gijbels (1996) for a comprehensive discussion.

Smoothing means a compromise between bias and variance. When the bandwidth h is larger, more observations are used so bias is higher and variance is lower, and the fitted curve is smoother. When the bandwidth h is small, only a small number of observations have influential weights so bias is low but variance is high. The choice of a kernel is usually not so crucial.

Smoothing Spline

The *smoothing spline* method finds an estimator $\hat{g}_\lambda(t)$ of $g(t)$ which minimizes the following penalized sum of squares (PSS)

$$PSS(\lambda) = \sum_{i=1}^{n}(y_i - g(t_i))^2 + \lambda \int \left(\frac{d^2 g(t)}{dt^2}\right)^2 dt, \qquad (2.49)$$

where λ is a smoothing parameter. It can be shown that the solution $\hat{g}_\lambda(t)$ which minimizes $PSS(\lambda)$ is a natural cubic spline with knots at the distinct times t_i, i.e., $\hat{g}_\lambda(t)$ is a cubic polynomial between successive t-values and has continuous first and second derivatives at the distinct times.

In (2.49), the first term is the residual sum of squares, which measures how close the fitted curve to the data, and the second term is a *roughness penalty*, which is large when $\hat{g}_\lambda(t)$ is "rough" (rapidly changing slope). Without the roughness penalty, the solution $\hat{g}_\lambda(t)$ of (2.49) simply interpolate the data. The smoothing parameter λ controls the tradeoff between the smoothness of the fitted curve and the faith with the data. Larger values of λ give more weight to the penalty term, leading to fitted curves with smaller variance but higher bias, or vice versa.

The foregoing smoothing methods are most commonly used in nonparametric regression models. Although it may not be obvious in some cases, many smoothers are *linear*, i.e., the estimators have the following linear form

$$\hat{g}_\rho(t) = \sum_{i=1}^{n} w_i(t, \rho) y_i \equiv A(\rho)\mathbf{y}, \qquad (2.50)$$

where ρ is a smoothing parameters (e.g., r, or h, or λ), $w_i(t, \rho)$ is the weight, $A(\rho)$ is an appropriate matrix with components $w_i(t, \rho)$, and $\mathbf{y} = (y_1, \cdots, y_n)$. The local polynomial, regression spline, smoothing spline are all linear smoothers. From (2.50), we have

$$E(\hat{g}_\rho(t)) = A(\rho)E(\mathbf{y}), \qquad Cov(\hat{g}_\rho(t)) = A(\rho)Cov(\mathbf{y})A(\rho)^T,$$

which can be used to derive the biases and variances of the fitted curves and to construct confidence bands for the fitted curves.

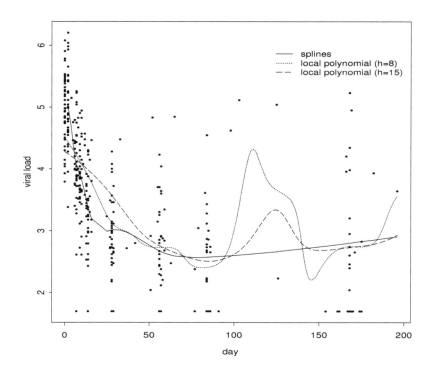

Figure 2.2 *Nonparametric smoothing curves based on spline and local polynomial methods (with bandwidths $h = 8$ and $h = 15$ respectively). Viral load is in $\log_{10}$ scale.*

Example 2.12 *AIDS data*

Consider the AIDS dataset in Chapter 1 (page 9). One can see from Figure 1.3 that the long-term viral load trajectories are complex and may not be modeled well by a parametric model. So we may try nonparametric or semiparametric models. Assume all the data are independent. Figure 2.2 shows nonparametric curves fitted by splines and local polynomial methods (with bandwidths $h = 8$ and $h = 15$ respectively). We can see that nonparametric models are very flexible in fitting complex longitudinal data. We can also see how the choice of bandwidth affects the fitted smooth curves.

Choosing Smoothing Parameters

Choosing the smoothing parameter ρ (e.g., r, or h, or λ) is crucial to the performance of a smoothing estimator. A general approach is to minimize the *mean square error (MSE)*

$$
\begin{aligned}
MSE(\hat{g}_\rho(t)) &= \sum_{i=1}^{n} E(\hat{g}_\rho(t_i) - g(t_i))^2 \\
&= \sum_{i=1}^{n} \left\{ [E(\hat{g}_\rho(t_i)) - g(t_i))]^2 + E[\hat{g}_\rho(t_i) - E(\hat{g}_\rho(t_i))]^2 \right\} \\
&= \text{bias}^2 + \text{variance},
\end{aligned}
$$

which is the sum of (squared) bias and variance. We should choose a smoothing parameter ρ with a tradeoff between the variance and bias to minimize the mean square error. In other words, smoothing parameter selection is a tradeoff between the goodness of fit and model complexity. The goodness of fit may be measured by the sum of squared errors (SSE):

$$
SSE(\rho) = \sum_{i=1}^{n}(y_i - \hat{y}_i)^2 = \sum_{i=1}^{n}(y_i - \hat{g}_\rho(t_i))^2.
$$

The model complexity can be measured by the trace of the smoother matrix $A(\rho)$, which roughly indicates the number of effective parameters in the model. One can also minimize a penalized sum of squares similar to (2.49). Another measure for goodness of fit is the log-likelihood.

The exact mean squared error is hard to compute since the true model is unknown. However, it can be approximated by a cross-validation or a AIC or a BIC criterion. For example, a leave-one-out *cross-validation* is defined as

$$
CV(\rho) = \frac{1}{n} \sum_{i=1}^{n}(y_i - \hat{g}_\rho^{-i}(t_i))^2,
$$

where $\hat{g}_\rho^{-i}(t_i)$ is a version of $\hat{g}_\rho(t_i)$ computed by leaving out the i-th data point (t_i, y_i). The cross-validation method selects ρ to minimize $CV(\rho)$, which balance the goodness of fit and the model complexity indirectly. Cross-validation methods are often computationally intensive. Alternatively, one can use standard model selection criteria such as the AIC or BIC methods to select ρ (see Section 2.7.1).

Semiparametric Models

A semiparametric model has both parametric components and nonparametric components. It is a compromise between a parametric model and a nonparametric model, and in the meantime it retains many nice features of both parametric and nonparametric models. For example, a semiparametric model is useful when some important variables, such as treatment or group effects, are best modeled parametrically, while some less important nuisance variables are modeled nonparametrically.

Let $\mathbf{x}_i = (x_{1i}, \cdots, x_{pi})^T$ be p covariates at time t_i. If the response y_i is known to be linearly related to $\mathbf{x}_i$, or if covariates $\mathbf{x}_i$ are best modeled in a linear fashion (e.g., $\mathbf{x}_i$ are discrete variables), we may consider a semiparametric regression model in which some residual variation may be modeled nonparametrically. A *semiparametric regression model* may be written as

$$y_i = \mathbf{x}_i^T \boldsymbol{\beta} + g(t_i) + e_i, \qquad (2.51)$$

where $g(t)$ is a unknown smooth function and e_i represents unexplained random error. In model (2.51), the first term on the right-hand side is parametric, which can in fact be any known parametric function such as a nonlinear function, while the second term is nonparametric.

Many of the estimating methods for nonparametric models can be extended to semiparametric models, including local polynomial methods, regression spline methods, and smoothing spline methods. There is an extensive literature on nonparametric and semiparametric models. Recent reviews of these methods can be found in Fan and Gijbels (1996), Ruppert, Wand, and Carroll (2003), Ramsay and Silverman (2005), and Wu and Zhang (2006), among others.

Nonparametric Regression Models with Non-Normal Responses

The foregoing nonparametric regression models are for continuous responses (normal responses) and for a single covariate (time). The ideas and approaches can be extended to nonparametric regression models with non-normal responses, such as binary responses and count responses, and multiple covariates, which are briefly described as follows.

For regression models with non-normal responses, generalized linear models are natural choices. A *nonparametric generalized linear model (GLM)* for cross-sectional data, with a single covariate time t_i, can be written as

$$h(E(y_i)) = g(t_i), \qquad i = 1, \cdots, n, \qquad (2.52)$$

where $h(\cdot)$ is a *known* link function, $g(t)$ is a *unknown* smooth function, and y_i follows a distribution in the exponential family. Estimation methods for model (2.52) are similar to those for normal responses. For example, for a regression splines or basis-based approach, we can approximate $g(t)$ in (2.52) by

$$\hat{g}_r(t) = \sum_{j=0}^{r} \gamma_j \psi_j(t), \qquad (2.53)$$

where $\psi_j(t)$'s are some basis functions.

For a smoothing spline approach, one maximizes the following penalized log-likelihood

$$PL(\lambda) = \sum_{i=1}^{n} l_i(y_i; g(t_i)) - \frac{1}{2}\lambda \int \left(\frac{d^2 g(u)}{du^2}\right)^2 du, \qquad (2.54)$$

where l_i is the log-likelihood for individual i. In (2.54), the first term represents faith to the data and the second term is a penalty function. Note that, when the response y_i is normal, the penalized log-likelihood (2.54) reduces to the penalized least-squares (2.49). The penalized log-likelihood (2.54) is a compromise between faith to the data and roughness of the fitted function, with the smoothing parameter λ controlling this compromise. The solution of (2.54) is again a cubic spline. Therefore, estimation methods for nonparametric GLMs are similar to those for normal response, with log-likelihood replacing residual sum of squares.

Nonparametric Regression Models with Multiple Covariates

Nonparametric regression models with one covariate can be extended to nonparametric models with multiple covariates. We describe one such extension in the framework of nonparametric GLMs for cross-sectional data. Let $x_1, \cdots, x_p$ be p covariates. We may consider the following *generalized additive model*

$$h(E(y)) = \eta(x_1, \cdots, x_p) = \beta_0 + g_1(x_1) + \cdots + g_p(x_p), \quad (2.55)$$

where $h(\cdot)$ is a known link function, and $\eta(x_1, \cdots, x_p)$ and $g_j(\cdot)$'s are unknown smooth functions. Hastie and Tibshirani (1990) provided a detailed discussion of generalized additive models. Note that, when the link function is the identity link $h(y) = y$, model (2.55) reduces to a nonparametric regression model with normal response and multiple covariates.

Estimation methods for model (2.55) are similar to those for models with a single covariate. For example, a basis-based approach estimator is given by

$$\hat{\eta}_r(x_1, \cdots, x_p) = \sum_{j=0}^{p} \sum_{k=1}^{r} \gamma_{jk} \psi_{jk}(x_j), \quad (2.56)$$

where $\{\psi_{j1}(x_j), \cdots, \psi_{jr}(x_j)\}$ are r basis functions for estimating $g_j(x_j)$, $j = 1, \cdots, p$, and γ_{jk} are coefficients to be estimated.

In a generalized additive model, we can allow some covariates to enter the model parametrically and other covariates to enter the model nonparametrically. This leads to the following semiparametric model or *partial linear model*:

$$h(E(y)) = \beta_0 + \beta_1 x_1 + \cdots + \beta_q x_q + g_{q+1}(x_{q+1}) + \cdots + g_p(x_p),$$

where $q < p$. The above partial linear model is useful, for example, when $x_1, \cdots, x_q$ are important covariates or categorical variables or are roughly linear in the model while the remaining covariates $(x_{q+1}, \cdots, x_p)$ may be nuisance factors. Hardle, Liang, and Gao (2000) provided a detailed discussion of partially linear models.

2.5.2 Nonparametric and Semiparametric Mixed Effects Models

The nonparametric and semiparametric regression models in the previous section are for cross-sectional or i.i.d. data. They can be extended to the analyses of longitudinal data or clustered data. One such extension is to introduce random effects in the models to account for within-individual correlation and between-individual variation in longitudinal or clustered data. Another approach is based on marginal GEE methods. In this section, we briefly discuss these extensions and focus on nonparametric and semiparametric mixed effects models, which may be viewed as a combination of nonparametric models and parametric mixed effects models.

Nonparametric Mixed Effects Models

Longitudinal data can be written as $\{(y_{ij}, t_{ij}), i = 1, \cdots, n; j = 1, \cdots, n_i\}$, where y_{ij} is the response for individual i at design time point t_{ij}. A nonparametric (population mean) model for the longitudinal data can be written as

$$y_{ij} = g(t_{ij}) + e_{ij}, \qquad i = 1, \cdots, n; \quad j = 1, \cdots, n_i, \qquad (2.57)$$

where $g(t)$ is a unknown smooth function and $e_{ij} = e_i(t_{ij})$ represents random noise. By introducing a random smooth process in model (2.57), we have the following *nonparametric mixed effects model* (or *nonparametric subject-specific regression model*)

$$\begin{aligned} y_{ij} &= g(t_{ij}) + u_i(t_{ij}) + e_{ij}, & (2.58) \\ u_i(t) &\sim GP(0, \gamma), & \mathbf{e}_i \sim N(0, R_i), \ i = 1, \cdots, n; j = 1, \cdots, n_i, \end{aligned}$$

where $u_i(t)$ is a random smooth process and assumed to follow a Gaussian process with mean 0 and covariance function $Cov(u_i(s), u_i(t)) = \gamma(s, t)$, denoted by $u_i(t) \sim GP(0, \gamma)$, and $\mathbf{e}_i = (e_{i1}, \cdots, e_{in_i})^T$. The random process $u_i(t)$ incorporates within-individual correlation and between-individual variation, similar to the roles played by random effects in a parametric mixed effects model. It also allows subject-specific inference.

Estimation methods for nonparametric mixed effects models are a combination of nonparametric methods described in the previous section and methods for parametric mixed effects models. In many cases, a nonparametric mixed effects model can be approximated by a parametric LME model, as shown below.

We first consider a regression spline approach. As discussed in Sections 2.5.1, regression splines are piecewise polynomials that are specified by a group of knots and some continuity conditions. They can be represented as linear combinations of regression spline bases, such as B-spline basis and truncated power basis (Shi, Weiss, and Taylor 1996; Rice and Wu 2001). The idea of a mixed effects regression spline method is to approximate (2.58) by

$$y_i(t_{ij}) \approx \sum_{k=0}^{r} \beta_k \psi_k(t_{ij}) + \sum_{k=0}^{q} b_{ik} \phi_k(t_{ij}) + e_i(t_{ij}), \qquad (2.59)$$

where $\Psi_r(t) = (\psi_1(t), \cdots, \psi_r(t))^T$ and $\Phi_q(t) = (\phi_1(t), \cdots, \phi_q(t))^T$ are known basis functions, $\boldsymbol{\beta} = (\beta_1, \cdots, \beta_r)^T$ are fixed coefficients, and $\mathbf{b}_i = (b_{i1}, \cdots, b_{iq})^T$ are random coefficients. We assume that $\mathbf{b}_i \sim N(0, D)$. The numbers r and q of basis functions control the smoothness and goodness of fit of the estimated function.

Let $\mathbf{x}_{ij} = \Phi_r(t_{ij})$ and $\mathbf{z}_{ij} = \Psi_r(t_{ij})$. Model (2.59) can be written as a standard LME model

$$y_{ij} = \mathbf{x}_{ij}^T \boldsymbol{\beta} + \mathbf{z}_{ij}^T \mathbf{b}_i + e_{ij}, \qquad (2.60)$$
$$\mathbf{b}_i \sim N(0, D), \qquad \mathbf{e}_i \sim N(0, R_i). \qquad (2.61)$$

So standard estimation methods for LME models can be used for approximate inference of the original nonparametric mixed effects model (2.58). Note that $Cov(u_i(s), u_i(t)) = \gamma(s, t) \approx \Phi_q(s)^T D \Phi_q(t)$. The basis functions can be any commonly used ones, and we often choose $\phi_j(t) = \psi_j(t)$ for simplicity. Choices of r and q are based on trade-off between goodness of fit and model complexity.

For a local polynomial method, we can approximate $g(t)$ and $u_i(t)$ in the nonparametric mixed effects model (2.58) using a Taylor approximation around a neighborhood of interest. That is, at a neighborhood of a fixed t, we have

$$g(t_{ij}) \approx g(t) + g^{(1)}(t)(t_{ij} - t) + \cdots + \frac{g^{(r)}(t)}{r!}(t_{ij} - t)^r = \mathbf{x}_{ij}^T \boldsymbol{\beta},$$

$$u_i(t_{ij}) \approx u_i(t) + u_i^{(1)}(t)(t_{ij} - t) + \cdots + \frac{u_i^{(q)}(t)}{q!}(t_{ij} - t)^q = \mathbf{z}_{ij}^T \mathbf{b}_i,$$

where $g^{(k)}(t)$ is the k-th derivative of $g(t)$ and similarly for $u_i^{(k)}(t)$, $\mathbf{x}_{ij} = (1, t_{ij} - t, \cdots, (t_{ij} - t)^r)^T$, $\mathbf{z}_{ij} = (1, t_{ij} - t, \cdots, (t_{ij} - t)^q)^T$, $\boldsymbol{\beta} = (g(t), g^{(1)}(t), \cdots, g^{(r)}(t))^T$, $\mathbf{b}_i = (u_i(t), u_i^{(1)}(t), \cdots, u_i^{(q)}(t))^T$. We often choose $r = q$ so $\mathbf{z}_{ij} = \mathbf{x}_{ij}$. Thus, within a neighborhood of t, we can again approximate the nonparametric mixed effects model (2.58) by a LME model:

$$y_{ij} = \mathbf{x}_{ij}^T \boldsymbol{\beta} + \mathbf{z}_{ij}^T \mathbf{b}_i + e_{ij}, \qquad (2.62)$$
$$\mathbf{b}_i \sim N(0, D), \qquad \mathbf{e}_i \sim N(0, R_i). \qquad (2.63)$$

Inference can be based on local likelihood methods (Tibshirani and Hastie 1987; Wu and Zhang 2006). Alternatively, Lin and Carroll (2000) considered a local polynomial kernel GEE method for clustered or longitudinal data. Wang, Carroll, and Lin (2005) considered efficient semiparametric marginal estimation for longitudinal or clustered data, and Wang (2003) considered marginal nonparametric kernel regressions which incorporate correlation within individuals.

Finally, we consider a smoothing splines method. Note that the nonparametric mixed effects model (2.58) can be approximated by (Wu and Zhang 2006)

$$\mathbf{y}_i \;=\; X_i\boldsymbol{\beta} + Z_i\mathbf{u}_i + \mathbf{e}_i, \tag{2.64}$$

$$\mathbf{u}_i \;\sim\; N(0, D), \qquad \mathbf{e}_i \sim N(0, R_i), \tag{2.65}$$

where $X_i = Z_i = (\mathbf{x}_{i1}, \cdots, \mathbf{x}_{in_i})^T$, $g(t_{ij}) = \mathbf{x}_{ij}^T\boldsymbol{\beta}$ and $u_i(t_{ij}) = \mathbf{x}_{ij}^T\mathbf{u}_i$, $\boldsymbol{\beta} = (g(t_1^*), \cdots, g(t_K^*))^T$ with t_j^*'s being all the distinct design time points and K being the total number of the distinct design time points, and $\mathbf{x}_{ij} = (x_{ij1}, \cdots, x_{ijK})^T$ with $x_{ijk} = 1$ if $t_{ij} = t_k^*$ and 0 otherwise. The cubic mixed effects smoothing splines estimators of $g(t)$ and $u_i(t)$ are defined as the minimizers of the following penalized log-likelihood criterion

$$\begin{aligned}
P(\lambda_1, \lambda_2) \;=\; & \sum_{i=1}^{n} \left\{ (\mathbf{y}_i - X_i\boldsymbol{\beta} - Z_i\mathbf{u}_i)^T R_i^{-1} (\mathbf{y}_i - X_i\boldsymbol{\beta} - Z_i\mathbf{u}_i) \right. \\
& \left. + \log|D| + \mathbf{u}_i^T D^{-1}\mathbf{u}_i + \log|R_i| \right\} \\
& + \lambda_1 \sum_{i=1}^{n} \int \left(\frac{d^2 u_i(t)}{dt^2} \right)^2 dt + \lambda_2 \int \left(\frac{d^2 g(t)}{dt^2} \right)^2 dt,
\end{aligned}$$

where λ_1 and λ_2 are smoothing parameters.

Semiparametric Mixed Effects Models

A semiparametric mixed effects model for longitudinal data introduces random effects in the parametric terms and the nonparametric terms. A general *semiparametric mixed effects model* can be written as

$$y_{ij} \;=\; \mathbf{x}_{ij}^T\boldsymbol{\beta} + \mathbf{z}_{ij}\mathbf{b}_i + g(t_{ij}) + u_i(t_{ij}) + e_{ij}, \tag{2.66}$$

$$i = 1, \cdots, n; j = 1, \cdots, n_i,$$

where $\mathbf{x}_{ij}$ and $\mathbf{z}_{ij}$ are known design vectors containing covariates, $\boldsymbol{\beta}$ contains fixed effects, $\mathbf{b}_i$ contains random effects, $g(t)$ is a unknown smooth function, $u_i(t)$ is a unknown random smooth process of time, and e_{ij} is the random error. We may assume that $\mathbf{b}_i \sim N(0, D)$, $\mathbf{e}_i \sim N(0, R_i)$, and $u_i(t)$ follows a Gaussian process with mean 0 and covariance function $\gamma(s, t)$.

For a specific application, it may not be necessary to include all the terms in (2.66). Dropping one or more terms in model (2.66) leading to various semiparametric mixed effects models. Inference can again be based on common smoothing methods, such as regression spline methods, local polynomial methods, and smoothing spline methods, with some modification. For example, Zhang et al. (1998) used a smoothing spline method and connected it to a LME model.

Semiparametric Nonlinear Mixed Effects Models

In semiparametric mixed effects model (2.66), the parametric components and nonparametric components enter the model in a linear fashion. In some applications, the parametric and/or nonparametric components may enter the model in a nonlinear fashion, leading to a semiparametric nonlinear mixed effects model. Specifically, a general *semiparametric nonlinear mixed effects (SNLME) model* may be written as follows:

$$y_{ij} = h(\boldsymbol{\beta}_{ij}, g_i(t_{ij})) + e_{ij}, \tag{2.67}$$

$$\boldsymbol{\beta}_{ij} = \mathbf{x}_{ij}^T \boldsymbol{\beta} + \mathbf{z}_{ij}^T \mathbf{b}_i, \tag{2.68}$$

$$g_i(t) = g(t) + u_i(t), \qquad i = 1, \cdots, n, \; j = 1, \cdots, n_i, \tag{2.69}$$

where $h(\cdot)$ is a known nonlinear function, $g(\cdot)$ and $u_i(\cdot)$ are unknown smooth fixed and random functions respectively, $\boldsymbol{\beta}$ contains fixed effects, $\mathbf{b}_i$ contains random effects, and e_{ij} is random error. We assume that $\mathbf{e}_i = (e_{i1}, \cdots, e_{in_i})^T \sim N(0, R_i)$, $\mathbf{b}_i \sim N(0, D)$, and $u_i(t)$ is a zero-mean Gaussian stochastic process with covariance function $\gamma(s, t)$.

Model (2.67) – (2.69) is very general in the sense that it contains many mixed effects models as special cases. For example, if the nonparametric component (2.69) is dropped, it reduces to a parametric NLME model. If the parametric component (2.68) is dropped and $h(\cdot)$ is linear, it reduces to a nonparametric mixed effects model. If $h(\cdot)$ is linear, it reduces to a semiparametric mixed effects model. It also includes the semiparametric model of Ke and Wang (2001) as a special case. Inference for model (2.67) – (2.69) can again be based on a combination of parametric and nonparametric methods. Wu and Zhang (2002) considered a basis-based approach and approximate the model by a parametric NLME model so standard parametric methods may be used. A similar approach is also considered in Liu and Wu (2007).

Lin and Zhang (1999) proposed the following very flexible *generalized additive mixed model*

$$h(E(\mathbf{y}_i | \mathbf{b}_i)) = \beta_0 + g_1(x_{i1}) + \cdots g_p(x_{ip}) + \mathbf{z}_i^T \mathbf{b}_i, \tag{2.70}$$

where $h(\cdot)$ is a known link function, $g_k(\cdot)$'s are unknown smooth functions, β_0 is an unknown parameter, $\mathbf{z}_i$ are covariates, and $\mathbf{b}_i$ are random effects. Lin and Zhang (1999) combined a smoothing spline method and the usual PQL method for GLMMs, leading to penalties for both the spline terms and the random effects which they called double PQL.

Example 2.13 *A semiparametric NLME model*

We return to the AIDS dataset in Chapter 1 (page 9). The parametric NLME models (2.33) or (2.34) (see page 56) may be used to fit viral load trajectories in the short term (say, first three months). For long-term viral load trajectories, however, a nonparametric or a semiparametric model may be more appropriate and more flexible. For example, for parametric NLME model (2.34), we may allow the second-phase viral decay rate λ_{2i} to vary over time nonparametrically, since viral load trajectories after the initial period seem complicated and may not be modeled parametrically. In other words, we may consider the following semiparametric NLME model for long-term HIV viral dynamics (Liu and Wu 2007)

$$y_{ij} = \log_{10}(P_{1i}e^{-\lambda_{1ij}t_{ij}} + P_{2i}e^{-\lambda_{2ij}t_{ij}}) + e_{ij}, \qquad (2.71)$$

$$P_{1i} = P_1 + b_{1i}, \quad \lambda_{1ij} = \lambda_1 + \beta_1 CD4_{ij} + b_{2i}, \quad P_{2i} = P_2 + b_{3i},$$

$$\lambda_{2ij} = w(t_{ij}) + h_i(t_{ij}), \qquad (2.72)$$

where $w(t_{ij})$ and $h_i(t_{ij})$ are nonparametric fixed and random smooth functions respectively.

We may consider linear combinations of natural cubic splines with percentile-based knots to approximate the nonparametric functions $w(t)$ and $h_i(t)$. Based on AIC and BIC criteria, we obtain the following approximation for λ_{2ij}:

$$\lambda_{2ij} \approx \beta_2 + \beta_3\,\psi_1(t_{ij}) + \beta_4\,\psi_2(t_{ij}) + b_{4i}, \qquad (2.73)$$

where $\psi_k(\cdot)$'s are basis functions. Therefore, a semiparametric NLME model is approximated by a parametric NLME model.

We fit the above semiparametric NLME model to the AIDS data in the entire study period (long term). To avoid very small estimates, which may be unstable, we standardize CD4 values and re-scale time to have a range of $[0,1]$. The resulting estimates (standard errors) are: $\hat{P}_1 = 11.69\,(0.19), \hat{\lambda}_1 = 62.29\,(4.09), \hat{\beta}_1 = 0.49\,(3.42), \hat{P}_2 = 6.53\,(0.79), \hat{\beta}_2 = -5.37\,(7.84), \hat{\beta}_3 = 11.25\,(12.01), \hat{\beta}_4 = 0.65\,(4.32)$. Liu and Wu (2007) showed that the above semiparametric NLME model provides more flexible fit for long-term viral load trajectories than the parametric NLME model.

2.6 Computational Strategies

The likelihood method is the standard approach for statistical inference in mixed effects models. It is also a fundamental component of Bayesian inference. As shown in previous sections, for GLMM and NLME models, the likelihoods involve intractable integrals with respect to the random effects since these models are *nonlinear* in the random effects. For LME models, the random effects can be integrated out from the likelihood since the likelihood is *linear* in the random effects, so analytic expressions of parameter estimates are available for LME models and the computation is much simpler. In other words, computational challenges mainly arise in GLMM, NLME, and frailty models. In this section we give a summary of the commonly used computational approaches for mixed effects models. We focus on NLME models for illustration. The methods for GLMM are similar, although their performance may be different (see Section 2.4.2). Similar methods may also be considered for frailty models (see Chapter 7).

Many estimation methods for mixed effects models have been proposed (Davidian and Giltinan, 1995, 2003; Jiang 2007; McCulloch, Searle, and Neuhaus 2008). Most of these approaches are in one of the following three categories: "exact" methods, EM algorithms, and approximate methods. These three approaches are briefly described as follows.

The so-called *"exact" methods* use the Gauss-Hermite quadrature method or Monte Carlo methods to approximate the likelihood of a mixed effects model, where the intractable integral in the likelihood is approximated by a summation. These methods can approximate the exact likelihood with arbitrary accuracy, in the expense of computing time. So they are called "exact" methods. They are also called *direct maximization methods* since they directly maximize the approximated likelihoods to find MLEs. These "exact" methods are often used when the random effects follow normal distributions and their dimensions are low, since the computation of these methods becomes very intensive or even infeasible when the dimensions of the integrals (or random effects) are not low.

The *Gauss-Hermite quadrature method* approximates the integration in the likelihood by a summation on a specified number of quadrature points for each dimension of the integration, assuming the random effects are normally distributed. It can approximate the integral to any practical degree of accuracy (so it is called an "exact" method here). However, when the dimension of the random effects (or the integral) is not small, this method involves a summation over a large number of quadrature points, so the computation can be extremely slow or even infeasible. The *adaptive quadrature* method, which uses fewer quadrature points, is therefore preferred. The Gauss-Hermite quadrature method is thus usually used for low dimension integrations. See Chapter 12 for a detailed description of the Gauss-Hermite quadrature method. Alternatively, one can use Monte Carlo methods to simulate a large sample of the random effects from the assumed distribution and then approximate the integral by the sum over simulated values.

The *EM algorithm* indirectly maximizes the likelihood by iterating between an E-step and an M-step until convergence, where the random effects are treated as "missing data". For GLMM and NLME models, computation of the E-step often requires numerical methods (e.g., Gauss-Hermite quadrature) or Monte Carlo methods or MCMC methods, leading to various types of EM algorithms such as the Monte Carlo EM algorithm (MCEM). The M-step may also be done sequentially, leading to variants of the EM algorithm such as the ECM algorithm (see Section 3.4.3). The EM algorithms may also be called *indirect maximization methods*.

The EM algorithm is general, stable, and conceptually simple. They may be used for almost any mixed effects models with any distributions and any dimensions of the random effects, so they are widely used. A drawback of the EM algorithms is intensive computation and slow convergence. McCulloch (1997) and McCulloch, Searle, and Neuhaus (2008) provided detailed discussions of Monte Carlo EM algorithms and Monte Carlo Newton-Raphson methods for GLMMs. Walker (1996) proposed a Monte Carlo EM algorithm for NLME models.

The *approximate methods* use Taylor expansions or Laplace approximations to approximate the model or likelihood. They avoid intractable integrals in the likelihoods so are computationally very efficient. However, unlike the foregoing "exact" methods, the approximations of these methods cannot be made arbitrary accurate, and the accuracies of approximations may depend on the richness of within-individual data, the types of the response (e.g., discrete or continuous), and other factors. For GLMMs and NLME models, currently the most widely used estimation methods are probably the approximate methods based on first-order Taylor expansions or Laplace approximations, such as the PQL method for GLMMs (Breslow and Clayton 1993) and the method of Lindstrom and Bates (Lindstrom and Bates 1990), because these methods are widely implemented in standard software such as R/Splus and SAS. The approximate methods based on Taylor expansions and Laplace approximations are asymptotically equivalent (Demidenko 2004), and they can be derived using different approaches.

The approximate methods may also be viewed as being based on one of the following two approaches:

- *linearization methods.* The idea is to *linearize* the nonlinear models using a first-order Taylor expansion. So the approach involves iteratively solving certain LME models. This can be done in different ways. The most popular procedure is perhaps that of Lindstrom and Bates (1990). A main advantage of this approach is that available methods for LME models can be readily incorporated.

- *direct approximation.* The idea is to directly approximate the integral in the likelihood using a Laplace approximation, which is a natural way to approximate an integral of the form $\int e^{-l(\mathbf{b})} d\mathbf{b}$ based on a Taylor expansion of $-l(\mathbf{b})$ about the value $\hat{\mathbf{b}}$ which maximizes $l(\mathbf{b})$ (see, e.g., Wolfinger 1993; Wolfinger and Lin 1997). This method also avoids integration with respect to the random effects. Lee, Nelder, and Pawitan (2006) proposed the so-called h-likelihood method which is an alternative implementation of the Laplace approximation.

There are other approximate methods. For example, we may consider higher-order Taylor or Laplace approximations, or approximate the first two moments and then use a GEE-type approach (Vonesh et al. 2002; Davidian and Giltinan 2003).

In addition to the above three approaches, other estimation methods are available, such as simulated maximum likelihood methods and simulated methods of moments (see, e.g., Jiang 1998; McCulloch et al. 2008). We do not discuss details of these methods here.

In the next few sub-sections, we provide more detailed descriptions of the fore-going three approaches, with a focus on NLME models as illustration (methods for GLMMs are similar).

2.6.1 "Exact" Methods

By "exact" methods, we mean that one can approximate the exact likelihoods as accurate as desired at the expense of greater computational intensity, so they are not true exact methods in the strict sense. True exact estimation methods are unavailable for GLMM and NLME models since their likelihoods generally cannot be obtained analytically due to the intractable integrations.

The *Gauss-Hermite quadrature method*, or *Gaussian quadrature method*, is a deterministic numerical integration technique which can be used to approximate intractable integrals with arbitrary accuracy. The idea is to approximate the target integral by a weighted average of the integrand evaluated at suitably chosen points over a grid, and the accuracy of the approximation increases with the number of grid points (quadrature points). See Chapter 12 for a detailed description. It is most useful for evaluation of low-dimensional integrals. So it can be used to approximate the likelihood of a GLMM or a NLME model when the dimension of the random effects is low.

Consider the NLME model (2.20) and (2.21) (page 53). It is convenient to re-parameterize the random effects as follows. Let $D = D^{1/2}(D^{1/2})^T$ be the Cholesky decomposition of the covariance matrix D for the random effects $\mathbf{b}_i$. We consider the following re-parameterization:

$$\mathbf{b}_i = D^{1/2}\mathbf{b}_i^*, \qquad \text{so } \mathbf{b}_i^* \sim N(0, I).$$

Let $f(\mathbf{y}_i | \boldsymbol{x}_i, \boldsymbol{\theta}, \boldsymbol{b}_i^*)$ be the density function of the response $\mathbf{y}_i$ given the random effects $\mathbf{b}_i^*$, and let $f(\boldsymbol{b}_i^*)$ be the density of $\boldsymbol{b}_i^*$. The likelihood to be evaluated is given by

$$L(\boldsymbol{\theta}|\boldsymbol{y}) = \prod_{i=1}^{n} L_i(\boldsymbol{\theta}|\boldsymbol{y}_i) = \prod_{i=1}^{n} \int f(\boldsymbol{y}_i|\boldsymbol{x}_i, \boldsymbol{\theta}, \boldsymbol{b}_i^*) f(\boldsymbol{b}_i^*) \, d\boldsymbol{b}_i^*. \qquad (2.74)$$

First, consider the case of a one dimensional random effect (i.e., a scalar random effect). The Gauss-Hermite approximation to $L_i(\boldsymbol{\theta}|\boldsymbol{y}_i)$ is given by

$$L_i(\boldsymbol{\theta}|\boldsymbol{y}_i) \approx L_{i,GH}(\boldsymbol{\theta}|\boldsymbol{y}_i) = \sum_{j=1}^{m} w_j f(\boldsymbol{y}_i|\boldsymbol{x}_i, \boldsymbol{\theta}, d_j), \qquad (2.75)$$

where $\{d_1, \cdots, d_m\}$ are the m grid (quadrature) points and w_j's are appropriate weights (see Chapter 12). Likelihood inference can then be based on the approximate likelihood $\prod_{i=1}^{n} L_{i,GH}(\boldsymbol{\theta}|\boldsymbol{y}_i)$. When the number m of the quadrature points is large enough, approximation (2.75) can be sufficiently accurate.

For multi-dimensional random effects, one can transform the integral into a series of one-dimensional integrals (Davidian and Gallant 1993; Pinheiro and Bates 1995). For example, for the NLME model (2.20) and (2.21) (page 53) with $R_i = \sigma^2 I$, we have

$$
\begin{aligned}
L_i(\boldsymbol{\theta}|\boldsymbol{y}_i) &= \int (2\pi\sigma^2)^{-\frac{q}{2}} |D|^{-\frac{1}{2}} \exp\left[-\frac{1}{2\sigma^2}(\boldsymbol{y}_i - \mathbf{u}_i(\boldsymbol{\beta}, \mathbf{b}_i))^T \right. \\
&\qquad \left. \times (\boldsymbol{y}_i - \mathbf{u}_i(\boldsymbol{\beta}, \mathbf{b}_i)) \right] \exp\left(-\frac{1}{2\sigma^2}\mathbf{b}_i^T D^{-1}\mathbf{b}_i \right) d\mathbf{b}_i \\
&= \int (2\pi\sigma^2)^{-\frac{q}{2}} \exp\left[-\frac{1}{2\sigma^2}\left(\boldsymbol{y}_i - \mathbf{u}_i(\boldsymbol{\beta}, D^{T/2}\mathbf{b}_i^*)\right)^T \right. \\
&\qquad \left. \times \left(\boldsymbol{y}_i - \mathbf{u}_i(\boldsymbol{\beta}, D^{T/2}\mathbf{b}_i^*)\right) \right] \exp\left(-\frac{1}{2\sigma^2}\mathbf{b}_i^{*T}\mathbf{b}_i^* \right) d\mathbf{b}_i^* \\
&\approx \sum_{j_1=1}^{m} \cdots \sum_{j_q=1}^{m} \left\{ \exp\left[-\frac{1}{2\sigma^2}\left(\boldsymbol{y}_i - \mathbf{u}_i(\boldsymbol{\beta}, D^{T/2}\mathbf{d}_j)\right)^T (\,\cdot\,) \right] \right. \\
&\qquad \left. \times \prod_{k=1}^{q} w_{jk} \right\},
\end{aligned}
$$

where $\mathbf{u}_i(\boldsymbol{\beta}, \mathbf{b}_i) = E(\boldsymbol{y}_i|\boldsymbol{\beta}, \mathbf{b}_i)$, and $\mathbf{d}_j = (d_{j_1}, \cdots, d_{j_q})^T$ and w_{jk} are the abscissas and weights for the one-dimensional Gaussian quadrature method based on the standard normal $N(0, 1)$ kernel.

The numerical effort of the Gaussian quadrature methods increases exponentially with the dimension of the integral, so the Gaussian quadrature methods are most useful for low-dimensional integrals. An improved method is the *adaptive Gauss-Hermite quadrature* which reduces the number of quadrature points and thus the computational effort (Liu and Pierce 1994; Pinheiro and Bates 1995). The idea is to center the grid of the random effects around their estimates that maximize the posterior density of the random effects, which leads to a great reduction in the number of grid points required to achieve a similar accuracy. In this case, use of one grid point in the adaptive Gaussian quadrature approximation leads to a Laplace approximation of the integral (see Section 2.6.3).

We can also use a Monte Carlo method to approximate the likelihood $L_i(\boldsymbol{\theta}|\boldsymbol{y}_i)$. Let $\{\tilde{\mathbf{b}}_i^{(1)}, \cdots, \tilde{\mathbf{b}}_i^{(m)}\}$ be a sample of size m simulated from the distribution $f(\mathbf{b}_i^*)$ of $\mathbf{b}_i^*$. Then, the likelihood $L_i(\boldsymbol{\theta}|\boldsymbol{y}_i)$ for individual i in (2.74) can be approximated by

$$L_i(\boldsymbol{\theta}|\boldsymbol{y}_i) \approx L_{i,MC}(\boldsymbol{\theta}|\boldsymbol{y}_i) = \sum_{j=1}^{m} f(\mathbf{y}_i|\boldsymbol{x}_i, \boldsymbol{\theta}, \tilde{\mathbf{b}}_i^{(j)}). \qquad (2.76)$$

This is straightforward since $f(\mathbf{b}_i^*)$ is known ($\mathbf{b}_i^* \sim N(0, I)$), but a very large m may be needed to provide a satisfactory approximation, especially when the dimension of the integral is high. In principle, the approximation can be arbitrary accurate if m is sufficiently large. Approximate likelihood estimation can then be based on $\prod_{i=1}^{n} L_{i,MC}(\boldsymbol{\theta}|\boldsymbol{y}_i)$. Alternatively, importance sampling methods can also be used (see Chapter 12), which is sometimes more efficient than the foregoing simple Monte Carlo method.

Note that the Gauss-Hermite quadrature method may be viewed as a deterministic version of the Monte Carlo integration method: in Gauss-Hermite quadrature method the samples and weights are fixed in advance while in Monte Carlo method they are random.

2.6.2 EM Algorithms

The *expectation-maximization (EM) algorithm* (Dempster, Laird, and Rubin, 1977) is an iterative algorithm for finding MLEs in the presence of "missing data". It iterates between an E-step, which computes the conditional expectation of the "complete-data" log-likelihood given the current parameter estimates and observed data, and an M-step, which maximizes the conditional expectation in the E-step to update the parameter estimates. At each iteration, the likelihood is non-decreasing. The algorithm iterates until convergence, and the final parameter estimates are the MLE or a local maximizer. The EM algorithm and extensions are very popular in likelihood estimation because they are very general and stable. A detailed description of the EM algorithm will be given in Chapter 3 (Section 3.4). McLachlan and Krishnan (1997) provided a comprehensive discussion. Here we focus on an Monte Carlo EM algorithm for NLME models as an illustration. The approach for GLMMs are similar.

Consider NLME model (2.20) and (2.21) (page 53). By treating the random effects $\mathbf{b}_i$ as "missing data", we have "complete data" $\{(\mathbf{y}_i, \mathbf{b}_i), i = 1, \cdots, n\}$, where $\{\mathbf{y}_i, i = 1, \cdots, n\}$ are the observed data. Let $\boldsymbol{\theta} = (\boldsymbol{\beta}, \boldsymbol{\eta}, D)$ denote all parameters. The "complete-data" log-likelihood for individual i is given by

$$l_i^{(c)}(\boldsymbol{\theta}|\mathbf{y}_i, \mathbf{b}_i) = \log f(\mathbf{y}_i|\mathbf{b}_i, \boldsymbol{\beta}, \boldsymbol{\eta}) + \log f(\mathbf{b}_i|D).$$

Begin with a starting value $\boldsymbol{\theta}^{(0)}$. At k-th EM iteration, the E-step computes the following conditional expectation

$$
\begin{aligned}
Q(\boldsymbol{\theta}|\boldsymbol{\theta}^{(k)}) &= E\left[\sum_{i=1}^{n} l_i^{(c)}(\boldsymbol{\theta}|\mathbf{y}_i, \mathbf{b}_i)\big|\mathbf{y}_i, \boldsymbol{\theta}^{(k)}\right] \\
&= E\left[\sum_{i=1}^{n} \left\{(\log f(\mathbf{y}_i|\mathbf{b}_i, \boldsymbol{\beta}, \boldsymbol{\eta}) + \log f(\mathbf{b}_i|D))\big|\mathbf{y}_i, \boldsymbol{\theta}^{(k)}\right\}\right],
\end{aligned}
$$

where the conditional expectation is evaluated with respect to the conditional distribution $f(\mathbf{b}_i|\mathbf{y}_i, \boldsymbol{\theta}^{(k)})$.

Unlike the EM algorithm for a LME model in Section 2.2.2, here the conditional expectation $Q(\boldsymbol{\theta}|\boldsymbol{\theta}^{(k)})$ does not have an analytic (closed-form) expression, because the NLME model is *nonlinear* in the random effects $\mathbf{b}_i$. To evaluate $Q(\boldsymbol{\theta}|\boldsymbol{\theta}^{(k)})$, we may consider numerical integration methods such as the Gauss-Hermite quadrature method or importance sampling methods when the dimension of the random effects is low. In the following, we consider a general Monte Carlo method to approximate the conditional expectation $Q(\boldsymbol{\theta}|\boldsymbol{\theta}^{(k)})$. The resulting EM algorithm is called a *Monte Carlo EM algorithm*.

The idea of the Monte Carlo EM algorithm is to simulate a large sample of the "missing data" $\mathbf{b}_i$, say $\{\mathbf{b}_i^{(1)}, \mathbf{b}_i^{(2)}, \cdots, \mathbf{b}_i^{(M)}\}$, from the conditional distribution $f(\mathbf{b}_i|\mathbf{y}_i, \boldsymbol{\theta}^{(k)})$ at k-th EM iteration. This sampling may be accomplished using a MCMC method such as the Gibbs sampler or a rejection sampling method (see Chapter 12) since

$$f(\mathbf{b}_i|\mathbf{y}_i, \boldsymbol{\theta}^{(k)}) \propto f(\mathbf{y}_i|\mathbf{b}_i, \boldsymbol{\theta}^{(k)})f(\mathbf{b}_i|D^{(k)}),$$

where $f(\mathbf{y}_i|\mathbf{b}_i, \boldsymbol{\theta}^{(k)})$ and $f(\mathbf{b}_i|D^{(k)})$ are known distributions. Then, the conditional expectation $Q(\boldsymbol{\theta}|\boldsymbol{\theta}^{(k)})$ can be approximated by the following empirical mean (average):

$$\widetilde{Q}(\boldsymbol{\theta}|\boldsymbol{\theta}^{(k)}) = \frac{1}{M}\sum_{j=1}^{M}\left[\log f(\mathbf{y}_i|\mathbf{b}_i^{(j)}, \boldsymbol{\beta}, \boldsymbol{\eta}) + \log f(\mathbf{b}_i^{(j)}|D)\right].$$

The M-step is then to maximize $\widetilde{Q}(\boldsymbol{\theta}|\boldsymbol{\theta}^{(k)})$ to produce an updated estimate

$\theta^{(k+1)}$. Thus, the M-step is like a complete-data maximization, so a standard optimization procedure such as the Newton-Raphson method can be used.

In principle, the Monte Carlo approximation in the E-step can be made arbitrary accurate by increasing the number of Monte Carlo sample M, but the computing time also grows rapidly as M increases. A good strategy is to increase the value of M as the iteration number k increases. The foregoing Monte Carlo EM algorithm can be computationally intensive, especially when the dimension of the random effects is high. We will discuss more details of such Monte Carlo EM algorithms in the context of missing data in Chapter 4. Note that, in an iterative algorithm such as the EM algorithm, the choice of starting values is important for GLMM and NLME models.

2.6.3 Approximate Methods

Numerical integration methods and Monte Carlo EM algorithms for likelihood estimation for GLMM and NLME models can be computationally very intensive and sometimes may even exhibit convergence difficulties, especially when the dimension of the integrals (or random effects) is high. Therefore, computationally much more efficient approximate methods for GLMM and NLME models are widely used and are implemented in standard statistical software such as SAS and R/Splus. The key idea of these approximate methods is based on Taylor series expansions or Laplace approximations. In this section, we describe these approaches from two viewpoints: linearization and direct approximation. We also briefly discuss other approaches. Again, we focus on NLME models for illustration, but similar methods can also be used in GLMMs (Breslow and Clayton 1993).

Linearization Methods

The idea of linearization methods is to take a first-order Taylor expansion about estimates of parameters and random effects, which leads to a "working" LME model. One then updates parameter estimates from this LME model, and iterates the algorithm until converge. An early procedure, called *first-order method*, was proposed by Beal and Sheiner (1982), who took a Taylor series expansion of the NLME model (2.20) and (2.21) (page 53) about $b_i = 0$, the mean of the random effects. An improved procedure, called *first-order conditional method*, was proposed by Lindstrom and Bates (1990), who took a Taylor series expansion about $b_i = \hat{b}_i$, the empirical Bayes estimate of the random effects. Lindstrom and Bates (1990) showed that their procedure performs better than that of Beal and Sheiner (1982). The procedures may also be derived using Laplace approximations (Wolfinger 1993; Wolfinger and Lin 1997). An advantage of the linearization procedures is that available methods for LME models, such as methods to speed up the EM algorithm and REML estimates, can be readily incorporated.

In the following, we follow the procedure of Lindstrom and Bates (1990) with some modification. We rewrite NLME model (2.20) and (2.21) (page 53) as a single equation

$$y_{ij} = u_{ij}(\mathbf{x}_i, \boldsymbol{\beta}, \mathbf{b}_i) + e_{ij}, \qquad j = 1, \ldots, n_i; \ i = 1, \ldots, n,$$

where $u_{ij}(\cdot)$ is a nonlinear function. Let $\mathbf{u}_i = (u_{i1}, \ldots, u_{in_i})^T$. At each iteration, denote the current estimates of $(\boldsymbol{\beta}, \mathbf{b}_i)$ by $(\widehat{\boldsymbol{\beta}}, \widehat{\mathbf{b}}_i)$, suppressing the iteration number, where $\widehat{\mathbf{b}}_i$ is the empirical Bayesian estimate of $\mathbf{b}_i$. The procedure of Lindstrom and Bates (1990) is equivalent to *iteratively* solving the following "working" LME model (Wolfinger 1993)

$$\widetilde{\mathbf{y}}_i \;=\; W_i \boldsymbol{\beta} + T_i \mathbf{b}_i + \mathbf{e}_i, \qquad\qquad (2.77)$$

where

$$\widetilde{\mathbf{y}}_i \;=\; \mathbf{y}_i - \mathbf{u}_i(\mathbf{x}_i, \widehat{\boldsymbol{\beta}}, \widehat{\mathbf{b}}_i) + W_i \widehat{\boldsymbol{\beta}} + T_i \widehat{\mathbf{b}}_i,$$

$$W_i \;=\; \left. \frac{\partial \mathbf{u}_i(\mathbf{x}_i, \boldsymbol{\beta}, \widehat{\mathbf{b}}_i)}{\partial \boldsymbol{\beta}^T} \right|_{\boldsymbol{\beta} = \widehat{\boldsymbol{\beta}}}, \qquad T_i = \left. \frac{\partial \mathbf{u}_i(\mathbf{x}_i, \widehat{\boldsymbol{\beta}}, \mathbf{b}_i)}{\partial \mathbf{b}_i^T} \right|_{\mathbf{b}_i = \widehat{\mathbf{b}}_i}.$$

At each iteration we obtain the updated estimates $(\widehat{\boldsymbol{\beta}}, \widehat{\mathbf{b}}_i)$ of the parameters and random effects from the LME model (2.77) using standard methods described in Section 2.2. Since the method is to iteratively solve LME models in which the random effects can be integrated out from the likelihood, it avoids the intractable integral in the likelihood of the NLME model and thus offers a substantial computational advantage.

Pinheiro and Bates (1995) conducted an extensive simulation study and concluded that the method of Lindstrom and Bates (1990) performs very well for NLME models. In general, the performance of linearization methods may depend on the number n_i of within-individual measurements and the within and between individual variations (Davidian and Giltinan 1995, 2003; Vonesh and Chinchilli 1997). Note that, unlike the "exact" methods in Section 2.6.1, linearization methods cannot be made arbitrary accurate.

Laplacian Approximations

Laplacian approximations are often used in Bayesian computation (e.g., Tierney and Kadane 1986). A first-order Laplace approximation can be written as

$$\int e^{kp(\mathbf{v})} d\mathbf{v} = \left(\frac{2\pi}{k} \right)^{\gamma/2} \cdot \left| \frac{\partial^2 p(\hat{\mathbf{v}})}{\partial \mathbf{v}^2} \right|^{-\frac{1}{2}} \cdot e^{kp(\hat{\mathbf{v}})} + O(k^{-1}), \qquad (2.78)$$

where $\mathbf{v}$ is a γ-dimension vector, $\hat{\mathbf{v}}$ maximizes $p(\mathbf{v})$, and $\partial^2 p(\hat{\mathbf{v}})/\partial \mathbf{v}^2 = \partial^2 p(\mathbf{v})/\partial \mathbf{v}^2 \big|_{\mathbf{v} = \hat{\mathbf{v}}}$. This approximation is obtained from a Taylor series expansion of $p(\mathbf{v})$ about the value $\hat{\mathbf{v}}$ which maximizes $p(\mathbf{v})$. Thus, Laplacian approximations can be used to approximate intractable integrals in the forms of $\int e^{kp(\mathbf{v})} d\mathbf{v}$, so they can be used for approximate estimations in NLME and GLMM models (Wolfinger 1993; Vonesh 1996; Wolfinger and Lin 1997; Lee et al. 2006).

Consider a GLMM or NLME model with conditional density $f(\mathbf{y}_i|\boldsymbol{x}_i, \boldsymbol{\beta}, \boldsymbol{\eta}, \boldsymbol{b}_i)$ given the random effects $\mathbf{b}_i$. Let $\mathbf{u}_i(\mathbf{x}_i, \boldsymbol{\theta}, \mathbf{b}_i) = E(\mathbf{y}_i|\boldsymbol{\theta}, \mathbf{b}_i)$. We follow the approach of Lee et al. (2006). Treating the random effects $\mathbf{b}_i$ as "missing data", the "complete-data" log-likelihood can be written as

$$l_c(\boldsymbol{\theta}) \equiv \sum_{i=1}^{n} l_c^{(i)}(\boldsymbol{\theta}, \mathbf{b}_i) \equiv \sum_{i=1}^{n} \{ \log f(\mathbf{y}_i|\boldsymbol{x}_i, \boldsymbol{\beta}, \boldsymbol{\eta}, \boldsymbol{b}_i) + \log f(\boldsymbol{b}_i|D) \}.$$

Taking $k = n_i$, $kp(\mathbf{v}) = l_c^{(i)}(\boldsymbol{\theta}, \mathbf{b}_i)$, $\gamma = \dim(\mathbf{b}_i)$, and $\mathbf{v} = \mathbf{b}_i$ in the Laplace approximation (2.78), we can approximate the observed-data log-likelihood $l_o^{(i)}(\boldsymbol{\theta})$ for individual i based on the following result (Lee et al. 2006)

$$
\begin{aligned}
l_o^{(i)}(\boldsymbol{\theta}) &= \left\{ l_c^{(i)}(\boldsymbol{\theta}, \mathbf{b}_i) - \frac{1}{2} \log \left| \frac{1}{2\pi} D\left(l_c^{(i)}(\boldsymbol{\theta}, \mathbf{b}_i), \mathbf{b}_i \right) \right| \right\}_{\mathbf{b}_i = \hat{\mathbf{b}}_i} + O(n_i^{-1}) \\
&\equiv p_{\hat{\mathbf{b}}_i}(l_c^{(i)}(\boldsymbol{\theta}, \mathbf{b}_i)) + O(n_i^{-1}),
\end{aligned}
$$

where $D(l, \boldsymbol{\xi}) = -\partial^2 l / \partial \boldsymbol{\xi}^2$ and $\hat{\mathbf{b}}_i = \hat{\mathbf{b}}_i(\mathbf{y}_i, \boldsymbol{\theta})$ solves $\partial l_c^{(i)}(\boldsymbol{\theta}, \mathbf{b}_i)/\partial \mathbf{b}_i = 0$. Thus, the observed-data log-likelihood $l_o(\boldsymbol{\theta}) = \sum_{i=1}^{n} l_o^{(i)}(\boldsymbol{\theta})$ for all individuals in the sample can be approximated as

$$
\begin{aligned}
l_o(\boldsymbol{\theta}) &= \sum_{i=1}^{n} \left\{ p_{\hat{\mathbf{b}}_i}(l_c^{(i)}(\boldsymbol{\theta}, \mathbf{b}_i)) + O(n_i^{-1}) \right\} \\
&\equiv p_{\hat{\mathbf{b}}}(l_c(\boldsymbol{\theta}, \mathbf{b})) + n\, O\left[\left(\min_i n_i \right)^{-1} \right] \\
&\approx p_{\hat{\mathbf{b}}}(l_c(\boldsymbol{\theta}, \mathbf{b})),
\end{aligned}
$$

where $p_{\hat{\mathbf{b}}}(l_c(\boldsymbol{\theta}, \mathbf{b})) = \sum_{i=1}^{n} p_{\hat{\mathbf{b}}_i}(l_c^{(i)}(\boldsymbol{\theta}, \mathbf{b}_i))$ is the first-order Laplacian approximation to $l_o(\boldsymbol{\theta})$. So the estimate $\tilde{\boldsymbol{\theta}}$ of $\boldsymbol{\theta}$ which maximizes $p_{\hat{\mathbf{b}}}(l_c(\boldsymbol{\theta}, \mathbf{b}))$ is an approximate MLE.

Specifically, for the NLME model (2.20) and (2.21) (page 53) with $R_i = \sigma^2 I$, the observed-data likelihood $L(\boldsymbol{\theta}|\mathbf{y})$ can be approximated as follows

$$
\begin{aligned}
L(\boldsymbol{\theta}|\mathbf{y}) \approx (2\pi\sigma^2)^{-\frac{N}{2}} |D|^{-\frac{n}{2}} \prod_{i=1}^{n} \Bigg\{ \left| \frac{\partial^2 q(\mathbf{y}_i, \boldsymbol{\theta}, \hat{\mathbf{b}}_i)}{\partial \mathbf{b}_i \partial \mathbf{b}_i^T} \right|^{-\frac{1}{2}} \\
\times \exp\left(-\frac{q(\mathbf{y}_i, \boldsymbol{\theta}, \hat{\mathbf{b}}_i)}{2\sigma^2} \right) \Bigg\},
\end{aligned}
$$

where $N = \sum_{i=1}^{n} n_i$ and

$$q(\mathbf{y}_i, \boldsymbol{\theta}, \hat{\mathbf{b}}_i) = (\mathbf{y}_i - \mathbf{u}_i(\mathbf{x}_i, \boldsymbol{\theta}, \mathbf{b}_i))^T (\mathbf{y}_i - \mathbf{u}_i(\mathbf{x}_i, \boldsymbol{\theta}, \mathbf{b}_i)) + \mathbf{b}_i^T D^{-1} \mathbf{b}_i.$$

In the above Laplacian approximation, it can be shown that, as $\min_i n_i$ grows faster than n, the function $p_{\hat{\mathbf{b}}}(l_c(\boldsymbol{\theta}, \mathbf{b}))$ approaches to the observed-data log-likelihood function $l_o(\boldsymbol{\theta})$. In other words, the approximation improves as the number of within individual measurements increases. Vonesh (1996) showed that the approximate estimates are consistent when *both* the number of measurements within each individual *and* the sample size go to infinite.

The Laplacian approximation for GLMMs corresponds to the *penalized quasi-likelihood (PQL)* (Breslow and Clayton 1993). It can also be obtained by iteratively solving a set of "working" LME models (Vonesh et al. 2002; Dean and Nielsen 2007). Lee, Nelder, and Pawitan (2006) proposed a *hierarchical likelihood* (h-likelihood) approach for approximate inference in GLMMs. The idea is essentially to use Laplace approximations for estimation of the mean parameters by integrating out the random effects, and for estimation of the variance components by integrating out the mean parameters and random effects.

We can also consider approximate inference for GLMM and NLME models using a GEE approach (see Chapter 10 for a detailed description of GEE methods). For example, we can approximate the first two moments of the models using a first-order Taylor expansion about $\mathbf{b}_i = 0$. We can then construct GEE-type equations using the approximate first two moments. For NLME models, the approximate mean is not of the GLM type and the approximate variance-covariance matrix is not a "working" covariance matrix as in a standard GEE method, but the GEE approach is quite general and can be broadly used.

Discussion

The approximate methods based on Taylor expansions or Laplace approximations are widely used because of their computational efficiencies and available software. The performance of these approximate methods may depend on

- the number n_i of within individual or cluster measurements and the sample size n;
- the discreteness of the response variable y;
- the magnitudes of the variations of the within-individual measurements and the between-individual measurements.
- some other factors such as nonlinearity of the models.

These approximate methods work very well for large values of n_i and n, continuous responses, and small variabilities of the within-individual measurements and the between-individual measurements. Pinheiro and Bates (1995) evaluated the performances of approximate methods and Gauss-Hermite quadrature methods for NLME models, and Joe (2008) investigated the methods for GLMMs. For NLME models, the approximate methods work very well even when the number n_i of within individual measurements is not large (Pinheiro and Bates 1995; Hartford and Davidian 2000; Ko and Davidian 2000). For binary responses with sparse within individual measurements, these approximate methods may not work well (Breslow and Lin 1995; Joe 2008). Higher order Taylor expansions or Laplace approximations may provide improvements (Solomon and Cox 1992; Raudenbush et al. 2000; Vonesh et al. 2002; Lee et al. 2006).

Lai and Shih (2003) and Lai, Shih, and Wong (2006) proposed hybrid methods that combine Laplacian approximations and Monte Carlo simulations for GLMM and NLME models. They proposed to use Monte Carlo approximations to the likelihoods for subjects with sparse observations and use Laplacian approximation for the likelihoods of other subjects that satisfy a certain diagnostic check on the adequacy of Laplacian approximations.

In all the approximate methods, standard errors are obtained by assuming that the approximations are exactly correct. Even when the approximate methods work well, standard errors may be under-estimated (Davidian and Giltinan 1995). More reliable standard errors may be obtained by bootstrap methods (Higgins, Davidian, and Giltinan 1997).

2.7 Further Topics

2.7.1 Model Selection and Further Topics

Model selection for mixed effects models includes selection of random effects and selection of covariates. Standard model selection methods, such as the AIC and BIC criteria and the likelihood ratio tests, can still be used for mixed effects models. However, special attention is required for mixed effects models since even the simple definition of sample size may not be clear (it involves both number of subjects and number of observations within each subject), so one should be careful about their asymptotic properties. Recently, Jiang et al. (2008) proposed so-called fence methods for mixed model selection. In practice, one can start with simple methods such as graphical tools. In this section, we briefly discuss some simple approaches which may be used in practice.

Selection of Random Effects

For longitudinal or clustered data, random effects are used to incorporate the correlation between the repeated observations within each individual and to address the variation of regression parameters across individuals or clusters. Thus, a simple method to select random effects is to check the variabilities of regression parameters across individuals. If a regression parameter varies substantially across individuals, a random effect may be introduced for this parameter, i.e., this parameter is allowed to be individual-specific. Otherwise, a random effect may not be necessary for this parameter.

More specifically, if there are sufficient repeated observations within each individual and the model is not complicated, one can fit a regression model to the repeated observations within each individual *separately*, using standard regression models for cross-sectional data. This leads to estimates of individual-specific parameters for each individual. We then compare these individual-specific parameter estimates across all individuals, either via graphical displays or via individual confidence intervals, and check the variabilities of these individual-specific estimates and decide the need of random effects. As an example, consider the children growth data in Section 1.3.4 (page 13). We can see from Figure 1.6 that the individual intercepts vary greatly across children but the individual slopes vary little across children. So we may consider a LME model with random intercept but fixed slope for modeling this dataset. Pinheiro and Bates (2002) showed more examples of this simple approach.

More formally, we may choose the random effects based on standard model selection criteria such as the *Akaike information criterion (AIC)* (Sakamoto, Ishiguro, and Kitagawa 1986), the *Bayesian information criterion (BIC)* (Schwarz 1978), and the *likelihood ratio test (LRT)*. The AIC and BIC are defined as

$$AIC = -2l(\hat{\boldsymbol{\theta}}) + 2p,$$
$$BIC = -2l(\hat{\boldsymbol{\theta}}) + p\log(N),$$

where $l(\hat{\boldsymbol{\theta}})$ is the maximized log-likelihood under the fitted model, $\hat{\boldsymbol{\theta}}$ is the MLE of $\boldsymbol{\theta}$, p is the number of parameters in the model, and N is the total number of observations used to fit the model. When comparing two models, the model with smaller AIC or BIC values is preferred.

When comparing two nested models, we may also consider the well-known likelihood ratio test (LRT), which may be used to test if one model fits significantly better than the other. The LRT statistic is defined as

$$T = -2\log(L_1(\hat{\boldsymbol{\theta}}_1)/L_2(\hat{\boldsymbol{\theta}}_2)),$$

where L_1 and L_2 are the likelihoods and $\hat{\boldsymbol{\theta}}_1$ and $\hat{\boldsymbol{\theta}}_2$ are MLEs of $\boldsymbol{\theta}$ under model I and model II respectively. Under the null hypothesis that the two models fit equally well and some *regularity conditions*, the LRT statistic asymptotically follows a χ^2-distribution with degrees of freedom being the difference in the numbers of parameters in the two models. One of the regularity conditions is that the parameters are not on the boundary of the parameter space, which may be violated in testing variance components as shown below.

Consider the LRT for testing if some or all random effects should be included in the model. Suppose that the random effects $\mathbf{b}_i \sim N(0, D)$, where for simplicity the covariance matrix D is assumed to be diagonal with elements $\mathbf{d} = (d_{11}, d_{22}, \cdots, d_{qq})$. These diagonal elements characterize the magnitudes of unexplained variation in the corresponding individual-specific parameters. For example, a large value of d_{jj} (say) indicates a great variation of the j-th component of $\boldsymbol{\beta}_i$ in (2.21) (page 53), suggesting a possible need of a random effect for this parameter. In general, we may use the LRT to test

$$H_0 : \mathbf{d}_s = \mathbf{0} \qquad \text{versus} \qquad H_1 : \mathbf{d}_s > \mathbf{0} \qquad (2.79)$$

where $\mathbf{d}_s$ is a subset of $\mathbf{d}$. If we rejects H_0, we should keep the corresponding random effects in the model; otherwise, no random effects are needed for the corresponding parameters.

For the testing problem (2.79), the regularity conditions for the asymptotics of the LRT do *not* hold, because the parameters $\mathbf{d}_s$ are on the boundary of the parameter space of $\mathbf{d}$ under H_0, where the parameter space for $\mathbf{d}$ is the positive orthant $\mathcal{D}^+ = \{\mathbf{d} : d_{jj} \geq 0, j = 1, \cdots, q\}$. Thus, the LRT statistic T does *not* follow a χ^2 distribution asymptotically. The testing problem (2.79) is an example of order-restricted tests or multivariate one-sided tests which have been well developed. Silvapulle and Sen (2004) and van Eeden (2006) provided recent review of the literature in order-restricted or constrained statistical inference.

Under some regularity conditions, asymptotically the LRT statistics for order-restricted or constrained hypotheses typically follow a *mixture of χ^2 variates* or a *weighted average of χ^2 variates*, called a $\bar{\chi}^2$ distribution, rather than a single χ^2 distribution as in the standard unrestricted cases. Specifically, consider testing the following order-restricted hypotheses

$$H_0 : R\boldsymbol{\theta} = \mathbf{0} \qquad \text{versus} \qquad H_1 : R\boldsymbol{\theta} \geq \mathbf{0}, \qquad (2.80)$$

where $\boldsymbol{\theta}$ denotes the parameters in the model $f(\mathbf{y}, \boldsymbol{\theta})$, and R is a known full-rank matrix of constants with order $r \times p$ $(r \leq p)$. The LRT statistic for testing H_0 versus H_1 is given by

$$LRT = 2 \left[\sup\{l(\boldsymbol{\theta}) : R\boldsymbol{\theta} \geq \mathbf{0}\} - \sup\{l(\boldsymbol{\theta}) : R\boldsymbol{\theta} = \mathbf{0}\} \right],$$

where $l(\boldsymbol{\theta}) = \log f(\mathbf{y}, \boldsymbol{\theta})$ is the log-likelihood. Then, under some regularity conditions (see Silvapulle and Sen 2004, page 146), the asymptotic null distribution of the LRT is the following $\bar{\chi}^2$ distribution

$$\lim_{n \to \infty} P(LRT \geq c \mid H_0) = \sum_{i=0}^{r} w_i \left(r, R\mathcal{I}(\boldsymbol{\theta}_0)^{-1} R^T \right) P(\chi_i^2 \geq c),$$

where c is any constant, $w_i(\cdot)$ is a (usually complicated) weight function (sometimes called level probability), χ_i^2 is the usual χ^2-distribution with degrees of freedom i, and $\mathcal{I}(\boldsymbol{\theta}) = -E(\partial^2 \log f(\mathbf{y}, \boldsymbol{\theta})/\partial\boldsymbol{\theta}^2)$ is the usual information matrix of the model.

For many order-restricted testing problems, exact p-values may be difficult to compute due to the complicated weight functions, especially in the presence of nuisance parameters. However, lower and upper bounds of the p-values or approximate p-values are relatively easy to obtain (e.g., Perlman 1969; Perlman and Wu 2002; Tamhane and Logan 2002, 2004; Silvapulle and Sen 2004). Constrained inference for LME models was considered in Stram and Lee (1994), Silvapulle 1997b, and Verbeke and Molenberghs (2000). Note that care must be taken for multivariate constrained tests to avoid potentially misleading conclusions (Silvapulle 1997a; Perlman and Wu 1999a).

Finally, note that an advantage of order-restricted tests (2.80) over unconstrained tests, such as the (two-sided) test $H_0 : R\boldsymbol{\theta} = \mathbf{0}$ versus $H_1 : R\boldsymbol{\theta} \neq \mathbf{0}$, is some power gain from incorporating known restrictions on parameter space into the inference.

Selection of Covariates

For covariate selection, we consider NLME models as illustration, since approaches for other mixed effects models are similar. In NLME modeling, an important objective is to understand how the individual-specific parameters $\boldsymbol{\beta}_i$ in the second stage (2.21) of a NLME model (see page 53) vary across individuals and whether some of this variation may be explained by individual characteristics represented by covariates $\mathbf{x}_i$. That is, one needs to determine an appropriate form of $h(\cdot)$ in (2.21):

$$\boldsymbol{\beta}_i = h(\mathbf{x}_i, \boldsymbol{\beta}, \mathbf{b}_i),$$

as in ordinary regression modeling. Once the second-stage model (2.21) is determined, the random effects represent the variations not explained by the covariates.

A simple approach to select covariates in (2.21) is as follows: (i) fit a NLME model (2.20) – (2.22) *without* any covariates in (2.21), and obtain empirical Bayes estimates of the random effects $\hat{\mathbf{b}}_i$; (ii) plot the components of $\hat{\mathbf{b}}_i$ against each covariate in $\mathbf{x}_i$; and (iii) identify important covariates from the plots (e.g., ones with systematic patterns) and plausible functional form of $h(\cdot)$ in model (2.21), since the estimates $\hat{\mathbf{b}}_i$ contain information about the covariates. This simple approach may be useful for initial screen when there is a large number of potential covariates, since in this case a formal variable selection method for NLME models may exhibit convergence problems. Wu and Wu (2002a) demonstrated such an approach. Formal variable selection methods include AIC/BIC criteria and LRT. Jiang et al. (2008) proposed an alternative formal model selection method and studies its theoretical properties.

Further Topics

There has been an explosion of research in mixed effects models in recent years. It is difficult to give a complete review. Here we briefly discuss a few selected topics.

In mixed effects models, we typically assume that the random effects follow a multivariate normal distribution $N(0, D)$. In practice, sometimes this assumption may not hold since the true random effects distribution may have heavier tails or skewed or do not follow any parametric distributions. Lai and Shih (2003) proposed a nonparametric estimation method for NLME models in which no distributional assumption is made for the random effects. They considered a nonparametric maximum likelihood method in which the likelihood function involves sums instead of integrals, so it offers substantial computational advantages in addition to the robustness of distributional assumption. They also developed asymptotic results for their proposed method. LME models with multivariate t-distributions in the random effects or in the random errors are discussed in Pinheiro et al. (2001) and Song et al. (2007).

To accommodate outliers in the data, Yeap and Davidian (2001) proposed robust methods for NLME models, and Sinha (2004) considered robust methods for GLMMs. We will give a detailed discussion of these robust methods in Chapter 9. Missing data, censored data, and measurement errors are common in longitudinal studies. Mixed effects models incorporating these incomplete observations are discussed in details in Chapters 4 – 6.

Model diagnostics for mixed effects models are important but the literature is somewhat limited. For LME and NLME models, model diagnostics can be based on the usual residuals (i.e., the differences between the observed values and fitted values), and normal Q-Q plots can be used to check normality assumptions (Davidian and Giltinan 1995; Pinheiro and Bates 2002). For GLMMs, Pearson or deviance residuals may be used (McCullagh and Nelder 1989). Recent research includes Tchetgen and Coull (2006), who derived a test for random effects distributions, and Waagepetersen (2006), who considered tests for normality. Further research in this area is needed.

2.7.2 Choosing a Mixed Effects Model and Method

We have reviewed various mixed effects models for longitudinal or clustered data. In practice, the choice of a mixed effects model should be based on the data and scientific objectives. For a mixed effects regression model, one would need to choose a response variable first. If the response is a continuous variable, one would naturally consider a LME model or a NLME model, which assumes normality for the response variable (or a transformation of the response). A LME model is often chosen empirically based on the observed data without theoretical justifications, while a NLME model is typically chosen based on some understanding of data-generation mechanisms, with some theoretical justifications, and a NLME model is often preferred if it can be derived and justified. If the response is binary or discrete, a GLMM can be considered. If the response is the time to an event of interest, a survival model can be used. If the observed longitudinal trajectories are complex, semiparametric or nonparametric mixed effects models are more flexible than parametric models and may thus be preferred.

Selections of random effects and covariates in a mixed effects model have been discussed in Section 2.7.1. In a preliminary analysis, random effects can be chosen based on the observed variations between individuals or clusters, although some of these variations may be partially explained by covariates. In data analysis, it is important to check if the final model indeed fits the observed data well. In other words, model diagnostics are important, which are often based on graphical tools such as residual plots.

We have also discussed a number of estimation methods for GLMM and NLME models. The choice of an estimation method mainly depends on computational burden. If computation is not an issue, one should choose exact or nearly exact methods. For GLMM and NLME models, if the dimension of random effects is small, one may consider nearly exact methods such as the Gauss-Hermite quadrature method. If the dimension of random effects is moderate, one may consider a Monte Carlo EM algorithm. If the dimension of random effects is moderate to large, computation is typically an issue, so computationally more efficient approximate methods are highly recommended. The performance of these approximate methods depends on the types of the response, the magnitudes of between and within individual variations, the nonlinearity of the models, and other factors. Thus, it would be desirable to use two or more methods and then compare the results. Alternatively, one may use an approximate method to obtain good starting values for more exact numerical or Monte Carlo iterative methods.

2.8 Software

Standard regression models for cross-sectional data, such as linear models, generalized linear models, and nonlinear models, are widely implemented in most statistical software, including R or Splus, SAS, and SPSS. For example, in R or Splus, the corresponding functions are **lm()**, **glm()**, and **nl()** respectively. The book by Venables and Ripley (2003) "Modern Applied Statistics with S-PLUS" (the *MASS* package) provided detailed descriptions. A function for fitting multi-level models in R is **bugs()**. Latest R implementations and packages can be found in the R webpage at *www.r-project.org/*.

Standard mixed effects models are also implemented in many statistical software. The following packages in Splus or R can be used to fit LME, NLME, and GLMM models respectively: function **lmer()** in the *lme4* package, functions **lme()** and **nlme()** in the *nlme* package, and function **glmmPQL()** in the *MASS* package. These packages are mainly based on the approximate methods of Lindstrom and Bates (1990) and Breslow and Clayton (1993). Standard errors are obtained assuming that the approximations are exactly correct. The book by Pinheiro and Bates (2002) provided a detailed description of fitting LME and NLME models, including many examples. The R function **glmm()** fits a GLMM with a random intercept using a normal mixing distribution computed by Gauss-Hermite integration. The R packages *mgcv* and *gamm* can be used to fit generalized additive models and generalized additive mixed models respectively. In Splus or R, the function **integrate()** or **integrate2()** can be used for numerical integration based on the adaptive quadrature method.

Current versions of SAS include **mixed**, **nlmixed**, and **glimmix** procedures for fitting LME, NLME, and GLMM models. An attractive feature of **proc nlmixed** in SAS is that it includes numerical integration method (the default is adaptive Gaussian quadrature) which allows "exact" likelihood computation, in addition to the approximate methods similar to *nlme* in Splus or R. The SAS **nlinmix** macro with the **expand=zero** option implements the first-order approximation (Beal and Sheiner 1982) using the GEE method, and the SAS **nlinmix** macro with the **expand=eblup** option implements the first-order conditional method (Lindstram and Bates 1990).

Other software for fitting mixed effects models include SPSS (the **mixed** command), STATA (the **xt** set of commands such as **xtmixed**), and HLM and ML-Win and others. In pharmacokinetics and pharmacodynamics (PK/PD) modeling, the *NONMEM* package for fitting NLME models is popular. For beginners, an easy software for fitting NLME models is WinNonmix.

When fitting nonlinear models, the choice of starting values is very important. Starting values may be obtained using some simple methods. One should try different starting values in a model fitting. Another potential problem for fitting nonlinear or generalized linear mixed models is convergence. Since iterative algorithms are used to fit these models, the algorithms sometime may not converge if the starting values are poorly chosen or if the sample size is not large but the number of parameters is large or other problems. When an algorithm does not converge, one can try different starting values or simplify the models to reduce the number of parameters.

CHAPTER 3

Missing Data, Measurement Errors, and Outliers

3.1 Introduction

In longitudinal studies, observed data are often incomplete since it is difficult to observe all data for every variable at each measurement time, especially if the studies last a long time. Incomplete data are also common in other types of studies, including cross-sectional studies. In many studies, data may also be censored or mis-measured. Specifically, the following problems are common in practice: (i) missing data: the data are completely missing, (ii) censored data: the true data are not observed but they are known to be in certain ranges, (iii) mis-measured data: the true data are not observed but their mis-measured versions are observed, (iv) outliers: the observed data may or may not be the true values but they are inconsistent with majority of the observed data. For last three cases, although the true data may not be observed, partial information is available for the true values so they may also be called incomplete data. To avoid confusion, we use the term *missing data* to refer to data that are completely missing and use the term *incomplete data* to refer to the more general cases, i.e., one of the above four cases. In this chapter, we review some commonly used general methods for incomplete data problems.

In the presence of missing data, it is important to choose appropriate methods for data analysis since simple or naive methods may lead to biased results. For example, if a subject withdraws from a study due to conditions related to the study, discarding or ignoring the missing data will lead to biased analyses. Thus, one should first check possible *reasons* of missing data or *how* the data are missing, i.e., the possible *missing data mechanisms*. Choices of appropriate methods for data analysis are then based on the missing data mechanisms. In practice it is often difficult to determine the missing data mechanisms, since the data analyst may not be the one who collects the data. In this case, one may analyze the data under several plausible missing data mechanisms and then check if the results differ substantially. This is called *sensitivity analysis*.

Censored data may be viewed as a special type of missing data in which the censored values are unobserved but are known to be in certain ranges. Censoring arises in studies where certain variables may have upper and/or low limits of detection. For example, in some AIDS studies viral loads may have low detection limits so that viral loads below the detection limits cannot be measured. Survival data or event time data are also examples of censored data. Due to the unique features of censored data, special methods are needed to analyze data with censoring.

Measurement errors in variables are common in practice. For example, blood pressure, fat intake in nutritional studies, and depression scores, may all be measured with errors, since it is usually difficult to measure these variables accurately. In the presence of measurement errors, the observed data are not the true values but mis-measured ones. If the observed data are measured with errors but are treated as true values, statistical analysis may be biased. For example, in regression models the covariates are usually treated as fixed or accurately measured. If the covariates are in fact measured with errors, statistical inference will be misleading, e.g., a significant covariate may seem to be non-significant. Thus, covariate measurement errors must be addressed for valid inference. To address measurement errors, we often require *replicates* or validation data. In longitudinal studies, the repeated observations within each individual may be viewed as replicates.

Outliers also arise frequently in practice. Outliers may be mis-measured values (e.g., recording errors) or may even be true values. Here we view outliers as incomplete data in a broad sense. It is known that maximum likelihood estimates are sensitive to outliers, i.e., a few outliers in the data may have great effects on the estimates such as misleading estimates and inflated standard errors. For multi-dimensional data, it may be difficult to detect outliers. Thus, we should perform robust analysis in which we can either assume heavy-tail distributions to accommodate outliers or use robust methods to downweight outliers.

There is an extensive literature on the analyses of incomplete data. For example, Little and Rubin (2002) provided an overview of missing data methods, Carroll et al. (2006) reviewed commonly used methods for measurement errors, and Maronna, Martin, and Yohai (2006) discussed recent developments of robust methods. However, a comprehensive review of incomplete data problems in mixed effects models seems unavailable, although incomplete data problems are especially common in the applications of these models.

In this chapter, we provide a review of general methods for incomplete data problems in regression models. We focus on basic concepts and approaches. Technical details and extensions to mixed effects models will be discussed in the following chapters.

3.2 Missing Data Mechanisms and Ignorability

3.2.1 Missing Data Mechanisms

When analyzing data with missing values, one first needs to consider possible missing data mechanisms. Rubin (1976) defined the following three missing data mechanisms:

- *missing completely at random (MCAR)*: missingness depends neither on the observed data nor on the unobserved data,

- *missing at random (MAR)*: missingness may depend on the observed data but *not* on the unobserved data,

- *missing not at random (MNAR)*: missingness may depend on the observed data *and* on the unobserved data.

For example, if a student did not hand in an assignment, the possibilities are (i) he forgot it, then the missing data is MCAR since the missingness is unrelated to any observed or missing data; (ii) he had done very well on all previous assignments so felt no need to do this assignment since not all assignments were counted, then the missing data is MAR since the missingness is related to his previous assignment marks (observed data) but not to the missing assignment; and (iii) he did not know how to do the assignment, then the missing data is MNAR since the missingness is related to the missing assignment (he would get a low mark if he did hand in the assignment). There may be other possibilities, but the missing data mechanism is either MCAR or MAR or MNAR.

When missing data are MCAR, we can view the individuals in the sample with completely observed data as a random (or representative) subsample of the population. Thus, the commonly used *complete-case (CC) method*, which discards all individuals with missing data, is still valid and leads to unbiased results. The only loss is efficiency, due to discarding some data, which leads to a smaller sample size. For multivariate data or longitudinal data, the loss of efficiency can be substantial since many individuals may have to be discarded. When the missing data are not MCAR (either MAR or MNAR), individuals with complete data cannot be treated as a random or representative subsample of the population, so the CC method may lead to biased results.

Note that MCAR is a very strong assumption, which may not be reasonable in many cases. In practice, MAR may be a more reasonable assumption, especially if one includes more observed variables in the analysis. In many cases, MNAR is a possibility, and such possibility should not be ignored as the missing data contain valuable information, i.e., the missingness is *informative*. Note that MNAR and MAR are not testable based on the observed data, so sensitivity analyses under various possible missing mechanisms are important. It is possible to test the MCAR assumption.

For missing covariates in regression models, MCAR means that the probability of missing a covariate value depends neither on the observed covariates/responses nor on the missing covariate. MAR means that the probability of missing a covariate value may depend on the observed values of the covariates or responses but not on the missing values of the covariates. MNAR means that the probability of missing a covariate value depends on the missing or unobserved values and may also on the observed covariates or responses.

In summary, if individuals with missing data differ from individuals with complete data in some *systematic* ways, the individuals with missing data cannot be viewed as a random or representative subsample of the population, so the missing data mechanism is unlikely to be MCAR. Statistical analysis with MCAR missing data is the simplest, but MCAR is also the strongest assumption.

3.2.2 Ignorability

In practice, it is often difficult to determine missing data mechanisms. If one chooses likelihood methods for data analysis, however, the missing data mechanisms can be ignored or the missingness is *ignorable* if the missing data are either MCAR or MAR and if the model parameters satisfy a reasonable assumption (Rubin 1976). The assumption on the parameters for ignorability is that the parameters in the model of interest and the parameters in the assumed missing data model (missing data mechanism) are *distinct* (i.e., the two models do not share any parameters). This distinctness assumption is often reasonable and will be assumed throughout the book. If the missing data are MNAR, the missing data mechanism is *non-ignorable*, and in this case we must incorporate the missing data mechanism in likelihood inference. Little and Rubin (2002) provided a more detailed discussion on ignorability.

Ignorability is a major advantage of likelihood methods for missing data problems. In many cases the MAR assumption may be realistic, and in these cases likelihood inference can proceed based on the *observed data*, ignoring the missing data mechanism. Other methods may not share this property. For example, if we use the generalized estimating equation (GEE) methods for inference, the missing data mechanism cannot be ignored even if the missing data are MAR (see Chapter 10).

To illustrate the concept of ignorability, consider missing covariates problems in regression models. For simplicity, we consider missing data in time-independent (e.g., baseline) covariates in a regression model for longitudinal data, without random effects. For individual i, let $\mathbf{y}_i = (y_{i1}, y_{i2}, \cdots, y_{im})^T$ be the response measurements and let $\mathbf{x}_i = (x_{i1}, x_{i2}, \cdots, x_{ip})^T$ be p time-independent covariates with possible missing data, $i = 1, 2, \cdots, n$. Let the regression model be $f(\mathbf{y}_i|\mathbf{x}_i, \boldsymbol{\beta}, \boldsymbol{\sigma})$, where the parameters $\boldsymbol{\beta}$ are regression coefficients and $\boldsymbol{\sigma}$ contains variance-covariance parameters.

When there are missing data in covariates $\mathbf{x}_i$, we should assume a model $f(\mathbf{x}_i|\boldsymbol{\alpha})$ for the covariates in order to address the missing data. Write $\mathbf{x}_i = (\mathbf{x}_{obs,i}, \mathbf{x}_{mis,i})$, where $\mathbf{x}_{obs,i}$ contains the observed components of $\mathbf{x}_i$ and $\mathbf{x}_{mis,i}$ contains the missing components of $\mathbf{x}_i$. Let

$$r_{ij} = \begin{cases} 1 & \text{if } x_{ij} \text{ is observed,} \\ 0 & \text{if } x_{ij} \text{ is missing,} \end{cases} \quad i = 1, \cdots, n, \ j = 1, \cdots, p. \quad , \quad (3.1)$$

be a missing data indicator. For simplicity, we assume that the r_{ij}'s are independent of each other, although in practice they may be correlated. To *approximate* a possible missing data mechanism, we assume a model for the missing data indicator $f(r_{ij}|\mathbf{x}_i, \boldsymbol{\phi})$, which describes how the probability of missing a covariate value depends on the observed and unobserved values.

If the missing data are MCAR, the value of r_{ij} does not depend on any observed or missing data, so we have

$$P(r_{ij} = 1) \equiv f(r_{ij}|\mathbf{x}_i, \boldsymbol{\phi}) = f(r_{ij}|\boldsymbol{\phi}).$$

In this case the missing data mechanism, or the model $f(r_{ij}|\boldsymbol{\phi})$, can be ignored in statistical inference. If the missing data are MAR, the value of r_{ij} does not depend on $\mathbf{x}_{mis,i}$ but may depend on $\mathbf{x}_{obs,i}$, so we have

$$P(r_{ij} = 1) \equiv f(r_{ij}|\mathbf{x}_i, \boldsymbol{\phi}) = f(r_{ij}|\mathbf{x}_{obs,i}, \boldsymbol{\phi}),$$

In this case, if the parameters $\boldsymbol{\phi}$ are distinct from the parameters $(\boldsymbol{\alpha}, \boldsymbol{\beta}, \boldsymbol{\sigma})$, the missing data mechanism can again be ignored in likelihood inference. The distinctness of the parameters is often a reasonable assumption in practice. Therefore, both MCAR and MAR are ignorable in likelihood inference, and we do not need to assume a missing data model for r_{ij}. In other words, under MCAR or MAR, statistical inference can simply be based on the *observed-data likelihood*

$$L_o(\boldsymbol{\alpha}, \boldsymbol{\beta}, \boldsymbol{\sigma}) = \prod_{i=1}^{n} \int f(\mathbf{y}_i|\mathbf{x}_i, \boldsymbol{\beta}, \boldsymbol{\sigma}) f(\mathbf{x}_i|\boldsymbol{\alpha}) \, d\mathbf{x}_{mis,i},$$

which leads to valid inference.

If the missing data are MNAR, the value of r_{ij} may depend on $\mathbf{x}_{mis,i}$ and $\mathbf{x}_{obs,i}$, so we have

$$P(r_{ij} = 1) \equiv f(r_{ij}|\mathbf{x}_i, \boldsymbol{\phi}) = f(r_{ij}|\mathbf{x}_{obs,i}, \mathbf{x}_{mis,i}, \boldsymbol{\phi}).$$

In this case the missing data mechanism is non-ignorable, so likelihood inference must incorporate the missing data model $f(r_{ij}|\mathbf{x}_i, \boldsymbol{\phi})$. In other words, inference should be based on the likelihood

$$L(\beta, \sigma, \alpha) = \prod_{i=1}^{n} \int f(\mathbf{y}_i | \mathbf{x}_i, \beta, \sigma) f(\mathbf{x}_i | \alpha) f(\mathbf{r}_i | \mathbf{x}_i, \phi) \, d\mathbf{x}_{mis,i},$$

where $\mathbf{r}_i = (r_{i1}, \cdots, r_{ip})$. The missing data model $f(r_{ij} | \mathbf{x}_i, \phi)$ is not testable based on the observed data, and different missing data models may fit the observed data equally well. Therefore, when the missing data are MNAR, it is important to perform *sensitivity analysis*, i.e., one assumes different missing data models and then checks if the resulting estimates are similar or not.

The parameters ϕ in the missing data model and the parameters α in the covariate model are usually not of primary inferential interest, i.e., they are *nuisance parameters*, since they are assumed just to address missing data. Therefore, we should avoid large or complicated models for $f(r_{ij} | \mathbf{x}_i, \phi)$ and $f(\mathbf{x}_i | \alpha)$ to avoid too many nuisance parameters. Too many nuisance parameters may affect precision of the main parameter estimates $\hat{\beta}$ and may lead to computational difficulties.

3.3 General Methods for Missing Data

3.3.1 Naive Methods

In this section we briefly review some naive methods which are commonly used for handling missing data in practice. These methods are very simple, so they are widely used. However, these naive methods often lead to inefficient or biased results, so they should be avoided whenever possible or be used when there is only a small portion of missing data.

The Complete-Case (CC) Method

The simplest and perhaps the most widely used method for missing data is the CC method, which simply discards all individuals with missing values. As discussed in Section 3.2, if the missing data are MCAR, individuals with missing values may be viewed as a random subsample of the original sample, so discarding these individuals will not lead to biased results but will lead to a loss of efficiency. When the missing data are not MCAR (i.e., MAR or MNAR), discarding individuals with missing values may lead to *biased* results since these individuals may differ from individuals with complete data in some systematic ways. Thus, the CC method is valid *only when* the missing data are MCAR.

Note that MCAR is a very strong assumption which is often invalid in practice, so the CC method often lead to biased results, but the bias may be small if the missing rate is low. For multivariate data or longitudinal data, the CC method may discard a large number of observations, so the loss of efficiency or bias may be substantial. Moreover, some data may be expensive to obtain, so discarding data is generally not recommended.

The Last-Value-Carried-Forward (LVCF) Method

A commonly used missing data method for longitudinal data is the so-called *last-value-carried-forward (LVCF)* method. The LVCF method imputes a missing value by the last observed value from the same individual. This method is very simple and does not discard any data. However, the LVCF method has at least two obvious drawbacks: i) if longitudinal trajectories vary greatly over time and the measurements are far apart, the imputed values can be very different from the (unobserved) true values, and ii) the missing data *uncertainty* is not taken into account. Carpenter et al. (2004) and Cook, Zeng, and Yi (2004) provided nice discussions of the LVCF method. Thus, the LVCF method is usually not recommended for missing data in longitudinal studies.

The Mean Imputation Method

Another simple imputation method is called the *mean-imputation* method, which imputes a missing value by the mean of observed data. The mean-imputation method is also very simple, but it shares similar drawbacks as the LVCF method, so it is generally not recommended.

Both the LVCF method and the mean imputation method are examples of *single imputation methods*, i.e., they impute a missing value by a single (guessed) value. There are other single imputation methods for missing data, such as baseline-value-carried-forward method and imputation methods based on conditional means or regression models. Some of these single imputation methods may produce unbiased estimates. However, a major drawback of these simple single imputation methods is that they fail to incorporate the *missing data uncertainty*. That is, if a value is missing, there is some uncertainty about the true value (i.e., the true value may take many possible values). Therefore, these simple single imputation methods may produce estimates with *under-estimated* standard errors (since they do not incorporate the missing data uncertainty), so statistical inference based on these single imputation methods may be misleading (e.g., p-values may be smaller than they should be), especially when the missing rate is not low.

For more reliable inference, the missing data uncertainty must be taken into account. The following approaches can be used to incorporate missing data uncertainty when an imputation method is used:

- impute each missing value by *several* possible values based on an imputation model, which is used to predict the missing value based on the observed data, and then combine the results to obtain a single overall inference. This approach is called the *multiple imputation* method, which will be described in details in Section 3.5.

- impute a missing value by a single predicted value based on an imputation model, and then *adjust the variances* of the estimates using theoretical results or resampling methods (e.g., bootstrap or jackknife) to reflect the missing data uncertainty.

Both approaches adjust the variances of parameter estimates to incorporate the uncertainty due to missing data, so they lead to more reliable inference. The choice of imputation models is important – one should try to choose a model that can best predict the missing data based on the observed data. We will further discuss these methods in Section 3.3.2.

There are other simple or naive methods for missing data. Little and Rubin (2002) provided a comprehensive overview and discussion.

3.3.2 Formal Methods

As noted in Section 3.3.1, commonly used simple or naive methods for missing data may lead to biased results and are generally not recommended, especially if the missing rate is not low. To address missing data appropriately, we should use more formal methods. These formal methods are often *model based*, i.e., they are often based on an assumed model for the data. In this section, we briefly review the following four formal methods:

- likelihood inference using EM algorithms,
- single imputation methods with variance adjustments,
- multiple imputation methods,
- Bayesian methods,
- weighted GEE methods.

In this book, we will mainly focus on EM algorithms for likelihood inference since likelihood methods are standard inferential tools for mixed effects models and have many desirable properties.

Likelihood Inference Using EM Algorithms

Likelihood methods are standard approaches for statistical inference, especially for mixed effects models. A major reason is that likelihood theory is well developed and maximum likelihood estimates (MLEs) have very attractive asymptotic properties, such as asymptotic normality and asymptotic most efficiency, under some regularity conditions. For missing data problems, likelihood methods are particularly appealing because both MCAR and MAR are ignorable in likelihood inference. Moreover, likelihood methods for mixed effects models allow *unbalanced* data in the response, i.e., if there are missing data in the responses with MCAR or MAR mechanisms, likelihood inference for mixed effects models can proceed in the usual way, ignoring the missing data.

In the case of missing covariates, if the missing data are MCAR or MAR, we only need to consider the observed data likelihood without specifying the missing data mechanisms. When the missing data are MNAR, the modification is straightforward since one only needs to incorporate a missing data model.

The EM algorithm is an iterative algorithm used to compute MLEs in the presence of missing data or unobservables. The EM algorithm is very general and stable, and it is guaranteed to converge to a local maxima. Therefore, EM algorithms are widely used for likelihood inference in the presence of missing data or unobservable quantities. We discuss details of the EM algorithm in Section 3.4.

Single Imputation Methods with Variance Adjustments

A single imputation method imputes a missing value by a single predicted value. The mean-imputation method and the LVCF method in Section 3.3.1 are both examples of single imputation methods. There are other single imputation methods, such as that based on regression models which use all observed data to predict the missing values. A well chosen imputation model may produce unbiased or approximately unbiased estimates. However, a major drawback of simple single imputation methods is that they fail to incorporate missing data uncertainty, so the variances or standard errors of the resulting estimates are under-estimated. Moreover, these methods may fail to lead to consistent estimates.

In some cases, one may be able to derive theoretical asymptotic standard errors which incorporate the missing data uncertainty (e.g., Schafer and Schenker 2000). However, these methods seem not generally applicable in other cases. More generally, we can obtain valid standard errors use well-known re-sample methods such as the bootstrap and jackknife to incorporate the missing data uncertainty (Rao and Shao 1992; Fay 1996; Shao 2002). Alternatively, multiple imputation methods appear to be more widely used since they are very general and are implemented in standard software.

Multiple Imputation Methods

Multiple imputation methods (Rubin 1987) impute *more than one* predicted values for *each* missing value to incorporate missing data uncertainty, where the imputations are generated based on a predictive distribution of the missing data given the observed data. This leads to several complete datasets. Each complete dataset is analyzed separately using standard complete-data methods as if there were no missing data. The results from the complete-data analyses are then combined. Once missing data are imputed, various analyses can be done by different analysts.

The major advantages of multiple imputation methods are i) missing data uncertainty is incorporated, ii) standard complete-data methods can be used, iii) the final results may be robust against the assumed imputation model when the missing rate is low since the observed data are not affected by the imputation model, and iv) different analyses can be performed on the multiply imputed dataset. Rubin (1987) provided the theory and justification of multiple imputation methods. More details of multiple imputation methods will be provided in Section 3.5.

Bayesian Methods

Bayesian methods incorporate prior information from similar studies, so they provide additional information for parameter estimates. Bayesian methods are particular appealing when there are too many parameters, which is often the case for missing data analyses, or when the sample size is not large. When non-informative priors are used, Bayesian methods are equivalent to likelihood methods (see Chapter 11). Ibrahim et al. (2002) showed a Bayes analysis for missing data in GLMs.

Bayesian methods are often implemented via Markov Chain Monte Carlo (MCMC) methods such as the Gibbs sampler, so they may be computationally intensive. There have been extensive developments of Bayesian methods in the last few decades, due to availability of modern computers and breakthrough in computational tools such as the MCMC methods which make Bayesian computation feasible. We will discuss Bayesian methods in more details in Chapter 11.

Weighted GEE

The EM algorithms, single or multiple imputation methods, and Bayesian methods are generally model-based and distributional assumptions are often required. So the results may be sensitive to the assumed models and distributions. Generalized estimating equations (GEE) methods are only based on the first two moments, without distributional assumptions. Moreover, if the mean structure is correctly specified, the mean parameters can be consistently estimated even if the covariance structure is mis-specified, and GEE estimates are asymptotically normal (Liang and Zeger 1986). Therefore, GEE methods are popular in the analyses of longitudinal data or clustered data.

In the presence of missing data, one can modify the GEE equations to weight the completely observed cases by the inverse of the missing probability, and the resulting estimates are still consistent (Robins et al. 1994, 1995). Recent work includes Ibrahim et al. (2005), Carpenter et al. (2006), and Yi (2008), among others. The weighted GEE method becomes less tractable when there are multiple missing variables and when the missing data patterns are non-monotone (see Table 3.1). For GEE methods the missing data mechanism MAR is not ignorable. Thus, if the missing data are not MCAR, one should incorporate the missing data mechanisms in GEE methods. GEE models and weighted GEE methods will be discussed in details in Chapter 10.

Other Methods

There are other methods for missing data analyses. Little and Rubin (2002) provided a comprehensive overview. For likelihood inference, analytic (or closed-form) MLEs can be obtained in some special cases, so the EM algorithm may not be needed. For example, for multivariate normal data or multinomial data with *monotone* missing data patterns (see Table 3.1), analytic expressions of MLEs can be obtained, since in this case the observed-data likelihood can be factored – see Little and Rubin (2002) for details and examples. Even if the missing data patterns are *non-monotone* (see Table 3.1), analytic expressions of MLEs may still be obtained under certain *conditional independence* assumptions, since in this case the observed-data likelihood can again be factored (Andersson and Perlman 1993; Perlman and Wu 1999). These conditional independence assumptions are often made in graphical models (Lauritzen 1996).

In most practical situations, however, the missing data patterns are non-monotone and certain conditional independence assumptions are not reasonable, so analytic solutions are unavailable and the EM algorithms should be used. In regression models with missing data, analytic expressions of MLEs are generally unavailable, even if the missing data patterns are monotone. Therefore, iterative algorithms such as the Newton-Raphson method and the EM algorithms are needed for maximum likelihood estimation.

3.3.3 Sensitivity Analysis

When the missing data are MNAR or non-ignorable, an assumed missing data model is *not* testable based on the *observed data*, since no data are available to verify the dependence of the missingness on the unobserved data. In fact, every MNAR model has an MAR counterpart in the sense that both models produce exactly the same fit to the observed data, but the two models may lead to different inferences (Molenberghs and Kenward 2007). Therefore, it is important to perform *sensitivity analysis* in which one considers several plausible missing data models and then checks the sensitivity of the results to the assumed models.

A simple approach for sensitivity analysis is to consider several plausible MNAR models which assume various dependencies of the missingness on the

Table 3.1 *Monotone and non-monotone missing data patterns*

ID	Monotone pattern				Non-monotone pattern			
	X_1	X_2	X_3	X_4	X_1	X_2	X_3	X_4
1	O	O	O	O	O	O	O	O
2	O	O	O	M	O	M	O	M
3	O	O	O	M	O	O	M	M
4	O	O	M	M	M	O	O	O
5	O	O	M	M	M	M	O	O
6	O	O	M	M	O	O	O	M
7	O	M	M	M	M	M	O	O
...								

Notation: ID – subject ID number, M – missing, O – observed.

unobserved and observed data. Then, one can check if the analysis results are robust to the assumed MNAR models. Alternatively, one could consider different distributions assumed for the models. Note that, parameter *identifiability* can be an issue for MNAR models, so one should not choose too complicated missing data models.

There are more formal methods for sensitivity analysis, such as local and global influence methods. Molenberghs and Kenward (2007) provided a detailed discussion on sensitivity analysis and provided some nice examples.

3.3.4 Selection Models versus Pattern-Mixture Models

There are two general model-based frameworks for missing data analyses: selection models and pattern-mixture models (Little and Rubin 2002). We briefly describe these two approaches using missing covariates in regression models as an illustration. We follow the notation and setup in Section 3.2.

The *selection model* approach factors the joint density of the data $(\mathbf{y}_i, \mathbf{x}_i, \mathbf{r}_i)$ as follows:

$$
\begin{aligned}
f(\mathbf{y}_i, \mathbf{x}_i, \mathbf{r}_i | \boldsymbol{\theta}, \boldsymbol{\phi}) &= f(\mathbf{y}_i, \mathbf{x}_i | \boldsymbol{\theta}) f(\mathbf{r}_i | \mathbf{y}_i, \mathbf{x}_i, \boldsymbol{\phi}) \\
&= f(\mathbf{y}_i | \mathbf{x}_i, \boldsymbol{\beta}, \boldsymbol{\sigma}) f(\mathbf{x}_i | \boldsymbol{\alpha}) f(\mathbf{r}_i | \mathbf{y}_i, \mathbf{x}_i, \boldsymbol{\phi}), \quad (3.2)
\end{aligned}
$$

where $\boldsymbol{\theta} = (\boldsymbol{\beta}, \boldsymbol{\sigma}, \boldsymbol{\alpha})$. In factorization (3.2), $f(\mathbf{y}_i | \mathbf{x}_i, \boldsymbol{\beta}, \boldsymbol{\sigma})$ is the regression model of interest, $f(\mathbf{x}_i | \boldsymbol{\alpha})$ is the assumed covariate model which is used to address missing data in covariates, and $f(\mathbf{r}_i | \mathbf{y}_i, \mathbf{x}_i, \boldsymbol{\phi})$ is the missing data model which specifies how the missingness of the covariates depends on the observed and missing data. If the missing data are MAR, we have

$$f(\mathbf{r}_i|\mathbf{y}_i, \mathbf{x}_i, \boldsymbol{\phi}) = f(\mathbf{r}_i|\mathbf{y}_i, \mathbf{x}_{obs,i}, \boldsymbol{\phi}).$$

If the missing data are MNAR, we have

$$f(\mathbf{r}_i|\mathbf{y}_i, \mathbf{x}_i, \boldsymbol{\phi}) = f(\mathbf{r}_i|\mathbf{y}_i, \mathbf{x}_{obs,i}, \mathbf{x}_{mis,i}, \boldsymbol{\phi}).$$

Therefore, the selection model approach is natural for likelihood inference, and it offers much flexibility for modeling the missing data mechanisms. A drawback is that the missing data model $f(\mathbf{r}_i|\mathbf{y}_i, \mathbf{x}_i, \boldsymbol{\phi})$ is not testable based on the observed data, so sensitivity analysis is required.

The *pattern-mixture model* approach factors the joint density of the data $(\mathbf{y}_i, \mathbf{x}_i, \mathbf{r}_i)$ as follows

$$\begin{aligned} f(\mathbf{y}_i, \mathbf{x}_i, \mathbf{r}_i|\boldsymbol{\theta}, \boldsymbol{\phi}) &= f(\mathbf{y}_i, \mathbf{x}_i|\mathbf{r}_i, \boldsymbol{\theta})f(\mathbf{r}_i|\boldsymbol{\phi}) \\ &= f(\mathbf{y}_i|\mathbf{x}_i, \boldsymbol{\beta}, \boldsymbol{\sigma})f(\mathbf{x}_i|\mathbf{r}_i, \boldsymbol{\alpha})f(\mathbf{r}_i|\boldsymbol{\phi}), \end{aligned} \quad (3.3)$$

assuming that $\mathbf{y}_i$ is independent of $\mathbf{r}_i$ given covariates $\mathbf{x}_i$. Factorization (3.3) indicates that a different covariate model is needed for each missing data pattern. An advantage of the pattern-mixture model is that we can avoid specifying how the missingness of the covariates depends on the observed and missing data, which is not testable based on the observed data, since the missing data model $f(\mathbf{r}_i|\boldsymbol{\phi})$ does not require such specification.

In this book we focus on the selection model approach for missing data problems since it appears to be more convenient for likelihood inference. Detailed discussions of pattern-mixture models can be found in Little (1993, 1994), Hogan and Laird (1997), and Little and Ruben (2002).

3.3.5 Choosing a Method for Missing Data

In statistical analyses with missing data, some naive methods such as the CC method and the LVCF method have been widely used due to their simplicity, but these naive methods are generally not recommended since they may lead to biased or misleading results. The consequences of using a naive method in practice depend on the fraction of missing data and missing data mechanisms. For example, if the fraction of missing data is low and the missing data mechanism is MCAR, the most commonly used CC method can lead to valid analysis, although it may not be efficient since some data are discarded. If the missing data mechanism is not MCAR, however, the CC method may lead to biased results. Note that MCAR is a very strong assumption and is often not valid in many cases. Therefore, a formal method for missing data should be used whenever possible in order to avoid biased results, especially when the missing rate is not low and the missing data mechanism is not MCAR.

The EM algorithms and multiple imputation methods are perhaps most widely used formal methods for missing data analyses. The EM algorithm has the advantages of generality and easy adaption to different missing data mechanisms and arbitrary missing data patterns, but it usually requires custom programming and it may be sensitive to the assumed models. Multiple imputation methods can take advantages of existing complete-data methods and software, and they are relatively robust against the assumed imputation models if the missing rate is not high. Thus, if the missing rate is not high, in a preliminary analysis one can use a general multiple imputation method to quickly impute missing data and then use standard methods and software for complete-data analysis. For example, one may use a multivariate normal model to create multiple imputations for missing data in continuous response and covariates in a regression model and then proceed standard regression analysis based on complete data. However, for a more thorough and formal analysis, creating proper multiple imputations for more complex models such as GLMM and NLME models may be challenging and requires custom programming, and combining multiple results from multiple imputations may not be trivial. If the assumed models hold, the EM algorithm may be preferred for a more formal analysis, which focuses on a specific analysis and may be viewed as a multiple imputation method with infinite many imputations. If different analyses are to be performed on the same dataset, as for many survey data, multiple imputation methods may be preferred.

For a single imputation method, one should make sure that the results are unbiased and missing data uncertainty is incorporated, which can be challenging in some cases. Weighted GEE methods are robust against distributional assumptions and may use standard software which allows weights. However, for complex models with non-MCAR missing data mechanisms and arbitrary missing data patterns, implementations of weighted GEE methods may not be straightforward. Moreover, weighted GEE methods are usually not the most efficient. In some sense, Bayesian methods are perhaps most general, and software such as WinBUGS is readily available for standard problems. However, for more complex models or problems, implementations of Bayesian methods for missing data can be computationally challenging or tedious.

3.4 EM Algorithms

3.4.1 Introduction

The expectation-maximization (EM) algorithm (Dempster, Laird, and Rubin 1977) is an iterative procedure used to compute maximum likelihood estimates in the presence of missing data or unobservables. It has become extremely popular in the last few decades, and it has also been widely used outside the missing data area. The popularity of the EM algorithm compared with other numerical methods such as Newton-Raphson methods is EM's superior stability properties and easy for implementation. Specifically, the advantages of the EM algorithm include (i) its convergence is stable and each iteration increases the likelihood, (ii) the M-step involves only complete-data maximization so it is often computationally straightforward, and (iii) it is very general and can be used in almost any maximum likelihood estimation problems with unknown quantities. A disadvantage is that the EM algorithm may be slow to converge for many problems. In recent years, there has been extensive research on the EM algorithm itself, such as methods to speed up convergence and extensions of the standard EM algorithm. McLachlan and Krishnan (1997) provided a comprehensive review of these developments.

An EM algorithm *iterates* between an E-step and an M-step as follows:

- *E-step*: computes the conditional expectation of the "complete-data" log-likelihood given the observed data and the current parameter estimates, where the "complete-data" contain both the observed data and the missing data,
- *M-step*: maximizes the conditional expectation in the E-step with respect to the unknown parameters to produce updated estimates of the parameters.

Given some starting values, we iterate between the E-step and the M-step until the parameter estimates converge. It can be shown that, under some reasonable regularity conditions, each EM iteration *increases* the likelihood, so the EM algorithm is guaranteed to converge to a local maxima (Wu 1983). Since the EM algorithm only converges to a local maxima, when the likelihood may have multiple modes, it is important to choose good starting values or try different starting values to make sure that the EM algorithm eventually converges to a global maxima.

In the following, we present some simple examples to illustrate the EM algorithms.

Example 3.1   *An EM algorithm for normal data*

Let $y_1, y_2, \cdots, y_n$ be an i.i.d. sample from normal distribution $N(\mu, \sigma^2)$, and let $\theta = (\mu, \sigma^2)$ be the unknown parameters. Suppose that $y_1, y_2, \cdots, y_r$ are observed, but $y_{r+1}, y_{r+2}, \cdots, y_n$ are missing, where $r < n$. Assume that the missing data are MAR or MCAR. Let $\mathbf{y}_{obs} = (y_1, y_2, \cdots, y_r)$ be the observed data and let $\mathbf{y}_{mis} = (y_{r+1}, y_{r+2}, \cdots, y_n)$ be the missing data. The "complete data" is then $\mathbf{y}_{com} = (\mathbf{y}_{obs}, \mathbf{y}_{mis}) = (y_1, y_2, \cdots, y_n)$. The observed data log-likelihood is given by

$$l_{obs}(\boldsymbol{\theta}) = -\frac{r}{2}\log(2\pi r\sigma^2) - \frac{1}{2}\sum_{i=1}^{r}\frac{(y_i - \mu)^2}{\sigma^2},$$

and the "complete-data" log-likelihood is given by

$$l_{com}(\boldsymbol{\theta}) = -\frac{n}{2}\log(2\pi n\sigma^2) - \frac{1}{2}\sum_{i=1}^{n}\frac{(y_i - \mu)^2}{\sigma^2}.$$

Let $\boldsymbol{\theta}^{(k)}$ be the parameter estimate from the $(k-1)$th EM iteration, $k = 1, 2, 3, \cdots$. At k-th EM iteration, the E-step computes the conditional expectation of the "complete-data" log-likelihood given the current parameter estimates $\boldsymbol{\theta}^{(k)}$ and the observed data $\mathbf{y}_{obs}$, i.e., the E-step computes

$$
\begin{aligned}
Q(\boldsymbol{\theta}|\boldsymbol{\theta}^{(k)}) &= E\left(l_{com}(\boldsymbol{\theta})|\boldsymbol{\theta}^{(k)}, \mathbf{y}_{obs}\right) \\
&= \left[-\frac{n}{2}\log(2\pi n\sigma^2) + \frac{n\mu^2}{2\sigma^2}\right] - \frac{1}{2\sigma^2}\left[E\left(\sum_{i=1}^{n}y_i^2|\boldsymbol{\theta}^{(k)}, \mathbf{y}_{obs}\right)\right. \\
&\qquad\qquad \left. -2\mu E\left(\sum_{i=1}^{n}y_i|\boldsymbol{\theta}^{(k)}, \mathbf{y}_{obs}\right)\right],
\end{aligned}
$$

where

$$E\left(\sum_{i=1}^{n}y_i|\boldsymbol{\theta}^{(k)}, \mathbf{y}_{obs}\right) = \sum_{i=1}^{r}y_i + (n-r)\mu^{(k)}, \tag{3.4}$$

$$E\left(\sum_{i=1}^{n}y_i^2|\boldsymbol{\theta}^{(k)}, \mathbf{y}_{obs}\right) = \sum_{i=1}^{r}y_i^2 + (n-r)(\mu^{(k)2} + \sigma^{(k)2}). \tag{3.5}$$

The M-step of the EM algorithm then updates the parameter estimates by maximizing $Q(\boldsymbol{\theta}|\boldsymbol{\theta}^{(k)})$ with respect to $\boldsymbol{\theta}$, which leads to

$$\mu^{(k+1)} = \frac{1}{n}E\left(\sum_{i=1}^{n}y_i|\boldsymbol{\theta}^{(k)}, \mathbf{y}_{obs}\right) = \frac{1}{n}\left[\sum_{i=1}^{r}y_i + (n-r)\mu^{(k)}\right],$$

$$
\begin{aligned}
(\sigma^{(k+1)})^2 &= \frac{1}{n}E\left(\sum_{i=1}^{n}y_i^2|\boldsymbol{\theta}^{(k)}, \mathbf{y}_{obs}\right) - (\mu^{(k+1)})^2 \\
&= \frac{1}{n}\left[\sum_{i=1}^{r}y_i^2 + (n-r)(\mu^{(k)2} + \sigma^{(k)2})\right] - (\mu^{(k+1)})^2.
\end{aligned}
$$

Iterate the E-step and the M-step (for $k = 1, 2, 3, \cdots$) until convergence, we obtain the following MLEs

$$\hat{\mu} = \frac{1}{r}\sum_{i=1}^{r}y_i, \qquad \hat{\sigma}^2 = \frac{1}{r}\sum_{i=1}^{r}y_i^2 - \hat{\mu}^2.$$

Note that in this example the EM algorithm is in fact not needed. We use it as an illustration of the EM algorithm, since it is simple and contains the essential ideas of the EM algorithm.

Table 3.2 *Estimates for the distress data based on different missing data methods*

Parameter	CC Method		LVCF Method		EM Algorithm	
	Estimate	S.E.	Estimate	S.E.	Estimate	S.E.
μ_1	1.07	0.060	1.13	0.045	1.13	0.044
μ_2	0.94	0.060	1.01	0.044	1.02	0.043
μ_3	0.84	0.057	0.91	0.043	0.89	0.041
μ_4	0.80	0.050	0.89	0.040	0.87	0.035
μ_5	0.75	0.055	0.87	0.042	0.83	0.039

Example 3.2 *Mental distress data*

Consider the GSI scores in the mental distress data described in Section 1.3.1 of Chapter 1 (see page 6). There are five measurements for each subject, and the proportions of missing data are 3.7%, 14%, 19.2%, 38.7%, and 36.2% at baseline, 3, 6, 18, and 60 months respectively. The missing data pattern is non-monotone. The missing data are assumed to be MAR for simplicity. We also assume that the GSI measurements at the five time points follow a multivariate normal distribution $N(\boldsymbol{\mu}, \Sigma)$, where $\boldsymbol{\mu} = (\mu_1, \cdots, \mu_5)$ and Σ is a 5×5 unstructured covariance matrix. We consider estimating $\boldsymbol{\mu}$ using the complete-case (CC) method, the last-value-carried-forward (LVCF) method, and the likelihood method base on the EM algorithm.

There are 261 subjects in the study, and 142 (54.4%) of them have missing data at one or more time points. So the CC method deletes 54.4% of the subjects, of which many have observed values, so the CC method is very inefficient and it may also be biased if the missing data are not MCAR. The LVCF method uses all observed data, but it imputes the last observed value for each missing value, so it ignores the missing data uncertainty and may also be biased. The EM algorithm also uses all observed data, and it leads to valid inference.

Table 3.2 gives the estimates of μ_j's and the associated standard errors. We see that the CC method is the worst, compared with EM estimates. Note that the performances of the CC and LVCF methods depend on the fractions of missing data. For example, the fractions of missing data are larger at times t_4 or t_5 than that at earlier times, so the estimates (standard errors) $\hat{\mu}_4$ or $\hat{\mu}_5$ based on the CC and LVCF methods are more biased than estimates ($\hat{\mu}_1$, $\hat{\mu}_2$, $\hat{\mu}_3$) at early time points.

3.4.2 An EM Algorithm for Missing Covariates

In this section, we further illustrate the EM algorithm for the missing covariate problem presented in Section 3.2 (page 99). We follow the notation in Section 3.2. Let the time-independent covariates for individual i be $\mathbf{x}_i$, which contains missing values. Since the covariates have missing data, we assume a covariate model $f(\mathbf{x}_i|\boldsymbol{\alpha})$ for likelihood inference. For completely observed covariates, no model assumptions are needed so these covariates are suppressed.

We consider the selection model approach for missing data and write the joint density for $(\mathbf{y}_i, \mathbf{x}_i, \mathbf{r}_i)$ as follows

$$f(\mathbf{y}_i, \mathbf{x}_i, \mathbf{r}_i|\boldsymbol{\theta}, \boldsymbol{\phi}) = f(\mathbf{y}_i|\mathbf{x}_i, \boldsymbol{\beta}, \boldsymbol{\sigma})f(\mathbf{x}_i|\boldsymbol{\alpha})f(\mathbf{r}_i|\mathbf{y}_i, \mathbf{x}_i, \boldsymbol{\phi}), \qquad (3.6)$$

where $\boldsymbol{\theta} = (\boldsymbol{\beta}, \boldsymbol{\sigma}, \boldsymbol{\alpha})$. For simplicity, we assume that the missing data in co-variates are ignorable (i.e., MCAR or MAR), so we can ignore the missing data model (mechanism) $f(\mathbf{r}_i|\mathbf{y}_i, \mathbf{x}_i, \boldsymbol{\phi})$ in the following likelihood inference. The likelihood for the observed data $\{(\mathbf{y}_i, \mathbf{x}_{obs,i}),\ i = 1, 2, \cdots, n\}$ is given by

$$L_{obs}(\boldsymbol{\theta}) = \prod_{i=1}^{n} \int [f(\mathbf{y}_i|\mathbf{x}_i, \boldsymbol{\beta}, \boldsymbol{\sigma})f(\mathbf{x}_i|\boldsymbol{\alpha})] \, d\mathbf{x}_{mis,i}, \qquad (3.7)$$

and the observed-data loglikelihood is $l_{obs}(\boldsymbol{\theta}) = \log L_{obs}(\boldsymbol{\theta})$. The observed-data likelihood $L_{obs}(\boldsymbol{\theta})$ generally does not have a closed form expression, except for linear models. So we use the EM algorithm to find the MLE of $\boldsymbol{\theta}$ as follows:

The "complete data" is

$$\{(\mathbf{y}_i, \mathbf{x}_i), i = 1, 2, \cdots, n\} = \{(\mathbf{y}_i, \mathbf{x}_{obs,i}, \mathbf{x}_{mis,i}), i = 1, 2, \cdots, n\},$$

and the "complete data" log-likelihood is given by

$$l_{com}(\boldsymbol{\theta}) = \sum_{i=1}^{n} [\log f(\mathbf{y}_i|\mathbf{x}_i, \boldsymbol{\beta}, \boldsymbol{\sigma}) + \log f(\mathbf{x}_i|\boldsymbol{\alpha})] . \qquad (3.8)$$

Let $\boldsymbol{\theta}^{(0)}$ be the starting value. At k-th EM iteration, $k = 0, 1, 2, \cdots$, the E-step computes the conditional expectation of the complete-data loglikelihood given the observed data and the current parameter estimates, i.e.,

$$\begin{aligned} Q(\boldsymbol{\theta}|\boldsymbol{\theta}^{(k)}) &= E\left(l_{com}(\boldsymbol{\theta})\big|\mathbf{y}_i, \mathbf{x}_{obs,i}, \boldsymbol{\theta}^{(k)}\right) \\ &= \sum_{i=1}^{n} \int \left[\log f(\mathbf{y}_i|\mathbf{x}_i, \boldsymbol{\beta}, \boldsymbol{\sigma}) + \log f(\mathbf{x}_i|\boldsymbol{\alpha})\right] \\ &\qquad \times f(x_{mis,i}|\mathbf{y}_i, \mathbf{x}_{obs,i}, \boldsymbol{\theta}^{(k)}) \, dx_{mis,i}. \end{aligned}$$

For linear models, computation of $Q(\boldsymbol{\theta}|\boldsymbol{\theta}^{(k)})$ is straightforward and reduces to computation of sufficient statistics. However, for nonlinear or generalized linear models, computation of $Q(\theta|\theta^{(k)})$ is usually tedious and may involve numerical or Monte Carlo methods.

The M-step of the EM algorithm is to maximize $Q(\theta|\theta^{(k)})$ with respect to θ to produce an updated estimate $\theta^{(k+1)}$. The M-step can usually be accomplished by standard optimization procedures such as the Newton-Raphson method. Iterating between the E-step and the M-step, we can show that the likelihood is increasing (or non-decreasing) at each iteration, so eventually the EM algorithm will converge to a, possibly local, maximum.

For mixed-effects models, we can treat the unobservable random effects as additional "missing covariates". Then the EM algorithm described above can be modified in a straightforward way. Specifically, consider a mixed-effects model with missing covariates (MAR). The observed data likelihood is now given by

$$L_{obs}^*(\theta) = \prod_{i=1}^{n} \int \int [f(\mathbf{y}_i|\mathbf{x}_i, \mathbf{b}_i, \beta, \sigma)f(\mathbf{x}_i|\alpha)f(\mathbf{b}_i|D)] \, d\mathbf{x}_{mis,i} \, d\mathbf{b}_i, \quad (3.9)$$

where $\theta = (\beta, \sigma, \alpha, D)$. The E-step of the EM algorithm at the k-th iteration can be written as

$$Q^*(\theta|\theta^{(k)}) = \sum_{i=1}^{n} \int \int \Big[\log f(\mathbf{y}_i|\mathbf{x}_i, \mathbf{b}_i, \beta, \sigma) + \log f(\mathbf{x}_i|\alpha) $$
$$+ \log f(\mathbf{b}_i|D) \Big] f(x_{mis,i}, \mathbf{b}_i|\mathbf{y}_i, \mathbf{x}_{obs,i}, \theta^{(k)}) \, d\mathbf{x}_{mis,i} \, d\mathbf{b}_i.$$

Computation of $Q^*(\theta|\theta^{(k)})$ becomes more tedious since it involves an intractable integral with a higher dimension. We will discuss more details in Chapter 4.

3.4.3 Properties and Extensions

In this section, we briefly discuss some properties of the standard EM algorithm and its extensions.

At the k-th iteration of the EM algorithm, we have

$$Q(\theta^{(k+1)}|\theta^{(k)}) \geq Q(\theta|\theta^{(k)}), \qquad k = 0, 1, 2, \cdots,$$

since $\theta^{(k+1)}$ maximizes $Q(\theta|\theta^{(k)})$. In fact, it can be shown that the observed data likelihood increases at each EM iteration, so the EM algorithm is guaranteed to convergence to a (possibly local) maximum (Wu 1983).

The rate of convergence of the EM algorithm depends on the fraction of missing information. Specifically, let

$$I_{obs}(\theta) = -\frac{\partial^2 l_{obs}(\theta)}{\partial\theta\partial\theta^T},$$

$$\mathcal{I}_{com}(\boldsymbol{\theta}) = -E\left(\frac{\partial^2 l_{com}(\boldsymbol{\theta})}{\partial\boldsymbol{\theta}\partial\boldsymbol{\theta}^T}\right), \quad \mathcal{I}_{mis}(\boldsymbol{\theta}) = -E\left(\frac{\partial^2 l_{mis}(\boldsymbol{\theta})}{\partial\boldsymbol{\theta}\partial\boldsymbol{\theta}^T}\right)$$

be the information matrix for the observed data, the expected information matrix for the complete data, and the expected information matrix for the missing data respectively, where all the expectations are taken given the observed data and $l_{mis}(\boldsymbol{\theta})$ is defined in a similar way as $l_{obs}(\boldsymbol{\theta})$. Then, the *missing information principle* says that the complete information equals to the sum of the observed information and the missing information, i.e.,

$$\mathcal{I}_{com}(\boldsymbol{\theta}) = I_{obs}(\boldsymbol{\theta}) + \mathcal{I}_{mis}(\boldsymbol{\theta}). \tag{3.10}$$

Let $\hat{\boldsymbol{\theta}}$ be the MLE of $\boldsymbol{\theta}$, and let

$$K(\hat{\boldsymbol{\theta}}) = \mathcal{I}_{com}(\hat{\boldsymbol{\theta}})^{-1}\mathcal{I}_{mis}(\hat{\boldsymbol{\theta}})$$

be the proportion of missing information. Then, it can be shown that

$$\boldsymbol{\theta}^{(k+1)} - \hat{\boldsymbol{\theta}} \approx K(\hat{\boldsymbol{\theta}})(\boldsymbol{\theta}^{(k)} - \hat{\boldsymbol{\theta}}),$$

and the *rate of convergence*

$$r = \lim_{k\to\infty} \frac{||\boldsymbol{\theta}^{(k+1)} - \hat{\boldsymbol{\theta}}||}{||\boldsymbol{\theta}^{(k)} - \hat{\boldsymbol{\theta}}||}$$

is the largest eigenvalue of $K(\hat{\boldsymbol{\theta}})$ (McLachlan and Krishnan 1997). Therefore, the rate of convergence of the EM algorithm depends on the proportion of missing information.

A major disadvantage of the EM algorithm is that it is often slow to converge. Various methods to speed up the EM algorithm have been proposed (e.g., Meng and van Dyk 1997, 1998; Liu, Rubin, and Wu 1998). One particularly popular approach is to introduce working parameters in the models to reduce the proportion of missing information, called a *parameter-expansion (PX) EM algorithm* (Liu, Rubin, and Wu 1998).

Note that the EM algorithm does not automatically produce the standard errors of the MLEs. In the presence of missing data, a well-known formula for computing the variance-covariance matrix of the MLEs is proposed by Louis (1982) based on the following observed information matrix

$$I_{obs}(\hat{\boldsymbol{\theta}}) = \mathcal{I}_{mis}(\hat{\boldsymbol{\theta}}) - E\left[S_{com}(\hat{\boldsymbol{\theta}})S_{com}(\hat{\boldsymbol{\theta}})^T | \mathbf{y}_{obs}\right],$$

where $S_{com}(\boldsymbol{\theta}) = \partial l_{com}(\boldsymbol{\theta})/\partial\boldsymbol{\theta}$. Thus, the variances of the components of $\hat{\boldsymbol{\theta}}$ can be estimated by the diagonal elements of $I_{obs}^{-1}(\hat{\boldsymbol{\theta}})$.

Many extensions of the standard EM algorithm have been proposed in the literature, which have greatly expanded the applicability of EM algorithms. For example, for missing data problems, the E-step is often difficult to compute since it typically involves intractable multi-dimensional integrals. A popular approach is to approximate the expectation in the E-step using Monte Carlo methods, leading to *Monte Carlo EM algorithms* (Wei and Tanner 1990; Booth and Hobert 1999; Fort and Moulines 2003; Caffo, Jank, and Jones 2005). Detailed examples of Monte Carlo EM algorithms can be found in Chapter 4. When the M-step is difficult to compute, we may consider a modified EM algorithm, called the *ECM algorithm* (Meng and Rubin 1993), in which the M-step is replaced by a sequence of constrained maximization (CM) steps, with each step being a maximization over a subset of parameters while holding others fixed. McLachlan and Krishnan (1997) and Little and Rubin (2002) provided a comprehensive review of various extensions of the standard EM algorithm.

3.5 Multiple Imputation

3.5.1 Introduction

Multiple imputation methods are also widely used for missing data problems, especially in sample survey. A multiple imputation method incorporates the missing data uncertainty by imputing *several* predicted values for each missing value. Standard methods for complete data can be used to analyze the imputed datasets. The advantages of multiple imputation methods are i) standard complete-data methods can be used for analysis; ii) missing data uncertainty is incorporated; iii) the imputed complete datasets can be used for various analyses, i.e., the models used for creating multiple imputation are not necessary the same as the models used for data analysis; and iv) software for multiple imputation is widely available.

A multiple imputation method consists of the following three steps:

STEP 1. Simulate m plausible values for *each* missing value based on an assumed imputation model, leading to m "complete datasets", where $m > 1$ is a positive integer (common choices are $m = 4, 5, 6$).

STEP 2. Each of the m "complete datasets" from Step 1 is analyzed *separately* using standard complete-data methods, leading to m analysis results.

STEP 3. The m results from Step 2 are *combined* in an appropriate way, leading to one overall final result.

In Step 1, the missing values are generated from an assumed imputation model, which is typically the predictive distribution of the missing data given the observed data. For example, for missing covariates in a regression model, we may generate $\mathbf{x}_{mis}$ from the predictive distribution $f(\mathbf{x}_{mis}|\mathbf{y}_i, \mathbf{x}_{obs})$. However, this step may not be easy since it requires integrating over unknown parameters. Simpler imputation methods may be used but the imputations may be *improper* in the sense that they may not provide valid frequentist inference (Little and Rubin 2002). A *proper* imputation may be achieved via a Bayesian framework (see Section 3.5.2). The number of imputation m does not have to be large – usually $m = 5$ or 6 imputations will lead to satisfactory results (Rubin 1987; Schafer 1997).

Step 2 is usually straightforward since we can just use available standard methods for complete data as if there were no missing data. For Step 3, formulas are available to appropriately combine the m results, either combining several estimates or combining several test statistics (Rubin 1987; Little and Rubin 2002). Some of the well-known formulas are given in Section 3.5.2.

Example 3.3   *Multiple imputation for normal data*

Suppose that $\{(x_{11}, x_{12}), (x_{21}, x_{22}), \cdots, (x_{n1}, x_{n2})\}$ is an i.i.d. sample from a bivariate normal distribution $N(\boldsymbol{\mu}, \Sigma)$, where

$$\boldsymbol{\mu} = \left(\begin{array}{c} \mu_1 \\ \mu_2 \end{array} \right), \qquad \Sigma = \left(\begin{array}{cc} \sigma_{11} & \sigma_{12} \\ \sigma_{21} & \sigma_{22} \end{array} \right),$$

and some data are missing. We consider a multiple imputation method to impute the missing data. For simplicity, we consider the case where all x_{i1} are observed but some x_{i2} are missing.

To generate imputations for the missing x_{i2}'s, we may consider the following predictive distribution of x_{i2} given x_{i1}:

$$(x_{i2}|x_{i1}, \boldsymbol{\theta}) \sim N((\mu_2 - \sigma_{12}\mu_1/\sigma_{11}) + (\sigma_{12}/\sigma_{11})x_{i1}, \quad \sigma_{22} - \sigma_{12}^2/\sigma_{11}),$$

where $\boldsymbol{\theta} = (\mu_1, \mu_2, \sigma_{11}, \sigma_{12}, \sigma_{22})$. Since $\boldsymbol{\theta}$ is *unknown*, we are unable to simulate x_{i2} from $f(x_{i2}|x_{i1}, \boldsymbol{\theta})$. A simple approach is to estimate $\boldsymbol{\theta}$ by $\hat{\boldsymbol{\theta}}$ using, say, the complete-case method, and then we substitute $\boldsymbol{\theta}$ by $\hat{\boldsymbol{\theta}}$ and draw the missing x_{i2} from distribution $f(x_{i2}|x_{i1}, \hat{\boldsymbol{\theta}})$. However, such an imputation method is *improper* since the *uncertainty* in estimating $\boldsymbol{\theta}$ is not propagated (Rubin 1987, Chapter 4). A *proper* imputation method should draw x_{i2} from $f(x_{i2}|x_{i1})$ via the data augmentation method (see Section 3.5.2). Note that another simple method is to draw x_{i2} from its marginal distribution $x_{i2} \sim N(\mu_2, \sigma_2)$, with the unknown parameters estimated by some naive methods, but such a method fails to incorporate the correlation between x_{i1} and x_{i2}.

3.5.2 Multiple Imputation Methods

In this section, we describe a general approach for generating *proper* multiple imputations that leads to valid frequentist inference. The method is implemented via a Bayesian framework, although a multiple imputation method is often a frequentist one.

We again consider missing covariates in regression models, as in the previous sections. Let $\boldsymbol{\beta}$ be a $(p \times 1)$ vector of parameters of interest, such as the regression coefficients. Let $\mathbf{y} = \{\mathbf{y}_1, \mathbf{y}_2, \cdots, \mathbf{y}_n\}$ be the responses and $\mathbf{x} = \{\mathbf{x}_1, \mathbf{x}_2, \cdots, \mathbf{x}_n\}$ be the covariates with missing values. Let $\hat{\boldsymbol{\beta}} = h_1(\mathbf{y}, \mathbf{x}_{obs}, \mathbf{x}_{mis})$ be the statistic that would be used to estimate $\boldsymbol{\beta}$ if there were no missing data, where h_1 is some known function. Let $V = \mathrm{Var}(\hat{\boldsymbol{\beta}}) = h_2(\mathbf{y}, \mathbf{x}_{obs}, \mathbf{x}_{mis})$ be the variance of $\hat{\boldsymbol{\beta}}$, where h_2 is some known function.

A simple multiple imputation method is to generate $\mathbf{x}_{mis}$ from the predictive distribution $f(\mathbf{x}_{mis}|\mathbf{y}, \mathbf{x}_{obs}, \tilde{\boldsymbol{\theta}})$, where the estimate $\tilde{\boldsymbol{\theta}}$ may be obtained using some naive methods such as the complete-case method. However, such a multiple imputation method is *improper* in the sense that it does not provide valid frequentist inference, since the uncertainty in estimating $\boldsymbol{\theta}$ is not propagated (Little and Rubin 2002). Frequency-valid multiple imputations can be obtained via *proper* imputations, which generates $\mathbf{x}_{mis}$ from the predictive distribution $f(\mathbf{x}_{mis}|\mathbf{y}, \mathbf{x}_{obs})$ and can be accomplished through the following Bayesian arguments.

Note that the predictive distribution of the missing data given the observed data can be written as

$$f(\mathbf{x}_{mis}|\mathbf{y}, \mathbf{x}_{obs}) = \int f(\mathbf{x}_{mis}|\mathbf{y}, \mathbf{x}_{obs}, \boldsymbol{\theta}) f(\boldsymbol{\theta}|\mathbf{y}, \mathbf{x}_{obs}) d\boldsymbol{\theta},$$

where the unknown parameters $\boldsymbol{\theta}$ contain $\boldsymbol{\beta}$. To generate proper multiple imputations of $\mathbf{x}_{mis}$ from the predictive distribution $f(\mathbf{x}_{mis}|\mathbf{y}, \mathbf{x}_{obs})$, we can first simulate $\boldsymbol{\theta}^*$ from $f(\boldsymbol{\theta}|\mathbf{y}, \mathbf{x}_{obs})$, and then simulate $\mathbf{x}_{mis}^*$ from $f(\mathbf{x}_{mis}|\mathbf{y}, \mathbf{x}_{obs}, \boldsymbol{\theta}^*)$, and then we iterate the procedure for a warm-in period until the simulated sequences stabilized.

Specifically, we can use the following *data augmentation* method (Tanner and Wong 1987; Schafer 1997; Little and Rubin 2002). Given a starting value $\boldsymbol{\theta}^{(0)}$, at iteration k ($k = 1, 2, 3, \cdots$), we first simulate

$$\mathbf{x}_{mis}^{(k)} \sim f(\mathbf{x}_{mis}|\mathbf{y}, \mathbf{x}_{obs}, \boldsymbol{\theta}^{(k-1)}) \propto f(\mathbf{y}|\mathbf{x}, \boldsymbol{\theta}^{(k-1)}) f(\mathbf{x}|\boldsymbol{\theta}^{(k-1)}),$$

which can be accomplished via rejection sampling methods since the density functions $f(\mathbf{y}|\mathbf{x}, \boldsymbol{\theta}^{(k-1)})$ and $f(\mathbf{x}|\boldsymbol{\theta}^{(k-1)})$ are known. Then we simulate

$$\boldsymbol{\theta}^{(k)} \sim f(\boldsymbol{\theta}|\mathbf{y}, \mathbf{x}_{obs}, \mathbf{x}_{mis}^{(k)}),$$

which is the posterior distribution of $\boldsymbol{\theta}$. Thus, we obtain a sequence of simulated values $\{(\mathbf{x}_{mis}^{(k)}, \boldsymbol{\theta}^{(k)}), k = 0, 1, 2, \cdots\}$, which is a Markov chain. The Markov chain will converge to the stationary distribution $f(\mathbf{x}_{mis}, \boldsymbol{\theta}|\mathbf{y}, \mathbf{x}_{obs})$ after a burn-in period. After a burn-in period, let the last values be $(\mathbf{x}_{mis}^*, \boldsymbol{\theta}^*)$. Then, we have

$$\boldsymbol{\theta}^* \sim f(\boldsymbol{\theta}|\mathbf{y}, \mathbf{x}_{obs}), \qquad \mathbf{x}_{mis}^* \sim f(\mathbf{x}_{mis}|\mathbf{y}, \mathbf{x}_{obs}).$$

That is, $\mathbf{x}_{mis}^*$ is a *proper* imputation of the missing $\mathbf{x}_{mis}$ from the predictive distribution $f(\mathbf{x}_{mis}|\mathbf{y}, \mathbf{x}_{obs})$. Repeating the foregoing procedure m times, we create m proper imputations $\{\mathbf{x}_{mis}^{(1)}, \cdots, \mathbf{x}_{mis}^{(m)}\}$ for the missing values $\mathbf{x}_{mis}$.

Once we generate m imputations, we have m "complete datasets":

$$\{\mathbf{y}, \mathbf{x}_{obs}, \mathbf{x}_{mis}^{(l)}\}, \qquad l = 1, 2, \cdots, m.$$

Let $\hat{\boldsymbol{\beta}}^{(l)} = h_1(\mathbf{y}, \mathbf{x}_{obs}, \mathbf{x}_{mis}^{(l)})$ be the estimate of $\boldsymbol{\beta}$ and $V^{(l)} = h_2(\mathbf{y}, \mathbf{x}_{obs}, \mathbf{x}_{mis}^{(l)})$ be the variance of $\hat{\boldsymbol{\beta}}^{(l)}$ based on the l-th "complete dataset", obtained using available complete-data methods. Following Rubin (1987), we can combine the m estimates and obtain the overall estimate of $\boldsymbol{\beta}$ as follows

$$\bar{\boldsymbol{\beta}} = \frac{1}{m} \sum_{l=1}^{m} \hat{\boldsymbol{\beta}}^{(l)}, \tag{3.11}$$

and the overall estimated variance of $\bar{\boldsymbol{\beta}}$

$$\widehat{Var}(\bar{\boldsymbol{\beta}}) = \left(1 + \frac{1}{m}\right) B + W, \tag{3.12}$$

where

$$B = \frac{1}{m-1} \sum_{l=1}^{m} (\hat{\boldsymbol{\beta}}^{(l)} - \bar{\boldsymbol{\beta}})^2, \qquad W = \frac{1}{m} \sum_{l=1}^{m} V^{(l)}$$

are the *between-imputation variance* and the *within-imputation variance* respectively. Therefore, in a multiple imputation method, the final overall estimate is simply the *average* of the m individual estimates from the m imputed "complete" datasets, and the overall variance is the *sum* of the between-imputation variance B and the within-imputation variance W. The between-imputation variance B reflects the missing data uncertainty.

Suppose that $(\bar{\boldsymbol{\beta}} - \boldsymbol{\beta})/\sqrt{W} \sim N(0, 1)$ approximately, assuming that $\boldsymbol{\beta}$ is a scalar for simplicity. It can be shown that (Rubin 1987)

$$\frac{\bar{\boldsymbol{\beta}} - \boldsymbol{\beta}}{\sqrt{\widehat{Var}(\bar{\boldsymbol{\beta}})}} \sim t(v) \tag{3.13}$$

approximately, where

$$v = (m-1)\left(1 + \frac{W}{(1-m^{-1})B}\right)^2.$$

Note that the value $r = (1 + m^{-1})B/W$ measures the relative increase in variance due to missing data.

For inference, hypothesis testing can be based on the following overall statistic

$$T = (\bar{\boldsymbol{\beta}} - \boldsymbol{\beta})^T \widehat{Var}(\bar{\boldsymbol{\beta}})^{-1} (\bar{\boldsymbol{\beta}} - \boldsymbol{\beta}).$$

Note that, due to the multiple imputations, the statistic T no longer follows a χ^2 distribution asymptotically. Thus Li et al. (1991) proposed the following overall statistic

$$T^* = \frac{T}{p(1 + r)}, \qquad (3.14)$$

where

$$r = \frac{1}{p}\left(1 + \frac{1}{m}\right)\text{tr}(BW^{-1})$$

is the average relative increase in variance due to missing data. Li et al. (1991) showed that the statistic T^* follows an $F(p, w)$ distribution asymptotically, where

$$w = 4 + (\tau - 4)\left(1 + \frac{1 - 2\tau^{-1}}{p}\right)^2, \qquad \tau = p(m - 1).$$

Thus, we can use the statistic T^* to perform an overall hypothesis testing for multiply imputed datasets. Alternatively, one can also combine the m individual test statistics from the m imputed "complete datasets", and then combine the m test statistics. See Meng and Rubin (1992) for such an approach.

Other Multiple Imputation Methods

There are many other multiple imputation methods in the literature. A common approach is to use regression models to create imputations for the missing data, in which the variable with missing data is treated as a response and other observed variables as predictors. For continuous variables with missing data, one may choose linear regression models. For binary variables with missing data, one may consider logistic regression models, while for categorical variables with more than two categories one may consider multinomial regression models or ordered logistic regression models.

Here we briefly describe one such approach based on *chained equations*, which creates imputations variable-by-variable (Raghunathan et al. 2001; van Buuren et al. 2006). Specifically, the idea is to specify an imputation model for each variable separately, using other variables as predictors. For each variable with missing data, an imputation is generated for the missing data based on the assumed imputation model, and then the imputed value is used in the imputation of the next variable. The method iterates the procedure using a Gibbs sampling approach until convergence. Repeating the algorithm m times to generate m imputations for each missing value. For this method, one should check the sensitivity of the order of the variables.

For many of multiple imputation methods in the literature, it may be difficult to check if these methods generate proper imputations, so further research is needed to study the validity of these methods. If the missing data rate is low, these multiple imputation methods may still produce reasonable results since a multiple imputation method only imputes the missing values while the observed values remain unchanged. When the missing data rate is moderate or high, the choice of imputation models is critical.

3.5.3 Examples

In the following we present two examples which further illustrate multiple imputation methods.

Example 3.4   (Example 3.3 continued)

Return to Example 3.3 (see page 118). We choose $m = 6$ imputations for each missing value. Let $\{x_{j2}^{(l)}, \ l = 1, 2, \cdots, 6\}$ be the 6 imputed values for each missing x_{j2} for some j's. Let $\{\mathbf{x}_i^{(l)} = (x_{i1}, x_{i2}^{(l)}), \ i = 1, 2, \cdots, n\}$ be the l-th "complete dataset", where $x_{i2}^{(l)} = x_{i2}$ if x_{i2} is observed, $l = 1, 2, \cdots, 6$. Let

$$\hat{\boldsymbol{\mu}}^{(l)} = \frac{1}{n} \sum_{i=1}^{n} \mathbf{x}_i^{(l)}, \qquad \hat{\Sigma}^{(l)} = \frac{1}{n-1} \sum_{i=1}^{n} (\mathbf{x}_i^{(l)} - \hat{\boldsymbol{\mu}}^{(l)})(\mathbf{x}_i^{(l)} - \hat{\boldsymbol{\mu}}^{(l)})^T$$

be respectively the estimates of $\boldsymbol{\mu}$ and Σ based on the l-th "complete dataset". Then, the overall estimate of $\boldsymbol{\mu}$ is given by

$$\bar{\boldsymbol{\mu}} = \frac{1}{6} \sum_{l=1}^{6} \hat{\boldsymbol{\mu}}^{(l)},$$

with overall estimated variance

$$\widehat{Var}(\bar{\boldsymbol{\mu}}) = \frac{7}{6} \left[\frac{1}{5} \sum_{l=1}^{6} (\hat{\boldsymbol{\mu}}^{(l)} - \bar{\boldsymbol{\mu}}) \right] + \frac{1}{6} \sum_{l=1}^{6} \hat{\Sigma}^{(l)},$$

which incorporates both the between imputation variation and the within imputation variation.

Example 3.5   *Multiple imputation for mental distress data (Example 3.2 continued)*

For the mental distress data, in Example 3.2 (page 113) we considered the CC method, the LVCF method, and the EM algorithm for estimating parameters μ_j's. We noted that the three methods give different estimates and the EM algorithm should be the only valid method among the three methods. Now we consider using a multiple imputation method to estimate the parameters. We consider $m = 3$ and $m = 6$ multiple imputations.

Table 3.3 shows the results of multiple imputations and the EM algorithm (we may view the EM algorithm as a multiple imputation method with infinite

Table 3.3 *Estimates for the distress data based on multiple imputations.*

Parameter	MI ($m = 3$)		MI ($m = 6$)		EM Algorithm	
	Estimate	S.E.	Estimate	S.E.	Estimate	S.E.
μ_1	1.13	0.044	1.13	0.045	1.13	0.044
μ_2	1.01	0.049	1.02	0.043	1.02	0.043
μ_3	0.88	0.040	0.89	0.042	0.89	0.041
μ_4	0.86	0.043	0.87	0.038	0.87	0.035
μ_5	0.85	0.050	0.84	0.048	0.83	0.039

MI: *multiple imputation, and m is the number of multiple imputations.*

many imputations). We see that even a small number of imputations $m = 3$ produces reasonable estimates, compared with EM estimates. A larger number of imputations $m = 6$ gives better estimates but the improvement is not substantial. Thus, in practice the number m of imputations in a multiple imputation method does not have to be large. Typically, $m = 4, 5$, and 6 are common choices.

Note also that the performance of a multiple imputation method depends on the fraction of missing data. For example, the fractions of missing data are larger at later measurement times t_4 or t_5 than that at earlier time points, so the standard errors of $\hat{\mu}_4$ or $\hat{\mu}_5$ based on the multiple imputation method are also larger (compared with EM estimates), which reflects the missing data uncertainty.

3.6 General Methods for Measurement Errors

3.6.1 Covariate Measurement Errors

In a standard regression model, covariates are usually assumed as fixed or accurately measured. In practice, however, some covariates may be measured with errors, such as blood pressure and CD4 counts. Thus, the observed covariate values are *not* their true values but mis-measured one. Here we call such observed data as incomplete data, but they are different from missing data. When data are missing, no information is available for the true values. When data are measured with errors, however, some information is available since we have the observed versions of the true values. Therefore, statistical methods for measurement errors are different from those for missing data, but the methods also share some similarities since data are incomplete in both cases.

If covariate measurement errors in regression models are ignored, parameter estimates may be biased, hypothesis testing may lose some power, and interesting features in the data may be masked (Carroll et al. 2006). Specifically, if covariate measurement errors are not addressed, true covariate effects may be undetected or under-estimated. Thus it is important to address covariate measurement errors in regression models.

To address measurement errors, we typically assume a model for the relationship between mis-measured values and true (but unobserved) values, called a *measurement error model*. There are two types of commonly used measurement error models:

- classical measurement error models,
- Berkson measurement error models,

which are briefly described as follows.

Let x_{ij} be the observed but possibly mis-measured covariate value and let x_{ij}^* be the unobserved true covariate value for individual i at measurement time t_{ij}, $i = 1, 2, \cdots, n$; $j = 1, 2, \cdots, n_i$. Let y_{ij} and $\mathbf{z}_{ij}$ be the response and accurately measured covariates for individual i at time t_{ij} respectively. The *classical measurement error model* assumes

$$x_{ij} = x_{ij}^* + e_{ij}, \qquad E(e_{ij}|x_{ij}^*) = 0, \tag{3.15}$$

i.e., the observed value equals to the true value plus measurement error. The *Berkson measurement error model* (Berkson 1950) assumes

$$x_{ij}^* = x_{ij} + e_{ij}, \qquad E(e_{ij}|x_{ij}) = 0, \tag{3.16}$$

i.e., the true value equals to the observed value plus measurement error. In classical measurement error models the variability of the observed values is larger than the variability of the true values, while in Berkson measurement error models the variability of the true values is larger than the variability of the observed values, assuming independence of e_{ij} and x_{ij}^* in (3.15) and independence of e_{ij} and x_{ij} in (3.16). In practice, both situations may arise, and the choice of the models is often not difficult given the specific application under consideration. We will further discuss this issue in Chapter 5. Also see Carroll et al. (2006) for a more detailed discussion of these models.

The measurement error is called *nondifferential* if

$$f(y_{ij}|x_{ij}, x_{ij}^*, \mathbf{z}_{ij}) = f(y_{ij}|x_{ij}^*, \mathbf{z}_{ij}),$$

i.e., given the true covariate x_{ij}^* and other covariates $\mathbf{z}_{ij}$, the observed covariate x_{ij} is independent of the response y_{ij} (i.e., x_{ij} contains no extra information about the response). In this case, we also call the observed value x_{ij} a *surrogate* of the true value x_{ij}^*. We will focus on nondifferential measurement error models.

To correct covariate measurement errors in regression models, we usually need the following additional information: i) *validation data*, in which some true covariates x_{ij}^* are observed; or ii) *replication data*, in which replicates of the observed covariates x_{ij} are available; or iii) *instrumental data*, in which another variable related to x_{ij} is observed. In longitudinal studies, fortunately the repeated observations with each individual may be viewed as replication data, so measurement errors in time-dependent covariates can be partially addressed in a longitudinal model without extra information. See Chapter 5 for more details.

3.6.2 General Methods for Measurement Errors

There has been extensive research in measurement error problems. Fuller (1987) gave an earlier review. For recent reviews, see Gustafson (2004) and Carroll et al. (2006).

Generally there are two common approaches for covariate measurement errors in regression models:

- *functional modeling*, where no distributional assumption is made for the true covariate x_{ij}^*. Commonly used methods include regression calibration and simulation extrapolation (SIMEX).

- *structural modeling*, where a model or distribution is typically assumed for the true covariate x_{ij}^*. Commonly used methods include likelihood methods and Bayesian methods.

Functional modeling is robust to misspecification of the covariate distribution for x_{ij}^*, but structural modeling may be more efficient if the model or distributional assumption holds. In particular, likelihood methods are often based on the structural modeling approach. Since we focus on likelihood inference for mixed effects models in this book, we will mainly consider the structural modeling approach in Chapter 5.

In functional modeling, the two most commonly used methods are perhaps the *regression calibration* method and the *simulation extrapolation (SIMEX)* method. The idea of a regression calibration method is to replace the unobserved true covariate x_{ij}^* by a regression of x_{ij}^* on the observed data (x_{ij}, z_{ij}) and estimate the parameters in this regression model, and then run a standard analysis. The standard errors in the standard analysis are then adjusted to account for the uncertainty in estimating the parameters in the regression of x_{ij}^* on (x_{ij}, z_{ij}), using the bootstrap or sandwich methods. SIMEX is a simulation-based method in which the measurement error generating process is imitated via Monte Carlo methods. Carroll et al. (2006) provided a detailed description of both methods. We will also show some examples in Chapter 5.

In structural modeling, the two most commonly used methods are perhaps the likelihood methods and Bayesian methods. For likelihood methods, the observed data likelihood function can be written as

$$L_{obs}(\boldsymbol{\theta}) \;=\; \prod_{i=1}^{n} \int f(\mathbf{y}_i | x_{ij}^*, \mathbf{z}_{ij}, \boldsymbol{\beta}) f(x_{ij} | x_{ij}^*, \mathbf{z}_{ij}, \boldsymbol{\alpha})$$

$$\times f(x_{ij}^* | \mathbf{z}_{ij}, \boldsymbol{\gamma}) \; dx_{ij}^*, \qquad (3.17)$$

where $\boldsymbol{\theta} = (\boldsymbol{\beta}, \boldsymbol{\alpha}, \boldsymbol{\gamma})$. In (3.17), $f(\mathbf{y}_i | x_{ij}^*, \mathbf{z}_{ij}, \boldsymbol{\beta})$ is often called the *outcome model*, which is of primary interest, $f(x_{ij} | x_{ij}^*, \mathbf{z}_{ij}, \boldsymbol{\alpha})$ is called the *error model*, and $f(x_{ij}^* | \mathbf{z}_{ij}, \boldsymbol{\gamma})$ is sometimes called the *exposure model* (Gustafson 2004). Due to the intractable multi-dimensional integral in (3.17), likelihood methods can be computationally intensive. We will provide details of the likelihood approach for mixed effects models with covariate measurement errors in Chapter 5. For Bayesian inference, one introduces prior distributions for the parameters and then conduct inference based on posterior distributions. We will discuss Bayesian methods in Chapter 11.

3.7 General Methods for Outliers

3.7.1 Outliers

An *outlier* is an observation which is distinctly different from the rest of the data or is quite inconsistent with majority of the data. Likelihood inference for mixed effects models is sensitive to outliers, i.e., a few outliers in the data can greatly affect the results and thus the conclusions. Thus it is important to address outliers.

For longitudinal data, there are two types of outliers: i) an outlier among the repeated measurements within an individual, although the individual as a unit may not be viewed as an outlier; and ii) an outlying individual who behaves very differently from the rest of the individuals in the sample. In many practical situations it may not be easy to distinguish between the two cases from the data. We will further discuss this issue in Chapter 9.

For multi-dimensional data, it is often difficult to detect outliers since graphical displays of the data may be difficult. Therefore, robust methods which either use heavy-tail distributions or downweight outlying observations are very valuable. A robust analysis may be compared with a standard analysis (without addressing outliers). If the results differ substantially, outliers may be present and robust analysis may be more reliable.

3.7.2 General Robust Methods

There are many robust methods in the literature. Comprehensive reviews and discussions are given in Huber (1981), Rousseeuw and Leroy (1987), and Maronna et al. (2006), among others. Here we briefly describe the following two common approaches:

- one approach is to use *heavy-tail distributions*, such as the t distributions, to replace typically assumed normal distributions in the models to accommodate possible outliers in the data;

- another approach is to bound or down-weight outlying observations using appropriate weight functions.

The basic ideas of these two approaches are reviewed below, with more details given in Chapter 9.

Normal distributions are standard assumptions in many regression models. In particular, in mixed effects models, the random effects are usually assumed to be normally distributed, and the within-individual errors in LME and NLME models are also assumed to be normal. It is well known that t-distributions have similar shapes as normal distributions but with *heavier tails*, which can be used to accommodate outliers. Thus, a common approach for robust inference is to replace the normal distributions assumed in the models by (multivariate) t-distributions. For example, in LME and NLME models, we may assume that the random effects $\mathbf{b}_i$ and/or the errors $\mathbf{e}_i$ follow multivariate t-distributions for robust inference. Lange et al. (1989) and Pinheiro et al. (2001) considered such an approach for mixed effects models. We will provide more details about this approach in Chapter 9.

Another common approach for robust inference, which is perhaps more widely used, is to bound or downweight outlying observations. Note that, in either maximum likelihood estimation or in GEE estimation, one typically solves a set of estimating equations to obtain parameter estimates. Thus, we can down-weight outlying observations in these estimating equations for robust inference.

Specifically, let $\mathbf{y}_i$ be a vector of responses for individual or cluster i, and let $\mathbf{x}_i$ be covariates in a regression model with regression coefficients β. Let

$$\boldsymbol{\mu}_i(\beta) \;=\; E(\mathbf{y}_i|\mathbf{x}_i, \beta), \quad R_i = Cov(\mathbf{y}_i), \quad \mathbf{r}_i(\beta) = R_i^{-1/2}(\mathbf{y}_i - \boldsymbol{\mu}_i(\beta)),$$

and $D_i(\beta) = \partial \boldsymbol{\mu}_i(\beta)/\partial \beta$. We can bound the influence of outlying responses and down-weight leverage points in the covariates $\mathbf{x}_i$ separately. This leads to the following estimating equation for robust estimate of β:

$$\sum_{i=1}^{n} \left\{ D_i^T(\beta) R_i^{-1/2} W_i(\mathbf{x}_i) \psi_i(\mathbf{r}_i(\beta)) \right\} = 0, \tag{3.18}$$

where ψ_i is a monotone function which bounds the influence of outlying data in the response $\mathbf{y}_i$, and $W_i(\cdot)$ is a weight function which downweights leverage points in the covariates $\mathbf{x}_i$. The resulting estimator is called an *M-estimator*. Note that, when $W_i(\mathbf{x}_i) = 1$ and $\psi_i(\mathbf{r}_i) = \mathbf{r}_i$, the estimate from equation (3.18) reduces to the standard MLE.

The function $\psi_i(\cdot)$ is often chosen to be the derivative of the Huber's ρ-function

$$\rho(u) \quad = \quad \left\{ \begin{array}{ll} u^2/2, & \text{for } |u| \leq k, \\ k|u| - u^2/2, & \text{otherwise} \end{array} \right. \tag{3.19}$$

so $\psi_i(u) = d\rho(u)/du$ is given by:

$$\psi_i(u) \quad = \quad \left\{ \begin{array}{ll} u & \text{if } |u| \leq k \\ k & \text{if } u > k \\ -k & \text{if } u < -k \end{array} \right. \tag{3.20}$$

which is bounded for large or small values of u, where k is a turning point. One can choose the robustness of the method by choosing appropriate values of the turning point k and the weight function $W_i(\cdot)$. It can be shown that the M-estimators are asymptotically normal. We will discuss the methods in greater details in Chapter 9.

3.8 Software

Software is available for handling missing data in some standard models, such as multivariate normal models and multinomial models. For other models, users often need to program their own functions, especially for likelihood inference based on EM algorithms. General purpose multiple imputation methods sometimes can be used to impute missing values in a dataset, even if the imputation models may be different from the analysis models, so software for multiple imputations may allow users to do less programming on their own for preliminary analysis. Horton and Kleinman (2007) provided a comparison for some missing data software.

Splus version 7.0 has a missing data library that can be used to perform EM algorithms and multiple imputations using data augmentation, based on the methods described in Schafer (1997). One can access the library using the Splus command **library(missing)**. Joseph Schafer also provides free software (macros for Splus), similar to the missing data library. SPSS version 12.0 also includes a missing data library (see von Hippel (2004) for a review).

SAS procedure **proc mi** in SAS/STAT can be used to create multiple imputations using several different imputation methods. In SAS, the first imputation step, the second complete-data analysis step, and final combining step can be done separately. Specifically, we can use **proc mi** to create imputations for missing data. Then, the resulting complete data can be analyzed using any of the SAS procedures (e.g., **proc glm**). Finally, the analysis results can be combined using **proc mianalyze**. In fact, each step can be performed using different software. For example, one can use SAS **proc mi** to create imputations and use Splus or R to perform analysis and combine results.

There are other missing data software. The Splus and R package MICE (Multiple Imputation by Chained Equation), available at *http://cran.r-project.org/*, can be used to perform a variety of multiple imputations using function **mice()**. Users can also program their own imputation functions in MICE. In Stata, ICE can be used to create multiple imputation using chained equations. Van Buuren (2006) provided a useful guide to software implementations for multiple imputation at website *http://www.multiple-imputation.com*. Weighted GEE methods may be implemented using software that allow for weights, such as SAS or STATA. Bayesian approaches may be fit using a package such as **WinBugs**.

STATA (*http://www.stata.com/merror/*) provides software for generalized linear measurement error models. The software provides the first implementation of regression calibration and SIMEX in a general-purpose statistical package. The software was written by Raymond Carroll, James Hardin, and Henrik Schmiediche. See Stata webpage for more details.

Various robust methods are also widely implemented. For example, the SAS procedure **proc robustreg** includes M-estimators for regression, Splus has the **lmRobMM()** function, SPSS has a robust procedure, and STATA has the **rreg** procedure. There is also a large number of R packages for robust data analysis (see R webpage).

CHAPTER 4

Mixed Effects Models with Missing Data

4.1 Introduction

In Chapter 3, we have reviewed some commonly used methods for *general* missing data problems. Missing data methods for regression models are reviewed in Little and Rubin (2002), Ibrahim et al. (2005), and Molenberghs and Kenward (2007), among others. These reviews mostly focus on models for cross-sectional data. In practice, missing data are especially common in longitudinal studies. In this chapter, we describe some statistical methods for various missing data problems in mixed effects models, including LME, GLMM, and NLME models. We will discuss missing data in survival models and frailty models in Chapter 7. We consider the following missing data problems in mixed effects models:

- missing covariates,
- missing responses,
- missing both covariates and responses,
- dropouts.

An advantage of the likelihood method for mixed effects models is that it allows *unbalanced* data in the response, i.e., the numbers and schedules of the response measurements can be different across individuals. In other words, if there are missing data in the response and the missing data mechanism is *ignorable* (e.g., MAR or MCAR), we can simply ignore these missing responses and proceed with likelihood inference. If the missing data in the response of a mixed effects model are non-ignorable, however, we must assume a missing data model and incorporate it in likelihood inference.

Missing data may also occur in covariates. There are two types of covariates in longitudinal studies: time-independent covariates and time-dependent covariates. *Time-independent covariates* are only measured once in a longitudinal study or their values do not change over time, such as baseline covariates or gender and race. *Time-dependent covariates* are measured repeatedly over time or their values change over time. For a mixed effects model with time-dependent covariates, standard model specifications require that all covariate values are available at the *same* times as the response measurements. However, this is often not the case in practice since some covariate values may not be available at the response measurement times. This leads to missing data in time-dependent covariates, which is a very common problem. Time-independent covariates may also have missing data.

In a mixed effects model with covariates, if the covariates have missing data, we must assume a model or distribution for the incompletely observed covariates in likelihood inference, even if the missing data are ignorable. When the missing data are non-ignorable, we should assume an additional missing data model, which describes how the missingness is related to unobserved data, and incorporate it in likelihood inference. For a continuous covariate we may consider a normal distribution for the covariate (perhaps after an appropriate transformation), while for a discrete or a binary covariate we may assume a multinomial distribution or a binomial distribution for the covariate. Likelihood inference with discrete covariates is often computationally simpler since the integration with respect to the missing covariates reduces to a summation. Mixed effects models contain *unobservable* random effects. For computation, we can treat these random effects as additional "missing data" (or "missing covariates").

In this chapter we mainly focus on likelihood methods for missing data in mixed effects models, since they are the standard inferential methods for mixed effects models. For mixed effects models with missing data, a typical approach is to specify the likelihood for all *observed data* and then conduct estimation and inference based on this observed-data likelihood. For such an approach, a major challenge is the implementation since the observed-data likelihood is often intractable. The EM algorithm is usually used for estimation, but it can be computationally very intensive and sometimes may offer convergence problems. Thus, computationally more efficient approximate methods have been proposed. These approximate methods are typically based on Taylor approximations or Laplace approximations, and they may offer substantial computational advantages over exact likelihood computation.

GEE methods and Bayesian methods for missing data problems will be discussed in Chapters 10 and 11 respectively.

4.2 Mixed Effects Models with Missing Covariates

In this section, we consider missing data in covariates for LME, GLMM, and NLME models, with ignorable and non-ignorable missing data mechanisms. We will consider both time-independent and time-dependent covariates, as well as continuous and discrete covariates. Note that a LME model may be viewed as a special case of a GLMM or a NLME model. GLMMs and NLME models can be written in a unified general form, and the missing data methods are similar for both models. So the materials in the following sections are presented in a general form. However, the performances of the methods for a GLMM and a NLME model can be different, due to different types of responses and assumed distributions.

4.2.1 Missing Data in Time-Independent Covariates

For missing covariates in regression models for cross-sectional data, Little and Rubin (2002), Ibrahim et al. (2005), and Molenberghs and Kenward (2007) provided comprehensive reviews. In this section, we focus on missing time-independent covariates in mixed effects models for longitudinal or clustered data, when the missing data mechanism is *ignorable* (e.g., when the missing data are either MAR or MCAR). We focus on the likelihood approach.

Let $\mathbf{y}_i = (y_{i1}, \ldots, y_{in_i})^T$ be the repeated measurements of the response for individual i, where y_{ij} is the response value for individual i at time t_{ij} (or cluster i and measurement j), $i = 1, \ldots, n$; $j = 1, \ldots, n_i$. Let $\mathbf{z}_i = (z_{i1}, \ldots, z_{ip})^T$ denote p time-independent covariates for individual i ($p \geq 1$). Suppose that each individual covariate in $\mathbf{z}_i$ has missing data with arbitrary missing patterns. With possible re-arrangements of the covariates, we can write $\mathbf{z}_i = (\mathbf{z}_{mis,i}, \mathbf{z}_{obs,i})$, where $\mathbf{z}_{mis,i}$ is a collection of the missing components of $\mathbf{z}_i$ and $\mathbf{z}_{obs,i}$ is a collection of the observed components of $\mathbf{z}_i$. For simplicity of presentation, we suppress covariates without missing data, since no distributional assumptions are needed for these completely observed covariates.

The Response Model

Let $f(\mathbf{y}_i|\mathbf{z}_i, \mathbf{b}_i, \boldsymbol{\beta}, \boldsymbol{\sigma})$ be the probability distribution (density) of the response $\mathbf{y}_i$ given random effects $\mathbf{b}_i$ and covariates $\mathbf{z}_i$, where $\boldsymbol{\beta}$ is a vector of mean parameters (fixed effects) and $\boldsymbol{\sigma}$ contains variance-covariance or dispersion parameters. Note that parameters $\boldsymbol{\sigma}$ are not necessary distinct from $\boldsymbol{\beta}$, as in some GLMMs. For simplicity, we assume that the responses $y_{i1}, y_{i2}, \cdots, y_{in_i}$ are conditionally independent given the random effects $\mathbf{b}_i$ – see Section 2.3.2 (page 52) and Section 2.4.2 (page 62) for some discussion about this assumption, but the methods can be applied to more general cases. We assume

$\mathbf{b}_i = (b_{i1}, \cdots, b_{is})^T$ i.i.d. $\sim N(0, D)$, and write $D = D(\boldsymbol{\eta})$ with $\boldsymbol{\eta}$ being the distinct parameters in D.

For a NLME model, $f(\mathbf{y}_i|\mathbf{z}_i, \mathbf{b}_i, \boldsymbol{\beta}, \boldsymbol{\sigma})$ is the density function of a multivariate normal distribution with a mean vector

$$\boldsymbol{\mu}_i = E(\mathbf{y}_i|\mathbf{z}_i, \mathbf{b}_i, \boldsymbol{\beta}) = g(\mathbf{z}_i, \boldsymbol{\beta}, \mathbf{b}_i),$$

where $g(\cdot)$ is a known nonlinear function, and a variance-covariance matrix $R_i = \text{Cov}(\mathbf{y}_i|\mathbf{z}_i, \mathbf{b}_i, \boldsymbol{\beta}) = \sigma^2 I$. Thus

$$f(\mathbf{y}_i|\mathbf{z}_i, \mathbf{b}_i, \boldsymbol{\beta}, \boldsymbol{\sigma}) = (2\pi)^{-\frac{n_i}{2}} R_i^{-\frac{1}{2}} \exp\left(-\frac{1}{2}(\mathbf{y}_i - \boldsymbol{\mu}_i)^T R_i^{-1}(\mathbf{y}_i - \boldsymbol{\mu}_i)\right),$$

where R_i can in fact be an arbitrary covariance matrix.

For a GLMM, we have

$$f(\mathbf{y}_i|\mathbf{z}_i, \mathbf{b}_i, \boldsymbol{\beta}, \boldsymbol{\sigma}) = \prod_{j=1}^{n_i} f(y_{ij}|\mathbf{z}_i, \mathbf{b}_i, \boldsymbol{\beta}, \boldsymbol{\sigma}),$$

where $f(y_{ij}|\mathbf{z}_i, \mathbf{b}_i, \boldsymbol{\beta}, \boldsymbol{\sigma})$ is the density function for a distribution in the exponential family, such as a binomial distribution or a Poisson distribution. In this case, the mean function can be written as

$$\mu_{ij} = E(y_{ij}|\mathbf{z}_i, \mathbf{b}_i, \boldsymbol{\beta}) = h(\mathbf{z}_i^T \boldsymbol{\beta} + \mathbf{w}_i^T \mathbf{b}_i),$$

where $h(\cdot)$ is a known link function and $\mathbf{w}_i$ may contain covariates. For a Binomial distribution or a Poisson distribution, the variance function is completely determined by the mean function. For example, for a logistic mixed effect model, the variance function is

$$Var(y_{ij}|\mathbf{z}_i, \mathbf{b}_i, \boldsymbol{\beta}) = E(y_{ij}|\mathbf{z}_i, \mathbf{b}_i, \boldsymbol{\beta})(1 - E(y_{ij}|\mathbf{z}_i, \mathbf{b}_i, \boldsymbol{\beta})),$$

and for a Poisson mixed effects model we have

$$Var(y_{ij}|\mathbf{z}_i, \mathbf{b}_i, \boldsymbol{\beta}) = E(y_{ij}|\mathbf{z}_i, \mathbf{b}_i, \boldsymbol{\beta}).$$

In both cases, the variance functions do not contain distinct parameters $\boldsymbol{\sigma}$.

The Covariate Models

Since covariates $\mathbf{z}_i$ have missing values, we assume a distribution $f(\mathbf{z}_i|\boldsymbol{\alpha})$ for $\mathbf{z}_i$ in likelihood inference. The parameters $\boldsymbol{\alpha}$ in $f(\mathbf{z}_i|\boldsymbol{\alpha})$ may be viewed as *nuisance parameters* since this covariate distribution is only used to address missing data and is not of primary inferential interest.

When the number p of covariates is large or when $\mathbf{z}_i$ contains both continuous and discrete covariates, we may write the joint distribution of $\mathbf{z}_i$ as a product of one-dimensional conditional distributions (Ibrahim et al. 1999):

$$f(\mathbf{z}_i|\boldsymbol{\alpha}) = f(z_{i1}|\boldsymbol{\alpha})f(z_{i2}|z_{i1}, \boldsymbol{\alpha}) \cdots f(z_{ip}|z_{i1}, \cdots, z_{i,p-1}, \boldsymbol{\alpha}). \tag{4.1}$$

Then, for each term on the right-hand side of (4.1), we may consider a standard regression model, such as a linear regression model for a continuous z_{ik} or a logistic regression model for a binary z_{ik}. This approach allows us to specify a joint distribution for $\mathbf{z}_i$ even if the individual covariates z_{ik}'s are of different types, and it also allows us to reduce the number of nuisance parameters in α since we can apply the usual variable selection to each model.

For example, if z_{ik} is a continuous variable with a roughly normal distribution (perhaps after an appropriate transformation), we may consider the following multiple linear regression model for $f(z_{ik}|z_{i1}, \cdots, z_{i,k-1}, \alpha)$

$$z_{ik} = \alpha_0 + \alpha_1 z_{i1} + \cdots + \alpha_{k-1} z_{i,k-1} + \epsilon_i, \qquad \epsilon_i \sim N(0, \alpha_k), \qquad (4.2)$$

i.e., given covariates $(z_{i1}, \cdots, z_{i,k-1})$, covariate z_{ik} is assumed to follow a normal distribution with mean $\eta_i = \alpha_0 + \alpha_1 z_{i1} + \cdots + \alpha_{k-1} z_{i,k-1}$ and variance α_k. If z_{ik} is a binary variable with values 1 or 0, we may consider the following logistic regression model for $f(z_{ik}|z_{i1}, \cdots, z_{i,k-1}, \alpha)$

$$\log \left(\frac{P(z_{ik} = 1)}{1 - P(z_{ik} = 1)} \right) = \alpha_0 + \alpha_1 z_{i1} + \cdots + \alpha_{k-1} z_{i,k-1} \qquad (4.3)$$

i.e., given covariates $(z_{i1}, \cdots, z_{i,k-1})$, covariate z_{ik} is assumed to follow a binomial distribution with mean $\mu_i = \exp(\eta_i)/(1 + \exp(\eta_i))$ and variance $\mu_i(1 - \mu_i)$. Then, we can delete non-significant terms in models (4.2) or (4.3) using a standard variable selection method such as the LRT or AIC/BIC criteria to reduce the number of parameters in models (4.2) or (4.3).

Sensitivity analysis should be performed to check if the results are sensitive to the order of the factorization in (4.1). This approach can also be used for multivariate regression models when the responses are of different types or follow different distributions.

The Observed-Data Likelihood

Likelihood inference is based on the *observed data* likelihood. Here the observed data are $\{(\mathbf{y}_i, \mathbf{z}_{obs,i}), \ i = 1, \ldots, n\}$, and the observed-data likelihood is given by

$$L_o(\boldsymbol{\theta}) = \prod_{i=1}^{n} L_{oi}(\boldsymbol{\theta}) \qquad (4.4)$$

$$= \prod_{i=1}^{n} \int\!\!\int f(\mathbf{y}_i|\mathbf{z}_i, \mathbf{b}_i, \boldsymbol{\beta}, \boldsymbol{\sigma}) f(\mathbf{b}_i|D(\boldsymbol{\eta})) f(\mathbf{z}_i|\alpha) \, d\mathbf{b}_i dz_{mis,i},$$

where $\boldsymbol{\theta} = (\alpha, \boldsymbol{\beta}, \boldsymbol{\sigma}, \boldsymbol{\eta})$ is the collection of all unknown parameters. Since the missing data mechanism is ignorable, we do not need to assume a missing data model. The observed-data likelihood $L_o(\boldsymbol{\theta})$ in (4.4) is often intractable since it involves a possibly high-dimensional integral and generally does not have an analytic or closed-form expression. Thus, a major challenge in likelihood inference for mixed-effects models with missing covariates is the evaluation of the intractable likelihood $L_o(\boldsymbol{\theta})$. However, we can follow the computational strategies described in Chapter 2 (Section 2.6, page 78) and consider numerical or Monte Carlo methods for "exact" likelihood estimation. In the following, we describe a Monte Carlo EM algorithm for maximum likelihood estimate of $\boldsymbol{\theta}$, as in Ibrahim et al. (1999) and Wu (2004).

A Monte Carlo EM Algorithm

By treating the unobservable random effects $\mathbf{b}_i$ as additional missing data, we have "complete data" $\{(\mathbf{y}_i, \mathbf{z}_i, \mathbf{b}_i), \ i = 1, \ldots, n\}$. The "complete-data" log-likelihood for all individuals can then be written as

$$
\begin{aligned}
l_c(\boldsymbol{\theta}) &= \sum_{i=1}^{n} l_c^{(i)}(\boldsymbol{\theta}; \mathbf{y}_i, \mathbf{z}_i, \mathbf{b}_i) \\
&= \sum_{i=1}^{n} \left\{ \log f(\mathbf{y}_i | \mathbf{b}_i, \mathbf{z}_i, \boldsymbol{\beta}, \sigma^2) + \log f(\mathbf{z}_i | \boldsymbol{\alpha}) + \log f(\mathbf{b}_i | D(\boldsymbol{\eta})) \right\}.
\end{aligned}
$$

Let $\boldsymbol{\theta}^{(t)}$ be the parameter estimates from the t-th EM iteration, $t = 1, 2, 3, \cdots$. The E-step for individual i at the $(t+1)$st EM iteration is given by

$$
\begin{aligned}
Q_i(\boldsymbol{\theta} | \boldsymbol{\theta}^{(t)}) &= E\left[l_c^{(i)}(\boldsymbol{\theta}; \mathbf{y}_i, \mathbf{z}_i, \mathbf{b}_i) \mid \mathbf{y}_i, \mathbf{z}_{obs,i}, \boldsymbol{\theta}^{(t)} \right] \\
&= \int \int \left\{ \log f(\mathbf{y}_i | \mathbf{z}_i, \mathbf{b}_i, \boldsymbol{\beta}, \sigma^2) + \log f(\mathbf{z}_i | \boldsymbol{\alpha}) \right. \\
&\qquad \left. + \log f(\mathbf{b}_i | D) \right\} f(\mathbf{z}_{mis,i}, \mathbf{b}_i | \mathbf{z}_{obs,i}, \mathbf{y}_i, \boldsymbol{\theta}^{(t)}) d\mathbf{b}_i dz_{mis,i} \\
&\equiv I_1 + I_2 + I_3. \qquad\qquad\qquad\qquad\qquad\qquad (4.5)
\end{aligned}
$$

The integral $Q_i(\boldsymbol{\theta} | \boldsymbol{\theta}^{(t)})$ is still intractable and does not have a closed form expression. However, since $Q_i(\boldsymbol{\theta} | \boldsymbol{\theta}^{(t)})$ is an *expectation* with respect to the conditional distribution $f(\mathbf{z}_{mis,i}, \mathbf{b}_i | \mathbf{z}_{obs,i}, \mathbf{y}_i, \boldsymbol{\theta}^{(t)})$, we can simulate many samples from the distribution $f(\mathbf{z}_{mis,i}, \mathbf{b}_i | \mathbf{z}_{obs,i}, \mathbf{y}_i, \boldsymbol{\theta}^{(t)})$ and then approximate $Q_i(\boldsymbol{\theta} | \boldsymbol{\theta}^{(t)})$ by an empirical mean, with the missing data $(\mathbf{z}_{mis,i}, \mathbf{b}_i)$ substituted by their simulated values.

To simulate samples of $(\mathbf{z}_{mis,i}, \mathbf{b}_i)$ from $f(\mathbf{z}_{mis,i}, \mathbf{b}_i | \mathbf{z}_{obs,i}, \mathbf{y}_i, \boldsymbol{\theta}^{(t)})$, we may use the *Gibbs sampler* (Gelfand and Smith 1990) to iteratively sample from lower dimensional *full conditionals* (see Chapter 12). That is, we can generate samples from $f(\mathbf{z}_{mis,i}, \mathbf{b}_i | \mathbf{z}_{obs,i}, \mathbf{y}_i, \boldsymbol{\theta}^{(t)})$ by iteratively sampling from the full conditionals $f(\mathbf{z}_{mis,i} | \mathbf{z}_{obs,i}, \mathbf{y}_i, \mathbf{b}_i, \boldsymbol{\theta}^{(t)})$ and $f(\mathbf{b}_i | \mathbf{z}_{mis,i}, \mathbf{z}_{obs,i}, \mathbf{y}_i, \boldsymbol{\theta}^{(t)})$ in turn. The resulting samples constitute a Markov chain which will converge to the stationary distribution $f(\mathbf{z}_{mis,i}, \mathbf{b}_i | \mathbf{z}_{obs,i}, \mathbf{y}_i, \boldsymbol{\theta}^{(t)})$. That is, after a *burn-in period*, we obtain a desired sample from $f(\mathbf{z}_{mis,i}, \mathbf{b}_i | \mathbf{z}_{obs,i}, \mathbf{y}_i, \boldsymbol{\theta}^{(t)})$. Repeating this process many times, we obtain many independent samples from

$f(\mathbf{z}_{mis,i}, \mathbf{b}_i | \mathbf{z}_{obs,i}, \mathbf{y}_i, \boldsymbol{\theta}^{(t)})$. See Chapter 12 (Appendix) for a detailed description of the Gibbs sampler and other MCMC methods.

To sample from the full conditionals, note that

$$f(\mathbf{z}_{mis,i} | \mathbf{z}_{obs,i}, \mathbf{y}_i, \mathbf{b}_i, \boldsymbol{\theta}^{(t)}) \quad \propto \quad f(\mathbf{z}_i | \boldsymbol{\theta}^{(t)}) f(\mathbf{y}_i | \mathbf{z}_i, \mathbf{b}_i, \boldsymbol{\theta}^{(t)}), \quad (4.6)$$

$$f(\mathbf{b}_i | \mathbf{z}_{mis,i}, \mathbf{z}_{obs,i}, \mathbf{y}_i, \boldsymbol{\theta}^{(t)}) \quad \propto \quad f(\mathbf{b}_i | \boldsymbol{\theta}^{(t)}) f(\mathbf{y}_i | \mathbf{z}_i, \mathbf{b}_i, \boldsymbol{\theta}^{(t)}), \quad (4.7)$$

so we only need to generate samples from the right-hand sides of (4.6) and (4.7), which can be accomplished using rejection sampling methods since the density functions on the right-hand sides of (4.6) and (4.7) are all known (see Chapter 12). If the density functions on the right-hand sides of (4.6) and (4.7) are log-concave, the adaptive rejection sampling method of Gilks and Wild (1992) can be used. Alternatively, integral $Q_i(\boldsymbol{\theta}|\boldsymbol{\theta}^{(t)})$ may be approximated using other MCMC methods or importance sampling methods. McCulloch (1997) discussed various approaches for GLMMs. We will discuss more details in Section 4.6.

For individual i, let

$$\left\{ (\tilde{\mathbf{z}}_{mis,i}^{(1)}, \tilde{\mathbf{b}}_i^{(1)}), (\tilde{\mathbf{z}}_{mis,i}^{(2)}, \tilde{\mathbf{b}}_i^{(2)}), \ldots, (\tilde{\mathbf{z}}_{mis,i}^{(m_t)}, \tilde{\mathbf{b}}_i^{(m_t)}) \right\}$$

denote a random sample of size m_t (large) generated from the distribution $f(\mathbf{z}_{mis,i}, \mathbf{b}_i | \mathbf{z}_{obs,i}, \mathbf{y}_i, \boldsymbol{\theta}^{(t)})$. Note that each $(\tilde{\mathbf{z}}_{mis,i}^{(j)}, \tilde{\mathbf{b}}_i^{(j)})$ depends on the EM iteration number t, which is suppressed throughout for notation simplicity. The E-step at the $(t+1)$st EM iteration can then be written as

$$Q(\boldsymbol{\theta}|\boldsymbol{\theta}^{(t)}) = \sum_{i=1}^{n} Q_i(\boldsymbol{\theta}|\boldsymbol{\theta}^{(t)})$$

$$\approx \sum_{i=1}^{n} \left\{ \frac{1}{m_t} \sum_{j=1}^{m_t} l_c(\boldsymbol{\theta}; \mathbf{y}_i, \mathbf{z}_{obs,i}, \tilde{\mathbf{z}}_{mis,i}^{(j)}, \tilde{\mathbf{b}}_i^{(j)}) \right\}$$

$$= \sum_{i=1}^{n} \sum_{j=1}^{m_t} \frac{1}{m_t} \log f(\mathbf{y}_i | \mathbf{z}_{obs,i}, \tilde{\mathbf{z}}_{mis,i}^{(j)}, \tilde{\mathbf{b}}_i^{(j)}, \boldsymbol{\beta}, \sigma^2)$$

$$+ \sum_{i=1}^{n} \sum_{j=1}^{m_t} \frac{1}{m_t} \log f(\mathbf{z}_{obs,i}, \tilde{\mathbf{z}}_{mis,i}^{(j)} | \boldsymbol{\alpha})$$

$$+ \sum_{i=1}^{n} \sum_{j=1}^{m_t} \frac{1}{m_t} \log f(\tilde{\mathbf{b}}_i^{(j)} | D)$$

$$\equiv Q^{(1)}(\boldsymbol{\beta}, \sigma^2 | \boldsymbol{\theta}^{(t)}) + Q^{(2)}(\boldsymbol{\alpha} | \boldsymbol{\theta}^{(t)}) + Q^{(3)}(D | \boldsymbol{\theta}^{(t)}). \quad (4.8)$$

The M-step of the Monte Carlo EM algorithm is to maximize $Q(\boldsymbol{\theta}|\boldsymbol{\theta}^{(t)})$ with respect to $\boldsymbol{\theta}$ to produce an updated estimate $\boldsymbol{\theta}^{(t+1)}$. From (4.8), the M-step is like a complete-data maximization, so standard optimization procedures for complete-data can be used, such as the Newton-Raphason method. A more computationally efficient method here would be to use the ECM algorithm (Meng and Rubin 1993), in which a sequence of maximizations is performed on the individual components of $\boldsymbol{\theta}$, holding the remain parameters fixed (see also Chapter 3).

Iterating between the E- and M-steps until convergence, we obtain an MLE of $\boldsymbol{\theta}$ or a local maximum of the observed-data likelihood.

Standard Errors

The EM algorithm does not automatically produce standard errors of the MLEs. To obtain the variance-covariance matrix of $\hat{\boldsymbol{\theta}}$, we may use the formula of Louis (1982) as follows. Let $\hat{\boldsymbol{\theta}}$ be the MLE of $\boldsymbol{\theta}$ at convergence. In Louis's formula (see Chapter 3, page 116), we can approximate the expectations by their corresponding Monte Carlo mean approximations and then approximate the observed-data information matrix by

$$\hat{I}(\hat{\boldsymbol{\theta}}) = -\ddot{Q}(\hat{\boldsymbol{\theta}}|\hat{\boldsymbol{\theta}}) - \sum_{i=1}^{n}\sum_{j=1}^{m_{t^*}}\frac{1}{m_{t^*}}S_{ij}(\hat{\boldsymbol{\theta}})S_{ij}^T(\hat{\boldsymbol{\theta}}) + \sum_{i=1}^{n}\dot{Q}_i(\hat{\boldsymbol{\theta}}|\hat{\boldsymbol{\theta}})\dot{Q}_i^T(\hat{\boldsymbol{\theta}}|\hat{\boldsymbol{\theta}}), \quad (4.9)$$

where t^* is the last EM iteration number, and

$$S_{ij}(\hat{\boldsymbol{\theta}}) = \left.\frac{\partial l_c(\boldsymbol{\theta}; \mathbf{y}_i, \mathbf{z}_{obs,i}, \tilde{\mathbf{z}}_{mis,i}^{(j)}, \tilde{\mathbf{b}}_i^{(j)})}{\partial \boldsymbol{\theta}}\right|_{\boldsymbol{\theta}=\hat{\boldsymbol{\theta}}},$$

$$\dot{Q}_i(\hat{\boldsymbol{\theta}}|\hat{\boldsymbol{\theta}}) = \frac{1}{m_{t^*}}\sum_{j=1}^{m_{t^*}}S_{ij}(\hat{\boldsymbol{\theta}}), \quad \ddot{Q}(\hat{\boldsymbol{\theta}}|\hat{\boldsymbol{\theta}}) = \sum_{i=1}^{n}\left(\frac{1}{m_{t^*}}\sum_{j=1}^{m_{t^*}}\left.\frac{\partial S_{ij}(\boldsymbol{\theta})}{\partial \boldsymbol{\theta}}\right|_{\boldsymbol{\theta}=\hat{\boldsymbol{\theta}}}\right).$$

Thus, the asymptotic variance-covariance matrix of $\hat{\boldsymbol{\theta}}$ can be approximated by $\hat{I}^{-1}(\hat{\boldsymbol{\theta}})$, and the asymptotic standard errors of $\hat{\boldsymbol{\theta}}$ can be estimated by the square roots of the diagonal elements of the matrix $\hat{I}^{-1}(\hat{\boldsymbol{\theta}})$. An alternative formula is given by (see, e.g., McLachlan and Krishnan 1997)

$$\widehat{Cov}(\hat{\boldsymbol{\theta}}) \approx \left[\sum_{i=1}^{n}\left(\frac{1}{m_t}\sum_{j=1}^{m_t}S_{ij}(\hat{\boldsymbol{\theta}})\right)\left(\frac{1}{m_t}\sum_{j=1}^{m_i}S_{ij}^T(\hat{\boldsymbol{\theta}})\right)\right]^{-1}, \quad (4.10)$$

which only uses the first derivatives so may be easier to compute.

Summary

In summary, the foregoing Monte Carlo EM algorithm proceeds as follows:

STEP 1. Obtain initial values for the parameters $\boldsymbol{\theta}^{(0)}$ and random effects $\mathbf{b}_i^{(0)}$. For example, these initial values can be the estimates based on the complete-case method (i.e., delete all incomplete data).

STEP 2. At the $(t+1)$th EM iteration, generate Monte Carlo samples of the missing covariates $\mathbf{z}_{mis,i}$ and the random effects $\mathbf{b}_i$ using the Gibbs sampler along with rejection sampling methods or using other MCMC methods or importance sampling methods;

STEP 3. At the $(t+1)$th EM iteration, obtain updated parameter estimate $\boldsymbol{\theta}^{(t+1)}$ in the M-step using standard complete-data optimization methods such as the Newton-Raphason method;

STEP 4. Iterate between Steps 2 and 3 until convergence (i.e., the changes in parameter estimates are smaller than a pre-specified threshold value such as 1%).

Some specific computational details and strategies, including sampling methods in the E-step, numbers of Monte Carlo samples in the E-step, and convergence of the EM algorithm, will be discussed in Section 4.6.

Note that, for LME models with missing covariates, the random effects $\mathbf{b}_i$ in $Q(\boldsymbol{\theta}|\boldsymbol{\theta}^{(t)})$ of (4.5) can be integrated out since LME models are linear in the random effects. Thus, in this case the integral in $Q(\boldsymbol{\theta}|\boldsymbol{\theta}^{(t)})$ of (4.5) has a lower dimension and the Monte Carlo EM algorithm is computationally much simpler. Note also that, when a covariate is *discrete* (or categorical), the integration with respect to this covariate reduces to a summation, so Monte Carlo sampling for this covariate is not needed (see Section 4.3.1 for an example). This also reduces computational burden.

4.2.2 Non-Ignorable Missing Covariates

In Section 4.2.1, the missing data mechanism is assumed to be ignorable. In practice, sometimes the missing covariates may be non-ignorable in the sense that the missingness may be related to the missing data. In this case, in order to avoid biased inference, we must assume a possible missing data mechanism or model and then incorporate it in likelihood inference. The estimation procedure then proceeds in a similar way, as in Ibrahim et al. (1999, 2001) and described as follows:

Let $\mathbf{r}_i = (r_{i1}, \ldots, r_{ip})^T$ be a $(p \times 1)$ vector of missing data indicators such that $r_{ij} = 1$ if the jth covariate is missing for individual i and $r_{ij} = 0$ otherwise, $j = 1, 2, \ldots, p; i = 1, 2, \cdots, n$. To model the missing data mechanism, we need to assume a distribution for $\mathbf{r}_i$, i.e., we need to assume how the probability of missing a covariate value may be related to the missing and observed values. The parameters in the missing data model are viewed as nuisance parameters since they are not of primary inferential interest, so we should avoid assuming a too complicated missing data model.

To reduce the number of nuisance parameters, we may write the joint distribution of $\mathbf{r}_i$ as a product of one-dimensional conditional distributions, similar to the approach for covariate models in Section 4.2.1. Since each r_{ij} is a binary variable, we naturally choose logistic regression models for the conditional distributions, which link the distribution of r_{ij} to the missing values $\mathbf{z}_{mis,i}$ and the observed covariates $\mathbf{z}_{obs,i}$. For example, we may consider the following missing data model

$$
\begin{aligned}
f(\mathbf{r}_i|\mathbf{z}_i, \boldsymbol{\phi}) &= f(r_{i1}|\mathbf{z}_i, \boldsymbol{\phi})f(r_{i2}|r_{i1}, \mathbf{z}_i, \boldsymbol{\phi}) \cdots f(r_{ip}|\bar{r}_{ip}, \mathbf{z}_i, \boldsymbol{\phi}), \\
\text{logit}[P(r_{ik} = 1)] &= \phi_{0k} + \cdots + \phi_{k-1,k}r_{i,k-1} + \phi_{kk}\mathbf{z}_{mis,i} \\
&\quad + \phi_{k,k+1}\mathbf{z}_{obs,i}, \qquad k = 2, 3, \cdots, p, \qquad (4.11)
\end{aligned}
$$

where $\bar{r}_{ik} = (r_{i1}, \cdots, r_{i,k-1})$ and $\boldsymbol{\phi} = (\phi_{01}, \cdots, \phi_{pp})$. In model (4.11), we allow the probability of missingness $P(r_{ij} = 1)$ to possibly depend on the missing and observed covariates. Thus we allow the missing data mechanism to be non-ignorable. We can also choose $P(r_{ij} = 1)$ to possibly depend on $\mathbf{y}_i$, i.e., consider a missing data model $f(\mathbf{r}_i|\mathbf{y}_i, \mathbf{z}_i, \boldsymbol{\phi})$.

Note that the missing data model (4.11) cannot be verified based on the *observed data* since the missing data $\mathbf{z}_{mis,i}$ are not observed. In other words, different missing data models may fit the observed data equally well. Therefore, as discussed in Chapter 3 (Section 3.3.3, page 107), sensitivity analyses based on different missing data models are important. Moreover, we should avoid to specify a large non-ignorable missing data model since such a model contains too many nuisance parameters and may become non-identifiable (Fitzmaurice et al. 1996). The "true" missing data mechanism may be very complex and unknown, and we only try to *approximate* it, so *simple* or parsimonious models are preferable, similar to the strategies for model selections in linear regression models.

Likelihood estimation can again be done using the EM algorithm. The procedure is similar to that in Section 4.2.1. The only modification is to incorporate the missing data model $f(\mathbf{r}_i|\mathbf{y}_i, \mathbf{z}_i, \boldsymbol{\phi})$ in the likelihood. The observed data are now $\{(\mathbf{y}_i, \mathbf{z}_{obs,i}, \mathbf{r}_i), \ i = 1, \ldots, n\}$. The "complete data" are $\{(\mathbf{y}_i, \mathbf{z}_i, \mathbf{b}_i, \mathbf{r}_i), i = 1, \ldots, n\}$, and the complete-data loglikelihood can then be written as

$$
\begin{aligned}
l_c(\boldsymbol{\theta}) &\equiv \sum_{i=1}^{n} l_c(\boldsymbol{\theta}; \mathbf{y}_i, \mathbf{z}_i, \mathbf{b}_i, \mathbf{r}_i) = \sum_{i=1}^{n} \Big\{ \log f(\mathbf{y}_i|\mathbf{z}_i, \boldsymbol{\beta}, \sigma^2, \mathbf{b}_i) \\
&\quad + \log f(\mathbf{z}_i|\boldsymbol{\alpha}) + \log f(\mathbf{b}_i|D(\boldsymbol{\eta})) + \log f(\mathbf{r}_i|\mathbf{y}_i, \mathbf{z}_i, \boldsymbol{\phi}) \Big\},
\end{aligned}
$$

where $\boldsymbol{\theta} = (\boldsymbol{\alpha}, \boldsymbol{\beta}, \sigma^2, \boldsymbol{\eta}, \boldsymbol{\phi})$ is the collection of all parameters.

At the $(t + 1)$st EM iteration, the E-step for individual i can be written as

$$
Q_i(\boldsymbol{\theta}|\boldsymbol{\theta}^{(t)}) = E\left[l_c(\boldsymbol{\theta}; \mathbf{y}_i, \mathbf{z}_i, \mathbf{r}_i, \mathbf{b}_i) \mid \mathbf{y}_i, \mathbf{z}_{obs,i}, \mathbf{r}_i, \boldsymbol{\theta}^{(t)}\right]
$$

$$= \int \int \left\{ \log f(\mathbf{y}_i | \mathbf{z}_i, \mathbf{b}_i, \boldsymbol{\beta}, \sigma^2) + \log f(\mathbf{z}_i | \boldsymbol{\alpha}) \right.$$
$$+ \log f(\mathbf{b}_i | D) + \log f(\mathbf{r}_i | \mathbf{y}_i, \mathbf{z}_i, \boldsymbol{\phi}) \Big\}$$
$$\times f(\mathbf{z}_{mis,i}, \mathbf{b}_i | \mathbf{z}_{obs,i}, \mathbf{y}_i, \mathbf{r}_i, \boldsymbol{\theta}^{(t)}) \ d\mathbf{b}_i \ d\mathbf{z}_{mis,i}.$$

As in Section 4.2.1, we can use the Gibbs sampler to generate samples from the conditional distribution $f(\mathbf{z}_{mis,i}, \mathbf{b}_i | \mathbf{z}_{obs,i}, \mathbf{y}_i, \mathbf{r}_i, \boldsymbol{\theta}^{(t)})$ by iteratively sampling from the full conditionals $f(\mathbf{z}_{mis,i} | \mathbf{b}_i, \mathbf{z}_{obs,i}, \mathbf{y}_i, \mathbf{r}_i, \boldsymbol{\theta}^{(t)})$ and $f(\mathbf{b}_i | \mathbf{z}_{mis,i}, \mathbf{z}_{obs,i}, \mathbf{y}_i, \mathbf{r}_i, \boldsymbol{\theta}^{(t)})$. Sampling the full conditionals via rejection sampling methods can be achieved by noting that

$$f(\mathbf{z}_{mis,i} | \mathbf{b}_i, \mathbf{z}_{obs,i}, \mathbf{y}_i, \mathbf{r}_i, \boldsymbol{\theta}^{(t)}) \quad \propto \quad f(\mathbf{z}_i | \boldsymbol{\alpha}^{(t)}) f(\mathbf{y}_i | \mathbf{z}_i, \mathbf{b}_i, \boldsymbol{\beta}^{(t)}, \sigma^{2(t)})$$
$$\times f(\mathbf{r}_i | \mathbf{y}_i, \mathbf{z}_i, \boldsymbol{\phi}^{(t)}),$$
$$f(\mathbf{b}_i | \mathbf{z}_{mis,i}, \mathbf{z}_{obs,i}, \mathbf{y}_i, \mathbf{r}_i, \boldsymbol{\theta}^{(t)}) \quad \propto \quad f(\mathbf{b}_i | D^{(t)}) f(\mathbf{y}_i | \mathbf{z}_i, \mathbf{b}_i, \boldsymbol{\beta}^{(t)}, \sigma^{2(t)}),$$

where the density functions on the right-hand sides are all known so a rejection sampling method can be implemented (see Chapter 12).

In summary, when the missing covariates are non-ignorable, likelihood inference is similar to the case when the missing covariates are ignorable. The only modification is to add an assumed model for the missing data mechanism in the likelihood. This missing data model introduces additional nuisance parameters and may cause parameter identifiability, and it is not testable based on the observed data. Therefore, simple missing data models are preferable and sensitivity analysis plays an important role. In fact, non-ignorable missing data models may be viewed as tools for *sensitivity analysis* since, as noted in Chapter 3 (Section 3.3.3, page 107), every MNAR (missing not at random) model has an MAR (missing at random) counterpart in the sense that both models produce exactly the same fit to the observed data, but the two models may lead to different inferences (Molenberghs and Kenward 2007).

4.2.3 Missing Data in Time-Dependent Covariates

In longitudinal studies, many covariates are measured over time along with the response measurements. Such covariates are called *time-dependent covariates*. In a mixed effects model with time-dependent covariates, standard model specifications require that all time-dependent covariate values are available at *each* time point where the response is measured. However, in practice this is often not the case for various reasons, which leads to missing data in time-dependent covariates. These types of missing data in time-dependent covariates are very common and are almost inevitable in a longitudinal study, especially if the study lasts a long time. Figure 4.1 shows viral load and CD4 measurements for four randomly selected subjects in an AIDS study. One can see that the viral

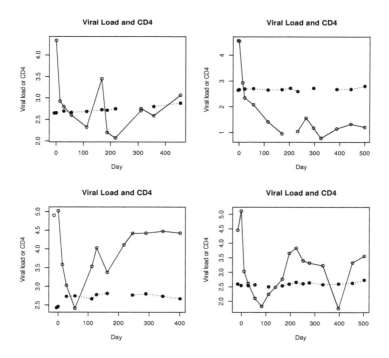

Figure 4.1 *Viral load and CD4 measurements from an AIDS study for four randomly selected subjects. Both viral load and CD4 values are* $\log_{10}$*-transformed. The solid lines with open circles are viral loads and the dotted lines with solid circles are CD4 values.*

load and CD4 values are sometimes not measured at the same times, which leads to missing time-dependent covariates if one variable is treated as the response and the other variable is treated as a covariate.

In this section, we consider missing time-dependent covariates in mixed effects models, including GLMM and NLME models, and we allow non-ignorable missing data mechanisms. For simplicity, we first focus on a single time-varying covariate, and then we discuss how to extend the method to more than one time-dependent covariates.

For a time-dependent covariate with missing data, we need to model the covariate process over time for likelihood inference. The general approach is similar to that in Sections 4.2.1 and 4.2.2, so we omit some details. Let z_{ij} be the covariate value for the i-th individual at time t_{ij}, $i = 1, \cdots, n; j = 1, \cdots, n_i$, and let $\mathbf{z}_i = (z_{i1}, \ldots, z_{in_i})^T$ be all the measurements of the covariate on individual i. Write $\mathbf{z}_i = (\mathbf{z}_{mis,i}, \mathbf{z}_{obs,i})$, where $\mathbf{z}_{mis,i}$ contains the missing components of $\mathbf{z}_i$ and $\mathbf{z}_{obs,i}$ contains the observed components of $\mathbf{z}_i$. Let $\mathbf{r}_i = (r_{i1}, \ldots, r_{in_i})^T$ be a vector of missing data indicators such that $r_{ij} = 1$ if z_{ij} is missing and $r_{ij} = 0$ otherwise.

In the absence of a theoretical model, we may model the covariate process empirically using a LME model:

$$z_i(t) = U_i(t)\,\boldsymbol{\alpha} + V_i(t)\,\mathbf{a}_i + \epsilon_i(t), \qquad i = 1, \ldots, n, \qquad (4.12)$$

where $\boldsymbol{\alpha}$ contains fixed parameters, $\mathbf{a}_i$'s are random-effects, $z_i(t)$, $U_i(t)$, $V_i(t)$, and $\epsilon_i(t)$ are the covariate value, design matrices, and random error for individual i at time t respectively. We assume that $\mathbf{a}_i$'s are i.i.d. $\sim N(\mathbf{0}, A)$, ϵ_{il} i.i.d. $\sim N(0, \delta^2 I)$, and $\mathbf{a}_i$ and $\epsilon_i = (\epsilon_{i1}, \ldots, \epsilon_{in_i})^T$ are independent, where A is a covariance matrix and δ^2 is the within-individual variance for the covariate process and $\epsilon_{il} = \epsilon_i(t_{il})$. We further assume that the errors and random effects in the covariate model (i.e., ϵ_i and $\mathbf{a}_i$) are independent of those in the response model (i.e., $\mathbf{e}_i$ and $\mathbf{b}_i$).

In covariate model (4.12), the covariate measurement times may be different from the response measurement times. At the response measurement time t_{ij}, the possibly unobserved covariate value can be viewed as

$$z_{ij}^* = z_i^*(t_{ij}) = U_{ij}\,\boldsymbol{\alpha} + V_{ij}\,\mathbf{a}_i,$$

where $U_{ij} = U_i(t_{ij})$ and $V_{ij} = V_i(t_{ij})$. In model fitting, we can simply fit the covariate model (4.12) to the *observed* covariate data.

The observed data are $\{(\mathbf{y}_i, \mathbf{z}_{obs,i}, \mathbf{r}_i), i = 1, \ldots, n\}$, and the observed data log-likelihood is given by

$$
\begin{aligned}
l_o(\boldsymbol{\theta}) \;=\; & \sum_{i=1}^{n} \int \int \int \Big\{ \log f(\mathbf{y}_i | \mathbf{z}_i, \mathbf{b}_i; \boldsymbol{\beta}, \sigma^2) + \log f(\mathbf{z}_i | \mathbf{a}_i; \boldsymbol{\alpha}, \delta^2) \\
& + \log f(\mathbf{a}_i | A) + \log f(\mathbf{b}_i | D) \\
& + \log f(\mathbf{r}_i | \mathbf{y}_i, \mathbf{z}_i, \boldsymbol{\phi}) \Big\}\, d\mathbf{z}_{mis,i}\, d\mathbf{a}_i\, d\mathbf{b}_i,
\end{aligned}
$$

where $\boldsymbol{\theta} = (\boldsymbol{\alpha}, \boldsymbol{\beta}, \sigma^2, \delta^2, A, D, \boldsymbol{\phi})$ is the collection of all parameters. The likelihood $l_o(\boldsymbol{\theta})$ involves an intractable integral with a high dimension, so the computation can be challenging.

Consider again a Monte Carlo EM algorithm for estimation. The "complete data" are now $\{(\mathbf{y}_i, \mathbf{z}_i, \mathbf{a}_i, \mathbf{b}_i, \mathbf{r}_i), i = 1, \ldots, n\}$. The E-step at the $(t + 1)$-th EM iteration $(t = 0, 1, \cdots)$ can be written as

$$
\begin{aligned}
Q(\boldsymbol{\theta} | \boldsymbol{\theta}^{(t)}) \;=\; & \int \int \int \Big\{ \log f(\mathbf{y}_i | \mathbf{z}_i, \mathbf{b}_i; \boldsymbol{\beta}, \sigma^2) + \log f(\mathbf{z}_i | \mathbf{a}_i; \boldsymbol{\alpha}, \delta^2) \\
& + \log f(\mathbf{a}_i | A) + \log f(\mathbf{b}_i | D) + \log f(\mathbf{r}_i | \mathbf{y}_i, \mathbf{z}_i, \boldsymbol{\phi}) \Big\} \\
& \times f(\mathbf{z}_{mis,i}, \mathbf{a}_i, \mathbf{b}_i | \mathbf{z}_{obs,i}, \mathbf{y}_i, \mathbf{r}_i, \boldsymbol{\theta}^{(t)})\, d\mathbf{a}_i\, d\mathbf{b}_i\, d\mathbf{z}_{mis,i},
\end{aligned}
$$

where $\theta^{(t)}$ is the parameter estimate from the t-th EM iteration. To approximate $Q(\theta|\theta^{(t)})$ using a Monte Carlo method, we need to simulate samples from the conditional distribution $f(\mathbf{z}_{mis,i}, \mathbf{a}_i, \mathbf{b}_i|\mathbf{z}_{obs,i}, \mathbf{y}_i, \mathbf{r}_i, \theta^{(t)})$. This can again be done using the Gibbs sampler method to break down a high dimensional complex density into several lower dimensional and more manageable ones (full conditionals), i.e., $f(\mathbf{z}_{mis,i}|\mathbf{z}_{obs,i}, \mathbf{y}_i, \mathbf{a}_i, \mathbf{b}_i, \mathbf{r}_i, \theta^{(t)})$, $f(\mathbf{a}_i|\mathbf{b}_i, \mathbf{z}_i, \mathbf{y}_i, \mathbf{r}_i, \theta^{(t)})$, and $f(\mathbf{b}_i|\mathbf{a}_i, \mathbf{z}_i, \mathbf{y}_i, \mathbf{r}_i, \theta^{(t)})$. Sampling from these full conditionals can be accomplished by rejection sampling methods by noting that

$$
\begin{aligned}
f(\mathbf{z}_{mis,i}|\mathbf{z}_{obs,i}, \mathbf{y}_i, \mathbf{a}_i, \mathbf{b}_i, \mathbf{r}_i, \theta^{(t)}) \quad &\propto \quad f(\mathbf{y}_i|\mathbf{z}_i, \mathbf{b}_i, \beta^{(t)}, \sigma^{2(t)}) \\
&\times f(\mathbf{z}_i|\mathbf{a}_i, \alpha^{(t)}, \delta^{2(t)}) \\
&\times f(\mathbf{r}_i|\mathbf{y}_i, \mathbf{z}_i, \phi^{(t)}), \\
f(\mathbf{a}_i|\mathbf{b}_i, \mathbf{z}_i, \mathbf{y}_i, \mathbf{r}_i, \theta^{(t)}) \quad &\propto \quad f(\mathbf{y}_i|\mathbf{z}_i, \mathbf{b}_i, \beta^{(t)}, \sigma^{2(t)}) \\
&\times f(\mathbf{a}_i|A^{(t)})f(\mathbf{z}_i|\mathbf{a}_i, \alpha^{(t)}, \delta^{2(t)}) \\
f(\mathbf{b}_i|\mathbf{a}_i, \mathbf{z}_i, \mathbf{y}_i, \mathbf{r}_i, \theta^{(t)}) \quad &\propto \quad f(\mathbf{b}_i|D^{(t)})f(\mathbf{y}_i|\mathbf{z}_i, \mathbf{b}_i, \beta^{(t)}, \sigma^{2(t)}),
\end{aligned}
$$

where the density functions on the right-hand sides are all known.

The missing data model $f(\mathbf{r}_i|\mathbf{y}_i, \mathbf{z}_i, \phi)$ can be chosen based on the discussion in Section 4.2.2. An alternative non-ignorable missing data model is to assume that the missingness depends on the unobserved random effects in the models. That is, we may consider the following non-ignorable missing data model

$$
\begin{aligned}
f(\mathbf{r}_i|\mathbf{a}_i, \mathbf{b}_i, \phi) \quad &= \quad f(r_{i1}|\mathbf{a}_i, \mathbf{b}_i, \phi)f(r_{i2}|r_{i1}, \mathbf{a}_i, \mathbf{b}_i, \phi)\cdots \\
&\times f(r_{in_i}|\bar{r}_{in_i}, \mathbf{a}_i, \mathbf{b}_i, \phi), \quad (4.13) \\
\text{logit}[P(r_{ij} = 1)] \quad &= \quad \phi_{0j} + \cdots + \phi_{j-1,j}r_{i,j-1} + \phi_a^T\mathbf{a}_i + \phi_b^T\mathbf{b}_i, \\
&\quad i = 1, 2, \cdots, n; \qquad j = 1, 2, \cdots, n_i,
\end{aligned}
$$

where $\bar{r}_{ij} = (r_{i1}, \cdots, r_{i,j-1})$ and $\phi = (\phi_{01}, \cdots, \phi_{pp}, \phi_a, \phi_b)$. The random effects $\mathbf{a}_i$ and $\mathbf{b}_i$ may represent summaries of the covariate and response processes, such as rates of change or histories of the covariate and response processes (Tsiatis, DeGruttola, and Wulfsohn 1995). The missing data model (4.13) sometimes is called a *shared-parameter model* (Wu and Carroll 1988). These models are also useful if we believe that the missingness is related to a latent process that governs the covariate and response processes.

When there are two or more time-dependent covariates with missing data, a simple method would be to model each covariate process separately, but this method may lose some efficiency if the covariates are highly correlated. A more efficient approach is to consider a multivariate longitudinal model for the covariates such as a multivariate LME model (Shah et al. 1997), or write the joint distribution of the covariates as a product of univariate conditional distributions, as in Section 4.2.1. A multivariate LME model can be converted into a univariate LME model by stacking all covariates. Then, the foregoing arguments can be applied to the univariate LME model. We will present such an example in Section 4.2.4 in the context of semiparametric and nonparametric mixed models.

4.2.4 Multivariate, Semiparametric, and Nonparametric Models

In practice, many longitudinal processes are very complex, so parametric models may not be flexible enough to model these processes. In Section 2.5 of Chapter 2, we discussed nonparametric and semiparametric mixed effects models and showed that these models are very flexible for modeling complex longitudinal processes. Thus, nonparametric and semiparametric mixed effects models are useful for modeling time-dependent covariates in the presence of missing data. Often, we need to model several covariate processes and these processes may be highly correlated, so a multivariate model would provide more efficient inference. In this section, we describe a multivariate semiparametric mixed effects model for jointly modeling several (continuous) covariate processes simultaneously.

Suppose that there are K time-dependent covariates. Let z_{ijk} be the k-th covariate value for individual i at time $t_{ij}, i = 1, \cdots, n; \ j = 1, \cdots, n_i; \ k = 1, \cdots, K$. We assume that the covariate processes have the same measurement schedules, i.e., assume $n_1 = n_2 = \cdots = m$, but we allow some of the covariate values to be missing. Let r_{ijk} be the missing data indicator such that $r_{ijk} = 1$ if z_{ijk} is observed and $r_{ijk} = 0$ otherwise. We define $\mathbf{z}_{ik} = (z_{i1k}, \ldots, z_{in_ik})^T$ and $\mathbf{z}_i = (\mathbf{z}_{i1}^T, \ldots, \mathbf{z}_{iK}^T)^T$. For the k-th covariate process, we consider the following semiparametric mixed-effects model

$$\mathbf{z}_{ik} = \mathbf{U}_{ik}^* \boldsymbol{\alpha}_k^* + \mathbf{V}_{ik}^* \mathbf{a}_{ik}^* + \mathbf{w}_k + \mathbf{h}_{ik} + \mathbf{e}_{ik}, \qquad i = 1, 2, \cdots, n, \quad (4.14)$$

where $\mathbf{U}_{ik}^*$ and $\mathbf{V}_{ik}^*$ are known design matrices, $\boldsymbol{\alpha}_k^*$ is a vector of population parameters, $\mathbf{a}_{ik}^*$ is a vector of random effects, $\mathbf{w}_k = (w_k(t_{i1}), \ldots, w_k(t_{in_i}))^T$, $\mathbf{h}_{ik} = (h_{ik}(t_{i1}), \ldots, h_{ik}(t_{in_i}))^T$, and $w_k(t)$ and $h_{ik}(t)$ are unknown *nonparametric* smooth fixed-effects and random-effects functions respectively. We assume that $\mathbf{e}_{ik} \sim N(\mathbf{0}, \delta_{kk}I)$, $\mathbf{a}_{ik} \sim N(\mathbf{0}, A_k)$, and $\mathbf{e}_{ik}, \mathbf{a}_{ik}^*, \mathbf{h}_{ik}$ are independent of each other.

Note that the semiparametric mixed-effects model (4.14) reduces to a nonparametric mixed-effects model when the parametric part $\mathbf{U}_{ik}^* \boldsymbol{\alpha}_k^* + \mathbf{V}_{ik}^* \mathbf{a}_{ik}^*$ in the model is dropped, and it reduces to a parametric LME model when the nonparametric part $\mathbf{w}_k + \mathbf{h}_{ik}$ in the model is dropped.

In Section 2.5 of Chapter 2, we reviewed some nonparametric smoothing methods. Here we focus on the basis-based approach, following Rice and Wu (2001). That is, we approximate the nonparametric smooth functions $w_k(t)$ and $h_{ik}(t)$ by linear combinations of some basis functions $\boldsymbol{\psi}_k(t) =$

$(\psi_{k0}(t), \psi_{k1}(t), \ldots, \psi_{kp_k}(t))^T$ and $\boldsymbol{\nu}_k(t) = (\nu_{k0}(t), \nu_{k1}(t), \ldots, \nu_{kq_k}(t))^T$
as follows:

$$w_k(t) \approx \tilde{w}_{p_k}(t) = \sum_{j=0}^{p_k-1} \eta_{jk} \, \psi_j(t) = \boldsymbol{\psi}_k(t)^T \, \boldsymbol{\eta}_k, \qquad (4.15)$$

$$h_{ik}(t) \approx \tilde{h}_{iq_k}(t) = \sum_{j=0}^{q_k-1} \xi_{ijk} \, \nu_j(t) = \boldsymbol{\nu}_k(t)^T \, \boldsymbol{\xi}_{ik}, \qquad (4.16)$$

where $\boldsymbol{\eta}_k$ and $\boldsymbol{\xi}_{ik}$ are vectors of unknown fixed and random coefficients respectively. Substituting $w_k(t)$ and $h_{ik}(t)$ by their approximations $\tilde{w}_{p_k}(t)$ and $\tilde{h}_{iq_k}(t)$, we can approximate model (4.14) by the following LME model for the k-th covariate process

$$\mathbf{z}_{ik} = \mathbf{U}_{ik}\boldsymbol{\alpha}_k + \mathbf{V}_{ik}\mathbf{a}_{ik} + \mathbf{e}_{ik}, \quad i = 1, 2, \cdots, n, \ k = 1, 2, \cdots, K, \ (4.17)$$

where $\boldsymbol{\alpha}_k = (\mathbf{a}_k^{*T}, \ \boldsymbol{\eta}_k^T)^T$ are fixed effects, $\mathbf{a}_{ik} = (\mathbf{a}_i^{*T}, \ \boldsymbol{\xi}_{ik}^T)^T$ are random effects, $\mathbf{U}_{ik} = (\mathbf{U}_{ik}^*, \Psi_{ik})$, $\mathbf{V}_{ik} = (\mathbf{V}_{ik}^*, \Gamma_{ik})$, Ψ_{ik} is a matrix whose (j, l)-th element is $\psi_{kl}(t_{ij})$, and Γ_k is a matrix whose (j, l)-th element is $\nu_{kl}(t_{ij})$.

To incorporate the correlation among the K covariates, we let Σ be the $K \times K$ covariance matrix for the K covariates, i.e., $Cov(\mathbf{z}_{i1}, \ldots, \mathbf{z}_{iK}) = \Sigma = (\delta_{ij})_{K \times K}$. Let $U_i = \text{diag}(\mathbf{U}_{i1}, \ldots, \mathbf{U}_{iK})$ be a block diagonal matrix with the k-th block being $\mathbf{U}_{ik}$, and let $V_i = \text{diag}(\mathbf{V}_{i1}, \ldots, \mathbf{V}_{iK})$ be a block diagonal matrix with the k-th block being $\mathbf{V}_{ik}$. Then, we can write all K covariate models in (4.17) as a *single* multivariate LME model

$$\mathbf{z}_i = U_i\boldsymbol{\alpha} + V_i\mathbf{a}_i + \mathbf{e}_i, \qquad i = 1, 2, \cdots, n, \qquad (4.18)$$

where $\boldsymbol{\alpha} = (\boldsymbol{\alpha}_1^T, \ldots, \boldsymbol{\alpha}_K^T)^T$ are fixed effects, $\mathbf{a}_i = (\mathbf{a}_{i1}^T, \ldots, \mathbf{a}_{iK}^T)^T$ are random effects, $\mathbf{a}_i \sim N(\mathbf{0}, A)$, and $\mathbf{e}_i \sim N(\mathbf{0}, \Sigma \otimes I_i)$.

LME Model (4.18) incorporates the correlation between the time-dependent covariates and the correlation between the repeated measurements within each individual. Thus, a multivariate parametric or semiparametric or nonparametric mixed-effects model can be written or approximated by a univariate LME model. Likelihood inference then proceeds as in the case for a univariate LME model, but computation can be tedious here since the covariance matrices have special structures and the number of parameters may increase substantially.

When some covariates are continuous and some are discrete, we can write their joint distribution as a product of conditional distributions. Specifically, suppose that $\mathbf{z}_i$ contains all continuous covariates and $\mathbf{x}_i$ contains all discrete covariates. We can write

$$f(\mathbf{x}_i, \mathbf{z}_i) = f(\mathbf{x}_i)f(\mathbf{z}_i \mid \mathbf{x}_i).$$

Then, a multivariate discrete model can be chosen for $f(\mathbf{x}_i)$, such as a multinomial or conditional logistic mixed effects model, and a multivariate LME model can be chosen for $f(\mathbf{z}_i \mid \mathbf{x}_i)$. Zeng and Cook (2007) proposed a multivariate model for longitudinal discrete data using a transitional model approach, which can also be used to model multivariate discrete longitudinal data.

4.3 Approximate Methods

A major issue in likelihood methods for mixed effects models with missing data is the intensive computation. As we have seen in Section 4.2, likelihood functions for the observed data typically involve high-dimensional and intractable integrals with no analytic expressions. Numerical integration techniques such as the Gauss-Hermite quadrature may be practically infeasible due to the high dimensionalities of the integrals. Monte Carlo EM algorithms or other Monte Carlo and stochastic methods can also be computationally very intensive and may even exhibit convergence problems. Therefore, computationally more efficient approximate methods are particularly valuable here. In Section 2.6.3 of Chapter 2, we described some computationally very efficient approximate methods based on Taylor or Laplace approximations. These approximate methods can be extended to mixed effects models with missing data. In this section, we describe these approximate methods.

Note that, in likelihood inference for mixed effects models, most of the computational difficulties arise for models that are *nonlinear* in the "unobservables" (i.e., the random effects and variables with missing data). In other words, most computational challenges arise in GLMM, NLME, and frailty models. For LME models, the random effects can be integrated out from the likelihoods since they are *linear* in the models – this greatly simplifies computation. When a mixed effects model is nonlinear in the random effects, however, one must integrate out the random effects in the likelihood, which leads to much of the computational difficulties.

Therefore, to avoid some of the major computational difficulties, one approach is to *linearize* the nonlinear models using Taylor series approximations. Such approaches have been widely used for complete data mixed effects models (e.g., Beal and Sheiner 1992; Breslow and Clayton 1993; Lindstrom and Bates 1990). Wu (2004) extended this approach to missing data problems. Alternatively, we can use Laplace approximations to directly approximate the intractable integrals in the likelihoods (Wolfinger 1993; Vonesh 1996; Lee et al. 2006). We discuss these approaches in the context of missing covariates in the following subsections.

As discussed in Section 2.6.3 of Chapter 2, the approximate methods based on Taylor approximations or Laplace approximations are asymptotically equivalent (Demidenko 2004), and the approximate estimates are consistent when *both* the number of measurements within each individual *and* the number of individuals go to infinite (Vonesh 1996). For complete-data NLME models, Pinheiro and Bates (1995) showed that these approximate methods perform very well. However, the approximate methods may perform less satisfactory for GLMMs, especially for binary responses such as logistic mixed effects models (Breslow and Lin 1995; Lin and Breslow 1996; Joe 2008). In the missing data cases, the performances of these approximate methods should be similar since the Monte Carlo EM algorithms are essentially based on complete-data models where the missing data are substituted by their simulated values.

4.3.1 Linearization

We focus on NLME models with missing covariates, but the approach can also be used in GLMMs with missing covariates. The idea is to take a first-order Taylor series expansion about some estimates of the random effects, which leads to a "working" LME model in which the random effects can be integrated out from the corresponding likelihood. The procedure is then iterated until convergence.

For complete-data NLME models, Beal and Sheiner (1982) took a first-order Taylor expansion about the means of the random effects, which are zeros. Lindstrom and Bates (1990) took a first-order Taylor expansion about the empirical Bayes estimates of the random effects, which are better estimates when the random effects variabilities are not small. Lindstrom and Bates (1990) showed that their method performs better than that of Beal and Sheiner (1982). Wu (2004) applied similar ideas to missing covariates in NLME models, as shown below.

In the following, we consider a NLME model with missing time-independent covariates and an ignorable missing data mechanism. Specifically, consider the following NLME model

$$y_{ij} = g_{ij}(\mathbf{z}_i, \boldsymbol{\beta}, \mathbf{b}_i) + e_{ij}, \qquad j = 1, \ldots, n_i; \ i = 1, \ldots, n, \qquad (4.19)$$

where $g_{ij}(\cdot)$ is a known nonlinear function, $\mathbf{e}_i \sim N(\mathbf{0}, R_i)$ with $R_i = \sigma^2 I_{n_i}$, and $\mathbf{b}_i \sim N(\mathbf{0}, D)$. Let $\mathbf{g}_i = (g_{i1}, \ldots, g_{in_i})^T$. To simplify the notation, we suppress iteration numbers. We begin with an initial estimate of $(\boldsymbol{\theta}, \mathbf{b}_i)$, and denote the estimate of $(\boldsymbol{\theta}, \mathbf{b}_i)$ at the current iteration by $(\widehat{\boldsymbol{\theta}}, \widehat{\mathbf{b}}_i)$. The linearization procedure of Lindstrom and Bates (1990) is equivalent to *iteratively* solving the following working LME model

$$\widetilde{\mathbf{y}}_i \ = \ X_i\boldsymbol{\beta} + T_i\mathbf{b}_i + \mathbf{e}_i, \quad i = 1, \ldots, n, \qquad (4.20)$$

where

$$\widetilde{\mathbf{y}}_i \ = \ \mathbf{y}_i - \mathbf{g}_i(\mathbf{z}_i, \widehat{\boldsymbol{\beta}}, \widehat{\mathbf{b}}_i) + X_i\widehat{\boldsymbol{\beta}} + T_i\widehat{\mathbf{b}}_i,$$

$$X_{ij} = \frac{\partial g_{ij}(\mathbf{z}_i, \boldsymbol{\beta}, \widehat{\mathbf{b}}_i)}{\partial \boldsymbol{\beta}^T}\Bigg|_{\boldsymbol{\beta}=\widehat{\boldsymbol{\beta}}}, \qquad T_{ij} = \frac{\partial g_{ij}(\mathbf{z}_i, \widehat{\boldsymbol{\beta}}, \mathbf{b}_i)}{\partial \mathbf{b}_i^T}\Bigg|_{\mathbf{b}_i=\widehat{\mathbf{b}}_i},$$

$X_i = (X_{i1}, \ldots, X_{in_i})^T, T_i = (T_{i1}, \ldots, T_{in_i})^T, \widetilde{\mathbf{y}}_i = (\widetilde{y}_{i1}, \ldots, \widetilde{y}_{in_i})^T$, and T_i is often independent of $\mathbf{z}_i$. Write $X_i = X_i(\mathbf{z}_i)$. When covariate $\mathbf{z}_i$ contains missing data, we can handle the missing covariates in LME model (4.20) at each iteration. For LME model (4.20), the random effects $\mathbf{b}_i$ in the likelihood can be integrated out, so the likelihood only involves an integration with respect to $\mathbf{z}_{mis,i}$ only, as shown below.

Note that

$$f(\mathbf{z}_{mis,i}, \mathbf{b}_i | \mathbf{z}_{obs,i}, \widetilde{\mathbf{y}}_{obs,i}, \widehat{\boldsymbol{\theta}}) = f(\mathbf{b}_i | \widetilde{\mathbf{y}}_i, \mathbf{z}_i, \widehat{\boldsymbol{\theta}}) f(\mathbf{z}_{mis,i} | \mathbf{z}_{obs,i}, \widetilde{\mathbf{y}}_{obs,i}, \widehat{\boldsymbol{\theta}}),$$

where

$$\widetilde{\mathbf{y}}_{obs,i} = \mathbf{y}_i - \mathbf{g}_i(\mathbf{z}_{obs,i}, \widehat{\boldsymbol{\beta}}, \widehat{\mathbf{b}}_i) + X_i(\mathbf{z}_{obs,i})\widehat{\boldsymbol{\beta}} + T_i\widehat{\mathbf{b}}_i.$$

Under the LME model (4.20), it can be shown that

$$[\mathbf{b}_i | \widetilde{\mathbf{y}}_i, \mathbf{z}_i, \widehat{\boldsymbol{\theta}}] \sim N(\widetilde{\mathbf{b}}_i, \widetilde{\Sigma}_i),$$

where

$$\widetilde{\Sigma}_i = (\widehat{\sigma}^{-2} T_i^T T_i + \widehat{D}^{-1})^{-1}, \qquad \widetilde{\mathbf{b}}_i = \widetilde{\Sigma}_i T_i^T (\widetilde{\mathbf{y}}_i - X_i(\mathbf{z}_i)\widehat{\boldsymbol{\beta}})/\widehat{\sigma}^2.$$

After some algebra, we can then integrate out $\mathbf{b}_i$ from the terms corresponding to I_1, I_2, and I_3 in (4.5) of Section 4.2.1 (page 136), and obtain the following results

$$\widetilde{I}_1 = -\frac{n_i}{2}\log(\sigma^2) - \frac{1}{2\sigma^2}\Bigg[tr(T_i^T T_i \widetilde{\Sigma}_i) + \int \Big(\widetilde{\mathbf{y}}_i - X_i(\mathbf{z}_i)\boldsymbol{\beta} - T_i\widetilde{\mathbf{b}}_i\Big)^T$$
$$\times \Big(\widetilde{\mathbf{y}}_i - X_i(\mathbf{z}_i)\boldsymbol{\beta} - T_i\widetilde{\mathbf{b}}_i\Big) f(\mathbf{z}_{mis,i}|\mathbf{z}_{obs,i}, \widetilde{\mathbf{y}}_{obs,i}, \widehat{\boldsymbol{\theta}})\, d\mathbf{z}_{mis,i}\Bigg],$$

$$\widetilde{I}_2 = \int \log f(\mathbf{z}_i|\boldsymbol{\alpha}) f(\mathbf{z}_{mis,i}|\mathbf{z}_{obs,i}, \widetilde{\mathbf{y}}_{obs,i}, \widehat{\boldsymbol{\theta}})\, d\mathbf{z}_{mis,i},$$

$$\widetilde{I}_3 = -\frac{1}{2}\log|D| - \frac{1}{2}tr(D^{-1}\widetilde{\Sigma}_i) - \frac{1}{2}\int \Big(\widetilde{\mathbf{b}}_i^T D^{-1}\widetilde{\mathbf{b}}_i\Big)$$
$$\times f(\mathbf{z}_{mis,i}|\mathbf{z}_{obs,i}, \widetilde{\mathbf{y}}_{obs,i}, \widehat{\boldsymbol{\theta}})\, d\mathbf{z}_{mis,i}.$$

Therefore, for LME model (4.20), the E-step of the Monte Carlo EM algorithm in Section 4.2.1 does not involve integration with respect to the random effects $\mathbf{b}_i$. That is, in the E-step we only need to simulate samples from the *lower* dimensional distribution $f(\mathbf{z}_{mis,i}|\mathbf{z}_{obs,i}, \widetilde{\mathbf{y}}_{obs,i}, \widehat{\boldsymbol{\theta}})$. This can again be done using the Gibbs sampler along with the rejection sampling methods by noting that

$$f(\mathbf{z}_{mis,i}|\mathbf{z}_{obs,i}, \widetilde{\mathbf{y}}_{obs,i}, \widehat{\boldsymbol{\theta}}) \propto f(\mathbf{z}_i|\widehat{\boldsymbol{\theta}}) f(\widetilde{\mathbf{y}}_i|\mathbf{z}_i, \widehat{\boldsymbol{\theta}}).$$

This greatly reduces computational burden since many computational difficulties, such as convergence problems, often arise from sampling the random effects (Gilks et al. 1996).

Suppose that $\{(\mathbf{z}_{mis,i}^{*(1)}, \ldots, \mathbf{z}_{mis,i}^{*(m_t)}\}$ is a sample of size m_t generated from $f(\mathbf{z}_{mis,i}|\mathbf{z}_{obs,i}, \widetilde{\mathbf{y}}_{obs,i}, \hat{\boldsymbol{\theta}})$ at the t-th EM iteration. Let

$$
\begin{aligned}
\mathbf{z}_i^{*(k)} &= \left(\mathbf{z}_{mis,i}^{*(k)}, \ \mathbf{z}_{obs,i}\right), \\
\widetilde{\mathbf{y}}_i^{*(k)} &= \mathbf{y}_i - \mathbf{g}_i(\mathbf{z}_i^{*(k)}, \hat{\boldsymbol{\beta}}, \widehat{\mathbf{b}}_i) + X_i(\mathbf{z}_i^{*(k)})\hat{\boldsymbol{\beta}} + T_i\widehat{\mathbf{b}}_i, \\
\widetilde{\mathbf{b}}_i^{*(k)} &= \widetilde{\Sigma}_i T_i^T(\widetilde{\mathbf{y}}_i^{*(k)} - X_i(\mathbf{z}_i^{*(k)})\hat{\boldsymbol{\beta}})/\hat{\sigma}^2, \quad k = 1, \ldots, m_t.
\end{aligned}
$$

The E-step for individual i can now be written as

$$
\begin{aligned}
\widetilde{Q}_i(\boldsymbol{\theta}|\hat{\boldsymbol{\theta}}) &= \left\{-\frac{n_i}{2}\log(\sigma^2) - \frac{1}{2\sigma^2}\left[tr(T_i^T T_i \widetilde{\Sigma}_i)\right.\right. \\
&\quad \left.\left. +\frac{1}{m_t}\sum_{k=1}^{m_t}\left(\widetilde{\mathbf{y}}_i^{*(k)} - X_i(\mathbf{z}_i^{*(k)})\boldsymbol{\beta} - T_i\widetilde{\mathbf{b}}_i^{*(k)}\right)^T(\cdot)\right]\right\} \\
&\quad +\left\{-\frac{1}{2}\log|D| - \frac{1}{2}tr(D^{-1}\widetilde{\Sigma}_i) - \frac{1}{2m_t}\sum_{k=1}^{m_t}\left(\widetilde{\mathbf{b}}_i^{*(k)T}D^{-1}\widetilde{\mathbf{b}}_i^{*(k)}\right)\right\} \\
&\quad +\frac{1}{m_t}\sum_{k=1}^{m_t}\log f\left(\mathbf{z}_i^{*(k)}|\boldsymbol{\alpha}\right).
\end{aligned}
\tag{4.21}
$$

The M-step then maximizes

$$
\widetilde{Q}(\boldsymbol{\theta}|\hat{\boldsymbol{\theta}}) = \sum_{i=1}^{n} \widetilde{Q}_i(\boldsymbol{\theta}|\hat{\boldsymbol{\theta}}),
$$

using standard complete-data optimization methods such as the Newton-Raphason method. Updated estimate for $\boldsymbol{\theta}$ can be obtained by maximizing $\widetilde{Q}(\boldsymbol{\theta}|\hat{\boldsymbol{\theta}})$ with respect to $\boldsymbol{\theta}$. Updated estimate of the random effect $\mathbf{b}_i$ is given by its empirical Bayes estimate

$$
\widehat{\mathbf{b}}_i = \hat{D}T_i^T\hat{V}_i^{-1}(\widetilde{\mathbf{y}}_i - X_i\hat{\boldsymbol{\beta}}),
$$

where $V_i = T_i D T_i^T + R_i$.

When the incompletely observed covariates are categorical or discrete, we can obtain *exact* expression for the E-step without using Monte Carlo approximation, as in Ibrahim et al. (1999). This further simplifies computation substantially. Specifically, suppose that covariate $\mathbf{z}_i$ is discrete. Note that, for any function $h(\cdot)$, we have

$$
\int h(\mathbf{z}_{mis,i}) \, f(\mathbf{z}_{mis,i}|\mathbf{z}_{obs,i}, \widetilde{\mathbf{y}}_{obs,i}, \hat{\boldsymbol{\theta}}) \, d\mathbf{z}_{mis,i} = \sum_{\mathbf{z}_{mis,i}(j)} h(\mathbf{z}_{mis,i}(j)) \, p_{ij}(\hat{\boldsymbol{\theta}}),
$$

where the sum $\sum_{\mathbf{z}_{mis,i}(j)}$ extends over all the possible values of the missing covariate components, with j indexing distinct covariate patterns, and

$$
\begin{aligned}
p_{ij}(\boldsymbol{\theta}) &= \Pr(\mathbf{z}_{mis,i}(j)|\mathbf{z}_{obs,i}, \tilde{\mathbf{y}}_i, \boldsymbol{\theta}) \\
&= \frac{f(\tilde{\mathbf{y}}_i|\mathbf{z}_{mis,i}(j), \mathbf{z}_{obs,i}, \boldsymbol{\beta}, \sigma^2) f(\mathbf{z}_{mis,i}(j), \mathbf{z}_{obs,i}|\boldsymbol{\alpha})}{\sum_{\mathbf{z}_{mis,i}(j)} f(\tilde{\mathbf{y}}_i|\mathbf{z}_i, \boldsymbol{\beta}, \sigma^2) f(\mathbf{z}_i|\boldsymbol{\alpha})}.
\end{aligned}
$$

Thus we can obtain an exact (analytic) expression for the E-step. This method can be easily extended to cases where the missing covariates contain both continuous and discrete components. In this case, we may partition the incompletely observed covariate vector into its continuous and categorical components. Then, the method described above can be used for the categorical components.

For complete-data LME models, efficient EM-type algorithms, which are extensions of the standard EM algorithm to accelerate convergence, have been developed (e.g., Meng and van Dyk 1997, 1998; van Dyk 2000). Since the approximate method in this section involves iteratively solving complete-data LME models, these efficient EM-type algorithms for LME models can be incorporated into the approximate method to further speed up the EM algorithm. We will provide such an example in Section 4.6.2.

4.3.2 Laplace Approximation

In this section, we use Laplace approximations to directly approximate the intractable integrals in the observed-data likelihoods, as in Section 2.6.3 of Chapter 2. Wolfinger (1993) and Vonesh (1996) considered such approach for complete-data NLME models, and Breslow and Clayton (1993) considered a closely related approach for complete-data GLMMs. Lee, Nelder, and Pawitan (2006) provided a systematic treatment of this type of approaches via so-called *hierarchical likelihood* or *h-likelihood*. In the following, we describe the method in the context of h-likelihood, which also produces restricted maximum likelihood estimation (REML) of the variance-covariance parameters. That is, it produces approximate MLEs of the mean parameters, and REMLs of the variance-covariance or dispersion parameters.

We illustrate the method for missing covariates in NLME models when the missing data mechanism is ignorable. Note that in the Laplace approximation (2.78) (page 86), the domain of variable $\mathbf{v}$ is the whole space $\mathbf{R}^\gamma$, where γ is the dimension of vector $\mathbf{v}$. Thus, to facilitate a Laplace approximation to the observed-data likelihood, we consider the following transformation (or scale) for the incompletely observed covariates $\mathbf{z}_i = (z_{i1}, \ldots, z_{ip})^T$:

$$
\mathbf{v}_i \equiv \mathbf{v}(\mathbf{z}_i) = (v_1(z_{i1}), \cdots, v_p(z_{ip})) \tag{4.22}
$$

such that, if the domain of the k-th component z_{ik} of $\mathbf{z}_i$ is $(-\infty, \infty)$ (the whole real line), no transformation is needed for z_{ik} (i.e., $v_k(z_{ik}) = z_{ik}$), but if the domain of z_{ik} is not the whole real line, we choose a monotone function $v_k(\cdot)$ so that the domain of $v_k(z_{ik})$ becomes the whole real line, $k = 1, \ldots, p$; $i = 1, \ldots, n$. For example, if z_{ik} is a covariate with positive domain (e.g., age or weight), we can choose the transformation or scale $v_k(z_{ik}) = \log(z_{ik})$. In fact, for a positive covariate z_{ik} we can always substitute z_{ik} by $w_{ik} = \exp(\log(z_{ik})) = \exp(v_k(z_{ik}))$.

Thus, the transformation or scale (4.22) is used to ensure that the domains of all the incompletely observed covariates are the whole real line. This transformation is necessary for the purpose of satisfactory Laplace approximations since Laplace approximations require the corresponding variables (i.e., $\mathbf{v}$ in (2.78)) to have unrestricted domains. After the unknown parameters in the models are estimated, we can transform $\mathbf{v}(\mathbf{z}_i)$ back to the original scale $\mathbf{z}_i$ if necessary. Lee et al. (2006) showed that the results are usually insensitive to the choice of scales or transformations. For the unobservable random effects $\mathbf{b}_i$, transformations or scales are not needed since we have assumed that $\mathbf{b}_i \sim N(0, D)$ so the domains of all the components of $\mathbf{b}_i$ are unrestricted.

Following Lee et al. (2006), the h-likelihood for individual i can be written as

$$h_i(\boldsymbol{\theta}, \mathbf{b}_i, \mathbf{v}_i) = \log f(\mathbf{y}_i|\mathbf{z}_i, \boldsymbol{\beta}, \sigma^2, \mathbf{b}_i) + \log f(\mathbf{b}_i|D) + \log f(\mathbf{v}_i|\boldsymbol{\alpha}),$$

where $\mathbf{v}_i = \mathbf{v}(\mathbf{z}_i)$. Let $(\tilde{\mathbf{b}}_i, \tilde{\mathbf{v}}_i)$ be the solution of equation

$$\frac{\partial h_i(\boldsymbol{\theta}, \mathbf{b}_i, \mathbf{v}_i)}{\partial(\mathbf{b}_i, \mathbf{v}_i)} = 0,$$

and let

$$D_i(h_i, \mathbf{b}_i, \mathbf{v}_i) = -\frac{\partial^2 h_i(\boldsymbol{\theta}, \mathbf{b}_i, \mathbf{v}_i)}{\partial(\mathbf{b}_i, \mathbf{v}_i)\partial(\mathbf{b}_i, \mathbf{v}_i)^T}.$$

Then, it can be shown that (Lee et al. 2006)

$$
p_{(\mathbf{b}_i, \mathbf{v}_i)}(h_i) \equiv p_{(\mathbf{b}_i, \mathbf{v}_i)}(h_i, \boldsymbol{\theta})
$$
$$
= \left[h_i(\boldsymbol{\theta}, \mathbf{b}_i, \mathbf{v}_i) - \frac{1}{2}\log\left|\frac{D_i(h_i, \mathbf{b}_i, \mathbf{v}_i)}{2\pi}\right| \right]\Bigg|_{(\mathbf{b}_i, \mathbf{v}_i) = (\tilde{\mathbf{b}}_i, \tilde{\mathbf{v}}_i)} \quad (4.23)
$$

is the first-order Laplace approximation to the observed-data (or marginal) log-likelihood $L_{oi}(\boldsymbol{\theta})$ (see (4.4) of Section 4.2.1, page 135) by *integrating out* the unobservables $(\mathbf{z}_{mis,i}, \mathbf{b}_i)$. That is, $p_{(\mathbf{b}_i, \mathbf{v}_i)}(h_i)$ is a first-order Laplace approximation of $L_{oi}(\boldsymbol{\theta})$ in (4.4). Therefore,

$$\tilde{L}_o(\boldsymbol{\theta}) \equiv \sum_{i=1}^{n} p_{(\mathbf{b}_i, \mathbf{v}_i)}(h_i, \boldsymbol{\theta}) \approx L_o(\boldsymbol{\theta})$$

is a first-order Laplace approximation to the observed-data (or marginal) log-likelihood $L_o(\boldsymbol{\theta})$ (see (4.4) of Section 4.2.1) by integrating out $(\mathbf{z}_{mis,i}, \mathbf{b}_i)$.

This approach avoids the intractable integration in the observed data likelihood $L_o(\boldsymbol{\theta})$ and offers a big computational advantage.

The above method can also be used to obtain REMLs of the variance parameters. To see this, we separate the mean parameters and variance-covariance parameters and write $\boldsymbol{\theta} = (\boldsymbol{\theta}_1, \boldsymbol{\theta}_2)$, where $\boldsymbol{\theta}_1$ contains the mean parameters and $\boldsymbol{\theta}_2$ contains the variance-covariance or dispersion parameters. Then we can use the approximate marginal likelihood $\tilde{L}_o(\boldsymbol{\theta})$ to make inference about the mean parameters $\boldsymbol{\theta}_1$. The resulting estimate $\hat{\boldsymbol{\theta}}_1$ is an approximate MLE of $\boldsymbol{\theta}_1$ based on the Laplace approximation to the observed-data likelihood by integrating out the random effects and missing covariates. For estimation of the dispersion parameters or the variance-covariance parameters $\boldsymbol{\theta}_2$, we can obtain REMLs by considering $p_{(\boldsymbol{\theta}_1, \mathbf{b}_i, \mathbf{v}_i)}(h_i)$, which is defined in a way similar to (4.23). That is, $p_{(\boldsymbol{\theta}_1, \mathbf{b}_i, \mathbf{v}_i)}(h_i)$ approximates the *restricted* log-likelihood by integrating out the mean parameters $\boldsymbol{\theta}_1$, the random effects $\mathbf{b}_i$, and the missing covariates $\mathbf{z}_{mis,i}$. See Lee et al. (2006) for a detailed discussion in the case of complete-data GLMMs.

The foregoing procedure can be iterated as follows:

STEP 1. Obtain initial estimates $\boldsymbol{\theta}^{(0)}$ and $(\mathbf{b}_i^{(0)}, \mathbf{v}_i^{(0)})$ of the parameters $\boldsymbol{\theta}$ and "missing data" $(\mathbf{b}_i, \mathbf{v}_i)$ respectively using a naive method;

STEP 2. At t-th iteration, given the current parameter estimates $\boldsymbol{\theta}^{(t)}$, update "missing data" estimates $(\mathbf{b}_i^{(t+1)}, \mathbf{v}_i^{(t+1)})$ by maximizing $h_i(\boldsymbol{\theta}^{(t)}, \mathbf{b}_i, \mathbf{v}_i)$ with respect to $(\mathbf{b}_i, \mathbf{v}_i)$, $i = 1, \ldots, n$;

STEP 3. Given the "missing data" estimates $(\mathbf{b}_i^{(t+1)}, \mathbf{v}_i^{(t+1)})$, update the parameter estimates $\boldsymbol{\theta}^{(t+1)}$ by maximizing

$$K(\boldsymbol{\theta}) \equiv \sum_{i=1}^{n} p_{(\mathbf{b}_i^{(t+1)}, \mathbf{v}^{(t+1)})}(h_i(\boldsymbol{\theta}, \mathbf{b}_i, \mathbf{v}_i))$$

with respect to $\boldsymbol{\theta}$. This step can be done in two or more steps if REML estimates of the variance-covariance parameters are desirable;

STEP 4. Iterating between Step 2 and Step 3 until convergence, we obtain the approximate MLE $\hat{\boldsymbol{\theta}}_{AP}$.

The estimates of the random effects obtained this way can be interpreted as empirical Bayes estimates.

At convergence, an approximate formula for the variance-covariance matrix of the approximate MLE $\hat{\boldsymbol{\theta}}_{AP}$ is given by

$$\text{Cov}(\hat{\boldsymbol{\theta}}_{AP}) = \left[-\frac{\partial^2 p_{(\mathbf{b}_i, \mathbf{v}_i)}(h_i(\boldsymbol{\theta}, \hat{\mathbf{b}}_i, \hat{\mathbf{v}}_i))}{\partial \boldsymbol{\theta} \partial \boldsymbol{\theta}^T} \right]_{\boldsymbol{\theta} = \hat{\boldsymbol{\theta}}_{AP}}^{-1} .$$

Note that if an incompletely observed covariate is discrete, the integration with respect to this covariate in $L_o(\boldsymbol{\theta})$ reduces to a summation, then the Laplace approximation with respect to this covariate is not needed. This reduces the dimension of the integral in $L_o(\boldsymbol{\theta})$ and thus simplifies the computation.

4.4 Mixed Effects Models with Missing Responses

Mixed effects models allow unbalanced data in the responses, i.e., the number of response measurements is allowed to vary across individuals or clusters. In other words, if there are missing data in the responses and the missingness is *ignorable*, inference for mixed effects models can proceed in the usual way as if there were no missing responses. Thus mixed effects models can be used to combine data from similar studies or clusters, and they can be used to model a wide varieties of longitudinal or clustered data.

In practice, however, missing data in the responses may be *non-ignorable* in the sense that the missingness may be related to the missing values. For example, if a subject drops out due to drug side-effects, the resulting missing response values may be non-ignorable. Sometimes, the dropout subjects may return to the study at a later time, so the missing data in the response may be *intermittent*. When a subject drops out from a longitudinal study, the corresponding time-dependent covariates will also be missing. Thus, missing response and missing covariates may arise simultaneously.

In this section, we consider mixed effects models with missing responses when the missingness is non-ignorable. Likelihood methods for mixed effects models with non-ignorable missing responses are conceptually the same as that for mixed effects models with nonignorable missing covariates, with straightforward modifications. For completeness, in the following we give an outline of the procedures. We focus on mixed effects models with non-igorable missing responses and missing covariates. The mixed effects models include LME, GLMM, and NLME models.

4.4.1 Exact Likelihood Inference

Let y_{ij} be the response value for individual i at time t_{ij} or cluster j, subject to nonignorable missing data, $i = 1, \ldots, n$; $j = 1, \ldots, n_i$. Let r_{ij} be the missing response indicator such that $r_{ij} = 1$ if y_{ij} is missing and $r_{ij} = 0$ if y_{ij} is observed. Let $\mathbf{z}_i = (z_{i1}, \ldots, z_{ip})^T$ be a collection of incompletely observed covariates for individual i, and let $\mathbf{v}_i$ be a collection of completely observed covariates. We define $\mathbf{y}_i = (y_{i1}, \ldots, y_{in_i})^T$ and define $\mathbf{r}_i$ similarly. Write $\mathbf{y}_i = (\mathbf{y}_{mis,i}, \mathbf{y}_{obs,i})$, where $\mathbf{y}_{mis,i}$ contains the missing components of $\mathbf{y}_i$ and

$\mathbf{y}_{obs,i}$ contains the observed components of $\mathbf{y}_i$, and write $\mathbf{z}_i = (\mathbf{z}_{mis,i}, \mathbf{z}_{obs,i})$ similarly. The missing covariates are assumed to be ignorable for simplicity of presentation. The observed data are $\{(\mathbf{y}_{obs,i}, \mathbf{z}_{obs,i}, \mathbf{v}_i, \mathbf{r}_i), \ i = 1, \ldots, n\}$. To allow for non-ignorable missing responses, we assume a model for the missing response indicators $\mathbf{r}_i$, denoted by $f(\mathbf{r}_i|\mathbf{y}_i, \mathbf{z}_i, \mathbf{v}_i, \boldsymbol{\phi})$, which links the probability of missingness to missing and observed values.

The observed data likelihood can be written as

$$L_o(\boldsymbol{\psi}) \;=\; \prod_{i=1}^{n} \int \int \int f(\mathbf{y}_i|\mathbf{z}_i, \mathbf{v}_i, \mathbf{b}_i, \boldsymbol{\beta}, \sigma^2) f(\mathbf{z}_i|\mathbf{v}_i, \boldsymbol{\alpha}) f(\mathbf{b}_i|D)$$
$$\times f(\mathbf{r}_i|\mathbf{y}_i, \mathbf{z}_i, \mathbf{v}_i, \boldsymbol{\phi}) \; d\mathbf{b}_i \; d\mathbf{z}_{mis,i} \; d\mathbf{y}_{mis,i},$$

where $\boldsymbol{\psi} = (\boldsymbol{\alpha}, \boldsymbol{\beta}, \sigma^2, \boldsymbol{\eta}, \boldsymbol{\phi})$ denotes the collection of all parameters. The observed data likelihood $L_o(\boldsymbol{\psi})$ involves an highly intractable integral. As in Ibrahim et al. (2001) and Stubbendick and Ibrahim (2003), here we consider a Monte Carlo EM algorithm for "exact" likelihood inference, incorporating missing responses and missing covariates *simultaneously*. The "complete data" are $\{(\mathbf{y}_i, \mathbf{z}_i, \mathbf{v}_i, \mathbf{b}_i, \mathbf{r}_i), \ i = 1, \ldots, n\}$, and the "complete-data" log-likelihood can be written as

$$l_c(\boldsymbol{\psi}) \;=\; \sum_{i=1}^{n} \Big\{ \log f(\mathbf{y}_i|\mathbf{z}_i, \mathbf{v}_i, \mathbf{b}_i, \boldsymbol{\beta}, \sigma^2) + \log f(\mathbf{z}_i|\mathbf{v}_i, \boldsymbol{\alpha})$$
$$+ \log f(\mathbf{b}_i|D(\boldsymbol{\eta})) + \log f(\mathbf{r}_i|\mathbf{y}_i, \mathbf{z}_i, \mathbf{v}_i, \boldsymbol{\phi}) \Big\}.$$

We may consider the following missing data model for the responses

$$f(\mathbf{r}_i|\mathbf{y}_i, \mathbf{z}_i, \mathbf{v}_i, \boldsymbol{\phi}) \;=\; f(r_{in_i}|\bar{r}_{in_i}, \bar{y}_{in_i}, \mathbf{z}_i, \mathbf{v}_i, \boldsymbol{\phi}_{n_i}) \times \cdots$$
$$\times f(r_{i2}|r_{i1}, \bar{y}_{i2}, \mathbf{z}_i, \mathbf{v}_i, \boldsymbol{\phi}_2) f(r_{i1}|y_{i1}, \mathbf{z}_i, \mathbf{v}_i, \boldsymbol{\phi}_1),$$

where $\boldsymbol{\phi} = (\boldsymbol{\phi}_1, \boldsymbol{\phi}_2, \ldots, \boldsymbol{\phi}_M)$, $M = \max_i\{n_i\}$, $\boldsymbol{\phi}_j = (\phi_1, \cdots, \phi_{M_j})$, $\bar{r}_{ij} = (r_{i1}, \cdots, r_{i,j-1})$, and $\bar{y}_{ij} = (y_{i1}, \cdots, y_{ij})$. We may assume, for example,

$$\text{logit}[P(r_{ij} = 1|\bar{r}_{ij}, \bar{y}_{ij}, \boldsymbol{\phi}_j)] = \phi_0 + \phi_1 r_{i,j-1} + \phi_2 y_{i1} + \cdots + \phi_{j+1} y_{ij}$$

for $j > 1$, and

$$\text{logit}[P(r_{i1} = 1|y_{i1}, \boldsymbol{\phi}_1)] = \phi_0 + \phi_2 y_{i1}.$$

Alternative missing data models can be specified in a similar way.

Let $\boldsymbol{\psi}^{(t)}$ be the parameter estimate from the t-th EM iteration, $t = 1, 2, \cdots$. The E-step for individual i at the $(t+1)$st EM iteration can be written as

$$Q_i(\boldsymbol{\psi}|\boldsymbol{\psi}^{(t)}) \;=\; E\left[l_c(\boldsymbol{\psi}|\mathbf{y}_i, \mathbf{z}_i, \mathbf{v}_i, \mathbf{r}_i, \mathbf{b}_i)|\mathbf{y}_{obs,i}, \mathbf{z}_{obs,i}, \mathbf{v}_i, \mathbf{r}_i, \boldsymbol{\psi}^{(t)}\right]$$
$$=\; \int \int \int \Big\{ \log f(\mathbf{y}_i|\mathbf{z}_i, \mathbf{v}_i, \mathbf{b}_i, \boldsymbol{\beta}, \sigma^2) + \log f(\mathbf{z}_i|\mathbf{v}_i, \boldsymbol{\alpha})$$

$$+ \log f(\mathbf{b}_i|D) + \log f(\mathbf{r}_i|\mathbf{y}_i, \mathbf{z}_i, \mathbf{v}_i, \boldsymbol{\phi}) \Big\}$$
$$\times f(\mathbf{y}_{mis,i}, \mathbf{z}_{mis,i}, \mathbf{b}_i|\mathbf{y}_{obs,i}, \mathbf{z}_{obs,i}, \mathbf{v}_i, \mathbf{r}_i, \boldsymbol{\psi}^{(t)})$$
$$\times d\mathbf{b}_i\, d\mathbf{z}_{mis,i} d\mathbf{y}_{mis,i}$$
$$\equiv \quad I_1 + I_2 + I_3 + I_4. \tag{4.24}$$

A Monte Carlo method can be used to approximate $Q_i(\boldsymbol{\psi}|\boldsymbol{\psi}^{(t)})$ in the E-step. That is, we generate m_t independent samples from the conditional distribution $f(\mathbf{y}_{mis,i}, \mathbf{z}_{mis,i}, \mathbf{b}_i|\mathbf{y}_{obs,i}, \mathbf{z}_{obs,i}, \mathbf{v}_i, \mathbf{r}_i, \boldsymbol{\psi}^{(t)})$ and then approximate the expectation $Q_i(\boldsymbol{\psi}|\boldsymbol{\psi}^{(t)})$ by its empirical mean, with missing data replaced by simulated values.

To generate samples from $f(\mathbf{y}_{mis,i}, \mathbf{z}_{mis,i}, \mathbf{b}_i|\mathbf{y}_{obs,i}, \mathbf{z}_{obs,i}, \mathbf{v}_i, \mathbf{r}_i, \boldsymbol{\psi}^{(t)})$, we can use the Gibbs sampler by iteratively sampling from the full conditionals $f(\mathbf{y}_{mis,i}|\mathbf{y}_{obs,i}, \mathbf{z}_i, \mathbf{v}_i, \mathbf{b}_i, \mathbf{r}_i, \boldsymbol{\psi}^{(t)})$, $f(\mathbf{z}_{mis,i}|\mathbf{z}_{obs,i}, \mathbf{y}_i, \mathbf{v}_i, \mathbf{b}_i, \mathbf{r}_i, \boldsymbol{\psi}^{(t)})$, and $f(\mathbf{b}_i|\mathbf{y}_i, \mathbf{z}_i, \mathbf{v}_i, \mathbf{r}_i, \boldsymbol{\psi}^{(t)})$. Sampling from these full conditionals can be done using rejection sampling methods by noting that

$$
\begin{aligned}
f(\mathbf{y}_{mis,i}|\mathbf{y}_{obs,i}, \mathbf{z}_i, \mathbf{v}_i, \mathbf{b}_i, \mathbf{r}_i, \boldsymbol{\psi}^{(t)}) \quad &\propto \quad f(\mathbf{y}_i|\mathbf{z}_i, \mathbf{v}_i, \mathbf{b}_i, \boldsymbol{\beta}^{(t)}, \sigma^{(t)2}) \\
&\quad \times f(\mathbf{r}_i|\mathbf{y}_i, \mathbf{z}_i, \mathbf{v}_i, \boldsymbol{\phi}^{(t)}), \\
f(\mathbf{z}_{mis,i}|\mathbf{z}_{obs,i}, \mathbf{y}_i, \mathbf{v}_i, \mathbf{b}_i, \mathbf{r}_i, \boldsymbol{\psi}^{(t)}) \quad &\propto \quad f(\mathbf{z}_i|\mathbf{v}_i, \boldsymbol{\alpha}^{(t)}) \\
&\quad \times f(\mathbf{y}_i|\mathbf{z}_i, \mathbf{v}_i, \mathbf{b}_i, \boldsymbol{\beta}^{(t)}, \sigma^{(t)2}) \\
&\quad \times f(\mathbf{r}_i|\mathbf{y}_i, \mathbf{z}_i, \mathbf{v}_i, \boldsymbol{\phi}^{(t)}), \\
f(\mathbf{b}_i|\mathbf{y}_i, \mathbf{z}_i, \mathbf{v}_i, \mathbf{r}_i, \boldsymbol{\psi}^{(t)}) \quad &\propto \quad f(\mathbf{b}_i|D^{(t)}) \\
&\quad \times f(\mathbf{y}_i|\mathbf{z}_i, \mathbf{v}_i, \mathbf{b}_i, \boldsymbol{\beta}^{(t)}, \sigma^{(t)2}),
\end{aligned}
$$

where the density functions on the right-hand sides are all known. Then, the M-step can be accomplished by standard complete-data optimization procedures, similar to that in Section 4.2.

The foregoing procedure can be extended to time-dependent covariates with non-ignorable missing data. Specifically, let z_{ij} be the covariate value for the i-th individual at time t_{ij}, $i = 1, \cdots, n; j = 1, \cdots, n_i$, and let $\mathbf{z}_i = (z_{i1}, \ldots, z_{in_i})^T$. Write $\mathbf{z}_i = (\mathbf{z}_{mis,i}, \mathbf{z}_{obs,i})$. Let $\mathbf{s}_i = (s_{i1}, \ldots, s_{in_i})^T$ be a vector of missing covariate indicators such that $s_{ij} = 1$ if z_{ij} is missing and $s_{ij} = 0$ otherwise. We can assume a mixed effects model for the covariate process, denoted by $f(\mathbf{z}_i|\mathbf{a}_i; \boldsymbol{\alpha})$, where $\mathbf{a}_i$'s are random effects with distribution $f(\mathbf{a}_i|A)$ and $\boldsymbol{\alpha}$ contains unknown fixed parameters. The observed data likelihood can then be written as

$$
\begin{aligned}
L_o(\boldsymbol{\psi}) \quad = \quad &\prod_{i=1}^{n} \int \int \int f(\mathbf{y}_i|\mathbf{z}_i, \mathbf{v}_i, \mathbf{b}_i, \boldsymbol{\beta}, \sigma^2) f(\mathbf{z}_i|\mathbf{a}_i, \boldsymbol{\alpha}) f(\mathbf{b}_i|D) f(\mathbf{a}_i|A) \\
&\quad \times f(\mathbf{r}_i|\mathbf{y}_i, \mathbf{z}_i, \mathbf{v}_i, \boldsymbol{\phi}_1) f(\mathbf{s}_i|\mathbf{z}_i, \boldsymbol{\phi}_2)\, d\mathbf{a}_i\, d\mathbf{b}_i\, d\mathbf{z}_{mis,i}\, d\mathbf{y}_{mis,i},
\end{aligned}
$$

where $f(\mathbf{s}_i|\mathbf{z}_i, \boldsymbol{\phi}_2)$ is the assumed missing data model for the missing co-variates. A Monte Carlo EM algorithm can again be used for estimation, but the computation becomes more intensive due to the high-dimensional and in-tractable integration.

For non-ignorable missing data, the assumed missing data models are not testable based on the observed data. So sensitivity analysis based on alter-native non-ignorable missing data models should be conducted. Diggle and Kenward (1994) and Little (1995) discussed various dropout models for lon-gitudinal data. For mixed effects models, an alternative missing data model is to link the missing probability to the random effects in the models, since these random effects characterize the individual-specific longitudinal processes and may also be viewed as summaries of the longitudinal trajectories. Thus, we may consider a missing data model $f(\mathbf{r}_i|\mathbf{b}_i, \boldsymbol{\phi})$, e.g.,

$$\text{logit}(P(r_{ij} = 1|\mathbf{b}_i, \boldsymbol{\phi}) = \phi_0 + \phi_1 r_{i,j-1} + \boldsymbol{\phi}_2^T \mathbf{b}_i.$$

Such a missing data model is sometimes called a *shared parameter model* (Wu and Carroll 1988; Little and Ruben 2002). In this case, the observed-data like-lihood can be written as

$$L_o^*(\boldsymbol{\psi}) = \prod_{i=1}^{n} \int \int f(\mathbf{y}_{obs,i}|\mathbf{z}_i, \mathbf{v}_i, \mathbf{b}_i, \boldsymbol{\beta}, \sigma^2) f(\mathbf{z}_i|\mathbf{v}_i, \boldsymbol{\alpha}) f(\mathbf{b}_i|D)$$

$$\times f(\mathbf{r}_i|\mathbf{b}_i, \boldsymbol{\phi}) \, d\mathbf{b}_i \, d\mathbf{z}_{mis,i}.$$

The foregoing Monte Carlo EM algorithm can be modified in a straightforward way to find the MLE of $\boldsymbol{\psi}$.

For models with non-ignorable missing data, there may be too many nuisance parameters, so identifiability of the parameters is an important issue (Fitzmau-rice et al. 1996). Since non-ignorable models can be complex, if there is not sufficient information in the data, the models can be non-identifiable in the sense that two different sets of parameters may lead to the same observed likeli-hood. In practice, one may check identifiability empirically. For example, if the Fisher information matrix is singular then the model may be non-identifiable, while if the Fisher information matrix is non-singular then the model is locally identifiable. As noted in Stubbendick and Ibrahim (2003), we may also check model identifiability by examing the convergence of the EM algorithm. If the model is non-identifiable, the EM may diverge quickly.

4.4.2 Approximate Likelihood Inference

A major issue in likelihood inference for mixed effects models with non-ignorable missing data is the computational challenge since the observed data likelihoods can be extremely intractable, as shown in Section 4.4.1. Therefore, computationally much more efficient approximate methods are highly valuable for these problems. These approximate methods are often based on Taylor or Laplace approximations, similar to that in Section 4.3. As an illustration, here we briefly describe a linearization method for the missing data problems in NLME models presented in Section 4.4.1.

Consider the following NLME model

$$
\begin{aligned}
y_{ij} &= g_{ij}(\mathbf{z}_i, \boldsymbol{\beta}, \mathbf{b}_i) + e_{ij}, \qquad j = 1, \ldots, n_i; \; i = 1, \ldots, n, \quad (4.25) \\
\mathbf{b}_i &\sim N(0, D), \qquad e_{ij} \text{ i.i.d.} \sim N(0, \sigma^2), \quad (4.26)
\end{aligned}
$$

where $g_{ij}(\cdot)$ is a known nonlinear function. To simplify the notation, we suppress the completely observed covariates $\mathbf{v}_i$, and denote the current estimate of ψ in the t-th EM iteration by $\widehat{\psi} = (\widehat{\alpha}, \widehat{\beta}, \widehat{D}, \widehat{\eta}, \widehat{\sigma}^2)$, suppressing the iteration number. Let $\mathbf{g}_i = (g_{i1}, \ldots, g_{in_i})^T$. Based on a first-order Taylor expansion around the current parameter and random effects estimates, the approximate method iteratively solves the following LME model

$$
\tilde{\mathbf{y}}_i = X_i \boldsymbol{\beta} + T_i \mathbf{b}_i + \mathbf{e}_i, \quad (4.27)
$$

where

$$
\begin{aligned}
\tilde{\mathbf{y}}_i &= \mathbf{y}_i - \mathbf{g}_i(\mathbf{z}_i, \widehat{\boldsymbol{\beta}}, \widehat{\mathbf{b}}_i) + X_i \widehat{\boldsymbol{\beta}} + T_i \widehat{\mathbf{b}}_i, \\
X_{ij} &= \left. \frac{\partial g_{ij}(\mathbf{z}_i, \boldsymbol{\beta}, \mathbf{b}_i)}{\partial \boldsymbol{\beta}^T} \right|_{(\widehat{\boldsymbol{\beta}}, \widehat{\mathbf{b}}_i)}, \qquad
T_{ij} = \left. \frac{\partial g_{ij}(\mathbf{z}_i, \boldsymbol{\beta}, \mathbf{b}_i)}{\partial \mathbf{b}_i^T} \right|_{(\widehat{\boldsymbol{\beta}}, \widehat{\mathbf{b}}_i)},
\end{aligned}
$$

$X_i \equiv X_i(\mathbf{z}_i) = (X_{i1}, \cdots, X_{in_i})^T$, $T_i \equiv T_i(\mathbf{z}_i) = (T_{i1}, \cdots, T_{in_i})^T$, and $\tilde{\mathbf{y}}_i = (\tilde{y}_{i1}, \cdots, \tilde{y}_{in_i})^T$.

Note that we have

$$
\begin{aligned}
f(\tilde{\mathbf{y}}_{mis,i}, \mathbf{z}_{mis,i}, \mathbf{b}_i | \tilde{\mathbf{y}}_{obs,i}, \mathbf{z}_{obs,i}, \mathbf{r}_i, \widehat{\psi}) &= f(\mathbf{b}_i | \tilde{\mathbf{y}}_i, \mathbf{z}_i, \widehat{\psi}) \\
&\times f(\tilde{\mathbf{y}}_{mis,i}, \mathbf{z}_{mis,i} | \tilde{\mathbf{y}}_{obs,i}, \mathbf{z}_{obs,i}, \mathbf{r}_i, \widehat{\psi}),
\end{aligned}
$$

where

$$
\tilde{\mathbf{y}}_{mis,i} = \mathbf{y}_{mis,i} - \mathbf{g}_i(\mathbf{z}_{mis,i}, \widehat{\boldsymbol{\beta}}, \widehat{\mathbf{b}}_i) + X_i(\mathbf{z}_{mis,i})\widehat{\boldsymbol{\beta}} + T_i(\mathbf{z}_{mis,i})\widehat{\mathbf{b}}_i,
$$

$\tilde{\mathbf{y}}_{obs,i}$ is defined similarly, and $\tilde{\mathbf{y}}_i = (\tilde{\mathbf{y}}_{mis,i}, \tilde{\mathbf{y}}_{obs,i})$. Under the LME model (4.27), it can be shown that

$$
[\mathbf{b}_i | \tilde{\mathbf{y}}_i, \mathbf{z}_i, \widehat{\psi}] \sim N(\tilde{\mathbf{b}}_i, \tilde{\Sigma}_i),
$$

where

$$
\tilde{\Sigma}_i = (\widehat{\sigma}^{-2} T_i^T T_i + \widehat{D}^{-1})^{-1}, \qquad
\tilde{\mathbf{b}}_i = \tilde{\Sigma}_i T_i^T (\tilde{\mathbf{y}}_i - X_i \widehat{\boldsymbol{\beta}})/\widehat{\sigma}^2.
$$

Then, we can integrate out $\mathbf{b}_i$ from $I_1 - I_4$ in (4.24) of Section 4.4.1 (page 156), and obtain the following results

$$
\tilde{I}_1 = -\frac{n_i}{2} \log(\sigma^2) - \frac{1}{2\sigma^2} \left[\text{tr}(T_i^T T_i \tilde{\Sigma}_i) + \int \left(\tilde{\mathbf{y}}_i - X_i \boldsymbol{\beta} - T_i \tilde{\mathbf{b}}_i \right)^T (\cdot) \right.
$$

$$\times f(\widetilde{\mathbf{y}}_{mis,i}, \mathbf{z}_{mis,i} | \widetilde{\mathbf{y}}_{obs,i}, \mathbf{z}_{obs,i}, \mathbf{r}_i, \widehat{\boldsymbol{\psi}}) \, d\widetilde{\mathbf{y}}_{mis,i} d\mathbf{z}_{mis,i} \Big],$$

$$\tilde{I}_2 = \int \log f(\mathbf{z}_i | \boldsymbol{\alpha}) f(\mathbf{z}_{mis,i} | \widetilde{\mathbf{y}}_i, \mathbf{z}_{obs,i}, \mathbf{r}_i, \widehat{\boldsymbol{\psi}}) d\mathbf{z}_{mis,i}$$

$$\tilde{I}_3 = -\frac{1}{2} \log |D| - \frac{1}{2} \text{tr}(D^{-1} \widetilde{\Sigma}_i) - \frac{1}{2} \int \left(\widetilde{\mathbf{b}}_i^T D^{-1} \widetilde{\mathbf{b}}_i \right)$$

$$\times f(\widetilde{\mathbf{y}}_{mis,i}, \mathbf{z}_{mis,i} | \widetilde{\mathbf{y}}_{obs,i}, \mathbf{z}_{obs,i}, \mathbf{r}_i, \widehat{\boldsymbol{\psi}}) \, d\widetilde{\mathbf{y}}_{mis,i} d\mathbf{z}_{mis,i},$$

$$\tilde{I}_4 = \int \log f(\mathbf{r}_i | \widetilde{\mathbf{y}}_i, \mathbf{z}_i, \boldsymbol{\phi})$$

$$\times f(\widetilde{\mathbf{y}}_{mis,i}, \mathbf{z}_{mis,i} | \widetilde{\mathbf{y}}_{obs,i}, \mathbf{z}_{obs,i}, \mathbf{r}_i, \widehat{\boldsymbol{\psi}}) \, d\widetilde{\mathbf{y}}_{mis,i} d\mathbf{z}_{mis,i}.$$

Thus, we have integrated out the random effects $\mathbf{b}_i$ from $\tilde{I}_1 - \tilde{I}_4$ in the E-step, and therefore greatly reduce the computational burden. The rest of the procedure is similar to that in Section 4.2.

Alternatively, a Laplace approximation may be considered. The procedure is similar to that in Section 4.3.2, so we omit the detail here.

4.5 Multiple Imputation Methods

4.5.1 Advantages and Issues of Multiple Imputation Methods

In previous sections, we have focused on likelihood methods or approximate likelihood methods for mixed effects models with missing data, implemented by Monte Carlo EM algorithms or Taylor/Laplace approximations. Likelihood methods are popular because MLEs are asymptotically most efficient and asymptotically normally distributed, if the assumed models hold and usual regularity conditions are satisfied. However, likelihood methods are specific to the assumed models, and in practice the assumed models may not hold exactly. Moreover, software for likelihood methods for specific problems may not be available, and general-purpose software may not be directly applicable to specific problems.

Multiple imputation methods, on the other hand, may be easier to use in practice since imputation models can be different from models used in data analyses and general-purpose software for multiple imputations is available and applicable to many specific problems (although they may not be optimal). In other words, once multiple imputations are created based on an imputation model, the resulting "complete datasets" can be used for a wide variety of analyses. For example, one may create multiple imputations based on a multivariate normal model and then analyze the imputed datasets using a logistic regression model (Little and Rubin 2002).

In fact, multiple imputation methods were first proposed for public use survey data, since the person who creates imputations for missing data can use auxiliary confidential and detailed information that may not be available to public or data analysts (Rubin 1996). Once the incomplete data are multiply imputed, data analysts can conduct a wide variety of statistical analyses based on the imputed datasets, although the analysts should not use a richer model than the model used for imputation (Schafer 1997; Little and Rubin 2002). Furthermore, multiple imputation methods are often more robust against the models used for imputation than the likelihood methods, especially when the missing rate is low. This is because the imputation model is only used to impute the missing values, while the observed data are not affected, so the impact may not be substantial if the missing rate is not high.

A multiple imputation method is closely related to the EM algorithm in that the EM algorithm "imputes" a missing value by a conditional mean in the E-step, while the multiple imputation method generates (or imputes) several possible values for a missing value from the imputation model, which is typically the predictive distribution of the missing data given the observed data (so it is a conditional distribution). Thus, computational approaches for multiple imputations are often similar to that for EM algorithms.

There may be many different approaches to create multiple imputations. However, one should check if the imputations are *proper*, since improper imputations lead to invalid inferences. Note that implementation of proper imputations can be computationally intensive. Proper imputations can be implemented in a Bayesian framework, although multiple imputation methods are quite general and are not necessary Bayesian.

Another main advantage of a multiple imputation method is that one can use standard complete-data methods to analyze the multiply imputed "complete datasets" using existing software. Moreover, methods to combine the complete-data analyses are often available.

4.5.2 Multiple Imputation for Mixed Effects Models with Missing Data

In this section, we consider multiple imputation methods for missing covariates in mixed effects models. For missing responses in mixed effects models, similar multiple imputation methods can be considered and the modification is straightforward. We focus on missing data in time-independent covariate z_i, with an ignorable missing data mechanism, following the same notation as in Section 4.2.1.

Let $f(y_i|z_i, b_i, \beta, \sigma)$ be the response model, $f(z_i|\alpha)$ be the covariate model, and $f(b_i|D)$ be the random effects model. Write $z_i = (z_{mis,i}, z_{obs,i})$. Let θ denote all parameters in the three models. We consider a Bayesian framework to generate proper multiple imputations. Specifically, let $f(\boldsymbol{\theta})$ be the prior distribution for the parameters in $\boldsymbol{\theta}$, and let

$$f(\boldsymbol{\theta}|\mathbf{y}_i, \mathbf{z}_i) \propto \int f(\boldsymbol{\theta}) f(\mathbf{y}_i|\mathbf{z}_i, \mathbf{b}_i, \boldsymbol{\beta}, \boldsymbol{\sigma}) f(\mathbf{z}_i|\boldsymbol{\alpha}) f(\mathbf{b}_i|D) \, d\mathbf{b}_i$$

be the posterior distribution. Multiple imputations for the missing data $\mathbf{z}_{mis,i}$ can be generated from the following posterior predictive distribution of the missing data given the observed data:

$$
\begin{aligned}
f(\mathbf{z}_{mis,i}|\mathbf{y}_i, \mathbf{z}_{obs,i}) &= \int\int f(\mathbf{x}_{mis,i}, \mathbf{b}_i|\mathbf{y}_i, \mathbf{z}_{obs,i}, \boldsymbol{\theta}) f(\boldsymbol{\theta}|\mathbf{y}_i, \mathbf{z}_{obs,i}) \, d\mathbf{b}_i \, d\boldsymbol{\theta} \\
&= \int\int f(\mathbf{z}_{mis,i}|\mathbf{y}_i, \mathbf{z}_{obs,i}, \boldsymbol{\theta}) f(\mathbf{b}_i|\mathbf{y}_i, \mathbf{z}_i, \boldsymbol{\theta}) \\
&\quad \times f(\boldsymbol{\theta}|\mathbf{y}_i, \mathbf{z}_{obs,i}) \, d\mathbf{b}_i \, d\boldsymbol{\theta}.
\end{aligned}
$$

To generate the desired imputations, we can use the *data augmentation* method (Tanner and Wong 1987). Specifically, we can *iteratively* sample from the conditional distributions $f(\mathbf{z}_{mis,i}|\mathbf{y}_i, \mathbf{z}_{obs,i}, \mathbf{b}_i, \boldsymbol{\theta})$, $f(\mathbf{b}_i|\mathbf{y}_i, \mathbf{z}_i, \boldsymbol{\theta})$, and $f(\boldsymbol{\theta}|\mathbf{y}_i, \mathbf{z}_i)$. Sampling from these conditional distributions can be done by rejection sampling methods, since we have

$$
\begin{aligned}
\mathbf{z}_{mis,i} &\sim f(\mathbf{z}_{mis,i}|\mathbf{y}_i, \mathbf{z}_{obs,i}, \mathbf{b}_i, \boldsymbol{\theta}) \propto f(\mathbf{y}_i|\mathbf{z}_i, \mathbf{b}_i, \boldsymbol{\theta}) f(\mathbf{z}_i|\boldsymbol{\theta}), \\
\mathbf{b}_i &\sim f(\mathbf{b}_i|\mathbf{y}_i, \mathbf{z}_i, \boldsymbol{\theta}) \propto f(\mathbf{y}_i|\mathbf{z}_i, \mathbf{b}_i, \boldsymbol{\theta}) f(\mathbf{b}_i|\boldsymbol{\theta}), \\
\boldsymbol{\theta} &\sim f(\boldsymbol{\theta}|\mathbf{y}_i, \mathbf{z}_i) \propto f(\boldsymbol{\theta}) f(\mathbf{y}_i|\mathbf{z}_i, \boldsymbol{\theta}).
\end{aligned}
$$

Iterating the above procedure, we obtain a sequence of simulated values for $(\mathbf{z}_{mis,i}, \mathbf{b}_i, \boldsymbol{\theta})$: $\{(\mathbf{z}_{mis,i}^{(k)}, \mathbf{b}_i^{(k)}, \boldsymbol{\theta}^{(k)}), k = 0, 1, 2, \cdots\}$, which is a Markov chain. After a burn-in period, the sequence will converge to the stationary distribution $f(\mathbf{z}_{mis,i}, \mathbf{b}_i, \boldsymbol{\theta}|\mathbf{y}_i, \mathbf{x}_{obs,i})$, so after burn-in we obtain a simulated value of $(\mathbf{z}_{mis,i}, \mathbf{b}_i, \boldsymbol{\theta})$ from the target distribution $f(\mathbf{z}_{mis,i}, \mathbf{b}_i, \boldsymbol{\theta}|\mathbf{y}_i, \mathbf{x}_{obs,i})$.

Let the last values in the sequence be $(\mathbf{z}_{mis,i}^*, \mathbf{b}_i^*, \boldsymbol{\theta}^*)$. Then, the value $\mathbf{z}_{mis,i}^*$ can be taken as a simulated value from the posterior predictive distribution $f(\mathbf{z}_{mis,i}|\mathbf{y}_i, \mathbf{z}_{obs,i})$, which is a marginal distribution of $f(\mathbf{z}_{mis,i}, \mathbf{b}_i, \boldsymbol{\theta}|\mathbf{y}_i, \mathbf{x}_{obs,i})$. Thus, $\mathbf{z}_{mis,i}^*$ can be used as a proper imputation of the missing data $\mathbf{z}_{mis,i}$. This imputation procedure is proper based on the definition of Rubin (1976).

Repeating the above procedure m times ($m > 1$), we create m proper multiple imputations, $\{\mathbf{z}_{mis,i}^{(1)}, \cdots, \mathbf{z}_{mis,i}^{(m)}\}$, for the missing values $\mathbf{z}_{mis,i}$. This process leads to m "complete datasets":

$$\{(\mathbf{y}, \mathbf{z}_{obs}, \mathbf{z}_{mis}^{(l)}), \, l = 1, 2, \cdots, m\}.$$

These m complete datasets can then be analyzed separately using any standard methods for complete data (these methods can be non-Bayesian), leading to m

sets of results. The m results are then combined for an overall inference, e.g., using the formulas given in Section 3.5.2 of Chapter 3 (page 118) to combine results.

4.5.3 Computational Issues and Other Methods

The computational methods for multiple imputation are similar to those for the Monte Carlo EM algorithm in Section 4.2, except that one has an additional layer of sampling from the posterior distribution $f(\theta|\mathbf{y}_i, \mathbf{z}_i)$. Thus the computational issues are similar to those for the Monte Carlo EM algorithms.

The data argumentation procedure described in the previous section can be computationally intensive. If the proportion of missing data is not too high, we may consider approximate methods to create multiple imputations. For example, we may use Taylor expansions to linearize a GLMM or NLME model, similar to that in Section 4.3, and then we generate multiple imputations from "working" LME models. This approximate method greatly reduces computational burden and still generate reasonable imputations for the missing data. Wu and Wu (2002b) considered a closely related approximate method where the multiple imputations were generated based on an approximate LME model.

Since the multiple imputation method described in the previous section is implemented via a Bayesian framework, one needs to choose prior distributions. In practice, we usually choose noninformative prior for θ, such as a uniform improper prior or a proper prior with noninformative hyperparameters. See Chapter 11 for choices of prior distributions for mixed effects models. Therefore, the results of multiple imputation methods are usually not very sensitive to the choice of priors.

There are other approaches to generate multiple imputations. For example, a simple approach would be to generate multiple imputations for $\mathbf{z}_{mis,i}$ from a predictive distribution where the unknown parameters and random effects are substituted by their corresponding estimates, i.e., generate an imputation as follows:

$$\mathbf{z}_{mis,i} \sim f(\mathbf{z}_{mis,i}|\mathbf{y}_i, \mathbf{z}_{obs,i}, \hat{\mathbf{b}}_i, \hat{\theta}) \propto f(\mathbf{y}_i|\mathbf{z}_i, \hat{\mathbf{b}}_i, \hat{\theta})f(\mathbf{z}_i|\hat{\theta}),$$

where $\hat{\mathbf{b}}_i$ and $\hat{\theta}$ are respectively the estimates of the random effects and parameters based on a simple method (say, the complete-case method). Note that, however, such a multiple imputation method is improper since the uncertainty in estimating the parameters and random effects are not incorporated in the generations of missing values.

One may also consider single imputation methods, e.g., impute $\mathbf{z}_{mis,i}$ by the mean $E(\mathbf{z}_{mis,i}|\mathbf{y}_i, \mathbf{z}_{obs,i}, \hat{\mathbf{b}}_i, \hat{\theta})$. However, in this case one should also adjust the standard errors of the resulting estimates to reflect the uncertainty of the missing data and the uncertainty in estimating the parameters and random effects, using (say) analytic methods or re-sampling methods.

4.6 Computational Strategies

For mixed effects models with missing data, a major problem with either the EM algorithms for likelihood inference or the multiple imputation methods is computational challenges, such as highly intensive computation and very slow or non-convergence. For Monte Carlo EM algorithms, computation in the E-steps typically is a combination of MCMC methods and rejection sampling methods, so they involve convergences of the MCMC methods within each EM iteration and the global convergence of the Monte Carlo EM algorithms. Although these computational intensive methods should work in theory, in practice many problems may arise. In this section, we discuss some common issues in more details.

We focus on the models and methods in Section 4.2.1 for illustration, i.e., we focus on missing time-independent covariates in NLME models with ignorable missing data mechanisms. Computational issues for other models and methods are often similar.

4.6.1 Sampling Methods

For Monte Carlo EM algorithms, a major step is to generate large numbers of Monte Carlo samples for the "missing data" in the E-step. Here we discuss several sampling methods to implement this, based on the models in Section 4.2.1 (page 133). We focus on rejection sampling methods, including the adaptive rejection sampling method and the multivariate rejection sampling method, and the importance sampling method. General ideas of these sampling methods can be found in Chapter 12.

For the Monte Carlo EM algorithm in Section 4.2.1, the E-step involves generating samples from the conditional distribution $f(\mathbf{z}_{mis,i}, \mathbf{b}_i | \mathbf{z}_{obs,i}, \mathbf{y}_i, \boldsymbol{\theta}^{(t)})$. One way to implement this, as described in Section 4.2.1, is to use the Gibbs sampler method by iteratively sampling from the lower dimensional full conditionals $f(\mathbf{z}_{mis,i} | \mathbf{z}_{obs,i}, \mathbf{y}_i, \mathbf{b}_i, \boldsymbol{\theta}^{(t)})$ and $f(\mathbf{b}_i | \mathbf{z}_{mis,i}, \mathbf{z}_{obs,i}, \mathbf{y}_i, \boldsymbol{\theta}^{(t)})$. This is a main advantage of the Gibbs sampler: We can break down the challenging problem of sampling from a high dimensional and intractable density into several more manageable problems of sampling from lower dimensional and easier-to-sample densities.

Sampling from the full conditionals can usually be accomplished by rejection sampling methods, since these full conditionals are proportional to products of known density functions. Specifically, if the density functions in the right-hand sides of (4.6) and (4.7) (page 137) are *log-concave* in appropriate parameters (i.e., the log-transformed density functions are concave in appropriate parameters), the *adaptive rejection algorithm* of Gilks and Wild (1992) may be used, as in Ibrahim et al. (1999). That is, if $f(\mathbf{z}_i|\boldsymbol{\theta}^{(t)})$ and $f(\mathbf{y}_i|\mathbf{z}_i, \mathbf{b}_i, \boldsymbol{\theta}^{(t)})$ in (4.6) are log-concave in each component of $\mathbf{z}_i$, and $f(\mathbf{b}_i|\boldsymbol{\theta}^{(t)})$ and $f(\mathbf{y}_i|\mathbf{z}_i, \mathbf{b}_i, \boldsymbol{\theta}^{(t)})$ in (4.7) are log-concave in each component of $\mathbf{b}_i$, we can use the adaptive rejection algorithm to sample from the full conditionals since the sum of log-concave densities is also log-concave. For the adaptive rejection algorithm, as long as the targeted density function is log-concave, we can generate desired samples even if the density function is very complicated (Gilks and Wild 1992).

The adaptive rejection sampling algorithm is widely used, especially in Gibbs sampling, but more general algorithms have also been proposed (Evans and Swartz 2000). For some density functions, the log-concavity may not be satisfied. Evans and Swartz (1998) extended Gilks and Wild's algorithm to more general cases. Evans and Swartz's method is an adaptive rejection method which does not require the log-concave restriction on the densities.

Alternatively, one may consider a *multivariate rejection sampling method* (Geweke 1996, section 3.2) to obtain desirable samples from the full conditionals in the Gibbs sampler, as in Booth and Hobert (1999). Specifically, suppose that we wish to generate samples from

$$f(\mathbf{b}_i|\mathbf{z}_{mis,i}, \mathbf{z}_{obs,i}, \mathbf{y}_i, \boldsymbol{\theta}^{(t)}) \propto f(\mathbf{b}_i|\boldsymbol{\theta}^{(t)}) f(\mathbf{y}_i|\mathbf{z}_i, \mathbf{b}_i, \boldsymbol{\theta}^{(t)}).$$

Let $f^*(\mathbf{b}_i) = f(\mathbf{y}_i|\mathbf{z}_i, \mathbf{b}_i, \boldsymbol{\theta}^{(t)})$ and $\tau = \sup_{\mathbf{u}}\{f^*(\mathbf{u})\}$. A random sample from $f(\mathbf{b}_i|\mathbf{z}_{mis,i}, \mathbf{z}_{obs,i}, \mathbf{y}_i, \boldsymbol{\theta}^{(t)})$ can then be obtained as follows by the multivariate rejection sampling method:

STEP 1. Sample $\mathbf{b}_i^*$ from $f(\mathbf{b}_i|\boldsymbol{\theta}^{(t)})$, and independently, sample w from the uniform(0,1) distribution;

STEP 2. If $w \leq f^*(\mathbf{b}_i^*)/\tau$, then accept $\mathbf{b}_i^*$, otherwise, go back to step 1.

Samples from $f(\mathbf{z}_{mis,i}|\mathbf{z}_{obs,i}, \mathbf{y}_i, \mathbf{b}_i, \boldsymbol{\theta}^{(t)})$ can be obtained in a similar way. Thus, the Gibbs sampler in conjunction with the multivariate rejection sampling method can also be used to obtain samples from $f(\mathbf{z}_{mis,i}, \mathbf{b}_i|\mathbf{z}_{obs,i}, \mathbf{y}_i, \boldsymbol{\theta}^{(t)})$. Note that sampling from $f(\mathbf{z}_i|\boldsymbol{\theta}^{(t)})$ and $f(\mathbf{b}_i|\boldsymbol{\theta}^{(t)})$ are often easy for many commonly assumed distributions such as a multivariate normal distribution, and the supremum τ can be found by standard optimization procedures. Booth and Hobert (1999) noted that in many cases the multivariate rejection sampling method can be fast even if the acceptance rate is low.

In the E-steps of Monte Carlo EM algorithms, we may also consider other sampling methods such as those based on Metropolis-Hastings algorithm (McCulloch 1997). However, some of these methods may generate *dependent* samples which may not lead to a straightforward assessment of Monte Carlo errors.

When the dimensions of the "missing data" $\mathbf{z}_{mis,i}$ or $\mathbf{b}_i$ are not small, the rejection sampling methods may be slow. In this case, we may use the *importance sampling methods* (see Chapter 12) to approximate the intractable integrals in the E-steps, where we may choose the importance functions to be a multivariate Student t density or a multivariate normal density whose mean and variance match the mode and curvature of $f(\mathbf{z}_{mis,i}, \mathbf{b}_i | \mathbf{z}_{obs,i}, \mathbf{y}_i, \boldsymbol{\theta}^{(t)})$.

Specifically, we may write

$$f(\mathbf{z}_{mis,i}, \mathbf{b}_i | \mathbf{z}_{obs,i}, \mathbf{y}_i, \boldsymbol{\theta}^{(t)}) = c \cdot \exp(q(\mathbf{z}_{mis,i}, \mathbf{b}_i)),$$

where c is the normalizing constant. Let $q^{(1)}(\mathbf{z}_{mis,i}, \mathbf{b}_i)$ and $q^{(2)}(\mathbf{z}_{mis,i}, \mathbf{b}_i)$ be the first and second derivatives of $q(\mathbf{z}_{mis,i}, \mathbf{b}_i)$ respectively. Let $(\widetilde{\mathbf{z}}_{mis,i}, \widetilde{\mathbf{b}}_i)$ be the solution of $q^{(1)}(\mathbf{z}_{mis,i}, \mathbf{b}_i) = 0$ which maximizes $q(\mathbf{z}_{mis,i}, \mathbf{b}_i)$. Then, the Laplace approximations of the mean and variance of $f(\mathbf{z}_{mis,i}, \mathbf{b}_i | \mathbf{z}_{obs,i}, \mathbf{y}_i, \boldsymbol{\theta}^{(t)})$ are $(\widetilde{\mathbf{z}}_{mis,i}, \widetilde{\mathbf{b}}_i)$ and $-(q^{(2)}(\widetilde{\mathbf{z}}_{mis,i}, \widetilde{\mathbf{b}}_i))^{-1}$ respectively. Suppose that

$$\left\{ (\widetilde{\mathbf{z}}_{mis,i}^{*(1)}, \widetilde{\mathbf{b}}_i^{*(1)}), \ldots, (\widetilde{\mathbf{z}}_{mis,i}^{*(m_t)}, \widetilde{\mathbf{b}}_i^{*(m_t)}) \right\}$$

is a random sample of size m_t generated from an importance function $h^*(\mathbf{z}_{mis,i}, \mathbf{b}_i)$, which is assumed to have the same support as $f(\mathbf{z}_{mis,i}, \mathbf{b}_i | \mathbf{z}_{obs,i}, \mathbf{y}_i, \mathbf{r}_i, \boldsymbol{\theta}^{(t)})$. Then we have the following approximation in the E-step

$$Q(\boldsymbol{\theta} | \boldsymbol{\theta}^{(t)}) \approx \sum_{i=1}^{n} \left\{ \frac{1}{m_t} \sum_{j=1}^{m_t} w_{ij}^{(t)} l_c \left(\boldsymbol{\theta}; \mathbf{y}_i, \mathbf{z}_{obs,i}, \widetilde{\mathbf{z}}_{mis,i}^{*(j)}, \widetilde{\mathbf{b}}_i^{*(j)} \right) \right\},$$

where

$$w_{ij}^{(t)} = \frac{f(\widetilde{\mathbf{z}}_{mis,i}^{*(j)}, \widetilde{\mathbf{b}}_i^{*(j)} | \mathbf{z}_{obs,i}, \mathbf{y}_i, \boldsymbol{\theta}^{(t)})}{h^*(\widetilde{\mathbf{z}}_{mis,i}^{*(j)}, \widetilde{\mathbf{b}}_i^{*(j)})}$$

are importance weights. Levine and Casella (2001) proposed an importance sampling scheme where we only need to generate random samples once.

For the foregoing sampling methods, the adaptive rejection method may be preferred when the appropriate densities are log-concave, while the multivariate rejection sampling method and the importance sampling method may be preferred in other cases. The adaptive and multivariate rejection sampling methods may be more efficient than the importance sampling methods when the dimensions of the missing data and random effects are not high and the sample size is not large. When the dimension of the integral in $Q(\boldsymbol{\theta} | \boldsymbol{\theta}^{(t)})$ in the E-step is high, however, rejection sampling methods can be inefficient, due to low acceptance rates. When the sample size is not small, the importance sampling methods may be more efficient than the rejection sampling methods, since in this case the importance function may closely resemble the target distribution.

For Monte Carlo EM algorithms, computation is often very intensive, and convergence problems such as very slow or non-convergence may arise. Therefore, computationally more efficient approximate methods based on Taylor or Laplace approximations are highly desirable in many problems.

4.6.2 Speed Up EM Algorithms

For many problems, the EM algorithm is known to be slow to converge. In particular, the Monte Carlo EM algorithms for mixed effects models with missing data can be very slow to converge, especially when the dimensions of the random effects and missing data are not small. In the literature, many EM-type algorithms have been proposed to speed up the convergence. These EM-type algorithms may substantially accelerate the EM algorithm while maintain the stability and simplicity of the standard EM algorithm (e.g., Meng and van Dyk 1997, 1998; Liu et al. 1998; van Dyk 2000). A popular idea of these procedures is to introduce working parameters in the models to reduce the fraction of missing information, and thus speed up the EM algorithm. In this section, we briefly describe one of these EM-type algorithms, called the *Parameter-Expansion (PX) EM algorithm* (Liu et al. 1998; van Dyk 2000).

The PX-EM algorithm introduces working parameters to the original model and then applies the standard EM algorithm to the expanded model rather than the original model. Liu et al. (1998) and van Dyk (2000) proposed PX-EM algorithms for complete-data LME models and showed that a PX-EM algorithm can be much faster than the standard EM algorithm, especially when the intra-individual variance is much larger than the random effects variability. Wu (2004) considered a Monte Carlo PX-EM algorithm for NLME models with missing data.

As an illustration, we consider a PX-EM algorithm for the approximate method described in Section 4.3.1 (page 148). Since the linearization method in Section 4.3.1 is equivalent to iteratively solving certain LME models, we can implement a PX-EM algorithm for these LME models within each iteration. Following Liu et al. (1998) and van Dyk (2000), the expanded model for LME model (4.20) (page 148) can be written as

$$\widetilde{\mathbf{y}}_i = X_i(\mathbf{z}_i)\boldsymbol{\beta} + T_i\,\Gamma\,\mathbf{b}_i + \mathbf{e}_i, \qquad i = 1, \ldots, n, \qquad (4.28)$$

where Γ is an $(s \times s)$ matrix containing working parameters. The implementation of the PX-EM algorithm is simple: the E-step is unchanged, while the

M-step is a simple modification of the original M-step by including a maximization over the working parameters Γ as well, which is straightforward for complete data. In other words, to implement the PX-EM algorithm, we only need to modify the original M-step by including the maximization over Γ.

Because the PX-EM algorithm is simply the standard EM algorithm applied to the expanded model (4.28), it retains the stability property of the standard EM algorithm. As argued in Meng and van Dyk (1997, 1998) and Liu et al. (1998), by introducing working parameters, the rate of convergence of an EM-type algorithm can be substantially improved while the stability property is retained.

Note that there can be various expansions for PX-EM implementations to an original model. In principle, we should expand the original model as much as possible in order to generate the fastest PX-EM algorithm (Liu et al. 1998; van Dyk 2000). For mixed effects models with sparse within-individual measurements, however, introducing too many working parameters may cause convergence problems. In such cases, we may want to reduce the number of working parameters if the cost of computation in the M-step outweigh the advantage of extra working parameters (Wu, 2004). When the intra-individual data are rich, on the other hand, we may try to expand the original model as much as possible, as long as the additional computation in the M-step is minimal.

Meng and van Dyk (1997, 1998) and van Dyk (2000) discussed other efficient EM-type methods for complete-data LME models based on alternative working parameters. One may also speed up the EM algorithms through alternative parameterizations of model parameters.

There are other ways to speed up the EM algorithm. For example, when the M-step is difficult to compute, we may consider the *Expectation-Conditional-Maximization (ECM) algorithm* (Meng and Rubin 1993). Note that an advantage of the EM algorithm is that its M-step involves only complete-data maximization. For many missing data problems, however, the M-step may be complicated since there may be too many parameters of different types. The complete-data maximization in the M-step becomes relatively simple if we condition on some parameters being estimated. This leads to the ECM algorithm.

Specifically, the idea of an ECM algorithm is to replace a complicated M-step of the EM algorithm with several computationally simpler conditional maximization (CM) steps, i.e., a sequence of constrained maximization steps, with each step being a maximization over a subset of parameters while holding other parameters fixed. The ECM algorithm shares the appealing properties of the standard EM algorithm (Meng and Rubin 1993).

4.6.3 Convergence

A major advantage of the EM algorithm is that the likelihood is guaranteed to increase (or non-decrease) at each EM iteration, so the EM algorithm will eventually converge to a (possibly local) maximum. For Monte Carlo EM algorithms, however, the likelihood is not guaranteed to increase at each EM iteration due to Monte Carlo errors in the E-step. However, under suitable regularity conditions, a Monte Carlo EM algorithm still converges to a maximum when the number of Monte Carlo samples increases as EM iteration increases (Chan and Ledolter 1995; Caffo, Jank, and Jones 2005).

When implementing the Monte Carlo E-step, it is necessary to choose the number m_t of Monte Carlo samples at the t-th EM iteration. Choosing a large value of m_t for all iterations may greatly increase computation burden. A good strategy is to increase m_t as the number t of EM iterations increases. Chan and Ledolter (1995) suggested to run Gibbs samplers to estimate the Monte Carlo variance based on several preliminary draws and then calculate the necessary value of m_t for a desired level of precision. It is obvious that larger values of m_t will result in more exact but slower computations. On the other hand, if the Monte Carlo error associated with $\hat{\theta}^{(t+1)}$ is not small, the $(t+1)$th iteration of an Monte Carlo EM algorithm is wasted because the EM step has been "swamped" by Monte Carlo error. Booth and Hobert (1999) proposed an automated method for choosing m_t in the context of complete-data GLM models. Their method can be extended as follows.

Let

$$Q^{(1)}(\theta|\hat{\theta}^{(t)}) = \frac{\partial Q(\theta|\hat{\theta}^{(t)})}{\partial \theta}, \qquad Q^{(2)}(\theta|\hat{\theta}^{(t)}) = \frac{\partial^2 Q(\theta|\hat{\theta}^{(t)})}{\partial \theta \partial \theta^T},$$

where $Q(\theta|\hat{\theta}^{(t)})$ is the quantity to be approximated in the E-step of a Monte Carlo EM algorithm. Let $\theta^{*(t+1)}$ be the solution of $Q^{(1)}(\theta|\hat{\theta}^{(t)}) = 0$. The conditional distribution of $[\theta^{(t+1)}|\theta^{(t)}]$ is approximately normal with mean $\theta^{*(t+1)}$ and a variance that can be estimated by

$$\widehat{\mathrm{Var}}\left(\theta^{(t+1)}|\theta^{(t)}\right) = Q^{(2)}(\theta^{*(t+1)}|\hat{\theta}^{(t)})^{-1}\widehat{\mathrm{Var}}\left(Q_i^{(1)}(\theta^{*(t+1)}|\hat{\theta}^{(t)})\right)$$
$$\times Q^{(2)}(\theta^{*(t+1)}|\hat{\theta}^{(t)})^{-1},$$

where

$$\widehat{\mathrm{Var}}\left(Q_i^{(1)}(\theta^{*(t+1)}|\hat{\theta}^{(t)})\right) = \frac{1}{m_t}\sum_{j=1}^{m_t}\left[w_{tj}\frac{\partial}{\partial \theta}\log f(\mathbf{y}, \mathbf{z}_{obs}, \tilde{\mathbf{z}}_{mis}^{(j)}, \tilde{\mathbf{b}}^{(j)}; \theta^{(t+1)})\right]$$
$$\times \left[w_{tj}\frac{\partial}{\partial \theta}\log f(\mathbf{y}, \mathbf{z}_{obs}, \tilde{\mathbf{z}}_{mis}^{(j)}, \tilde{\mathbf{b}}^{(j)}; \theta^{(t+1)})\right]^T,$$

and w_{tj} are importance weights which are all set equal to 1 when rejection sampling methods are used. Then, after the $(t + 1)$th EM iteration, we may construct an approximate $100(1 - \alpha)\%$ confidence ellipsoid for $\boldsymbol{\theta}^{*(t+1)}$ based on the above normal approximation. The EM step is swamped by Monte Carlo error if the previous value $\boldsymbol{\theta}^{(t)}$ lies in the confidence ellipsoid, and in that case we need to increase m_t, e.g., replace m_t by $m_t + m_t/k$ for some positive constant k. Note that this method of choosing m_t is completely automated.

To assess the convergence of the Gibbs sampler or Markov chains in the E-step, we may use standard graphical tools such as time-series plots and autocorrelations, and determine the burn-in or warm-up iterations in Gibbs sampler based on some preliminary draws (Gelman et al. 2003).

Assessing the convergence of a Monte Carlo EM algorithm sometimes may not be easy, due to Monte Carlo errors. We may, for example, claim that the Monte Carlo EM algorithm has converged if $||\boldsymbol{\theta}^{(t+1)} - \boldsymbol{\theta}^{(t)}|| < c$ for several consecutive EM iterations, where c is a pre-specified small positive value. One may also plot the likelihood at each EM iteration to check its behavior. It is also very important to choose good starting values, especially for nonlinear models.

4.7 Examples

Example 4.1 *A GLMM with missing data*

We consider the AIDS dataset in Section 1.3.2 of Chapter 1 to illustrate the methods presented in this chapter. In this study, viral load is repeatedly measured over time after the initiation of an anti-HIV treatment. Viral loads generally declined in the early period and some viral loads even dropped below the detection limit of 100, but most viral loads rebounded later in the study. It is of interest to explore if the events of viral loads dropping below the detection limit are associated with covariates. Previous studies showed that some baseline covariates such as the total complement levels (CH50), tumor necrosis factor (TNF), and CD4 cell counts may be associated with viral load trajectories.

Let y_{ij} be a binary response such that $y_{ij} = 1$ if the viral load for patient i is below the detection limit at time t_{ij} and $y_{ij} = 0$ otherwise. We follow Noh, Wu, and Lee (2009) and consider the following GLMM (a logistic regression model with random effects):

$$\text{logit}(P(y_{ij} = 1)) = (\beta_1 + b_{1i}) + (\beta_2 + b_{2i})z_{i1} + \beta_3 z_{i2} + \beta_4 z_{i3} + \beta_5 t_{ij}, \quad (4.29)$$

where $z_{i1} = CH50_i$, $z_{i2} = TNF_i$, and $z_{i3} = CD4_i$ are baseline covariate values for individual i, $\boldsymbol{\beta} = (\beta_1, \cdots, \beta_5)^T$ are fixed parameters, $\mathbf{b}_i = (b_{1i}, b_{2i})^T \sim N(0, D)$ are random effects, and D is an unrestricted covariance matrix, $i = 1, \cdots, n; \ j = 1, \cdots, n_i$.

In this study, the baseline values of CH50 (z_1), TNF (z_2), and CD4 (z_3) contain 19.0%, 16.7%, and 20.5% missing data respectively. To address these missing data in likelihood inference, we assume the following model for the covariates and write the joint covariate distribution as a product of three one-dimensional conditional distributions

$$f(z_{i1}, z_{i2}, z_{i3}|\alpha) = f(z_{i3}|z_{i1}, z_{i2}, \alpha_3)f(z_{i2}|z_{i1}, \alpha_2)f(z_{i1}|\alpha_1). \qquad (4.30)$$

For the three univariate conditional distributions, we may consider the following linear normal models, which appear to fit the observed covariate data reasonably well:

$$(z_{i3}|z_{i1}, z_{i2}, \alpha_3) \sim N(\alpha_{30} + \alpha_{31}z_{i1} + \alpha_{32}z_{i2}, \ \alpha_{33}), \qquad (4.31)$$

$$(z_{i2}|z_{i1}, \alpha_2) \sim N(\alpha_{20} + \alpha_{21}z_{i1}, \ \alpha_{22}), \quad (z_{i1}|\alpha) \sim N(\alpha_{10}, \ \alpha_{11}), \qquad (4.32)$$

where $\alpha_3 = (\alpha_{30}, \alpha_{31}, \alpha_{32}, \alpha_{33})$, $\alpha_2 = (\alpha_{20}, \alpha_{21}, \alpha_{22})$, and $\alpha_1 = (\alpha_{10}, \alpha_{11})$ are parameters for the covariate models, which are often viewed as nuisance parameters. Model selection methods such as AIC or BIC criteria may be used to reduce the number of nuisance parameters in the above linear normal models. Sensitivity analysis should be performed to check if the estimates of the main parameter β depend on the order of the covariates in factorizations (4.30). To avoid very small or very large estimates, which may be unstable, we standardize all the covariate values in the analysis.

It is possible that the missing data may be non-ignorable, although such a missing data mechanism is not testable based on the observed data. It is useful to assume a non-ignorable missing data model for sensitivity analysis. Let $r_{ik} = 1$ if z_{ik} is missing and $r_{ik} = 0$ if z_{ik} is observed, $k = 1, 2, 3$. We consider the following simple model for a non-ignorable missing data mechanism:

$$\log(P(r_{ik} = 1|z_{ik}, \phi)) = \phi_0 + \phi_1 z_{ik}, \qquad k = 1, 2, 3,$$

where $\phi = (\phi_0, \phi_1)^T$. We assume that r_{i1}, r_{i2}, r_{i3} are mutually independent for simplicity (to reduce the number of nuisance parameters).

The "complete-data" density functions (in log scale) in the above models are given by

$$\log f(\mathbf{y}_i|\mathbf{b}_i, \mathbf{z}_i, \beta) = \sum_{j=1}^{n_i} \{y_{ij} \log(P(y_{ij} = 1))$$

$$+ (1 - y_{ij}) \log(1 - P(y_{ij} = 1))\},$$

$$\log f(\mathbf{b}_i|D) = -\frac{1}{2} \log(2\pi D) - \frac{1}{2}\mathbf{b}_i^T D^{-1}\mathbf{b}_i,$$

$$\log f(\mathbf{z}_i|\alpha) = -\frac{1}{2} \log(2\pi\alpha_{33}) - \frac{1}{2\alpha_{33}}(z_{i3} - \alpha_{30} - \alpha_{31}z_{i1} - \alpha_{32}z_{i2})^2$$

$$-\frac{1}{2} \log(2\pi\alpha_{22}) - \frac{1}{2\alpha_{22}}(z_{i2} - \alpha_{20} - \alpha_{21}z_{i1})^2$$

$$-\frac{1}{2}\log(2\pi\alpha_{11}) - \frac{1}{2\alpha_{11}}(z_{i1} - \alpha_{10})^2,$$

$$\log f(\mathbf{r}_i|\mathbf{z}_i, \boldsymbol{\phi}) = \sum_{k=1}^{3}\{r_{ik}\log(P(y_{ij} = 1))$$

$$+(1 - r_{ik})\log(1 - P(y_{ij} = 1))\}.$$

These density functions are used in the computation.

In this example, the total dimension of the unobservables $(\mathbf{z}_{mis,i}, \mathbf{b}_i)$ is 8, so evaluation of the observed-data likelihood $L_o(\boldsymbol{\theta})$ using a numerical integration method or a Monte Carlo EM algorithm can be computationally intensive and may even offer convergence problems. Thus we consider an approximate method based on a Laplace approximation (i.e., the h-likelihood method described in Section 4.3.2), denoted by AP, which is computationally much more efficient. For comparison purpose, we also consider the complete-case (CC) method which discards all incomplete covariates and ignores the missing data and mechanism.

Table 4.1 *Parameter estimates for the GLMM (4.29) with missing covariates*

Parameter	CC method		AP method	
	Estimate	S.E.	Estimate	S.E.
β_1	−4.82	0.60	−5.03	0.55
β_2	0.46	0.53	0.25	0.39
β_3	−0.18	0.36	−0.14	0.27
β_4	0.36	0.31	0.49	0.29
β_5	0.03	0.005	0.03	0.004
d_{11}	1.57		1.97	
d_{12}	−0.81		−0.83	
d_{22}	0.97		1.17	

S.E. stands for standard error, and d_{ij} are elements of matrix D.

The resulting estimates are shown in Table 4.1 for parameters β and D (Noh, Wu, and Lee 2009). We see that the CC estimates and the AP estimates differ substantially, especially for the parameters $(\beta_2, \beta_3, \beta_4)$ which are associated with the missing covariates. This indicates the importance of addressing missing covariates in the model. Noh, Wu, and Lee (2009) conducted a simulation study and demonstrated that the AP method works well. The estimates based on the AP method show that the three baseline covariates do not appear significantly associated with viral loads dropping below the detection limit, but time is a significant predictor. The estimates of the diagonal elements of covariance matrix D indicate that there are large between-individual variations in the parameters and imply the need to include these random effects in the model.

Example 4.2 *A NLME model with missing data*

In AIDS studies, another interesting problem is to model HIV viral dynamics in the early period during an anti-HIV treatment, since the initial viral decay rate may reflect the efficacy of the treatment (Ding and Wu 2001). Based on previous studies (Wu and Ding 1999; Wu and Wu 2002), the following NLME model has been used to model HIV viral dynamics in the early period (e.g., first three months):

$$y_{ij} = \log_{10}(P_{1i}e^{-\lambda_{1ij}t_{ij}} + P_{2i}e^{-\lambda_{2i}t_{ij}}) + e_{ij}, \tag{4.33}$$

$$\log(P_{1i}) = \beta_1 + b_{1i}, \quad \lambda_{1ij} = \beta_2 + \beta_3 z_{1ij} + \beta_4 z_{2ij} + b_{2i}, \tag{4.34}$$

$$\log(P_{2i}) = \beta_5 + b_{3i}, \quad \lambda_{2i} = \beta_6 + b_{4i}, \tag{4.35}$$

where y_{ij} is the $\log_{10}$-transformation of the viral load measurement for the i-th patient at j-th measurement time t_{ij} (the log-transformation is used to make the data more normally distributed and to stabilize the variances), λ_{1ij} and λ_{2i} represent two-phase individual-specific viral decay rates, P_{1i} and P_{2i} are the individual-specific baseline values, β_j's are fixed effects, b_{ji}'s are random effects, and z_{1ij} (CD4) and z_{2ij} (CD8) are time-dependent covariates for the i-th patient at j-th measurement time t_{ij}, $i = 1, \cdots, n$; $j = 1, \cdots, n_i$. We assume that the within-individual errors e_{ij} i.i.d. $\sim N(0, \sigma^2)$ and the random effects $\mathbf{b}_i = (b_{1i}, b_{2i}, b_{3i}, b_{4i})^T \sim N(0, D)$, where D is unrestricted.

The covariates (z_{1ij}, z_{2ij}) in the model are time dependent and may be missing at the response measurement times. In order to address the missing covariates, we need to model the covariate processes. Note that

$$f(z_{1ij}, z_{2ij}|\boldsymbol{\alpha}) = f(z_{1ij}|z_{2ij}, \boldsymbol{\alpha})f(z_{2ij}|\boldsymbol{\alpha}). \tag{4.36}$$

Due to the large between-individual variations in these covariates, we consider the following LME models to empirically model the covariate processes and incorporate the between-individual variations:

$$\begin{aligned} z_{1ij} = {}& (\alpha_{10} + a_{i10}) + (\alpha_{11} + a_{i11})z_{2ij} + (\alpha_{12} + a_{i12})t_{ij} \\ & + (\alpha_{13} + a_{i13})t_{ij}^2 + \epsilon_{1i}, \end{aligned} \tag{4.37}$$

$$\begin{aligned} z_{2ij} = {}& (\alpha_{20} + a_{i20}) + (\alpha_{21} + a_{i21})t_{ij} \\ & + (\alpha_{22} + a_{i22})t_{ij}^2 + \epsilon_{2i}, \end{aligned} \tag{4.38}$$

where $\boldsymbol{\alpha} = (\alpha_{10}, \cdots, \alpha_{22})^T$ contains fixed parameters, $\mathbf{a}_i^{(1)} = (a_{i10}, \cdots, a_{i13})^T$ $\sim N(0, A_1)$ and $\mathbf{a}_i^{(2)} = (a_{i20}, \cdots, a_{i22})^T \sim N(0, A_2)$ are random effects, and $\epsilon_{1i} \sim N(0, \alpha_{14})$ and $\epsilon_{2i} \sim N(0, \alpha_{23})$ are within-individual covariate measurement errors.

The "complete-data" density functions in all the models are given by

$$
\log f(\mathbf{y}_i|\mathbf{z}_i, \boldsymbol{\beta}, \sigma^2, \mathbf{b}_i) = \sum_{j=1}^{n_i} \Bigg\{ -\frac{1}{2}\log(2\pi\sigma^2)
$$
$$
-\frac{1}{2\sigma^2}\big[y_{ij} - \log_{10}(P_{1i}e^{-\lambda_{1ij}t_{ij}}
$$
$$
+P_{2i}e^{-\lambda_{2i}t_{ij}})\big]^2 \Bigg\},
$$

$$
\log f(\mathbf{b}_i|D) = -\frac{1}{2}\log(2\pi D) - \frac{1}{2}\mathbf{b}_i^T D^{-1}\mathbf{b}_i,
$$

$$
\log f(\mathbf{z}_i|\mathbf{a}_i, \boldsymbol{\alpha}) = \sum_{j=1}^{n_i} \Bigg\{ -\frac{1}{2}\log(2\pi\alpha_{14})
$$
$$
-\frac{1}{2\alpha_{14}}\big[z_{1ij} - (\alpha_{10} + a_{i10}) - (\alpha_{11} + a_{i11})z_{2ij}
$$
$$
-(\alpha_{12} + a_{i12})t_{ij} - (\alpha_{13} + a_{i13})t_{ij}^2\big]^2
$$
$$
-\frac{\log(2\pi\alpha_{23})}{2} - \frac{1}{2\alpha_{23}}\big[z_{i2} - (\alpha_{20} + a_{i20})
$$
$$
-(\alpha_{21} + a_{i21})t_{ij} - (\alpha_{22} + a_{i22})t_{ij}^2\big]^2 \Bigg\},
$$

$$
\log f(\mathbf{a}_i|A) = -\frac{1}{2}\log(2\pi A_1) - \frac{1}{2}\mathbf{a}_i^{(1)T} A_1^{-1}\mathbf{a}_i^{(1)}
$$
$$
-\frac{1}{2}\log(2\pi A_2) - \frac{1}{2}\mathbf{a}_i^{(2)T} A_1^{-1}\mathbf{a}_i^{(2)}.
$$

These density functions are used in the computation.

In this example, the total dimension of the unobservables $(\mathbf{a}_i^{(1)}, \mathbf{a}_i^{(2)}, \mathbf{b}_i)$ is 11, which is very high. Thus, evaluation of the observed-data likelihood $L_o(\boldsymbol{\theta})$ using numerical or Monte Carlo methods can be computationally very challenging! Therefore, we again follow Noh, Wu, and Lee (2009) and use the approximate method (AP) for likelihood estimation (see Section 4.3.2), and compare it with the complete-case (CC) method, which ignores all missing data.

The resulting estimates are given in Table 4.2 for the main parameters $\boldsymbol{\beta}$. For sensitivity analysis, we consider the two possible orders of the covariates in factorization (4.36). We see that the CC method and the AP method give different results. The results based on the AP method should be more reliable, as confirmed in a simulation study in Noh, Wu, and Lee (2009). Note that the results are not sensitive to the order of the covariates in factorizations (4.36). Based on the AP method (order 1), we see that the first viral decay rate λ_{1ij} may change over time and may be significantly associated with the time-varying CD4 val-

Table 4.2 *Parameter estimates of the NLME model with missing data*

Parameter	CC method Estimate	CC method S.E.	AP method (order1) Estimate	AP method (order1) S.E.	AP method (order2) Estimate	AP method (order2) S.E.
β_1	12.48	0.20	13.71	0.19	13.75	0.20
β_2	0.48	0.020	0.75	0.020	0.74	0.020
β_3	0.047	0.016	0.058	0.012	0.062	0.014
β_4	−0.023	0.014	−0.015	0.010	−0.013	0.009
β_5	7.91	0.26	8.35	0.25	8.31	0.25
β_6	0.039	0.003	0.047	0.003	0.045	0.003

S.E.: standard error. Order1: $f(z_1, z_2) = f(z_1|z_2)f(z_2)$. *Order2:* $f(z_1, z_2) = f(z_2|z_1)f(z_1)$.

ues (estimate of β_3 is significant at 5% level) but may not be associated with the time-varying CD8 values (estimate of β_4 is not significant).

Example 4.3 *A Multiple Imputation Method*

Consider the study on mental distress described in Section 1.3.1 of Chapter 1 (page 6). In this study, nearly all time-dependent variables have missing data. Multiple imputation methods can be used to impute the missing data, leading to several complete datasets. These complete datasets can be analyzed in various ways using standard complete-data methods, and the results can then be combined for an overall inference. Note that the models for data analyses do not have to be the same as the models for creating multiple imputations, and the observed data are not affected by multiple imputations. See Schafer (1997) and Little and Ruben (2002) for detailed discussions on the choices of imputation models and analysis models.

As an illustration, here we focus on the variable "depression". To generate multiple imputations for missing data in depression scores, we assume the following LME model for generating multiple imputations

$$y_{ij} = \beta_1 + b_{1i} + (\beta_2 + b_{2i})t_{ij} + e_{ij}, \qquad (4.39)$$
$$(b_{1i}, b_{2i})^T \sim N(0, D), \qquad e_{ij} \text{ i.i.d. } \sim N(0, \sigma^2), \qquad (4.40)$$

where y_{ij} is the depression score for individual i at measurement time j, $i = 1, 2, \cdots, n$; $j = 1, \cdots, J$. In the following, we use model (4.39) to create multiple imputations for the missing data in y_{ij}'s, following the approach described in Schafer and Yucel (2002). For simplicity, we assume ignorable missing data, and we let the analysis model to be the same as the imputation model.

Table 4.3 *Parameter estimates based on the multiple imputation method and the CC method*

Method	Parameter	Estimate	S.E.	Parameter	Estimate	S.E.
CC Method	β_1	1.40	0.052	β_2	–0.39	0.053
MI Method	β_1	1.41	0.048	β_2	–0.45	0.071

S.E.: standard error

We consider a Bayesian framework to create proper multiple imputations, as described in Section 4.5 (page 159). This leads to m "complete datasets". We fit model (4.39) to each complete dataset and obtain parameter estimates, and then we combine the m sets of estimates using the formulas in Section 3.5.2 (page 118) to obtain overall estimates.

Table 2.1 lists the overall estimates for the parameters in model (4.39) based on $m = 3$ imputations, and compares the results based on the complete-case (CC) method. Table 2.1 shows that the estimates based on the multiple imputation method differ from the estimates based on the CC method, especially for the parameter β_2 which is associated with time. This indicates the importance of addressing missing data. Note that the standard error for $\hat{\beta}_2$ is larger based on the MI method than that based on the CC method, possibly reflecting the uncertainty of missing data. Figure 4.2 shows some diagnostic plots for the Gibbs sampling used to create proper multiple imputations. We see that the Markov chain for sampling β_2 converges faster than that for β_1. We took 500 runs as a burn-in period.

We may consider generating multiple imputations for missing data in other variables *separately* based on models similar to model (4.39). A better multiple imputation model, however, is to incorporate the correlations between the variables and generate multiple imputations using a *multivariate* LME model (see (8.24) in Section 8.6.1 in Chapter 8, page 282) to impute all missing data in all the variables *simultaneously*. See Schafer and Yucel (2002) for such an approach. Once each missing value is imputed by several values based on a multiple imputation method, one obtains several complete datasets. These complete datasets can then be analyzed in various ways using standard complete-data methods. For example, some variables may be used as covariates while other variables may be used as responses, although when generating multiple imputations all variables may be treated as responses in the multivariate imputation model. Standard regression methods can then be used to analyzed each of the "complete datasets" separately, and the results can then be combined.

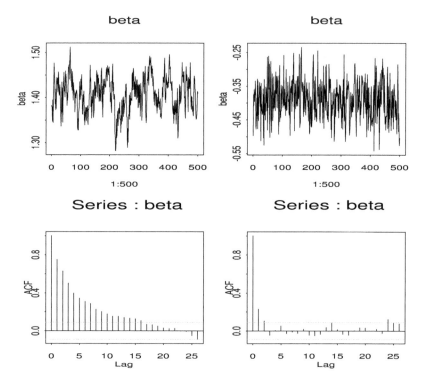

Figure 4.2 *Diagnostic plots for the Gibbs sampling used in the multiple imputation method. The figures in the left column shows time series and autocorrelation function (ACF) plots for sampling β_1, and the figures in the right column shows time series and autocorrelation plots for sampling β_2.*

CHAPTER 5

Mixed Effects Models with Measurement Errors

5.1 Introduction

In Section 3.6 of Chapter 3, we briefly described covariate measurement error problems in regression models and briefly reviewed several general approaches to address measurement errors. It is known that the naive method which ignores covariate measurement errors in regression models may produce biased and misleading results. To visually illustrate this, Figure 5.1 shows how the estimates of regression coefficients may be biased if measurement errors in a covariate is not addressed, based on an artificial dataset generated from the simple linear regression $y = \beta_0 + \beta_1 x + e$. To formally address measurement errors, we typically need to have validation data or replication data (Carroll et al. 2006). In practice, these validation data or replication data may not be available. However, for longitudinal studies the repeated measurements within each individual may be viewed as replication data, which allows us to partially address measurement errors in time-dependent covariates.

To address covariate measurement errors, we often assume an error structure or a measurement error model. Let z_{ij} be the observed covariate value for individual i at time t_{ij} (or replication j), with possible measurement errors, and let z_{ij}^* be the corresponding unobserved true covariate value. As discussed in Section 3.6 of Chapter 3, the following two measurement error models are commonly used (Carroll et al. 2006): the *classical measurement error model* assumes that

$$z_{ij} = z_{ij}^* + e_{ij}, \qquad E(e_{ij}|z_{ij}^*) = 0, \tag{5.1}$$

and the *Berkson measurement error model* (Berkson 1950) assumes that

$$z_{ij}^* = z_{ij} + e_{ij}, \qquad E(e_{ij}|z_{ij}) = 0, \tag{5.2}$$

where e_{ij} is the measurement error for individual i at time t_{ij} (or replication j).

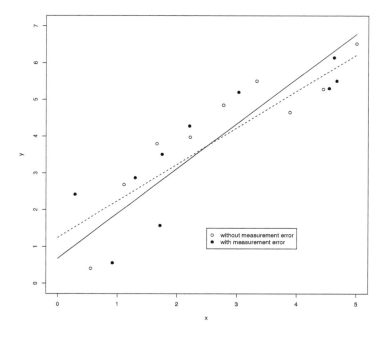

Figure 5.1 *The effect of covariate measurement error on fitted least square lines for simple linear regression model* $y = \beta_0 + \beta_1 x + e$. *The solid line is the least square line for x without measurement error, while the dotted line is the least square line for x with measurement error.*

In the classical measurement error model (5.1), the variability of the observed covariates z_{ij} is larger than the variability of the true covariates z_{ij}^*, i.e., var$(z_{ij}) \geq$ var(z_{ij}^*), while in the Berkson measurement error model (5.2), the variability of the observed covariates z_{ij} is less than the variability of the true covariates z_{ij}^*, i.e., var$(z_{ij}) \leq$ var(z_{ij}^*). This may help us to decide which measurement error model to use in practice. Carroll et al. (2006) provided the following suggestions: if an error-prone covariate is measured uniquely to an individual, such as blood pressure measurements, we can use a classical measurement error model; on the other hand, if all individuals in a group are given the same value of the error-prone covariate but the true covariate value is particular to an individual, such as the true exposure of an individual to air pollution in a city of people exposed to the same level of air pollution, we can use a Berkson measurement error model.

Once a measurement error model is chosen, there are two modeling approaches: (i) *functional modeling* approach, which makes no or minimal distributional assumptions for the unobserved true covariates; (ii) *structural modeling* approach, which assumes a model or a distribution for the unobserved true covariates. Functional modeling methods include regression calibration methods and simulation extrapolation (SIMEX) methods. They are usually robust to mis-specification of the true covariate distribution. Structural modeling methods include likelihood methods and Bayesian methods. They are usually more efficient if the assumed covariate model is correctly specified. For a time-dependent covariate in a longitudinal model, an assumed covariate model can be verified based on the observed data, so the structural modeling approach is often a good choice.

In this chapter, we consider covariate measurement error problems in mixed effects models. Since the likelihood method is the standard inferential approach for mixed effects models, we focus on likelihood methods to address measurement errors, assuming classical measurement error models. We also assume that the observed covariate is a *surrogate* of the unobserved true covariate, i.e., the measurement error is assumed to be *nondifferential* (see Section 3.6.1, page 123). In other words, given the true covariate z_{ij}^* and other covariates, the observed covariate z_{ij} is assumed to be independent of the response in the regression model of interest.

5.2 Measurement Error Models and Methods

In this section, we discuss commonly used measurement error models and methods in the context of mixed effects models for longitudinal data. There is an extensive literature in measurement error models and methods. Here we focus on the essential ideas and some of the commonly used ones.

5.2.1 Measurement Error Models

We first consider classical measurement error models for mis-measured covariates in mixed effects models, with a focus on likelihood methods. For likelihood methods or structural modeling methods in general, one typically assumes a model for the unobserved true covariates z_{ij}^*. In the following, we consider several choices for modeling the true covariates.

For simplicity of presentation, we consider a single time-dependent continuous covariate z_{ij}, with possible measurement errors. Extensions to more than one covariates are straightforward. Let y_{ij} be the response value for the i-th individual at time t_{ij}, $i = 1, \ldots, n$, $j = 1, \ldots, n_i$. Let z_{il} be the observed covariate value and z_{il}^* be the corresponding unobserved "true" covariate value for the i-th individual at time u_{il}, $i = 1, \ldots, n$, $l = 1, \ldots, m_i$. Note that we allow the covariate measurement times u_{il} to possibly differ from the response measurement times t_{ij}, so we use different notations for the covariate and the response measurement times (see Figure 4.1 on page 142 for an example). In other words, we allow missing data in the covariates, assuming MAR missing mechanism. Such situations arise frequently in longitudinal studies since many covariate values may not be available at the response measurement times. Let $\mathbf{y}_i = (y_{i1}, \ldots, y_{in_i})^T$ and $\mathbf{z}_i = (z_{i1}, \ldots, z_{im_i})^T$. We suppress the accurately measured covariates throughout since no distributional assumptions are needed for these covariates.

We write the response mixed effects model as $f(\mathbf{y}_i|\mathbf{z}_i, \mathbf{b}_i, \boldsymbol{\beta}, \boldsymbol{\sigma})$, given random effects $\mathbf{b}_i$, where $\boldsymbol{\beta}$ contains mean parameters and $\boldsymbol{\sigma}$ contains variance-covariance parameters (for some GLMMs, parameters $\boldsymbol{\sigma}$ may overlap with parameters $\boldsymbol{\beta}$, see Chapter 2). We assume $\mathbf{b}_i \sim N(0, B)$, where B is a unstructured covariance matrix. When covariate $\mathbf{z}_i$ is measured with error, we assume that the response depends on the unobserved true covariate value $\mathbf{z}_i^*$ rather than the observed but mis-measured covariate value $\mathbf{z}_i$, i.e., we assume a response model $f(\mathbf{y}_i|\mathbf{z}_i^*, \mathbf{b}_i, \boldsymbol{\beta}, \boldsymbol{\sigma})$. We also assume that $\mathbf{z}_i$ is a surrogate of $\mathbf{z}_i^*$ or that the measurement error is nondifferential, so

$$f(\mathbf{y}_i|\mathbf{z}_i, \mathbf{z}_i^*, \mathbf{b}_i, \boldsymbol{\beta}, \boldsymbol{\sigma}) = f(\mathbf{y}_i|\mathbf{z}_i^*, \mathbf{b}_i, \boldsymbol{\beta}, \boldsymbol{\sigma}).$$

A classical measurement error model can be written as

$$z_{il} = z_{il}^* + \epsilon_{il}, \qquad E(\epsilon_{il}|z_{il}^*) = 0, \qquad i = 1, \ldots, n, \ l = 1, \ldots, m_i, \quad (5.3)$$

where the repeated measurements $z_{i1}, \cdots, z_{im_i}$ may be viewed as replicates. For a time-independent covariate, we may write $z_{il} = z_i^* + \epsilon_{il}$ where z_{il}'s are true replicates of z_i^*. Although in this chapter we focus on time-dependent covariates, most of the ideas and methods can be applied to time-independent covariates, if replicates are available.

To model the true time-dependent covariate z_{ij}^*, we may consider a mixed effects model to incorporate between-individual variation and within-individual correlation, assuming the covariate values change smoothly over time. For example, we may consider the following LME model to address measurement errors

$$\mathbf{z}_i = U_i\boldsymbol{\alpha} + V_i\mathbf{a}_i + \boldsymbol{\epsilon}_i \equiv \mathbf{z}_i^* + \boldsymbol{\epsilon}_i, \qquad (5.4)$$

where U_i and V_i are known design matrices, $\boldsymbol{\alpha}$ contains unknown fixed parameters, $\mathbf{a}_i$ are random effects, and $\boldsymbol{\epsilon}_i = (\epsilon_{i1}, \cdots, \epsilon_{im_i})$ represent covariate measurement errors. We assume that $\mathbf{a}_i$ i.i.d. $\sim N(\mathbf{0}, A)$, $\boldsymbol{\epsilon}_i$ i.i.d. $\sim N(\mathbf{0}, R_i)$, and $\mathbf{a}_i$ and $\boldsymbol{\epsilon}_i$ are independent, where A and R_i are unknown covariance matrices. We often choose $R_i = \delta^2 I_{m_i}$, i.e., the within-individual covariate measurements are assumed to be conditionally independent given the random effects.

This assumption may be reasonable when the within-individual repeated measurements are far apart, and it reduces the number of nuisance parameters. The value of δ^2 reflects the magnitude of the measurement error. We further assume that ϵ_i and $\mathbf{a}_i$ are independent of $\mathbf{e}_i$ and $\mathbf{b}_i$, where $\mathbf{e}_i$ s are the within individual random errors in the response model.

In the measurement error model (5.4), the unobserved true covariate $\mathbf{z}_i^*$ is modeled as

$$\mathbf{z}_i^* = U_i\boldsymbol{\alpha} + V_i\mathbf{a}_i. \tag{5.5}$$

Given the observed covariate data $\{\mathbf{z}_i, i = 1, \cdots, n\}$, the assumed measurement error model (5.4) can be fitted using standard methods for LME models. Therefore, one can obtain parameter estimate $\hat{\boldsymbol{\alpha}}$ and the empirical Bayes estimates $\hat{\mathbf{a}}_i$, and obtain an estimate of the unobserved true covariate

$$\hat{\mathbf{z}}_i^* = U_i\hat{\boldsymbol{\alpha}} + V_i\hat{\mathbf{a}}_i,$$

as well as the associated standard errors.

Example 5.1 *A measurement error model for CD4 count*

In AIDS studies, CD4 counts are known be measured with errors. To address measurement errors in CD4 counts, we may treat the repeated measurements of CD4 over time as "replicates" and fit an empirical mixed effects model to the observed CD4 data. For the AIDS dataset described in Chapter 1, we consider the following quadratic polynomial LME model based on AIC/BIC model selection criteria:

$$z_{ij} = \alpha_0 + a_{0i} + (\alpha_1 + a_{1i})t_{ij} + (\alpha_2 + a_{2i})t_{ij}^2 + \epsilon_{ij} = z_{ij}^* + \epsilon_{ij}, \tag{5.6}$$

where z_{ij} is the observed CD4 value for individual i at time t_{ij}, $\boldsymbol{\alpha} = (\alpha_0, \alpha_0, \alpha_0)^T$ are fixed parameters, $\mathbf{a}_i = (a_{0i}, a_{1i}, a_{2i})^T$ are random effects, z_{ij}^* is the true but unobserved CD4 value, and ϵ_{ij} is the measurement error. We assume that $\mathbf{a}_i$ are i.i.d. $\sim N(0, A)$ and ϵ_{ij} s are i.i.d. $\sim N(0, \delta^2)$.

Figure 5.2 shows the fitted CD4 curves and the observed CD4 data for four randomly selected subjects. We can see that the LME model (5.6) fits the observed CD4 data reasonably well. If we assume that CD4 counts change smoothly over time, the LME model can be used to partially address measurement errors in CD4 counts, so we may assume that the estimated true CD4 value for subject i at time t_{ij} is $\hat{z}_{ij}^* = \hat{\alpha}_0 + \hat{a}_{0i} + (\hat{\alpha}_1 + \hat{a}_{1i})t_{ij} + (\hat{\alpha}_2 + \hat{a}_{2i})t_{ij}^2$. More complicated measurement error models can also be considered, as discussed below.

Measurement error model (5.4) is often an empirical model attempting to capture the main features of the true covariate trajectory. It includes polynomial regression models with individual-specific regression parameters. It can also

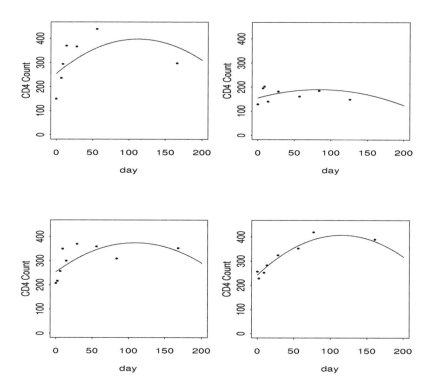

Figure 5.2 *A LME measurement error model for CD4 counts: Observed data versus fitted curves for 4 randomly selected subjects.*

be used to approximate semiparametric or nonparametric mixed effects models (see Section 2.5 of Chapter 2). Therefore, model (5.4) is very flexible for modeling complex longitudinal processes. However, model (5.4) assumes that the random effects in the model do not change over time, i.e., it assumes that the between-individual variations remain constant over time, which may be restrictive in some cases in practice, so more flexible models sometimes may be needed.

To allow the longitudinal trend to vary over time, or to allow a "wiggly" empirical representation of within-individual trajectory, Taylor, Cumberland and Sy (1994), Wang and Taylor (2001), and Xu and Zeger (2001) considered the following measurement error model

$$\mathbf{z}_i^* = U_i \boldsymbol{\alpha} + V_i \mathbf{a}_i + \mathbf{w}_i, \tag{5.7}$$

where U_i and V_i are known design matrices, $\boldsymbol{\alpha}$ contains fixed parameters, $\mathbf{a}_i$ contains random effects, $\mathbf{w}_i = (w_{i1}, \cdots, w_{im_i})^T$, $w_{il} = w_i(t_{il})$, and $w_i(t)$ is a mean-zero stochastic process such as an *integrated Ornstein-Uhlenbeck (IOU) process* or a stationary Gaussian process. Model (5.7) allows the longitudinal trend to vary with time and induces a within-subject autocorrelation structure that may be thought of as arising from evolving fluctuations in the covariate process. Note that a high-order polynomial LME model (5.4) may be similar to model (5.7) but is computationally more manageable.

To allow for missing data (ignorable) in the time-varying covariates, or to allow for different measurement schedules for the covariates and the response processes, we recast model (5.4) in continuous time, assuming covariate values change smoothly over time,

$$z_i(t) = \mathbf{u}_i^T(t)\boldsymbol{\alpha} + \mathbf{v}_i^T(t)\mathbf{a}_i + \epsilon_i(t), \qquad i = 1, \ldots, n,$$

where $z_i(t)$, $\mathbf{u}_i(t)$, $\mathbf{v}_i(t)$, and $\epsilon_i(t)$ are the observed covariate value, design vectors, and measurement error for individual i at time t respectively. At the response measurement time t_{ij}, which may be different from the covariate measurement times u_{il}, the unobserved true covariate value can be viewed as $z_{ij}^* = \mathbf{u}_{ij}^T\boldsymbol{\alpha} + \mathbf{v}_{ij}^T\mathbf{a}_i$, where $\mathbf{u}_{ij} = \mathbf{u}_i(t_{ij})$ and $\mathbf{v}_{ij} = \mathbf{v}_i(t_{ij})$.

The Berkson measurement error model (5.2) corresponds to regression calibration methods, in which the unobserved true covariate $\mathbf{z}_i^*$ in the response model is substituted by an estimate based on a regression of $\mathbf{z}_i^*$ on $\mathbf{z}_i$ (see Section 5.3.2). Berkson measurement error models can be useful in some applications. For example, suppose that all people in a small city are exposed to the same level of air population, measured by monitoring stations in the city, but the true (actual) individual exposures may vary around the observed values. In this case, a Berkson measurement error model may be a good choice to address the measurement error.

Wang (2004) proposed a second-order least square method, which requires no distributional assumptions, for nonlinear models with covariate measurement errors based on a Berkson model. The main idea is as follows. Let

$$\rho_i(\boldsymbol{\theta}) = \big\{ (y_{ij} - E(y_{ij}|\mathbf{z}_i, \boldsymbol{\theta}),\ y_{ij}y_{ik} - E(y_{ij}y_{ik}|\mathbf{z}_i, \boldsymbol{\theta})),$$
$$i = 1, \cdots, n;\ 1 \le j \le k \le n_i \big\}^T.$$

The second-order least square method finds $\boldsymbol{\theta}$ to minimize

$$R(\boldsymbol{\theta}) = \sum_{i=1}^n \rho_i(\boldsymbol{\theta})^T W_i(\mathbf{z}_i)\rho_i(\boldsymbol{\theta}),$$

where $W_i(\mathbf{z}_i)$ is a weighting matrix such as the inverse of a covariance matrix. The method involves evaluations of $E(\mathbf{z}_i^*|\mathbf{z}_i, \boldsymbol{\theta})$ and the first two moments, so a Berkson measurement error model fits in the framework. When the first two moments are difficult to evaluate, Monte Carlo methods such as the importance sampling method can be used for approximating these moments. Wang (2004) established the consistency and asymptotic normality of the resulting estimators.

For a regression model with covariate measurement errors, it is often assumed that measurement errors are either classical or Berkson. In practice, sometimes the covariates may be contaminated by a mixture of the two errors. Carroll, Delaigle, and Hall (2007) considered such problems and proposed a non-parametric estimator of a regression function.

5.2.2 Measurement Error Methods

Comprehensive reviews of covariate measurement error models and methods are provided in Fuller (1987), Gustafson (2004), and Carroll et al. (2006). Commonly used methods for covariate measurement errors in regression models include

- regression calibration methods,
- simulation extrapolation (SIMEX) methods,
- likelihood methods,
- Bayesian methods,

among others. In the above four approaches, regression calibration methods and SIMEX methods make minimal assumptions on the distributions of the unobserved true covariates, so they are more robust to the misspecifications of the covariate distributions. Likelihood methods and Bayesian methods make strong distributional assumptions on the unobserved true covariates, so they are more efficient if the covariate distributions are correctly specified. In the following, we discuss regression calibration methods and likelihood methods in some details, due to their popularity, and briefly mention the other methods.

Regression calibration methods, which include Berkson error models, attempt to model the distribution of the unobserved true covariate z^* given the observed covariate z and other observed covariates x, and then replace the unobserved true covariates z^* by an estimated value of $E(z^*|z, x)$, the regression of z^* on (z, x). After z^* is approximated by an estimate of $E(z^*|z, x)$, one then performs a standard analysis as if there were no covariate measurement error. Carroll et al. (2006) provided a detailed discussion of the methods.

Specifically, consider a mixed effects model $f(y_i|z_i, x_i, b_i, \beta, \sigma)$, where b_i is a vector of random effects, covariates z_i are measured with errors, and covariates x_i are measured without errors. Let z_i^* be the unobserved true covariate value corresponding to the observed value $\mathbf{z}_i$. To address measurement errors in covariates $\mathbf{z}_i$, a regression calibration method proceeds as follows:

STEP 1. Assume a regression model of $\mathbf{z}_i^*$ on $(\mathbf{z}_i, \mathbf{x}_i)$, say $E(\mathbf{z}_i^*|\mathbf{z}_i, \mathbf{x}_i, \boldsymbol{\alpha})$ with $\boldsymbol{\alpha}$ being the unknown regression parameters, and obtain parameter estimate $\hat{\boldsymbol{\alpha}}$ using standard estimation methods for regression models;

STEP 2. Replace $\mathbf{z}_i^*$ in the response model by its estimate $E(\mathbf{z}_i^*|\mathbf{z}_i, \mathbf{x}_i, \hat{\boldsymbol{\alpha}})$, and then perform a standard analysis on the approximate response model

$$f(\mathbf{y}_i \mid E(\mathbf{z}_i^*|\mathbf{z}_i, \mathbf{x}_i, \hat{\boldsymbol{\alpha}}), \mathbf{b}_i, \boldsymbol{\beta}, \boldsymbol{\sigma}) \approx f(\mathbf{y}_i|\mathbf{z}_i^*, \mathbf{b}_i, \boldsymbol{\beta}, \boldsymbol{\sigma});$$

STEP 3. Adjust the standard errors of the parameter estimates in the response model to reflect the *uncertainty* in the estimation of the covariate model in Step 1, using methods such as the bootstrap methods or sandwich methods.

The regression calibration method may be applicable to almost any regression models with covariate measurement errors. It is also closely related to the so-called *two-step approach* (see Section 5.3). A naive two-step approach is similar to the above three-step regression calibration method but without step 3, so it may under-estimate standard errors of parameter estimates in the response model, similar to a naive single-imputation method for missing data. Section 5.3.2 presents an application of regression calibration methods to NLME models with covariate measurement errors. Note that the Berkson error model is the most famous example of a regression calibration method.

Likelihood methods for covariate measurement errors are based on the observed-data *joint* likelihood of the response model and the covariate model. Let $\boldsymbol{\theta}_1$ denote all parameters in the response model and $\boldsymbol{\theta}_2$ denote all parameters in the covariate model. The joint likelihood for all observed data can be written as

$$L(\boldsymbol{\theta}_1, \boldsymbol{\theta}_2) = \prod_{i=1}^{n} f(\mathbf{y}_i|\mathbf{z}_i^*, \boldsymbol{\theta}_1) f(\mathbf{z}_i; \boldsymbol{\theta}_2). \tag{5.8}$$

MLE of $(\boldsymbol{\theta}_1, \boldsymbol{\theta}_2)$ can be obtained by maximizing the joint likelihood $L(\boldsymbol{\theta}_1, \boldsymbol{\theta}_2)$. When the response and covariates are modeled by mixed effects models respectively, the observed-data joint likelihood is given by

$$L(\boldsymbol{\theta}_1, \boldsymbol{\theta}_2) = \prod_{i=1}^{n} \int \int f(\mathbf{y}_i|\mathbf{z}_i^*, \mathbf{b}_i, \boldsymbol{\theta}_1) f(\mathbf{z}_i|\mathbf{a}_i, \boldsymbol{\theta}_2) \, d\mathbf{b}_i \, d\mathbf{a}_i, \tag{5.9}$$

where $\mathbf{b}_i$ and $\mathbf{a}_i$ are random effects in the response and covariate models. When the models are nonlinear in the random effects, the integral in the joint likelihood (5.9) can be intractable, so likelihood methods can be computationally intensive. An EM algorithm can be used for likelihood estimation, treating the random effects as "missing data". Section 5.4 provides more details of the likelihood methods for mixed effects models with covariate measurement errors.

Simulation extrapolation (SIMEX) methods are simulation-based methods of estimating and reducing bias due to measurement errors. They are often used in situations where the measurement error generating process can be imitated on a computer via Monte Carlo methods. More detailed discussion of this approach can be found in Carroll et al. (2006). Li and Lin (2003) and Yi (2008) are examples of recent developments of this approach.

Bayesian methods assume prior distributions for the parameters, in addition to distributional assumptions for the response and covariate models. We will discuss Bayesian methods in more details in Chapter 11.

Other measurement error methods include *instrumental variables, score function* methods, *quasi-likelihood* methods, nonparametric and semiparametric methods, etc. Carroll et al. (2006) provided a comprehensive overview of various methods for covariate measurement errors in nonlinear models. Zidek et al. (1998) considered a different approach for measurement errors in GLMMs. They specify a standard GLMM for the observed data and then use various Taylor series expansions to approximate the underlying true-data model.

5.2.3 Bias Analysis

When covariates in a regression model are measured with errors, statistical analysis ignoring the measurement errors may be biased. To see this, in the following we follow Wang et al. (1998) and briefly discuss bias analysis for GLMMs with covariate measurement errors. Details can be found in Wang et al. (1998), Wang et al. (2000), and Carroll et al. (2006).

Consider the following GLMM

$$g(E(y_{ij}|\mathbf{b}_i)) = \beta_0 + z_{ij}^*\beta_1 + \mathbf{x}_{ij}^T\beta_2 + \mathbf{w}_{ij}^T\mathbf{b}_i, \qquad (5.10)$$
$$i = 1, \cdots, n; \ j = 1, \cdots, n_i,$$

where $g(\cdot)$ is a known link function, z_{ij} is the observed covariate value with measurement error, z_{ij}^* is the corresponding unobserved true covariate value, $\mathbf{x}_{ij}$ and $\mathbf{w}_{ij}$ are accurately measured covariates, and $\mathbf{b}_i$ are random effects. In GLMM (5.10), the response y_{ij} is related to the true but unobserved covariate value z_{ij}^* rather than the observed but mis-measured covariate value z_{ij}.

By assuming the following classical measurement error model

$$z_{ij} = z_{ij}^* + \epsilon_{ij}, \qquad (5.11)$$

we can convert the GLMM (5.10) into a new GLMM in which y_{ij} is related to the observed covariate value z_{ij}. Then, analytic expressions for bias and bias-correction can be found by comparing the parameters in the two GLMMs. This can be done by noting that the classical measurement error model (5.11) can be converted to the following regression calibration model:

$$z_{ij}^* = \gamma_0 + \gamma_1 z_{ij} + e_{ij}^*.$$

Then, we can plug in z_{ij}^* in the GLMM (5.10) and obtain a new GLMM. Since the analytic expressions are somewhat complicated, we omit the details here. Wang et al. (1998) also noted that a naive application of regression calibration is not suitable for GLMMs with covariate measurement errors and they proposed a SIMEX approach. On the other hand, Buonaccorsi et al. (2000) noted that a regression calibration method is suitable for LME models with covariate measurement errors.

For hypothesis testing problems in regression models, Brunner and Austin (2009) demonstrated how the type I error rate can be greatly inflated when covariate measurement errors are ignored. If covariate measurement errors are not addressed, the inflated type I error can be very high, so a unimportant covariate in a regression model may appear to be highly significant. Their work shows the importance of addressing covariate measurement errors in regression models.

5.3 Two-Step Methods and Regression Calibration Methods

5.3.1 Two-Step Methods

For covariate measurement error problems in regression models, a commonly used simple approach is the so-called *two-step method*: in step 1 the "true" unobserved covariate values are estimated based on observed data or external data; and in step 2 the mis-measured covariates in the response model are simply substituted by their estimated values from the first step. Then, statistical inference for the response model proceeds as if the estimated covariate values are their true values. Such a naive two-step method is similar to the naive single-imputation methods for missing data (see Chapter 3), and is closely related to regression calibration methods.

A major advantage of the above naive two-step method is its simplicity. In some cases the naive two-step method may even produce approximately unbiased estimates of the main parameters in the response model, if the covariate model is appropriately chosen. However, as with naive single imputation methods for missing data, a major drawback of the naive two-step method is that it fails to incorporate the *uncertainty* in the estimation of the true covariate values in the first step, so the standard errors of the main parameter estimates may be under-estimated or unreliable.

In the likelihood framework, suppose that $\hat{\boldsymbol{\theta}}_2$ is an estimate of $\boldsymbol{\theta}_2$ in the covariate model $f(\mathbf{z}_i; \boldsymbol{\theta}_2)$ based on the observed covariate data $\{\mathbf{z}_i, i = 1, \cdots, n\}$, and suppose that $\hat{\mathbf{z}}_i^*$ is the corresponding estimate of the unobserved true covariate $\mathbf{z}_i^*$. Let $f(\mathbf{y}_i | \mathbf{z}_i^*, \boldsymbol{\theta}_1)$ be the response model which links the response $\mathbf{y}_i$ to the true covariate $\mathbf{z}_i^*$. The naive two-step method maximizes

$$L(\boldsymbol{\theta}_1, \hat{\boldsymbol{\theta}}_2) = \prod_{i=1}^{n} f(\mathbf{y}_i | \hat{\mathbf{z}}_i^*, \boldsymbol{\theta}_1) f(\mathbf{z}_i; \hat{\boldsymbol{\theta}}_2),$$

as a function of $\boldsymbol{\theta}_1$ to obtain MLE of the main parameters $\boldsymbol{\theta}_1$. Gong and Sammaniego (1981) called the resulting estimate of $\boldsymbol{\theta}_1$ a *pseudo-maximum likelihood estimate*. Buonaccorsi et al. (2000) derived some analytic results for this method. Ogden and Tarpey (2006) considered a similar two-step approach and adjusted standard errors of parameter estimates using bootstrap methods.

There are several ways to adjust the standard errors of main parameter estimates based on a simple two-step method:

- *re-sample methods*, such as the *bootstrap* methods;
- *joint inference* for the response and covariate models, such as that based on the joint likelihood;
- *approximate analytic methods* which incorporate the uncertainty in covariate estimation through theoretical arguments, such as that based on large-sample or asymptotic results;
- *multiple imputation methods*, which impute the unobserved true covariates by several plausible values based on reasonable imputation models so they incorporate the uncertainty of the unobserved true covariate values.

In these approaches, re-sample methods and multiple imputation methods can be computationally intensive, and approximate analytic methods may not be feasible or may be difficult in some cases. A joint likelihood method is appealing, as it provides the most efficient estimates if the models are correctly specified, but it can also be computationally intensive. In Section 5.3.2 we present a parametric bootstrap method. In Section 5.4 we describe a joint likelihood method.

5.3.2 A Two-Step Method for NLME Models with Measurement Errors

For NLME models with measurement errors in time-dependent covariates, Higgins, Davidian, and Giltinan (1997) proposed a two-step method and suggested to adjust standard errors of the main parameter estimates using a parametric bootstrap method. Their method is a nice application of the regression calibration method for mixed effects models with covariate measurement errors. We briefly describe the method as follows.

Consider the following NLME model with covariate $\mathbf{z}_i = (z_{i1}, \cdots, z_{im_i})^T$, which is measured with error,

$$
\begin{aligned}
y_{ij} &= g(z_{ij}^*, \boldsymbol{\beta}, \mathbf{b}_i) + e_{ij}, \quad i = 1, \cdots, n; \ j = 1, \cdots, n_i, \\
e_{ij} &\ \text{i.i.d.} \ \sim N(0, \sigma^2), \qquad \mathbf{b}_i \sim N(0, B),
\end{aligned}
\tag{5.12}
$$

where $g(\cdot)$ is a known nonlinear function, z_{ij}^* is the unobserved true covariate value with the corresponding observed version being z_{ij}, and $\mathbf{b}_i$ are random effects. To address covariate measurement errors, consider the following classical measurement error model

$$
\begin{aligned}
\mathbf{z}_i &= U_i \boldsymbol{\alpha} + V_i \mathbf{a}_i + \boldsymbol{\epsilon}_i \equiv \mathbf{z}_i^* + \boldsymbol{\epsilon}_i, \tag{5.13} \\
\mathbf{a}_i &\sim N(\mathbf{0}, A), \qquad \boldsymbol{\epsilon}_i \sim N(0, \delta^2 I).
\end{aligned}
$$

The two-step method in Higgins, Davidian, and Giltinan (1997) proceeds as follows. In the first step, we fit the covariate LME model (5.13) to the observed covariate data using a standard method for LME models and obtain an estimate of the unobserved true covariate

$$
\hat{\mathbf{z}}_i^* = \hat{E}(\mathbf{z}_i^* | \mathbf{z}_i) = U_i \hat{\boldsymbol{\alpha}} + V_i \hat{\mathbf{a}}_i,
$$

which can be interpreted as an empirical Bayes estimate. In the second step, we consider the following "working" NLME model

$$
y_{ij} \approx g(\hat{z}_{ij}^*, \boldsymbol{\beta}, \mathbf{b}_i) + e_{ij}^* \equiv E\left(y_{ij} \mid \hat{E}(\mathbf{z}_i^* | \mathbf{z}_i), \boldsymbol{\beta}, \mathbf{b}_i\right) + e_{ij}^*, \tag{5.14}
$$

in which the unobserved true covariate $\mathbf{z}_i^*$ is replaced by its estimate $\hat{\mathbf{z}}_i^*$. Estimates of the main parameters in the response NLME model (5.12) are then based on the "working" NLME model (5.14), using standard methods for NLME models.

To adjust the standard errors of the main parameter estimates $\hat{\boldsymbol{\beta}}$ to incorporate the uncertainty in the estimation of the covariate model (5.13), Higgins et al. (1997) proposed the following *parametric bootstrap method*:

STEP 1. Generate covariate values z_{ij} based on the covariate model (5.13), in which the unknown parameters are substituted by their estimates, the unknown random effects $\mathbf{a}_i$ are simulated from $N(\mathbf{0}, \hat{A})$, and the unknown error $\boldsymbol{\epsilon}_i$ are simulated from $N(\mathbf{0}, \hat{\delta}^2 I)$.

STEP 2. Generate responses y_{ij} based on the response model (5.12), in which the unknown parameters are substituted by their estimates, the unknown random effects $\mathbf{b}_i$ are simulated from $N(\mathbf{0}, \hat{B})$, and the unknown error $\mathbf{e}_i$ are simulated from $N(\mathbf{0}, \hat{\sigma}^2 I)$.

STEP 3. Based on the generated dataset from Steps 1 and 2 (the generated dataset is called *bootstrap dataset*), fit the NLME model (5.12) using the foregoing two-step method and obtain parameter estimates.

Repeating the above procedure B times (say, $B = 1000$), we can obtain estimated standard errors for the main parameters (say, β) from the sample covariance matrix across the B bootstrap datasets. This parametric Bootstrap method produces more reliable estimates of the standard errors for the main parameter estimates than the naive two-step method, if the assumed models are correct. The method can be computationally intensive.

The above two-step approach is a regression calibration method. It is very general and can be applied to other mixed effects models such as GLMMs with covariate measurement errors. The method performs well for NLME models based on simulation results in Higgins et al. (1997).

5.4 Likelihood Methods

5.4.1 Joint Likelihood

In regression calibration methods or two-step methods, the parameters in the response and covariate models are estimated separately, so standard errors of the parameter estimates in the response model need to be adjusted separately to reflect the uncertainty in the estimation of the covariate model. In this section, we consider a joint likelihood method which estimates all parameters *simultaneously*. The method incorporates the uncertainty in estimating the covariate and response model parameters simultaneously, and thus produces valid parameter estimates and standard errors. Moreover, the resulting MLEs are asymptotically most efficient if the assumed models hold and certain regularity conditions are satisfied.

The joint likelihood method is a *structural modeling* approach, in which one assumes a parametric model for the unobserved true covariates. For mixed effects models with error-prone time-dependent covariates, we can select the response and covariate models based on the observed data. In the following, we focus on the classical measurement error model (5.4), and follow the models and notation in Section 5.2 (page 179).

Let the response model be $f(\mathbf{y}_i|\mathbf{z}_i^*, \mathbf{b}_i, \beta, \sigma)$ and the covariate model be $f(\mathbf{z}_i|\mathbf{a}_i, \alpha)$, given the random effects $(\mathbf{a}_i, \mathbf{b}_i)$. Let $\theta = (\alpha, \beta, \sigma, R, A, B)$ be the collection of all unknown parameters. The *joint log-likelihood* for the observed data $\{(\mathbf{y}_i, \mathbf{z}_i), i = 1, \ldots, n\}$ can be written as

$$
\begin{aligned}
l_o(\theta) &\equiv \sum_{i=1}^{n} l_o^{(i)}(\theta) \\
&= \sum_{i=1}^{n} \log \int \int f(\mathbf{y}_i|\mathbf{z}_i^*, \mathbf{b}_i, \beta, \sigma) f(\mathbf{z}_i|\mathbf{a}_i, \alpha) f(\mathbf{a}_i|A) f(\mathbf{b}_i|B) \, d\mathbf{a}_i \, d\mathbf{b}_i
\end{aligned}
$$

$$= \sum_{i=1}^{n} \log \int \int f(\mathbf{y}_i|\mathbf{a}_i, \mathbf{b}_i, \boldsymbol{\theta}) f(\mathbf{z}_i|\mathbf{a}_i, \boldsymbol{\alpha}) f(\mathbf{a}_i|A) \, f(\mathbf{b}_i|B) \, d\mathbf{a}_i \, d\mathbf{b}_i,$$

where $f(\cdot)$ denotes a generic density function.

The log-likelihood $l_o(\boldsymbol{\theta})$ is intractable and generally does not have a closed-form expression when the models are nonlinear in the random effects $(\mathbf{a}_i, \mathbf{b}_i)$, such as GLMM and NLME models. Numerical integration methods such as the Gauss-Hermit quadrature is often infeasible here since the dimension of the random effects $(\mathbf{a}_i, \mathbf{b}_i)$ is usually not low. Therefore, in the next section we consider a Monte Carlo EM (MCEM) algorithm to find the MLEs of the parameters in $\boldsymbol{\theta}$, similar to the MCEM algorithm in Chapter 4.

5.4.2 Estimation Based on Monte Carlo EM Algorithms

By treating the unobservable random effects $\mathbf{a}_i$ and $\mathbf{b}_i$ as "missing data", we have "complete data" $\{(\mathbf{y}_i, \mathbf{z}_i, \mathbf{a}_i, \mathbf{b}_i), \ i = 1, \ldots, n\}$. The "complete data" log-likelihood is given by

$$
\begin{aligned}
l_c(\boldsymbol{\theta}) &= \sum_{i=1}^{n} l_c^{(i)}(\boldsymbol{\theta}) \\
&= \sum_{i=1}^{n} \Big[\log f(\mathbf{y}_i|\mathbf{a}_i, \mathbf{b}_i; \boldsymbol{\theta}) + \log f(\mathbf{z}_i|\mathbf{a}_i; \boldsymbol{\alpha}) \\
&\qquad\qquad + \log f(\mathbf{a}_i; A) + \log f(\mathbf{b}_i; B) \Big].
\end{aligned}
$$

Let $\boldsymbol{\theta}^{(t)}$ be the parameter estimates from the t-th EM iteration, $t = 1, 2, \cdots$. The E-step for individual i at the $(t + 1)$-th EM iteration computes the conditional expectation of the complete-data log-likelihood given the observed data and current parameter estimates:

$$
\begin{aligned}
Q_i(\boldsymbol{\theta}|\boldsymbol{\theta}^{(t)}) &= E(l_c^{(i)}(\boldsymbol{\theta})|\mathbf{y}_i, \mathbf{z}_i; \boldsymbol{\theta}^{(t)}) \\
&= \int \int \Big[\log f(\mathbf{y}_i|\mathbf{a}_i, \mathbf{b}_i; \boldsymbol{\theta}) + \log f(\mathbf{z}_i|\mathbf{a}_i; \boldsymbol{\alpha}) \\
&\qquad\quad + \log f(\mathbf{a}_i; A) + \log f(\mathbf{b}_i; B) \Big] f(\mathbf{a}_i, \mathbf{b}_i|\mathbf{y}_i, \mathbf{z}_i; \boldsymbol{\theta}^{(t)}) \, d\mathbf{a}_i \, d\mathbf{b}_i, \\
&\equiv I_1 + I_2 + I_3 + I_4,
\end{aligned}
$$

where the expectation is taken with respect to the conditional distribution $f(\mathbf{a}_i, \mathbf{b}_i|\mathbf{y}_i, \mathbf{z}_i; \boldsymbol{\theta}^{(t)})$.

The E-step can be evaluated using Monte Carlo methods. Specifically, we may use the Gibbs sampler to generate samples from the conditional distribution $f(\mathbf{a}_i, \mathbf{b}_i|\mathbf{y}_i, \mathbf{z}_i; \boldsymbol{\theta}^{(t)})$ by iteratively sampling from the full conditionals

$f(\mathbf{a}_i|\mathbf{y}_i, \mathbf{z}_i, \mathbf{b}_i; \boldsymbol{\theta}^{(t)})$ and $f(\mathbf{b}_i|\mathbf{y}_i, \mathbf{z}_i, \mathbf{a}_i; \boldsymbol{\theta}^{(t)})$. To sample from the full conditionals, note that

$$f(\mathbf{a}_i|\mathbf{y}_i, \mathbf{z}_i, \mathbf{b}_i; \boldsymbol{\theta}^{(t)}) \propto f(\mathbf{z}_i|\mathbf{a}_i; \boldsymbol{\theta}^{(t)}) \, f(\mathbf{a}_i; \boldsymbol{\theta}^{(t)}) \, f(\mathbf{y}_i|\mathbf{z}_i, \mathbf{a}_i, \mathbf{b}_i; \boldsymbol{\theta}^{(t)}),$$
$$f(\mathbf{b}_i|\mathbf{y}_i, \mathbf{z}_i, \mathbf{a}_i; \boldsymbol{\theta}^{(t)}) \propto f(\mathbf{b}_i; \boldsymbol{\theta}^{(t)}) \, f(\mathbf{y}_i|\mathbf{z}_i, \mathbf{a}_i, \mathbf{b}_i; \boldsymbol{\theta}^{(t)}),$$

where the density functions on the right-hand-sides are known. So samples from each of the full conditionals can be generated using rejection sampling methods or importance sampling methods.

After generating a large random sample from the distribution $f(\mathbf{a}_i, \mathbf{b}_i|\mathbf{y}_i, \mathbf{z}_i; \boldsymbol{\theta}^{(t)})$, the conditional expectation $Q(\boldsymbol{\theta}|\boldsymbol{\theta}^{(t)})$ in the E-step can be approximated by its empirical mean, with the random effects substituted by their simulated values. Then, the M-step is like a complete-data maximization, so standard complete-data optimization procedures may be used to update the parameter estimates. At convergence of the EM algorithm, an approximate variance-covariance matrix of $\hat{\boldsymbol{\theta}}$ is given by

$$\text{Cov}(\hat{\boldsymbol{\theta}}) \approx \left[\sum_{i=1}^{n} E(\mathbf{s}_c^{(i)}|\mathbf{y}_i, \mathbf{z}_i; \hat{\boldsymbol{\theta}}) \, E(\mathbf{s}_c^{(i)}|\mathbf{y}_i, \mathbf{z}_i; \hat{\boldsymbol{\theta}})^T \right]^{-1},$$

where $\mathbf{s}_c^{(i)} = \partial l_c^{(i)}(\boldsymbol{\theta})/\partial\boldsymbol{\theta}$ and the expectation can also be approximated by Monte Carlo methods.

The computational issues associated with the foregoing MCEM are similar to that in Chapter 4, but the MCEM here may exhibit more convergence problems such as very slow or non-convergence, since sampling the completely unobserved random effects $(\mathbf{a}_i, \mathbf{b}_i)$ in the E-step may offer more convergence problems in the Gibbs sampler. Therefore, computationally more efficient approximate methods, similar to that in Chapter 4, are very valuable here.

Parameter identifiability may be an issue in some cases. We discussed parameter identifiability in mixed effects models in Chapter 2, some of which may also be applicable in the current problems. In general, we should avoid building too large models which may contain too many parameters. For example, the covariate models are secondary and the parameters in the covariate models can be viewed as nuisance parameters. Thus, we should avoid building large or complicated covariate models, or we may impose some restrictions on the variance-covariance matrices, such as diagonal matrices, to reduce the number of parameters. In practice, if the MCEM algorithm diverges quickly, there may be problems in parameter identifiability.

5.5 Approximate Methods

For measurement error models in this chapter, the MCEM can be computationally very intensive and may exhibit convergence problems since sampling the random effects may lead to inefficient Gibbs sampler and may lead to a high degree of auto-correlation and lack of convergence in the Gibbs sampler. Thus, computationally more efficient approximate methods are particularly desirable.

In this section we consider two computationally very efficient approximate methods based on Taylor or Laplace approximations. Although these approximate methods are similar to that in Chapter 4 for missing data problems, the computational advantages of these approximate methods are more substantial in the measurement error problems since here we can completely eliminate any integration in likelihood computation and thus completely avoid any Monte Carlo approximations. We briefly describe these approximate methods in the following sections and skip some details since these details are similar to those in Chapter 4. As illustration, we focus on NLME models with measurement errors in time-dependent covariates and follow the notation in previous sections. Most of the general ideas and approaches apply to other mixed effects models as well.

5.5.1 Linearization

As one can see from previous sections, a mixed effects model with covariate measurement error can be converted to a new mixed effects model, if the covariate process is modeled using another mixed effects model. Thus, we can *linearize* or approximate the new mixed effects model by taking a first-order Taylor expansion around estimates of the parameters and random effects and obtain a "working" LME model, similar to Lindstrom and Bates (1990) and the approach described in Chapter 4. For the working LME model, the random effects can be integrated out in the E-step of an EM algorithm, and analytic or closed-form expressions for the E-step and the M-step can be obtained. One then iterates the procedure until convergence. This leads to a computationally very efficient EM algorithm.

Specifically, consider the NLME model (5.12) (see page 189) with covariate measurement errors. We can combine the NLME model for the response and the classical measurement error LME model (5.4) (see page 180) for the error-prone and time-dependent covariate by substituting the true covariates in the response model (5.12) with the assumed model (5.4), which leads to the following new NLME model

$$y_{ij} = g_{ij}(\boldsymbol{\alpha}, \boldsymbol{\beta}, \mathbf{a}_i, \mathbf{b}_i) + e_{ij}, \qquad i = 1, \ldots, n, \ j = 1, \ldots, n_i, \quad (5.15)$$

where $g_{ij}(\cdot)$ is a known nonlinear function. Let $\mathbf{g}_i = (g_{i1}, \ldots, g_{in_i})^T$.

In the following iterative algorithm, denote the estimate of $(\boldsymbol{\theta}, \mathbf{a}_i, \mathbf{b}_i)$ in the current iteration by $(\tilde{\boldsymbol{\theta}}, \tilde{\mathbf{a}}_i, \tilde{\mathbf{b}}_i)$, where $\boldsymbol{\theta}$ denotes all unknown parameters, and

$$\tilde{\mathbf{a}}_i = E(\mathbf{a}_i | \mathbf{y}_i, \mathbf{z}_i; \tilde{\boldsymbol{\theta}}), \qquad \tilde{\mathbf{b}}_i = E(\mathbf{b}_i | \mathbf{y}_i, \mathbf{z}_i; \tilde{\boldsymbol{\theta}}),$$

suppressing the iteration number. Taking a first-order Taylor expansion of the nonlinear function $g_{ij}(\alpha, \beta, a_i, b_i)$ around the current mean parameter estimates $\tilde{\alpha}$ and $\tilde{\beta}$ and the random effects estimates $\tilde{a}_i$ and $\tilde{b}_i$, we obtain the following working LME model

$$\tilde{\mathbf{y}}_i = W_i\,\boldsymbol{\alpha} + X_i\,\boldsymbol{\beta} + H_i\,\mathbf{a}_i + T_i\,\mathbf{b}_i + \mathbf{e}_i, \qquad (5.16)$$

where

$$
\begin{aligned}
\tilde{\mathbf{y}}_i &= \mathbf{y}_i - \mathbf{g}_i(\tilde{\alpha},\tilde{\beta},\tilde{\mathbf{a}}_i,\tilde{\mathbf{b}}_i) + W_i\,\tilde{\alpha} + X_i\,\tilde{\beta} + H_i\,\tilde{\mathbf{a}}_i + T_i\,\tilde{\mathbf{b}}_i, \\
\mathbf{w}_{ij} &= \frac{\partial g_{ij}(\alpha,\beta,\mathbf{a}_i,\mathbf{b}_i)}{\partial\alpha}, \qquad \mathbf{x}_{ij} = \frac{\partial g_{ij}(\alpha,\beta,\mathbf{a}_i,\mathbf{b}_i)}{\partial\beta}, \\
\mathbf{h}_{ij} &= \frac{\partial g_{ij}(\alpha,\beta,\mathbf{a}_i,\mathbf{b}_i)}{\partial\mathbf{a}_i}, \qquad \mathbf{t}_{ij} = \frac{\partial g_{ij}(\alpha,\beta,\mathbf{a}_i,\mathbf{b}_i)}{\partial\mathbf{b}_i},
\end{aligned}
$$

$W_i = (\mathbf{w}_{i1},\dots,\mathbf{w}_{in_i})^T$, $X_i = (\mathbf{x}_{i1},\dots,\mathbf{x}_{in_i})^T$, $H_i = (\mathbf{h}_{i1},\dots,\mathbf{h}_{in_i})^T$, $T_i = (\mathbf{t}_{i1},\dots,\mathbf{t}_{in_i})^T$, with all the partial derivatives being evaluated at the current estimates $(\tilde{\alpha}, \tilde{\beta}, \tilde{\mathbf{a}}_i, \tilde{\mathbf{b}}_i)$. The approximate method is then to *iteratively* solve the working LME model (5.16) using standard methods for LME models, such as an EM algorithm.

After some algebra, analytic or closed-form expressions for the E-step and M-step of an EM algorithm can be obtained for the LME model (5.16) (Liu and Wu 2009). This avoids Monte Carlo approximations in the E-step and iterative procedures in the M-step, and thus offers substantial computational advantages.

5.5.2 Laplace Approximation

The approximate method in the previous section linearizes the nonlinear models and then works on the resulting linear models to simplify computation. Alternatively, we can directly approximate the intractable integral in the observed-data log-likelihood $l_o(\boldsymbol{\theta})$ of the original nonlinear model using a first-order Laplace approximation, which also greatly simplifies computation. This approach is similar to that in Chapter 4 for missing data problems, but the method is simpler here because direct Laplace approximation can be used without any transformations. We follow the procedures in Lee et al. (2006) for Laplace approximations.

Let $\omega = \{\omega_i \equiv (\mathbf{a}_i, \mathbf{b}_i),\ i = 1,\dots,n\}$ be the collection of all random effects, and let q be the dimension of ω_i. Let $N_i = n_i + m_i$ be the total number of observed responses and covariates for individual i. Let $\hat{\omega}$ be the solution to the equation

$$\frac{\partial l_c(\boldsymbol{\theta}, \omega)}{\partial\omega} = \mathbf{0},$$

where $l_c(\boldsymbol{\theta}, \boldsymbol{\omega}) \equiv l_c(\boldsymbol{\theta})$ is the "complete-data" log-likelihood in (5.4.2). As in Chapter 4, we can approximate the observed-data log-likelihood $l_o^{(i)}(\boldsymbol{\theta})$ for individual i as follows (Lee et al. 2006)

$$
\begin{aligned}
l_o^{(i)}(\boldsymbol{\theta}) &= \left\{ l_c^{(i)}(\boldsymbol{\theta}, \boldsymbol{\omega}_i) - \frac{1}{2} \log \left| \frac{1}{2\pi} D(l_c^{(i)}(\boldsymbol{\theta}, \boldsymbol{\omega}_i), \boldsymbol{\omega}_i) \right| \right\}_{\boldsymbol{\omega}_i = \hat{\boldsymbol{\omega}}_i} + O(N_i^{-1}) \\
&\equiv p_{\hat{\boldsymbol{\omega}}_i}(l_c^{(i)}(\boldsymbol{\theta}, \boldsymbol{\omega}_i)) + O(N_i^{-1}),
\end{aligned}
$$

where $D(l, \boldsymbol{\xi}) = -\partial^2 l / \partial \boldsymbol{\xi}^2$. Thus, the observed-data log-likelihood $l_o(\boldsymbol{\theta}) = \sum_{i=1}^{n} l_o^{(i)}(\boldsymbol{\theta})$ can be approximated as

$$
\begin{aligned}
l_o(\boldsymbol{\theta}) &= \sum_{i=1}^{n} \left\{ p_{\hat{\boldsymbol{\omega}}_i}(l_c^{(i)}(\boldsymbol{\theta}, \boldsymbol{\omega}_i)) + O(N_i^{-1}) \right\} \\
&\equiv p_{\hat{\boldsymbol{\omega}}}(l_c(\boldsymbol{\theta}, \boldsymbol{\omega})) + nO\left[\left(\min_i N_i \right)^{-1} \right] \approx p_{\hat{\boldsymbol{\omega}}}(l_c(\boldsymbol{\theta}, \boldsymbol{\omega})),
\end{aligned}
$$

where

$$
\begin{aligned}
p_{\hat{\boldsymbol{\omega}}}(l_c(\boldsymbol{\theta}, \boldsymbol{\omega})) &= \sum_{i=1}^{n} p_{\hat{\boldsymbol{\omega}}_i}(l_c^{(i)}(\boldsymbol{\theta}, \boldsymbol{\omega}_i)) \\
&= \sum_{i=1}^{n} \left\{ l_c^{(i)}(\boldsymbol{\theta}, \boldsymbol{\omega}_i) - \frac{1}{2} \log \left| \frac{1}{2\pi} D(l_c^{(i)}(\boldsymbol{\theta}, \boldsymbol{\omega}_i), \boldsymbol{\omega}_i) \right| \right\}_{\boldsymbol{\omega}_i = \hat{\boldsymbol{\omega}}_i}.
\end{aligned}
$$

As $\min_i\{N_i\}$ grows faster than n, the function $p_{\hat{\boldsymbol{\omega}}}(l_c(\boldsymbol{\theta}, \boldsymbol{\omega}))$ approaches to the observed-data log-likelihood function $l_o(\boldsymbol{\theta})$. It can be shown that $p_{\hat{\boldsymbol{\omega}}}(l_c(\boldsymbol{\theta}, \boldsymbol{\omega}))$ is the first-order Laplace approximation to the observed-data log-likelihood $l_o(\boldsymbol{\theta})$ (Lee et al. 2006). This is equivalent to integrating out the random effects $\boldsymbol{\omega}$ in the likelihood $l_o(\boldsymbol{\theta})$ using the first-order Laplace approximation. Thus, the estimate $\hat{\boldsymbol{\theta}}_{AP}$ of $\boldsymbol{\theta}$, which maximizes $p_{\hat{\boldsymbol{\omega}}}(l_c(\boldsymbol{\theta}, \boldsymbol{\omega}))$ with respect to $\boldsymbol{\theta}$, is an approximate MLE of $\boldsymbol{\theta}$.

Similarly, we can obtain restricted maximum likelihood (REML) estimates of the variance-covariance parameters by integrating out the mean parameters and the random effects simultaneously using the first-order Laplace approximation to the observed-data log-likelihood $l_o(\boldsymbol{\theta})$. Specifically, to obtain REML estimates of the variance-covariance parameters, we can split $\boldsymbol{\theta}$ into two parts: $\boldsymbol{\theta} = (\boldsymbol{\theta}_1, \boldsymbol{\theta}_2)$, where $\boldsymbol{\theta}_1$ contains the mean parameters and $\boldsymbol{\theta}_2$ contains the variance-covariance parameters. Then we use the above approach to integrate out the mean parameters and random effects using Laplace approximations as follows.

Let $(\hat{\boldsymbol{\omega}}, \hat{\boldsymbol{\theta}}_1)$ be the solution to the equation

$$
\frac{\partial l_c(\boldsymbol{\theta}_1, \boldsymbol{\theta}_2, \boldsymbol{\omega})}{\partial (\boldsymbol{\omega}, \boldsymbol{\theta}_1)} = 0.
$$

Then, it can be shown that

$$
p_{(\hat{\boldsymbol{\omega}},\hat{\boldsymbol{\theta}}_1)}(l_c(\boldsymbol{\theta}, \boldsymbol{\omega})) \;=\; \sum_{i=1}^{n} \Big\{ l_c^{(i)}(\boldsymbol{\theta}_2, \hat{\boldsymbol{\theta}}_1, \hat{\boldsymbol{\omega}}_i)
$$
$$
-\frac{1}{2}\log\left| \frac{1}{2\pi} D\left(l_c^{(i)}(\boldsymbol{\theta}_2, \hat{\boldsymbol{\theta}}_1, \hat{\boldsymbol{\omega}}_i), (\hat{\boldsymbol{\theta}}_1, \hat{\boldsymbol{\omega}}_i) \right) \right| \Big\}
$$

is the first-order Laplace approximation to the observed-data log-likelihood $l_o(\boldsymbol{\theta})$ by integrating out the mean parameters $\boldsymbol{\theta}_1$ and the random effects $\boldsymbol{\omega}$ simultaneously (Lee et al. 2006), where $h(a, \hat{b}) \equiv h(a, b)|_{b=\hat{b}}$. Thus, by maximizing $p_{(\hat{\boldsymbol{\omega}},\hat{\boldsymbol{\theta}}_1)}(l_c(\boldsymbol{\theta}_1, \boldsymbol{\theta}_2, \boldsymbol{\omega}))$ with respect to $\boldsymbol{\theta}_2$, we obtain an approximate REML of the variance-covariance parameters $\boldsymbol{\theta}_2$.

For NLME models, the Taylor approximation and the Laplace approximation are asymptotically equivalent (Demidenko 2004). It can be shown that the approximate estimates are consistent only when *both* the number of individuals n *and* the number of within individual measurements n_i go to infinity. For finite samples, simulation results show that these approximate methods perform well for normal responses (e.g., NLME models) but may perform less satisfactory for discrete responses (e.g., GLMMs) (Pinheiro and Bates 1995; Liu and Wu 2007; Joe 2008). When the approximate methods do not perform well for GLMMs, higher order approximations such as a second-order Laplace approximation may improve the performance.

In practice the performance of these approximate methods depends on a number of factors, including

- the discreteness of the response,
- the degree of nonlinearity of the models,
- the variabilities of the within-individual and between-individual measurements,
- the sample sizes (both the number of individuals and the number of measurements within each individual).

Thus, their finite-sample performance should be evaluated on a case-by-case basis. In some situations one can use these approximate methods for initial analyses and then use more exact methods for the final analysis. The approximate methods can also be used to produce excellent starting values for EM algorithms or other iterative algorithms, since the choice of starting values is very important for nonlinear models.

5.6 Measurement Error and Missing Data

In many longitudinal studies, missing data and measurement errors often arise simultaneously. For example, in many longitudinal studies some subjects may drop out early, leading to missing data in both responses and time-dependent covariates, in addition to measurement errors in some covariates before subjects drop out. Thus, statistical methods are required to address measurement errors, missing data, and dropouts *simultaneously*.

For time-dependent covariates with both measurement errors and missing data, if the missing data mechanism is ignorable, the models and methods for measurement errors presented in Sections 5.4 – 5.5 can mostly be used with little modification. Note that, however, when covariates are measured with errors, one should be careful assuming the MAR missing data mechanism since the observed covariates may contain measurement errors and the true covariates may be unobserved (they are treated as observed values in the absence of measurement errors). Yi, Liu, and Wu (2009) provided some discussion on this issue. When the missing covariates are nonignorable, some modifications are required to incorporate the nonignorable missing data mechanisms. These modifications are often straightforward since we only need to incorporate a missing data model in the likelihood.

In the following sections, we briefly discuss mixed effects models with covariate measurement errors and nonignorable missing data in the covariates and in the responses. We omit the details since the procedures are conceptually the same as previous ones.

5.6.1 Measurement Errors and Missing Data in Covariates

We consider likelihood inference for a mixed effects model with a time-dependent covariate z_{ij}, which has both measurement errors and non-ignorable missing data. Since the missing data mechanism is non-ignorable, we need to assume a missing data model and incorporate it in the likelihood for valid inference. Let $\mathbf{r}_i = (r_{i1}, \ldots, r_{in_i})^T$ be a vector of missing data indicators for individual i such that $r_{ij} = 1$ if z_{ij} is missing and 0 otherwise. We follow the notation in previous sections, and write $\mathbf{z}_i = (\mathbf{z}_{mis,i}, \mathbf{z}_{obs,i})$, where $\mathbf{z}_{mis,i}$ contains the missing components of $\mathbf{z}_i$ and $\mathbf{z}_{obs,i}$ contains the observed components of $\mathbf{z}_i$. The observed-data joint log-likelihood can be written as

$$l_o(\boldsymbol{\theta}) = \sum_{i=1}^{n} \left[\int \int \int f(\mathbf{y}_i|\mathbf{z}_i^*, \ \mathbf{b}_i, \boldsymbol{\beta}, \boldsymbol{\sigma}) \, f(\mathbf{z}_i|\mathbf{a}_i, \boldsymbol{\alpha}) f(\mathbf{a}_i|A) \right.$$

$$\left. \times f(\mathbf{b}_i|B) f(\mathbf{r}_i|\mathbf{z}_i, \mathbf{y}_i; \boldsymbol{\eta}) \, d\mathbf{a}_i \, d\mathbf{b}_i \, d\mathbf{z}_{mis,i} \right], \quad (5.17)$$

where $\boldsymbol{\theta}$ contains all parameters and $\mathbf{z}_i^*$ is the unobserved true covariate whose observed version is $\mathbf{z}_i$. The non-ignorable missing data model $f(\mathbf{r}_i|\mathbf{z}_i, \mathbf{y}_i; \boldsymbol{\eta}) = f(\mathbf{r}_i|\mathbf{z}_{mis,i}, \mathbf{z}_{obs,i}, \mathbf{y}_i; \boldsymbol{\eta})$ links the missingness of the covariates to both missing and observed data.

Likelihood estimation then proceeds in a way similar to that in Chapter 4. For example, we may use a Monte Carlo EM algorithm to obtain MLE of $\boldsymbol{\theta}$, which involves sampling from the distribution $f(\mathbf{z}_{mis,i}, \mathbf{a}_i, \mathbf{b}_i | \mathbf{z}_{obs,i}, \mathbf{y}_i, \boldsymbol{\theta}^{(t)})$ at t-th EM iteration. This can again be done by using the Gibbs sampler along with rejection sampling methods or using importance sampling or other MCMC methods. We omit the details here since the procedures are similar to the ones presented in Chapter 4. Such Monte Carlo EM algorithms can be computationally very intensive and may exhibit convergence problems. Therefore, computationally more efficient approximate methods based on Taylor or Laplace approximations, similar to that in Section 5.4, are extremely valuable here and are highly recommended.

For non-ignorable missing covariate problems, when the covariates are also measured with errors, it is reasonable to assume that the missingness of the covariates may be related to the *true* but unobserved covariate values rather than the observed but mis-measured covariate values. In this case, if the covariate process is modeled using a LME model, as in Section 5.4, the random effects or some functions of the random effects in the LME model can be used to summarize the unobserved true covariate process or the history of the true covariate process. This suggests that we may consider the following non-ignorable missing data model

$$f(\mathbf{r}_i | \mathbf{z}_i; \boldsymbol{\eta}) = f(\mathbf{r}_i | \mathbf{a}_i, \mathbf{b}_i; \boldsymbol{\eta}). \tag{5.18}$$

That is, the missingness of the covariates is related to the unobserved true covariate and response values through the corresponding random effects $(\mathbf{a}_i, \mathbf{b}_i)$ in the covariate and response models respectively. Thus, we again have a shared parameter model (Wu and Carroll 1988; Little and Rubin 2002).

Under the missing data model (5.18), the observed-data loglikelihood can be written as

$$l_o^*(\boldsymbol{\theta}) = \sum_{i=1}^{n} \left[\int \int f(\mathbf{y}_i | \mathbf{a}_i, \mathbf{b}_i, \boldsymbol{\theta}) f(\mathbf{z}_{obs,i} | \mathbf{a}_i, \boldsymbol{\alpha}) \right.$$
$$\left. \times f(\mathbf{a}_i | A) f(\mathbf{b}_i | B) f(\mathbf{r}_i | \mathbf{a}_i, \mathbf{b}_i; \boldsymbol{\eta}) d\mathbf{a}_i \, d\mathbf{b}_i \right]. \tag{5.19}$$

Note that the integration in the likelihood (5.19) has a lower dimension than that in the likelihood (5.17), so the computation is simpler. In this case, for the approximate method based on Taylor expansions similar to that in Section 5.5.1, closed form expressions of the E- and M-step of the EM algorithm can be obtained. The approximate method based on Laplace approximations is also similar to that in Section 5.5.2 and is straightforward. Therefore, the computation for both exact and approximate likelihood inference under the missing data model (5.18) is simpler than that based on the missing data model in (5.17).

Since non-ignorable missing data models are not testable based on the observed data, it is a good strategy to consider different missing data models and then check if the main parameter estimates in the response model are sensitive to the assumed missing data models. The foregoing two types of missing data models are good choices for sensitivity analysis.

5.6.2 Measurement Errors in Covariates and Missing Data in Responses

In a mixed effects model for longitudinal data, measurement errors in covariates and non-ignorable missing data in the response is also a common problem in practice. For example, dropouts are extremely common in longitudinal studies. When a subject drops out, the response values of the subject from the time of dropout are missing, and such missing data are likely to be non-ignorable. Note that, if the responses are missing with an ignorable missing data mechanism, the missing data in the responses can be ignored since a mixed effects model allows unbalanced data in the response. When the responses are non-ignorably missing, however, we must assume a missing data model for the responses and incorporate it in likelihood inference. The approach is conceptually the same as that for non-ignorable missing covariates in Section 5.6.1. For completeness, we briefly describe the approach below.

Let $\mathbf{y}_i = (y_{i1}, \dots, y_{in_i})^T$ be the responses with possible non-ignorable missing data. Let $\mathbf{r}_i = (r_{i1}, \dots, r_{in_i})^T$ be a vector of missing response indicators for individual i such that $r_{ij} = 1$ if y_{ij} is missing and 0 otherwise. We write $\mathbf{y}_i = (\mathbf{y}_{mis,i}, \mathbf{y}_{obs,i})$. As in Section 5.6.1, we may consider two types of missing data models for sensitivity analysis, i.e. we may assume that the probability of missing data $P(r_{ij} = 1)$ (i) depends on the observed and missing values; or (ii) depends on the random effects in the models. That is, in case (i) we can assume a distribution $f(\mathbf{r}_i | \mathbf{y}_i, \mathbf{z}_i; \boldsymbol{\eta})$ for $\mathbf{r}_i$, The log-likelihood for the observed data $\{(\mathbf{y}_{obs,i}, \mathbf{z}_i, \mathbf{r}_i), \ i = 1, \dots, n\}$ can then be written as

$$l_o(\boldsymbol{\theta}) = \sum_{i=1}^{n} \log \int \int \int \left[f(\mathbf{y}_i | \mathbf{z}_i^*, \mathbf{b}_i; \boldsymbol{\beta}, \sigma^2) \, f(\mathbf{z}_i | \mathbf{a}_i; \boldsymbol{\alpha}) f(\mathbf{a}_i; A) \right.$$
$$\left. \times f(\mathbf{b}_i; B) \, f(\mathbf{r}_i | \mathbf{y}_i, \mathbf{z}_i; \boldsymbol{\eta}) \right] d\mathbf{y}_{mis,i} \, d\mathbf{a}_i \, d\mathbf{b}_i. \qquad (5.20)$$

Alternatively, as in Section 5.5.2, we may assume that the missingness of the responses is related to the random effects $\mathbf{a}_i$ and $\mathbf{b}_i$ in the response and covariate models, since the observed covariates and responses may be measured with errors. For example, we may assume that

$$\text{logit}[P(r_{ij} = 1 | \mathbf{a}_i, \mathbf{b}_i; \boldsymbol{\eta})] = \eta_0 + \eta_1 r_{i,j-1} + \boldsymbol{\eta}_2^T \mathbf{a}_i + \boldsymbol{\eta}_3^T \mathbf{b}_i.$$

The log-likelihood for the observed data $\{(\mathbf{y}_{obs,i}, \mathbf{z}_i, \mathbf{r}_i), \ i = 1, \dots, n\}$ can then be written as

$$
\begin{aligned}
l_o^*(\boldsymbol{\theta}) &= \sum_{i=1}^{n} \log \int \int \int \left[f(\mathbf{y}_i|\mathbf{z}_i^*, \mathbf{b}_i;\, \boldsymbol{\beta},\, \sigma^2) f(\mathbf{z}_i|\mathbf{a}_i;\, \boldsymbol{\alpha}) \right. \\
&\qquad \left. \times f(\mathbf{a}_i;\, A) f(\mathbf{b}_i;\, B)\, f(\mathbf{r}_i|\mathbf{a}_i,\, \mathbf{b}_i;\, \boldsymbol{\eta}) \right] d\mathbf{y}_{mis,i}\, d\mathbf{a}_i\, d\mathbf{b}_i \\
&= \sum_{i=1}^{n} \log \int \int \left[f(\mathbf{y}_{obs,i}|\mathbf{a}_i,\, \mathbf{b}_i;\, \boldsymbol{\beta},\, \sigma^2) f(\mathbf{z}_i|\mathbf{a}_i;\, \boldsymbol{\alpha}) \right. \\
&\qquad \left. \times f(\mathbf{a}_i;\, A) f(\mathbf{b}_i;\, B)\, f(\mathbf{r}_i|\mathbf{a}_i,\, \mathbf{b}_i;\, \boldsymbol{\eta}) \right] d\mathbf{a}_i\, d\mathbf{b}_i. \qquad (5.21)
\end{aligned}
$$

Note that for the log-likelihood $l_o^*(\boldsymbol{\theta})$ in (5.21) the missing responses $\mathbf{y}_{mis,i}$ are integrated out, which is not possible for the log-likelihood $l_o(\boldsymbol{\theta})$ in (5.20) where the missingness is directly related to the missing response values.

Exact likelihood inference can be carried out using Monte Carlo EM algorithms, as in the previous sections, but again they are computationally very intensive. So computationally more efficient approximate methods, similar to the ones in the previous sections, are highly valuable here. Note that the computation for models in (5.21) is simpler than that for models in (5.20) since the dimension of the integral in (5.21) is lower.

The foregoing methods can be extended to mixed effects models with missing covariates, missing responses, *and* covariate measurement errors. Such an extension is straightforward conceptually but the computation becomes even more challenging. As an example, consider a mixed effects models with non-ignorable missing covariates, non-ignorable missing responses, and covariate measurement errors. Let $\mathbf{r}_i$ and $\mathbf{s}_i$ be the missing covariate indicators and the missing response indicators respectively. Then, the observed data log-likelihood can be written as

$$
\begin{aligned}
l_o(\boldsymbol{\theta}) &= \sum_{i=1}^{n} \log \int \int \int \int \left[f(\mathbf{y}_i|\mathbf{z}_i^*, \mathbf{b}_i;\, \boldsymbol{\beta},\, \sigma^2)\, f(\mathbf{z}_i|\mathbf{a}_i;\, \boldsymbol{\alpha}) f(\mathbf{a}_i;\, A) \right. \\
&\qquad \times f(\mathbf{b}_i;\, B)\, f(\mathbf{r}_i|\mathbf{y}_i,\, \mathbf{z}_i;\, \boldsymbol{\eta}) \\
&\qquad \left. \times f(\mathbf{s}_i|\mathbf{y}_i,\, \mathbf{z}_i;\, \boldsymbol{\psi}) \right] d\mathbf{y}_{mis,i}\, d\mathbf{z}_{mis,i}\, d\mathbf{a}_i\, d\mathbf{b}_i.
\end{aligned}
$$

The computation for exact likelihood inference based on the above log-likelihood can be extremely intensive and may be practically infeasible! Therefore, computationally more efficient approximate methods based on Taylor or Laplace approximations are extremely valuable for such problems and may even be the only choice.

Note that, for the above problems, there may be too many model parameters, so parameter or model identifiability can become a major issue. A possible solution is to simplify secondary models, such as the missing data models and the covariate models, to reduce the number of nuisance parameters. Alternatively, we may impose certain restrictions on some parameter space, e.g., assume that the covariance matrices are diagonal.

Example 5.2 *Covariate measurement errors and missing responses*

We return to Example 4.2 in Chapter 4. In this study, some response values were missing, and the missingness may be informative or nonignorable. Thus, a sensitivity analysis with a non-ignorable missing data model is useful. Moreover, some covariates such as CD4 counts were measured with errors. Here we follow Liu and Wu (2007) and consider a joint inference for a NLME model with covariate measurement errors and nonignorable missing responses.

In this study, the long-term viral load trajectories (response profiles) appear to be complicated, so Liu and Wu (2007) considered a semiparametric NLME model in which the second phase viral decay rate is modeled nonparametrically to incorporate complicated viral load trajectories. Specifically, they considered the following semiparametric NLME model

$$y_{ij} = \log_{10}(P_{1i}e^{-\lambda_{1ij}t_{ij}} + P_{2i}e^{-\lambda_{2ij}t_{ij}}) + e_{ij}, \quad (5.22)$$

$$\log(P_{1i}) = \beta_1 + b_{1i}, \quad \lambda_{1ij} = \beta_2 + \beta_3 z_{ij}^* + b_{2i}, \quad (5.23)$$

$$\log(P_{2i}) = \beta_4 + b_{3i}, \quad \lambda_{2ij} = w(t_{ij}) + h_i(t_{ij}), \quad (5.24)$$

where y_{ij} is the $\log_{10}$-transformation of the viral load measurement for patient i at time t_{ij}, P_{1i} and P_{2i} are baseline values, λ_{1ij} and λ_{2ij} are the first and second phases of viral decay rates respectively and are allowed to vary over time, z_{ij} is the observed but mis-measured CD4 value and z_{ij}^* is the corresponding unobserved true CD4 value for patient i at time t_{ij}, $(\beta_1, \beta_2, \beta_3, \beta_4)$ are fixed parameters, (b_{1i}, b_{2i}, b_{3i}) are random effects, and $w(t_{ij})$ and $h_i(t_{ij})$ are nonparametric smooth fixed and random functions respectively. We assume that e_{ij} i.i.d. $\sim N(0, \delta^2)$.

As discussed in Section 2.5 of Chapter 2, we can use linear combinations of natural cubic splines with percentile-based knots to approximate the nonparametric functions $w(t)$ and $h_i(t)$ respectively. Based on AIC and BIC criteria, we obtained the following model for λ_{2ij} in (5.24):

$$\lambda_{2ij} \approx \beta_5 + \beta_6\,\psi_1(t_{ij}) + \beta_7\,\psi_2(t_{ij}) + b_{4i}, \quad (5.25)$$

where β_j's are fixed parameters, $\psi_j(\cdot)$'s are basis functions, and b_{4i} is a random effect. We assume that $\mathbf{b}_i = (b_{1i}, b_{2i}, b_{3i}, b_{4i})^T \sim N(0, D)$. See Example 2.13 (page 77) in Chapter 2 for more details.

To address measurement errors in CD4, we model the CD4 process by the following empirical LME model, which may be viewed as a classical measurement error model,

$$z_{ij} = (\alpha_1 + a_{i1}) + (\alpha_2 + a_{i2})\,u_{ij} + (\alpha_3 + a_{i3})\,u_{ij}^2 + \epsilon_{ij}, \quad (5.26)$$

Table 5.1 *Parameter estimates (standard errors) for the models in the example.*

Method	α_1	α_2	α_3	β_1	β_2	β_3	β_4	β_5	β_6	β_7
NV	–	–	–	11.72	65.71	0.84	6.87	–2.58	8.66	–1.90
	–	–	–	(.2)	(3.8)	(3.2)	(.6)	(5.5)	(8.9)	(3.1)
LK	–.42	4.15	–3.75	11.72	67.08	1.52	6.97	–1.83	7.75	–2.54
	(.1)	(.5)	(.6)	(.2)	(5.2)	(6.2)	(.7)	(5.8)	(8.8)	(3.5)
AP	–.43	4.21	–3.78	11.70	66.97	1.50	6.96	–1.90	7.86	–2.63
	(.1)	(.6)	(.6)	(.2)	(4.4)	(5.8)	(.6)	(5.5)	(7.9)	(3.0)

Estimates of variance components: $\hat{\delta} = 0.35$, $\hat{\sigma} = 0.51$.

$$= z_{ij}^* + \epsilon_{ij},$$

where u_{ij} is the CD4 measurement time for the i-th individual and j-th measurement, $\alpha = (\alpha_1, \alpha_2, \alpha_3)^T$ are fixed parameters, $\mathbf{a}_i = (a_{i1}, a_{i2}, a_{i3})^T$ are random effects, and ϵ_{ij} represent measurement error, $i = 1, \cdots, n$, $j = 1, \cdots, m_i$. We assume that ϵ_{ij} i.i.d. $\sim N(0, \sigma^2)$ and $\mathbf{a}_i \sim N(0, A)$. For possibly nonignorable missing data in viral loads due to (say) informative dropouts, we define $r_{ij} = 1$ if y_{ij} is observed and $r_{ij} = 0$ otherwise, and we consider the following simple nonignorable missing data model

$$\text{logit}[P(r_{ij} = 1 | \boldsymbol{\eta})] = \eta_1 + \eta_2 z_{ij} + \eta_3 y_{ij}, \quad i = 1, \cdots, n; \ j = 1, \cdots, n_i,$$

where the r_{ij}'s are assumed to be independent in order to reduce the number of nuisance parameters.

We estimate the model parameters using a *naive* method (NV), which ignores measurement errors and missing data, the likelihood method (LK) based on a Monte Carlo EM algorithm, and an approximate method (AP) based on a first-order Taylor approximation. Table 5.1 presents the resulting parameter estimates and standard errors (from Liu and Wu, 2007). We see that the likelihood method (LK) and the approximate method AP) give similar parameter estimates, but the naive method (NV) may severely under-estimate the covariate effect (i.e., the estimate of parameter β_3).

The results in Table 5.1 indicates that the approximate method based on a Taylor expansion performs very well here, compared to the "exact" likelihood method, since both produce similar estimates. This example also shows that it is important to incorporate covariate measurement errors and missing data in likelihood inference in order to avoid biased results.

CHAPTER 6

Mixed Effects Models with Censoring

6.1 Introduction

In many longitudinal studies, such as studies on environmental pollution and infectious diseases, measurements of some variables may be subject to a detection limit, i.e., a certain threshold value below or above which the measurements are not quantifiable. In other words, some data are left or right *censored*, so their values are not observed. In this chapter, we consider mixed effects models for censored longitudinal data.

In some studies, the detection limits may vary from laboratory to laboratory and may change over time as new laboratory methods are implemented. For example, in HIV/AIDS studies viral loads may be subject to left censoring due to a detection limit so viral load values below this limit cannot be measured or quantified. Figure 1.3 in Chapter 1 (page 11) and Figure 6.1 in this chapter (page 209) show viral load data in an AIDS study where the detection limit is 100 (or 2 in $\log_{10}$ scale). Because the proportion of censored data may not be small in many longitudinal studies, failure to account for the censoring in statistical analysis may lead to significant biases in the parameter estimates and inference (Hughes 1999).

For AIDS data with left censoring, commonly used naive methods include imputing the censored values by the detection limit or half the detection limit or other simple imputation procedures (Paxton et al. 1997). These naive methods, however, may lead to biased results, as demonstrated in Hughes (1999). Moreover, these naive single imputation methods ignore the uncertainty of the censored values. Therefore, one should not use these naive methods in statistical inference, and more appropriate formal methods described in this chapter should be considered.

Censored data may be viewed as special missing data since, although the true values are not observed, there is some information available for the censored values: the censored values are known to be smaller or larger than a known number. This is different from usual missing data for which one knows nothing about the unobserved true values. Censored data may also be viewed as non-ignorable missing data since the missingness (censoring) is related to the unobserved values (the censored values).

Censored data are perhaps more well known in the context of survival analysis. Mixed effects models with censored responses are similar to parametric frailty models or parametric survival models with random effects. For Cox proportional hazards models, we model the hazard function rather than the mean response and we allow the hazard functions to be nonparametric. We will discuss survival models in Chapter 7.

In the presence of censoring, valid likelihood inference must incorporate the censoring mechanism. Hughes (1999) proposed a Monte Carlo EM algorithm for LME models with censored responses. Fitzgerald et al. (2002) extended Hughes' method to NLME models with censored data. In the context of survival data, Pettitt (1986) considered mixed effects models with right-censored data and discussed some special cases. Wu (2002) considered NLME models with both censored data and covariate measurement errors. In this chapter, we focus on likelihood methods for mixed effects models with censored responses.

6.2 Mixed Effects Models with Censored Responses

In this section, we consider mixed effects models for longitudinal data with left censored responses, based on an EM algorithm for likelihood estimation. The method can be extended to right censored responses or "doubly censored" responses in a straightforward way.

Let y_{ij} be the response for individual i at time t_{ij}, subject to left censoring, $i = 1, \ldots, n; \ j = 1, \ldots, n_i$. In the presence of left censoring, the observed value of the response y_{ij} can be written as (q_{ij}, c_{ij}), where q_{ij} is the observed value and c_{ij} is the censoring indicator such that y_{ij} is observed if $c_{ij} = 0$ and y_{ij} is left censored if $c_{ij} = 1$, i.e.,

$$y_{ij} = \begin{cases} q_{ij} & \text{if } c_{ij} = 0, \\ \leq d & \text{if } c_{ij} = 1, \end{cases} \tag{6.1}$$

where d is a known constant such as a detection limit. Let $\mathbf{z}_i$ be a collection of covariates. Denote $\mathbf{y}_i = (y_{i1}, \ldots, y_{in_i})^T, \mathbf{q}_i = (q_{i1}, \ldots, q_{in_i})^T$, and $\mathbf{c}_i = (c_{i1}, \ldots, c_{in_i})^T$. The observed data are $\{(\mathbf{q}_i, \mathbf{c}_i, \mathbf{z}_i), \ i = 1, \ldots, n\}$.

Let $f(\mathbf{y}_i | \mathbf{z}_i, \mathbf{b}_i, \beta, \sigma)$ be the density function of a mixed effects model, given random effects $\mathbf{b}_i$, where β contains mean parameters and σ contains variance-covariance parameters. We assume that $y_{i1}, \ldots, y_{in_i}$ are conditionally independent given the random effects $\mathbf{b}_i$, and we assume that $\mathbf{b}_i \sim N(0, B)$, where

B is a unknown covariance matrix. Let $f(\cdot)$ denote a generic density function and $F(\cdot)$ denote the corresponding cumulative density function (cdf). Conditioning on the random effects $\mathbf{b}_i$, a detectable measurement y_{ij} contributes $f(y_{ij}|\mathbf{z}_i, \mathbf{b}_i, \boldsymbol{\beta}, \boldsymbol{\sigma})$ in the likelihood, whereas a non-detectable measurement contributes

$$F(d\,|\mathbf{z}_i, \mathbf{b}_i, \boldsymbol{\beta}, \boldsymbol{\sigma}) \equiv P(Y_{ij} < d\,|\mathbf{z}_i, \mathbf{b}_i, \boldsymbol{\beta}, \boldsymbol{\sigma})$$

in the likelihood, where Y_{ij} is the random version of y_{ij}. Let $\boldsymbol{\theta}$ be the collection of all unknown parameters. The likelihood for the *observed data* $\{(\mathbf{q}_i, \mathbf{c}_i, \mathbf{z}_i),\ i = 1, \ldots, n\}$ can be written as

$$L_o(\boldsymbol{\theta}) = \prod_{i=1}^{n} \int \left\{ \prod_{j=1}^{n_i} (f(y_{ij}|\,\mathbf{z}_i, \mathbf{b}_i, \boldsymbol{\theta}))^{1-c_{ij}}\, (F(d\mid \mathbf{z}_i, \mathbf{b}_i, \boldsymbol{\theta}))^{c_{ij}} \right\}$$
$$\times f(\mathbf{b}_i|B)\, d\mathbf{b}_i,$$

which generally does not have an analytic expression.

For LME models with censored responses, Hughes (1999) proposed a Monte Carlo EM algorithm to find the MLEs of the parameters $\boldsymbol{\theta}$. His method can be extended to GLMM and NLME models with censored data. The idea is as follows. Treating the non-detectable components in $\mathbf{y}_i$ and the random effects $\mathbf{b}_i$ as "missing data", we have *complete data* $\{(\mathbf{y}_i, \mathbf{z}_i, \mathbf{b}_i), i = 1, 2, \cdots, n\}$. The complete-data log-likelihood can then be written as

$$l_c(\boldsymbol{\theta}) = \sum_{i=1}^{n} l_c^{(i)}(\boldsymbol{\theta}) = \sum_{i=1}^{n} [\log f(\mathbf{y}_i|\mathbf{z}_i, \mathbf{b}_i, \boldsymbol{\theta}) + \log f(\mathbf{b}_i|\boldsymbol{\theta})]\,.$$

The E-step for the i-th observation at the $(k+1)$-th EM iteration can be written as

$$Q_i(\boldsymbol{\theta}|\boldsymbol{\theta}^{(k)}) = E\left\{ l_c^{(i)}(\boldsymbol{\theta}) \mid \mathbf{q}_i, \mathbf{c}_i, \mathbf{z}_i; \boldsymbol{\theta}^{(k)} \right\}, \quad k = 1, 2, \cdots \tag{6.2}$$

To implement this E-step, we can use Monte Carlo methods to approximate the conditional expectation $Q_i(\boldsymbol{\theta}|\boldsymbol{\theta}^{(k)})$, similar to the approaches presented in previous chapters.

In the next section, we follow Hughes (1999) and describe a Monte Carlo EM algorithm for LME models in which *analytic* expressions for the E-step and the M-step can be obtained. Then in the following sections we extend the method to GLMM and NLME models for which the computation becomes more intensive since analytic expressions are no longer available. These approaches are conceptually similar to the Monte Carlo EM algorithms in Chapter 4 for non-ignorable missing responses, since censored data can be considered as special non-ignorable missing data.

6.2.1 LME Models

Consider the following standard LME model

$$\begin{aligned}
\mathbf{y}_i &= X_i\boldsymbol{\beta} + Z_i\mathbf{b}_i + \mathbf{e}_i, \qquad i = 1, 2, \cdots, n, \\
\mathbf{b}_i &\sim N(0, B), \qquad \mathbf{e}_i \sim N(0, \sigma^2 I),
\end{aligned} \tag{6.3}$$

where the design matrices X_i and Z_i contain covariates $\mathbf{z}_i$. Suppose that the response $\mathbf{y}_i$ is subject to left censoring. Let $\boldsymbol{\theta} = (\boldsymbol{\beta}, B, \sigma^2)$ be all unknown parameters. In this section, we follow Hughes (1999) and find MLE of $\boldsymbol{\theta}$ based on a Monte Carlo EM algorithm.

For LME model (6.3), the "complete-data" log-likelihood for individual i can be written as

$$\begin{aligned}
l_c^{(i)}(\boldsymbol{\theta}) &= \left\{ -\frac{n_i}{2}\log(2\pi\sigma^2) - \frac{1}{2\sigma^2}(\mathbf{y}_i - X_i\boldsymbol{\beta} - Z_i\mathbf{b}_i)^T(\mathbf{y}_i - X_i\boldsymbol{\beta} - Z_i\mathbf{b}_i) \right\} \\
&\quad + \left\{ -\frac{1}{2}\log(2\pi|B|) - \frac{1}{2}\mathbf{b}_i^T B^{-1}\mathbf{b}_i \right\}.
\end{aligned}$$

To compute $Q_i(\boldsymbol{\theta}|\boldsymbol{\theta}^{(k)}) = E\left[l_c^{(i)}(\boldsymbol{\theta}) \mid \mathbf{q}_i, \mathbf{c}_i, \mathbf{z}_i; \boldsymbol{\theta}^{(k)}\right]$ in the E-step of the EM algorithm at the k-th iteration, we only need to compute the following conditional expectations of the sufficient statistics for $\boldsymbol{\beta}$, B, and σ^2 respectively:

$$E(\mathbf{y}_i|\mathbf{q}_i, \mathbf{c}_i, \mathbf{z}_i; \boldsymbol{\theta}^{(k)}), \quad E(\mathbf{b}_i\mathbf{b}_i^T|\mathbf{q}_i, \mathbf{c}_i, \mathbf{z}_i, \boldsymbol{\theta}^{(k)}), \quad E(\mathbf{e}_i^T\mathbf{e}_i|\mathbf{q}_i, \mathbf{c}_i, \mathbf{z}_i, \boldsymbol{\theta}^{(k)}).$$

To compute the above conditional expectations of the sufficient statistics, note that

$$\begin{aligned}
f(\mathbf{b}_i|\mathbf{q}_i, \mathbf{c}_i, \mathbf{z}_i, \boldsymbol{\theta}^{(k)}) &= \int_{R_i} f(\mathbf{b}_i|\mathbf{y}_i, \mathbf{z}_i, \boldsymbol{\theta}^{(k)})d\mathbf{y}_i \\
f(\mathbf{e}_i|\mathbf{q}_i, \mathbf{c}_i, \mathbf{z}_i, \boldsymbol{\theta}^{(k)}) &= \int_{R_i} f(\mathbf{e}_i|\mathbf{y}_i, \mathbf{z}_i, \boldsymbol{\theta}^{(k)})d\mathbf{y}_i,
\end{aligned}$$

where R_i is the set of $\mathbf{y}_i$ consistent with the observed data $(\mathbf{q}_i, \mathbf{c}_i)$. That is, the distributions with censored data can be obtained by averaging the corresponding distributions with uncensored data over the values of $\mathbf{y}_i$ that are consistent with the observed censoring patterns. Therefore, we have

$$\begin{aligned}
E(\mathbf{b}_i\mathbf{b}_i^T|\mathbf{q}_i, \mathbf{c}_i, \mathbf{z}_i; \boldsymbol{\theta}^{(k)}) &= \int_{R_i} E(\mathbf{b}_i\mathbf{b}_i^T|\mathbf{y}_i, \mathbf{z}_i; \boldsymbol{\theta}^{(k)})f(\mathbf{y}_i|\mathbf{q}_i, \mathbf{c}_i, \mathbf{z}_i; \boldsymbol{\theta}^{(k)})d\mathbf{y}_i, \\
E(\mathbf{e}_i^T\mathbf{e}_i|\mathbf{q}_i, \mathbf{c}_i, \mathbf{z}_i; \boldsymbol{\theta}^{(k)}) &= \int_{R_i} E(\mathbf{e}_i^T\mathbf{e}_i|\mathbf{y}_i, \mathbf{z}_i; \boldsymbol{\theta}^{(k)})f(\mathbf{y}_i|\mathbf{q}_i, \mathbf{c}_i, \mathbf{z}_i; \boldsymbol{\theta}^{(k)})d\mathbf{y}_i,
\end{aligned}$$

where

$$\begin{aligned}
E(\mathbf{b}_i\mathbf{b}_i^T|\mathbf{y}_i, \mathbf{z}_i; \boldsymbol{\theta}^{(k)}) &= \hat{\mathbf{b}}_i\hat{\mathbf{b}}_i^T + D^{(k)} - D^{(k)}Z_i^T W_i Z_i D^{(k)}, \\
E(\mathbf{e}_i^T\mathbf{e}_i|\mathbf{y}_i, \mathbf{z}_i; \boldsymbol{\theta}^{(k)}) &= \hat{\mathbf{e}}_i^T\hat{\mathbf{e}}_i + \hat{\sigma}^{2(k)}(n_i - \hat{\sigma}^{2(k)}\mathrm{tr}(W_i)),
\end{aligned}$$

$$\hat{\mathbf{b}}_i = D^{(k)} Z_i^T W_i (\mathbf{y}_i - X_i \boldsymbol{\beta}^{(k)}),$$
$$\hat{\mathbf{e}}_i = (I_i - Z_i D^{(k)} Z_i^T W_i)(\mathbf{y}_i - X_i \boldsymbol{\beta}^{(k)}),$$
$$V_i = Z_i D Z_i^T + \sigma^2 I,$$
$$W_i = V_i^{-1}.$$

Note that the above conditional expectations are obtained based on the fact that $(\mathbf{b}_i, \mathbf{y}_i)$ follows a multivariate normal distribution, so these conditional expectations are obtained using well-known results for multivariate normal distributions.

Computation of the foregoing conditional expectations in the E-step is challenging due to the intractable integrals. Hughes (1999) proposed to use a Monte Carlo method to approximate these conditional expectations in the E-step. The idea is to simulate samples of the censored responses $\mathbf{y}_i$ from the conditional distribution $f(\mathbf{y}_i | \mathbf{q}_i, \mathbf{c}_i, \mathbf{z}_i; \boldsymbol{\theta}^{(k)})$ at EM iteration k, and then approximate the conditional expectations by their corresponding empirical means. Specifically, suppose that $\{\mathbf{y}_i^{(j)}, \ j = 1, \cdots, m_k\}$ are m_k simulated values of the censored components in $\mathbf{y}_i$ generated from distribution $f(\mathbf{y}_i | \mathbf{q}_i, \mathbf{c}_i, \mathbf{z}_i; \boldsymbol{\theta}^{(k)})$, where the observed (non-censored) components of $\mathbf{y}_i$ remain unchanged. Then, we have

$$E(\mathbf{y}_i | \mathbf{q}_i, \mathbf{c}_i, \mathbf{z}_i; \boldsymbol{\theta}^{(k)}) \quad \approx \quad \frac{1}{m_k} \sum_{j=1}^{m_k} \mathbf{y}_i^{(j)}, \tag{6.4}$$

$$E(\mathbf{b}_i \mathbf{b}_i^T | \mathbf{q}_i, \mathbf{c}_i, \mathbf{z}_i; \boldsymbol{\theta}^{(k)}) \quad \approx \quad \frac{1}{m_k} \sum_{j=1}^{m_k} E(\mathbf{b}_i \mathbf{b}_i^T | \mathbf{y}_i^{(j)}, \mathbf{z}_i; \boldsymbol{\theta}^{(k)}), \tag{6.5}$$

$$E(\mathbf{e}_i^T \mathbf{e}_i | \mathbf{q}_i, \mathbf{c}_i, \mathbf{z}_i; \boldsymbol{\theta}^{(k)}) \quad \approx \quad \frac{1}{m_k} \sum_{j=1}^{m_k} E(\mathbf{e}_i^T \mathbf{e}_i | \mathbf{y}_i^{(j)}, \mathbf{z}_i; \boldsymbol{\theta}^{(k)}). \tag{6.6}$$

The above approximations can be made arbitrary accurate by increase the number of Monte Carlo samples m_k. Typically we choose m_k to increase with the EM iteration number k.

To simulate samples from the conditional distribution $f(\mathbf{y}_i | \mathbf{q}_i, \mathbf{c}_i, \mathbf{z}_i; \boldsymbol{\theta})$ in the E-step, we can first simulate samples from the multivariate normal distribution $f(\mathbf{y}_i | \mathbf{z}_i; \boldsymbol{\theta})$, i.e., $N(X_i \boldsymbol{\beta}, V_i)$, and then reject those samples that do not fit the observed censoring pattern, but this procedure may have a very low acceptance rate. Hughes (1999) suggested to use the Gibbs sampler by simulating from univariate conditional distributions. Breslaw (1994) showed how to use the Gibbs sampler to sample from the truncated normal distribution.

The M-step of the Monte Carlo EM algorithm can be written as

$$\boldsymbol{\beta}^{(k+1)} \quad = \quad \sum_{i=1}^{n} \left(X_i^T W_i X_i \right)^{-1} X_i^T W_i E(\mathbf{y}_i | \mathbf{q}_i, \mathbf{c}_i, \mathbf{z}_i; \boldsymbol{\theta}^{(k)}), \tag{6.7}$$

$$D^{(k+1)} = \sum_{i=1}^{n} E(\mathbf{b}_i \mathbf{b}_i^T | \mathbf{q}_i, \mathbf{c}_i, \mathbf{z}_i; \boldsymbol{\theta}^{(k)})/n, \qquad (6.8)$$

$$\sigma^{2(k+1)} = \sum_{i=1}^{n} E(\mathbf{e}_i^T \mathbf{e}_i | \mathbf{q}_i, \mathbf{c}_i, \mathbf{z}_i; \boldsymbol{\theta}^{(k)}) \Big/ \sum_{i=1}^{n} n_i, \quad k = 1, 2, \cdots (6.9)$$

Iterating between the above E- and M-steps until convergence, we obtain the MLE of $\boldsymbol{\theta}$. Approximate standard errors of the parameter estimates can be obtained based on a modified version of Louis (1982) formula as follows:

$$\text{Var}(\hat{\boldsymbol{\beta}}) = \left[\sum_{i=1}^{n} X_i^T V_i^{-1} X_i - X_i^T V_i^{-1} U_i V_i^{-1} X_i \right]^{-1},$$

where

$$U_i = \text{Var}(\mathbf{y}_i - X_i \boldsymbol{\beta} \mid \mathbf{q}_i, \mathbf{c}_i, \mathbf{z}_i, \hat{\boldsymbol{\theta}}).$$

Based on the fitted model, we can obtain predicted values for the unobserved responses as follows. At convergence of the EM algorithm, the empirical Bayes estimate of the random effects $\mathbf{b}_i$ is given by

$$\hat{\mathbf{b}}_i = D Z_i^T W_i (E(\mathbf{y}_i | \mathbf{q}_i, \mathbf{c}_i, \mathbf{z}_i, \hat{\boldsymbol{\theta}}) - X_i \hat{\boldsymbol{\beta}}).$$

Then, at any time point t_{ij}, the predicted value of a unobserved (censored) y_{ij} can be obtained by substituting the parameters and random effects by their estimates, i.e.,

$$\hat{\mathbf{y}}_i = X_i \hat{\boldsymbol{\beta}} + Z_i \hat{\mathbf{b}}_i.$$

An approximate variance for the predicted value is given by

$$\widehat{\text{Var}}(\hat{\mathbf{y}}_i) \approx Z_i \hat{B} Z_i^T + \hat{\sigma}^2 I,$$

and an approximate confidence interval for a unobserved (censored) y_{ij} is then given by

$$\hat{y}_{ij} \pm z_{\alpha/2} (\widehat{Var}(\hat{y}_{ij}))^{1/2},$$

where $z_{\alpha/2}$ is the $1 - \alpha/2$ percentile of the standard normal distribution $N(0, 1)$.

In the above approach we have assumed that the models hold for censored values. In some applications, however, the assumed models may not hold for censored values. In this case, alternative approaches are required.

Example 6.1 *LME models with censored responses*

Return to the AIDS dataset in Section 1.3.2 of Chapter 1. As an illustration of the method in this section, we focus on viral load data in the first 90 days, and empirically model the data using the following LME model, which is a

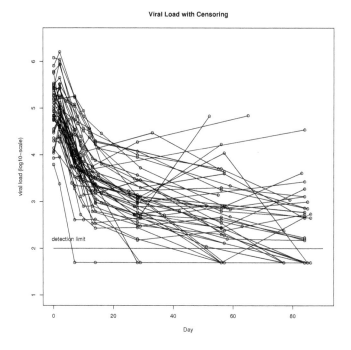

Figure 6.1 *Viral loads trajectories (in log_{10} scale) in the first three months. The viral loads have a lower detection limit of $log_{10}(100) = 2$. Viral loads below the detection limit are substituted by half the limit (i.e., $log_{10}(50) = 1.69$).*

quadratic polynomial with random coefficients:

$$y_{ij} = (\beta_1 + b_{1i}) + (\beta_2 + b_{2i})t_{ij} + (\beta_3 + b_{3i})t_{ij}^2 + e_{ij}, \quad (6.10)$$

$$(b_{1i}, b_{2i}, b_{3i})^T \sim N(0, D), \qquad e_{ij} \text{ i.i.d. } \sim N(0, \sigma^2), \quad (6.11)$$

where y_{ij} is the log_{10}-transformation of the viral load for individual i at time t_{ij}, $i = 1, \cdots, n$; $j = 1, \cdots, n_i$, β_j's are fixed parameters, and b_{ji}'s are random effects. In this study, viral loads have a lower detection limit of 100, which is 2 after the log_{10}-transformation. Figure 6.1 shows the viral load data, in which the viral loads below the detection limit are left censored and are substituted by half the detection limit (i.e., $log_{10}(50) = 1.69$).

We consider two estimation methods: the naive method, which simply imputes censored viral loads by half the detection limit, and the likelihood method, which is based on the Monte Carlo EM algorithm of Hughes (1999), as described in this section. The Monte Carlo EM algorithm is considered converged if the maximum relative change in parameters between successive EM itera-

Table 6.1 *Parameter estimates for the LME model with censoring*

Method	Par.	Est.	S.E.	Par.	Est.	S.E.	Par.	Est.	S.E.
Naive	β_1	4.77	0.087	β_2	−6.97	0.323	β_3	5.09	0.335
Likelihood	β_1	4.78	0.083	β_2	−7.15	0.363	β_3	5.01	0.409

Par.: parameter, Est.: estimate, S.E.: standard error.

tions is less than 0.001. The initial number of Monte Carlo samples in the EM algorithm is chosen to be 500, and it is doubled if the absolute change in any parameter is less than the estimated Monte Carlo standard error for that parameter, as suggested by Hughes (1999). The burn-in period for the Gibbs sampler is 25 iterations, since the sampler burns in quickly in this example.

Table 6.1 shows the estimation results. We see that the two methods give different results. In particular, the likelihood method produces larger standard errors of the parameter estimates, which reflects the *uncertainty* of the censored values. On the other hand, the naive method ignores the uncertainty of the censored values so it under-estimates the standard errors of the parameter estimates, and it may also give biased estimation. The results based on the Monte Carlo EM algorithm should be more reliable than that based on the naive method.

6.2.2 GLMM and NLME Models

We can extend the method for LME models in Section 6.2.1 to GLMM and NLME models with censored responses. For GLMM and NLME models, the parameters and random effects are *nonlinear* in the models, so two difficulties arise:

- the Monte Carlo E-step becomes more complicated since the random effects cannot be integrated out,
- analytic expressions for the E-step and the M-step are no longer available, so iterative algorithms are needed.

Therefore, although the extension from LME models to GLMM and NLME models is conceptually straightforward, the computation becomes much more challenging, as described below.

For GLMM and NLME models with censored responses, the E-step of a Monte Carlo EM algorithm for likelihood estimation is to compute the following conditional expectation of the "complete data" loglikelihood given the observed data and parameter estimates at the current iteration:

$$Q_i(\boldsymbol{\theta}|\boldsymbol{\theta}^{(k)}) = \int\int \left[\log f(\mathbf{y}_i|\mathbf{z}_i, \mathbf{b}_i, \boldsymbol{\theta}^{(k)}) + \log f(\mathbf{b}_i|\boldsymbol{\theta}^{(k)})\right]$$
$$\times f(\mathbf{y}_i, \mathbf{b}_i|\mathbf{q}_i, \mathbf{c}_i, \mathbf{z}_i; \boldsymbol{\theta}^{(k)})\, d\mathbf{y}_{cen,i}\, d\mathbf{b}_i,$$

where $\mathbf{y}_{cen,i}$ contains the censored components of $\mathbf{y}_i$ and k is the EM iteration number. Since the density function $f(\mathbf{y}_i|\mathbf{z}_i, \mathbf{b}_i, \boldsymbol{\theta}^{(k)})$ is nonlinear in the random effects $\mathbf{b}_i$ for both GLMM and NLME models, we cannot integrate out the unobservable random effects $\mathbf{b}_i$ in $Q_i(\boldsymbol{\theta}|\boldsymbol{\theta}^{(k)})$, so we are unable to obtain closed-form expressions similar to those in Section 6.2.1. However, we can use Monte Carlo methods to simulate the "missing data" $(\mathbf{y}_{cen,i}, \mathbf{b}_i)$ from the conditional distribution $f(\mathbf{y}_i, \mathbf{b}_i|\mathbf{q}_i, \mathbf{c}_i, \mathbf{z}_i; \boldsymbol{\theta}^{(k)})$, and then approximate $Q_i(\boldsymbol{\theta}|\boldsymbol{\theta}^{(k)})$ by an empirical mean based on many simulated values of the missing data, as in Chapter 4. This simulation step can again be done by Gibbs sampler along with rejection sampling methods or the importance sampling method or other MCMC methods. We will show some details in Section 6.3.

Exact likelihood estimation for GLMM and NLME models with censored data based on Monte Carlo EM algorithms can be computationally intensive. However, as in Chapter 4, computationally more efficient approximate methods based on Taylor or Laplace approximations can be developed. For example, for the linearization procedure based on a first-order Taylor approximation, we iteratively solve a working LME model, so the analytic expressions in Section 6.2.1 for LME models can be used at each iteration, which greatly simplifies computation. We will show the details in Section 6.3 for a more complex but similar problem.

6.2.3 Imputation Methods

Single Imputation Methods

For censored longitudinal data, several naive single imputation methods are commonly used in practice. The simplest imputation method is to impute the censored values by the detection limit or half the detection limit. An obvious drawback of this naive single imputation method is that it fails to incorporate the *uncertainty* of the censored and unobserved values, in addition to possible biases resulted from the single imputed value (the detection limit or half the detection limit).

Another single imputation method is to impute the censored values based on a regression model. For example, we may generate an imputation for a censored response from its predictive distribution $f(\mathbf{y}_i|\mathbf{z}_i, \hat{\mathbf{b}}_i, \hat{\boldsymbol{\theta}})$, where the estimates $(\hat{\mathbf{b}}_i, \hat{\boldsymbol{\theta}})$ can be obtained from some simple method, e.g., from a model where

the censored values are substituted by half the detection limit. This imputation method again fails to incorporate the uncertainty of the censored values.
One may tempt to create multiple imputations based on model $f(\mathbf{y}_i|\mathbf{z}_i, \hat{\mathbf{b}}_i, \hat{\boldsymbol{\theta}})$.
However, such a multiple imputation method is usually improper (Rubin 1987),
since it fails to incorporate the uncertainty in estimating the random effects and
the parameters.

More appropriate single imputation methods can be based on regression models and then *adjust* the variances or standard errors of the parameter estimates
to reflect the uncertainty of the censored values. Note that the imputation
model, or the regression model used to create imputations, should incorporate the censoring information, i.e., the imputed values for the censored data
should not be larger than the detection limit if the data are left censored. An
assumption for such an approach is that the model continues to hold for censored values. The uncertainty of the censored values can also be incorporated
via the following proper multiple imputation method.

Multiple Imputation Methods

In the following, we briefly describe a multiple imputation method to generate
imputations for censored values $\mathbf{y}_{cen,i}$. As in Section 3.5 of Chapter 3, we can
generate multiple imputations via a Bayesian framework. Specifically, multiple
imputations can be generated from the following predictive distribution of the
censored values $\mathbf{y}_{cen,i}$ given the observed values $(\mathbf{y}_{obs,i}, \mathbf{c}_i, \mathbf{z}_i)$:

$$f(\mathbf{y}_{cen,i}|\mathbf{y}_{obs,i}, \mathbf{c}_i, \mathbf{z}_i) = \iiint f(\mathbf{y}_i|\mathbf{z}_i, \mathbf{c}_i, \mathbf{b}_i, \boldsymbol{\theta})f(\mathbf{b}_i|B)$$
$$\times f(\boldsymbol{\theta}|\boldsymbol{\eta})\ d\mathbf{y}_{cen,i}\ d\mathbf{b}_i\ d\boldsymbol{\theta},$$

where $\boldsymbol{\eta}$ contains hyper-parameters for the prior distribution of $\boldsymbol{\theta} \equiv$
$(\boldsymbol{\beta}, \sigma^2, B)$, $\mathbf{c}_i$ is a vector of censoring indicators as defined earlier, and $\mathbf{z}_i$ contains observed covariates.

Let $\boldsymbol{\eta}_0$ be the initial values of the hyper-parameters $\boldsymbol{\eta}$. One can use the following procedure to implement the multiple imputation method

- Generate a sample $\boldsymbol{\theta}^*$ from $f(\boldsymbol{\theta}|\boldsymbol{\eta}) = f(\boldsymbol{\beta}|\boldsymbol{\eta})f(\sigma^2|\boldsymbol{\eta})f(B|\boldsymbol{\eta})$, assuming
independence of the parameters;
- Generate a sample $\mathbf{b}_i^*$ from $f(\mathbf{b}_i|B^*)$, where B^* is a component in $\boldsymbol{\theta}^*$.
- Generate a sample $\mathbf{y}_{cen,i}^*$ from the following distribution via rejection sampling methods

$$f(\mathbf{y}_i|\mathbf{z}_i, \mathbf{c}_i, \mathbf{b}_i^*, \boldsymbol{\theta}^*) \propto f(\mathbf{y}_i|\mathbf{z}_i, \mathbf{b}_i^*, \boldsymbol{\theta}^*)f(\mathbf{c}_i|\mathbf{y}_i, \mathbf{z}_i, \mathbf{b}_i^*, \boldsymbol{\theta}^*).$$

Iterating the above procedure, we obtain a Markov chain

$$\left\{ (\mathbf{y}_{cen,i}^{*(k)}, \boldsymbol{\theta}^{*(k)}, \mathbf{b}_i^{*(k)}),\ k = 0, 1, 2, \cdots \right\},$$

which has a stationary distribution. After a burn-in period, we may view the last value $\mathbf{y}^*_{cen,i}$ as an imputation generated from the predictive distribution $f(\mathbf{y}_{cen,i}|\mathbf{y}_{obs,i}, \mathbf{c}_i, \mathbf{z}_i)$.

Repeating this process m times, we generate m "complete datasets". Each of these m complete datasets can be analyzed separately as if there were no censored values. The m analysis results are then combined to form an overall result. See Chapters 3 and 4 for details of similar procedures. Note that this multiple imputation method may be computationally more intensive than the Monte Carlo EM algorithm, since it adds another layer of sampling for the parameters.

Discussion

The likelihood method based on the EM algorithm and the multiple imputation method in previous sections can be extended to mixed effects models with censoring in covariates. When covariates are censored, the approaches are similar to that for censored responses, with some modifications. Censored covariates may arise, for example, when the censored response in a regression model is treated as a covariate in another regression model. In AIDS studies, sometimes we may treat CD4 as the response and viral load as a covariate, but viral load often has a lower detection limit. In this case, we can model the CD4 process using a mixed effects model, with the censored viral loads as a covariate.

Both the EM algorithm and the multiple imputation method are computationally intensive. To reduce computational burden, one may consider approximate methods which are computationally much more efficient. For example, for the multiple imputation method, we may use a first-order Taylor expansion to linearize a NLME model and then generate imputations based on the resulting working LME model. This can greatly reduce computational burden since (i) for LME models the random effects can be integrated out so sampling the random effects may not be needed; and (ii) closed-form expressions are available for LME models so some iterative algorithms may be avoided. In the meantime, such an approximate method may still generate reasonable imputations. Note that, for a multiple imputation method, the observed values remain unchanged throughout. Thus, the choice of imputation models only affects the imputed values, so if the proportion of censored values is not large, approximate imputation models should perform well.

For both the likelihood method based on the EM algorithm and the multiple imputation method in previous sections, we have assumed that the models hold for both observed and censored values. This assumption is not testable based on the observed data. In some applications, the assumed models may not hold for censored values, although they may fit the observed data well, since the censored values may be viewed as somewhat "unusual". One may consider models with mixture distributions, in which a different distribution is assumed for the censored values. Such an approach may be used for sensitivity analysis.

6.3 Mixed Effects Models with Censoring and Measurement Errors

In many longitudinal studies, we often need to address censored responses and covariate measurement error *simultaneously*. For example, in the studies of HIV viral dynamics, we may introduce CD4 cell counts to partially explain the variation in viral load trajectories. Since viral load often has a lower detection limit while CD4 is typically measured with errors, we need to address censored viral loads (response) and measurement errors in CD4 (covariate) simultaneously, since ignoring either one of them may lead to biased results. In this section, we discuss *joint inference* for mixed effects models with censored responses and covariate measurement errors.

We focus on a *joint likelihood* method and use a Monte Carlo EM algorithm for estimation. The idea is similar to that in Chapter 5 for joint inference of mixed effects models with non-ignorable missing responses and covariate measurement errors. One only needs to incorporate the censoring information and treats the censored responses as special non-ignorable missing responses. In the following sections, we first consider LME models with censored responses and covariate measurement errors, in which some analytic results can be obtained, following Wu (2004). Then, we extend the method to GLMM and NLME models with censored response and covariate measurement errors, following Wu (2002). For simplicity, we focus on a single time-dependent and error-prone covariate.

6.3.1 LME Models

In this section, we consider a LME model with censoring in the response and measurement errors in time-dependent covariates, following Wu (2004). Let z_{ik} be the observed but possibly mis-measured covariate value for individual i at time u_{ik}, $i = 1, \cdots, n$; $k = 1, \cdots, m_i$. To address measurement errors in covariate z_{ik}, we consider a classical measurement error model and model the covariate process by the following LME model

$$
\begin{aligned}
z_{ik} &= \mathbf{u}_{ik}^T \boldsymbol{\alpha}_i + \epsilon_{ik} \equiv z_{ik}^* + \epsilon_{ik}, & (6.12) \\
\boldsymbol{\alpha}_i &= \boldsymbol{\alpha} + \mathbf{a}_i, & i = 1, \cdots, n; \; k = 1, \cdots, m_i, \\
\mathbf{a}_i &\sim N(0, A), & \boldsymbol{\epsilon}_i \sim N(0, \delta^2 I),
\end{aligned}
$$

where $\mathbf{u}_{ik}$ is a $p \times 1$ design vector, $\boldsymbol{\alpha}_i = (\alpha_{i1}, \cdots, \alpha_{ip})^T$ is a $p \times 1$ vector of individual-specific regression parameters, ϵ_{ik} is the measurement error for individual i at time u_{ik} and $\epsilon_i = (\epsilon_{i1}, \dots, \epsilon_{im_i})^T$, $\alpha = (\alpha_1, \cdots, \alpha_p)^T$ is a $p \times 1$ vector of unknown fixed parameters (fixed effects), $\mathbf{a}_i$ are random effects, A is a unknown $p \times p$ covariance matrix, and δ^2 is the unknown variance for within-individual measurements. We can interpret

$$z_{ik}^* \equiv \mathbf{u}_{ik}^T \boldsymbol{\alpha}_i$$

as the "true" covariate value for individual i at time u_{ik}. In covariate model (6.12), we allow the covariate observation times u_{ik} to possibly differ from the response observation times t_{ij}, i.e., we allow missing data in the covariate where the missing data mechanism is assumed to be ignorable.

Note that, at a given time, the covariate trajectory up to that time may be summarized by the individual-specific parameters $\boldsymbol{\alpha}_i$. For example, if we fit a straight line to the covariate process, i.e., $\mathbf{u}_{ik} = (1, u_{ik})^T$, then parameters $\boldsymbol{\alpha}_i$ represents the individual-specific intercept and slope of the covariate process. When the observed covariates z_{ij}'s are measured with errors, we may be interested in the relationship between the response and the true covariate values z_{ij}^* or the past history or summaries of the regularized covariate values. That is, the response is assumed to be related to the covariate through the individual-specific parameter $\boldsymbol{\alpha}_i$ or the random effects $\mathbf{a}_i$. We assume that $\mathbf{z}_i$ is a surrogate for $\boldsymbol{\alpha}_i$.

Let $h(\boldsymbol{\alpha}_i)$ be a known linear function which summarizes the covariate process. For the response process, we consider the following LME model

$$
\begin{aligned}
y_{ij} &= \mathbf{x}_{ij}^T \boldsymbol{\beta}_i + e_{ij}, \qquad i = 1, \cdots, n; \quad j = 1, \cdots, n_i, \qquad (6.13) \\
\boldsymbol{\beta}_i &= \mathbf{d}(\mathbf{z}_i^*, \boldsymbol{\beta}, \mathbf{b}_i) \equiv \mathbf{d}(h(\boldsymbol{\alpha}_i), \boldsymbol{\beta}, \mathbf{b}_i), \\
\mathbf{b}_i &\sim N(\mathbf{0}, B), \qquad \mathbf{e}_i \sim N(\mathbf{0}, \sigma^2 I),
\end{aligned}
$$

where $\mathbf{x}_{ij}$ is a $s \times 1$ design vector, $\boldsymbol{\beta}_i = (\beta_{i1}, \cdots, \beta_{is})^T$ is a $(s \times 1)$ vector of individual-specific regression parameters, $\boldsymbol{\beta} = (\beta_1, \cdots, \beta_r)^T$ is a $(r \times 1)$ vector of population parameters (fixed effects), $\mathbf{d}(\cdot)$ is a s-dimensional known linear function, $\mathbf{e}_i = (e_{i1}, \cdots, e_{in_i})^T$, σ^2 is the unknown within individual variance, I is the identity matrix, B is a $s \times s$ covariance matrix, and $\mathbf{b}_i = (b_{i1}, \cdots, b_{is})^T$ is a vector of random effects.

A common form for the linear function $\mathbf{d}(\cdot)$ is

$$\mathbf{d}(h(\boldsymbol{\alpha}_i), \boldsymbol{\beta}, \mathbf{b}_i) = D_{ij} \boldsymbol{\beta} + V_i \mathbf{b}_i,$$

where the matrix D_{ij} depends on $\boldsymbol{\alpha}_i$ and other covariates such as time and V_i is a matrix of 0's and 1's. We assume that ϵ_i is independent of $\mathbf{a}_i$, e_i is independent of ϵ_i, and $\mathbf{b}_i$ is independent of e_i and $\mathbf{a}_i$. In model (6.13), we link the individual-specific regression parameters $\boldsymbol{\beta}_{ij}$ to the error-prone covariate through the function $h(\boldsymbol{\alpha}_i)$ which summarizes the true covariate process. Note that the parameters of the unrestricted covariance matrices A and B are identifiable only if $n > \max\{1 + p(p+1)/2, \ 1 + s(s+1)/2\}$.

In the presence of censoring in the response and measurement error in the covariate, we consider a joint likelihood based on the covariate and response models (6.12) and (6.13). The observed data are $\{(\mathbf{q}_i, \mathbf{c}_i, \mathbf{z}_i), \ i = 1, \ldots, n\}$ and the observed-data joint likelihood can be written as

$$
L_o(\boldsymbol{\theta}) \;=\; \prod_{i=1}^{n} \left[\int \int \left\{ \prod_{j=1}^{n_i} \left(f(y_{ij}|z_{ij}^*, \boldsymbol{\alpha}_i, \mathbf{b}_i) \right)^{1-c_{ij}} \right. \right.
$$

$$
\times \left. \left(F(d \mid z_{ij}^*, \boldsymbol{\alpha}_i, \mathbf{b}_i) \right)^{c_{ij}} \right\}
$$

$$
\times \left\{ \prod_{k=1}^{m_i} f(z_{ik}|\boldsymbol{\alpha}_i) \right\} f(\boldsymbol{\alpha}_i) f(\mathbf{b}_i) d\boldsymbol{\alpha}_i d\mathbf{b}_i \Bigg], \quad (6.14)
$$

where

$$
(y_{ij}|z_{ij}^*, \boldsymbol{\alpha}_i, \mathbf{b}_i) \;\sim\; N(\mathbf{x}_{ij}^T\boldsymbol{\beta}_{ij}, \sigma^2) \equiv N(\mathbf{x}_{ij}^T\mathbf{d}(h(\boldsymbol{\alpha}_i), \boldsymbol{\beta}, \mathbf{b}_i), \ \sigma^2)
$$

$$
(z_{ik}|\boldsymbol{\alpha}_i) \;\sim\; N(\mathbf{u}_{ik}^T\boldsymbol{\alpha}_i, \delta^2) \equiv N(z_{ik}^*, \ \delta^2).
$$

We consider an EM algorithm for computing the MLEs of the parameters. The E-step for the i-th observation at the $(k+1)$st EM iteration can be written as

$$
Q_i(\boldsymbol{\theta}|\boldsymbol{\theta}^{(k)}) \;=\; E\Big[\big\{ \log f(\mathbf{y}_i|\boldsymbol{\alpha}_i, \mathbf{b}_i, \boldsymbol{\theta}) + \log f(\mathbf{z}_i|\boldsymbol{\alpha}_i, \boldsymbol{\theta})
$$

$$
+ \log f(\boldsymbol{\alpha}_i|\boldsymbol{\theta}) + \log f(\mathbf{b}_i|\boldsymbol{\theta}) \big\} \mid \mathbf{q}_i, \mathbf{c}_i, \mathbf{z}_i; \boldsymbol{\theta}^{(k)} \Big].
$$

As in Section 6.2, to carry out the E-step, we only need to compute the following conditional expectations of the sufficient statistics (also see the Appendix):

$$
E(\mathbf{y}_i|\mathbf{q}_i, \mathbf{c}_i, \mathbf{z}_i; \boldsymbol{\theta}^{(k)}), \quad E(\boldsymbol{\alpha}_i|\mathbf{q}_i, \mathbf{c}_i, \mathbf{z}_i, \boldsymbol{\theta}^{(k)}), \quad E(\boldsymbol{\alpha}_i\boldsymbol{\alpha}_i^T|\mathbf{q}_i, \mathbf{c}_i, \mathbf{z}_i, \boldsymbol{\theta}^{(k)}),
$$

$$
E(\mathbf{b}_i|\mathbf{q}_i, \mathbf{c}_i, \mathbf{z}_i, \boldsymbol{\theta}^{(k)}), \quad E(\mathbf{b}_i\mathbf{b}_i^T|\mathbf{q}_i, \mathbf{c}_i, \mathbf{z}_i, \boldsymbol{\theta}^{(k)}).
$$

To compute these quantities, we note that

$$
E(\boldsymbol{\alpha}_i|\mathbf{q}_i, \mathbf{c}_i, \mathbf{z}_i; \boldsymbol{\theta}^{(k)}) \;=\; \int_{R_i} E(\boldsymbol{\alpha}_i|\mathbf{y}_i, \mathbf{z}_i; \boldsymbol{\theta}^{(k)}) f(\mathbf{y}_i|\mathbf{q}_i, \mathbf{c}_i, \mathbf{z}_i; \boldsymbol{\theta}^{(k)}) d\mathbf{y}_i,
$$

$$
E(\boldsymbol{\alpha}_i\boldsymbol{\alpha}_i^T|\mathbf{q}_i, \mathbf{c}_i, \mathbf{z}_i; \boldsymbol{\theta}^{(k)}) \;=\; \int_{R_i} E(\boldsymbol{\alpha}_i\boldsymbol{\alpha}_i^T|\mathbf{y}_i, \mathbf{z}_i; \boldsymbol{\theta}^{(k)})
$$

$$
\times f(\mathbf{y}_i|\mathbf{q}_i, \mathbf{c}_i, \mathbf{z}_i; \boldsymbol{\theta}^{(k)}) d\mathbf{y}_i,
$$

where R_i is the set of $\mathbf{y}_i$ consistent with the observed data $(\mathbf{q}_i, \mathbf{c}_i)$. Similar expressions hold for $E(\mathbf{b}_i|\mathbf{q}_i, \mathbf{c}_i, \mathbf{z}_i; \boldsymbol{\theta}^{(k)})$ and $E(\mathbf{b}_i\mathbf{b}_i^T|\mathbf{q}_i, \mathbf{c}_i, \mathbf{z}_i; \boldsymbol{\theta}^{(k)})$.

The Monte Carlo EM algorithm then proceeds by generating $\mathbf{y}_i$ from the conditional distribution $f(\mathbf{y}_i|\mathbf{q}_i, \mathbf{c}_i, \mathbf{z}_i; \boldsymbol{\theta}^{(k)})$ using the Gibbs sampler, and then approximate the expectation in the E-step by an empirical mean, as in Section 6.2. These computations involve evaluating the conditional expectations: $E(\alpha_i | \mathbf{y}_i, \mathbf{z}_i; \boldsymbol{\theta}^{(k)})$, $E(\alpha_i \alpha_i^T | \mathbf{y}_i, \mathbf{z}_i; \boldsymbol{\theta}^{(k)})$, $E(\mathbf{b}_i | \mathbf{y}_i, \mathbf{z}_i; \boldsymbol{\theta}^{(k)})$, and $E(\mathbf{b}_i \mathbf{b}_i^T | \mathbf{y}_i, \mathbf{z}_i; \boldsymbol{\theta}^{(k)})$. To obtain these conditional expectations, note that $(\alpha_i, \mathbf{y}_i, \mathbf{z}_i)$ and $(\mathbf{b}_i, \mathbf{y}_i, \mathbf{z}_i)$ follow multivariate normal distributions. Thus the conditional expectations can be obtained using well-known results for multivariate normal distributions (see the Appendix).

The M-step of the Monte Carlo EM algorithm is like a complete-data maximization and is thus straightforward. Let $\hat{\boldsymbol{\theta}}$ denote the estimate of $\boldsymbol{\theta}$ at the convergence of the Monte Carlo EM algorithm. The observed information matrix is given by

$$
\begin{aligned}
I(\mathbf{q}_i, \mathbf{c}_i, \mathbf{z}_i; \hat{\boldsymbol{\theta}}) \;=\;& E_{\boldsymbol{\theta}} \left(\frac{\partial^2 l_c(\boldsymbol{\theta})}{\partial \boldsymbol{\theta} \partial \boldsymbol{\theta}^T} \;\middle|\; \mathbf{q}_i, \mathbf{c}_i, \mathbf{z}_i \right) \Bigg|_{\boldsymbol{\theta}=\hat{\boldsymbol{\theta}}} \\
& - E_{\boldsymbol{\theta}} \left(S_c(\hat{\boldsymbol{\theta}}) S_c^T(\hat{\boldsymbol{\theta}}) \;\middle|\; \mathbf{q}_i, \mathbf{c}_i, \mathbf{z}_i \right),
\end{aligned}
$$

where

$$
S_c(\hat{\boldsymbol{\theta}}) = \frac{\partial l_c(\boldsymbol{\theta})}{\partial \boldsymbol{\theta}} \Big|_{\boldsymbol{\theta}=\hat{\boldsymbol{\theta}}}.
$$

This information matrix can be computed using the conditional expectations derived in the Appendix. The estimate of the asymptotic covariance matrix of $(\hat{\alpha}, \hat{\beta})$ is the upper $(p+r) \times (p+r)$ block of $I^{-1}(\mathbf{q}_i, \mathbf{c}_i, \mathbf{z}_i; \hat{\boldsymbol{\theta}})$.

6.3.2 GLMM and NLME Models

The method for LME models in Section 6.3.1 can be extended to GLMM and NLME models with censored response and covariate measurement errors. The extension is relatively straightforward but the computation can be much more challenging. In this section, we follow Wu (2002) and focus on NLME models, but the method can also be applied to GLMMs.

Consider a NLME model with an error-prone time-dependent covariate. We write the NLME model as a hierarchical two-stage model (Davidian and Giltinan 1995), i.e., the first stage models the within-individual data and specifies the intra-individual variation

$$
y_{ij} \;=\; g(\mathbf{t}_{ij}, \boldsymbol{\beta}_{ij}) + e_{ij}, \qquad \mathbf{e}_i \sim N(\mathbf{0}, \sigma^2 I), \tag{6.15}
$$
$$
j = 1, \cdots, n_i, \; i = 1, \cdots, n,
$$

while the second stage introduces covariates and random effects to account for inter-individual variation

$$
\boldsymbol{\beta}_{ij} \;=\; \mathbf{d}(z_{ij}^*, \boldsymbol{\beta}, \mathbf{b}_i), \qquad \mathbf{b}_i \;\; \text{i.i.d.} \sim N(\mathbf{0}, B), \tag{6.16}
$$

where $g(\cdot)$ is a known nonlinear function, $\mathbf{t}_{ij}$ is a vector containing independent variables such as time, $\boldsymbol{\beta}_{ij} = (\beta_{ij1}, \cdots, \beta_{ijs})^T$ are individual-specific regression parameters, $\boldsymbol{\beta} = (\beta_1, \cdots, \beta_r)^T$ are population parameters or fixed effects, $\mathbf{d}(\cdot)$ is a s-dimensional known linear function, σ^2 is the unknown within individual variance, I is the identity matrix, $\mathbf{b}_i$'s are random effects, and B is an unknown covariance matrix. We assume that $\mathbf{e}_i = (e_{i1}, \cdots, e_{in_i})^T$, $\mathbf{b}_i$, $\boldsymbol{\epsilon}_i$, and $\boldsymbol{\alpha}_i$ are independent of each other. The dependence of $\boldsymbol{\beta}_{ij}$ on $\mathbf{t}_{ij}$ through the function $\mathbf{d}(\cdot)$ is suppressed in the models. We also assume that the within-individual errors e_{ij} are conditionally independent given the random effects $\mathbf{b}_i$, but the method adapt readily to relaxation of this assumption.

As in Section 6.3.1, we consider the following classical measurement error model for the error-prone covariate

$$
\begin{aligned}
z_{ik} &= \mathbf{u}_{ik}^T \boldsymbol{\alpha}_i + \epsilon_{ik} \equiv z_{ik}^* + \epsilon_{ik}, & (6.17)\\
\boldsymbol{\alpha}_i &= \boldsymbol{\alpha} + \mathbf{a}_i, & i = 1, \cdots, n; \quad k = 1, \cdots, m_i,\\
\boldsymbol{\alpha}_i &\sim N(\boldsymbol{\alpha}, A), & \boldsymbol{\epsilon}_i \sim N(0, \delta^2 I),
\end{aligned}
$$

where ϵ_{ik} represents measurement error and other notation is the same as that in Section 6.3.1.

Let $\boldsymbol{\theta} = (\boldsymbol{\beta}, \boldsymbol{\alpha}, A, B, \sigma, \delta)$ be the collection of all unknown parameters. The *joint likelihood* for the observed data $\{(q_i, \mathbf{c}_i, \mathbf{z}_i),\ i = 1, \ldots, n\}$ can be written as follows

$$
\begin{aligned}
L_o(\boldsymbol{\theta}) = \prod_{i=1}^{n} &\left[\int \int \left\{ \prod_{j=1}^{n_i} (f(y_{ij}|\boldsymbol{\alpha}_i, \mathbf{b}_i))^{1-c_{ij}} (F(d\,|\,\boldsymbol{\alpha}_i, \mathbf{b}_i))^{c_{ij}} \right\} \right.\\
&\left. \times \left\{ \prod_{k=1}^{m_i} f(z_{ik}|\boldsymbol{\alpha}_i) \right\} f(\boldsymbol{\alpha}_i) f(\mathbf{b}_i)\, d\boldsymbol{\alpha}_i\, d\mathbf{b}_i \right],
\end{aligned}
$$

where

$$
f(y_{ij}|\boldsymbol{\alpha}_i, \mathbf{b}_i) = (2\pi\sigma^2)^{-1/2} \exp\left[-\frac{\{y_{ij} - g\left(\mathbf{t}_{ij}, \mathbf{d}(\mathbf{u}_{ij}^T \boldsymbol{\alpha}_i, \boldsymbol{\beta}, \mathbf{b}_i)\right)\}^2}{2\sigma^2} \right],
$$

$$
F(d\,|\,\boldsymbol{\alpha}_i, \mathbf{b}_i) = \int_{-\infty}^{d} f(y_{ij}|\boldsymbol{\alpha}_i, \mathbf{b}_i)\, dy_{ij},
$$

$$
f(z_{ik}|\boldsymbol{\alpha}_i) = (2\pi\delta^2)^{-1/2} \exp\left\{ -\frac{1}{2\delta^2} (z_{ij} - \mathbf{u}_{ik}^T \boldsymbol{\alpha}_i)^2 \right\},
$$

$$
f(\boldsymbol{\alpha}_i) = (2\pi|A|)^{-1/2} \exp\left\{ -\frac{1}{2} (\boldsymbol{\alpha}_i - \boldsymbol{\alpha})^T A^{-1} (\boldsymbol{\alpha}_i - \boldsymbol{\alpha}) \right\},
$$

$$
f(\mathbf{b}_i) = (2\pi|B|)^{-1/2} \exp\left\{ -\frac{1}{2} \mathbf{b}_i^T B^{-1} \mathbf{b}_i \right\}.
$$

We again consider a Monte Carlo EM algorithm to obtain the MLE of $\boldsymbol{\theta}$. The

E-step at the k-th EM iteration computes

$$Q(\boldsymbol{\theta}|\boldsymbol{\theta}^{(k)}) \;\; = \;\; \sum_{i=1}^{n} E\Big[\big(\log f(\mathbf{y}_i|\mathbf{z}_i, \boldsymbol{\alpha}_i, \mathbf{b}_i, \boldsymbol{\theta}) + \log f(\mathbf{z}_i|\boldsymbol{\alpha}_i, \boldsymbol{\theta})$$

$$+ \log f(\boldsymbol{\alpha}_i|\boldsymbol{\theta}) + \log f(\mathbf{b}_i|\boldsymbol{\theta})\big) \mid \mathbf{q}_i, \mathbf{c}_i, \mathbf{z}_i; \boldsymbol{\theta}^{(k)}\Big].$$

Since the expectation in $Q(\boldsymbol{\theta}|\boldsymbol{\theta}^{(k)})$ is difficult to evaluate analytically, we can use a Monte Carlo method to generate large samples of the "missing data" $(\mathbf{y}_{cen,i}, \boldsymbol{\alpha}_i, \mathbf{b}_i)$ from the conditional distribution $f(\mathbf{y}_i, \boldsymbol{\alpha}_i, \mathbf{b}_i \mid \mathbf{q}_i, \mathbf{c}_i, \mathbf{z}_i; \boldsymbol{\theta}^{(k)})$, and then approximate $Q(\boldsymbol{\theta}|\boldsymbol{\theta}^{(k)})$ by an empirical mean. This can again be done using the Gibbs sampler by iteratively sampling from the following full conditionals

$$f(\mathbf{y}_{cen,i}|\mathbf{q}_i, \mathbf{c}_i, \mathbf{z}_i, \boldsymbol{\alpha}_i, \mathbf{b}_i; \boldsymbol{\theta}^{(k)}),$$
$$f(\boldsymbol{\alpha}_i \mid \mathbf{y}_i, \mathbf{z}_i, \mathbf{b}_i; \boldsymbol{\theta}^{(k)}), \qquad f(\mathbf{b}_i \mid \mathbf{y}_i, \mathbf{z}_i, \boldsymbol{\alpha}_i, \boldsymbol{\theta}^{(k)}),$$

where sampling from the full conditionals can be done using rejection sampling methods. Other MCMC or importance sampling methods can also be used. The procedures are similar to that in Chapter 5. However, such an Monte Carlo EM algorithm can be computationally very intensive and may offer convergence problems. Therefore, in the following we describe a computationally more efficient approximate method based on a first-order Taylor approximation (linearization), following Wu (2002).

We can write the NLME model (6.15) and (6.16) as a single equation as follows:

$$y_{ij} = g_{ij}(\boldsymbol{\beta}, \mathbf{b}_i) + e_{ij} \tag{6.18}$$

where $g_{ij}(\cdot)$ is a nonlinear function and the covariates are suppressed in the expression. Let $\mathbf{g}_i = (g_{i1}, \ldots, g_{in_i})^T$. In the following iterative algorithm, denote the current estimate of $(\boldsymbol{\beta}, \mathbf{b}_i)$ by $(\hat{\boldsymbol{\beta}}, \hat{\mathbf{b}}_i)$. The linearization step yields the following working LME model

$$\mathbf{w}_i \;\; = \;\; \tilde{X}_i\boldsymbol{\beta} + \tilde{Z}_i\mathbf{b}_i + \mathbf{e}_i \tag{6.19}$$

$$= \;\; \tilde{X}_{i(1)}\boldsymbol{\beta}_{-r} + \beta_r\tilde{X}_{i(2)}\mathbf{u}_{ij}^T\boldsymbol{\alpha}_i + \tilde{Z}_i\mathbf{b}_i + \mathbf{e}_i, \tag{6.20}$$

where

$$\mathbf{w}_i \;\; = \;\; \mathbf{y}_i - \mathbf{g}_i(\hat{\boldsymbol{\beta}}, \hat{\mathbf{b}}_i) + \tilde{X}_i\hat{\boldsymbol{\beta}} + \tilde{Z}_i\hat{\mathbf{b}}_i,$$

$$\tilde{X}_{ij} \;\; = \;\; \frac{\partial g_{ij}(\boldsymbol{\beta}, \mathbf{b}_i)}{\partial \boldsymbol{\beta}^T}\Big|_{(\hat{\boldsymbol{\beta}}, \hat{\mathbf{b}}_i)}, \qquad \tilde{Z}_{ij} = \frac{\partial g_{ij}(\boldsymbol{\beta}, \mathbf{b}_i)}{\partial \mathbf{b}_i^T}\Big|_{(\hat{\boldsymbol{\beta}}, \hat{\mathbf{b}}_i)},$$

$\tilde{X}_i = (\tilde{X}_{i1}^T, \ldots, \tilde{X}_{in_i}^T)^T$, $\tilde{Z}_i = (\tilde{Z}_{i1}^T, \ldots, \tilde{Z}_{in_i}^T)^T$, $\mathbf{w}_i = (w_{i1}, \ldots, w_{in_i})^T$, $\tilde{X}_{i(1)}$ is the first $(r-1)$ columns of $\tilde{X}_i$, $\tilde{X}_{i(2)}$ is the last column of $\tilde{X}_i$, and $\boldsymbol{\beta}_{-r} = (\beta_1, \cdots, \beta_{r-1})^T$.

The approximate method is to iteratively solve the LME model (6.19) or (6.20), so in each iteration we can use the closed-form results in Section 6.3.1 for LME models where the computation is much simpler. Specifically, in the E-step of the k-th EM iteration, we only need to compute the following conditional expectations

$$E(\boldsymbol{\alpha}_i|\mathbf{q}_i, \mathbf{c}_i, \mathbf{z}_i, \boldsymbol{\theta}^{(k)}), \qquad E(\boldsymbol{\alpha}_i\boldsymbol{\alpha}_i^T|\mathbf{q}_i, \mathbf{c}_i, \mathbf{z}_i, \boldsymbol{\theta}^{(k)}),$$
$$E(\mathbf{b}_i|\mathbf{q}_i, \mathbf{c}_i, \mathbf{z}_i, \boldsymbol{\theta}^{(k)}), \qquad E(\mathbf{b}_i\mathbf{b}_i^T|\mathbf{q}_i, \mathbf{c}_i, \mathbf{z}_i, \boldsymbol{\theta}^{(k)}).$$

To compute these quantities, we can use a Monte Carlo method to generate $\mathbf{y}_{cen,i}$ from $f(\mathbf{y}_i|q_i, c_i, \mathbf{z}_i, \boldsymbol{\theta}^{(k)})$ and then approximate the conditional expectations by their empirical means, as in Section 6.3.2. This greatly simplifies the computation. Recall that in the Monte Carlo EM algorithm for exact likelihood estimation, which is described earlier, we need to sample $(\mathbf{y}_{cen,i}, \boldsymbol{\alpha}_i, \mathbf{b}_i)$ from the conditional distribution $f(\mathbf{y}_i, \boldsymbol{\alpha}_i, \mathbf{b}_i \mid \mathbf{q}_i, \mathbf{c}_i, \mathbf{z}_i; \boldsymbol{\theta}^{(k)})$. This sampling step can be a big computational burden and may lead to many computational difficulties such as non-convergence or very slow convergence. The above approximate method simplifies the computation by avoiding sampling the high-dimensional random effects $(\boldsymbol{\alpha}_i, \mathbf{b}_i)$ since sampling these random effects often causes convergence problems.

Under the working LME model (6.19) or (6.20), closed-form expressions of the updated parameter estimates in the M-step can be written as follows (Wu 2002):

$$\hat{\boldsymbol{\alpha}}^{(k+1)} = \frac{1}{n}\sum_{i=1}^{n} E(\boldsymbol{\alpha}_i|\mathbf{q}_i, \mathbf{c}_i, \mathbf{z}_i, \hat{\boldsymbol{\theta}}^{(k)}),$$

$$\hat{\boldsymbol{\beta}}^{(k+1)} = \left[\sum_{i=1}^{n}\sum_{j=1}^{n_i} E\left(\tilde{X}_{ij}^T\tilde{X}_{ij}\Big|\mathbf{q}_i, \mathbf{c}_i, \mathbf{z}_i, \hat{\boldsymbol{\theta}}^{(k)}\right)\right]^{-1}$$
$$\times \sum_{i=1}^{n}\sum_{j=1}^{n_i} E\left[\tilde{X}_{ij}^T(w_{ij} - \tilde{Z}_{ij}\mathbf{b}_i)\Big|\mathbf{q}_i, \mathbf{c}_i, \mathbf{z}_i, \hat{\boldsymbol{\theta}}^{(k)}\right],$$

$$\hat{A}^{(k+1)} = \frac{1}{n}\sum_{i=1}^{n} E\left[(\boldsymbol{\alpha}_i - \hat{\boldsymbol{\alpha}})(\boldsymbol{\alpha}_i - \hat{\boldsymbol{\alpha}})^T|\mathbf{q}_i, \mathbf{c}_i, \mathbf{z}_i, \hat{\boldsymbol{\theta}}^{(k)}\right],$$

$$\hat{B}^{(k+1)} = \frac{1}{n}\sum_{i=1}^{n} E(\mathbf{b}_i\mathbf{b}_i^T|\mathbf{q}_i, \mathbf{c}_i, \mathbf{z}_i, \hat{\boldsymbol{\theta}}^{(k)}),$$

$$\hat{\delta}^{2(k+1)} = \frac{1}{N_1}\sum_{i=1}^{n}\sum_{k=1}^{m_i} E\left[(z_{ik} - \mathbf{u}_{ik}^T\boldsymbol{\alpha}_i)^2|\mathbf{q}_i, \mathbf{c}_i, \mathbf{z}_i, \hat{\boldsymbol{\theta}}^{(k)}\right],$$

$$\hat{\sigma}^{2(k+1)} = \frac{1}{N_2}\sum_{i=1}^{n}\sum_{j=1}^{n_i} E\left[(w_{ij} - \tilde{X}_{ij}\boldsymbol{\beta} - \tilde{Z}_{ij}\mathbf{b}_i)^2|\mathbf{q}_i, \mathbf{c}_i, \mathbf{z}_i, \hat{\boldsymbol{\theta}}^{(k)}\right],$$

$$k = 1, 2, \cdots,$$

where $N_1 = \sum_{i=1}^{n} m_i$, $N_2 = \sum_{i=1}^{n} n_i$. Note that such closed-form expressions are unavailable for exact likelihood estimation in which iterative algorithms are required to update the parameter estimates. Iterating the above E-step and M-step until convergence, we obtain an approximate MLE (or local maximum) of $\boldsymbol{\theta}$.

Approximate methods based on Laplace approximations can also be developed. In this case, we can directly approximate the observed-data likelihood $L_o(\boldsymbol{\theta})$ using a first-order Laplace approximation, and thus completely avoid any Monte Carlo simulations.

6.4 Mixed Effects Models with Censoring and Missing Data

Censoring and Missing data

In many longitudinal studies, censored data and missing data may arise simultaneously. For example, in addition to some data being censored, subjects may also drop out and covariates and responses may have missing data. In other words, in some applications, in addition to the responses being censored, the responses may have other types of non-ignorable missing data, and covariates may have missing data as well. In fact, in practice censoring, dropouts, missing data, and measurement errors may all arise simultaneously. Statistical analyses failing to address any one of these problems may lead to biased results. Therefore, it is important to address all these incomplete data problems *simultaneously*.

In Section 6.3, we address censored response and covariate measurement errors simultaneously. In this section, we address censored response and missing data simultaneously. We focus on mixed effects models with censoring and non-ignorable missing data in the response, including informative dropouts. The methods for mixed effects models with censored responses and non-ignorable missing data in the covariates are similar and thus are omitted here.

Let $\mathbf{r}_i = (r_{i1}, \ldots, r_{in_i})^T$ be a vector of missing response indicators such that $r_{ij} = 1$ if y_{ij} is missing and 0 otherwise, $i = 1, 2, \cdots, n$. For monotone dropouts, i.e., subjects never return to the study once they dropout, the $\mathbf{r}_i$'s have a monotone pattern. Let $\mathbf{c}_i = (c_{i1}, \ldots, c_{in_i})^T$ be the censoring indicator. We write $\mathbf{y}_i = (\mathbf{y}_{cen,i}, \mathbf{y}_{mis,i}, \mathbf{y}_{obs,i})$, where $\mathbf{y}_{cen,i}$, $\mathbf{y}_{mis,i}$, and $\mathbf{y}_{obs,i}$ correspond to the censored, missing, and observed components of $\mathbf{y}_i$ respectively. Let $\mathbf{z}_i$ be the completely observed covariates.

As in previous sections, we write the mixed effects response model as $f(\mathbf{y}_i|\mathbf{z}_i, \mathbf{b}_i, \boldsymbol{\beta}, \boldsymbol{\sigma})$, where $\boldsymbol{\beta}$ is a $r \times 1$ vector of fixed-effects parameters and $\boldsymbol{\sigma}$ contains parameters characterizing within individual variation, and $\mathbf{b}_i$ is a $s \times 1$ vector of random-effects for individual i. We assume that $\mathbf{b}_i \sim N(\mathbf{0}, D)$, where D is an unstructured covariance matrix. Given $\mathbf{b}_i$, we assume that $y_{i1}, \cdots, y_{in_i}$ are conditionally independent. We consider non-ignorable missing responses, i.e., the distribution of $\mathbf{r}_i$ may depend on the unobserved values $(\mathbf{y}_{i,mis}, \mathbf{y}_{i,cen})$ or on the unobservable random effects $\mathbf{b}_i$.

Let $\boldsymbol{\theta}$ be the collection of all parameters. All the observed data are

$$\{(\mathbf{y}_{obs,i}, \mathbf{z}_i, \mathbf{c}_i, \mathbf{r}_i), \ i = 1, \ldots, n\}.$$

Consider a non-ignorable missing data model $f(\mathbf{r}_i|\mathbf{y}_i, \mathbf{z}_i, \mathbf{c}_i, \boldsymbol{\phi})$ where the missingness of the response may depend on the missing or censored values $(\mathbf{y}_{i,mis}, \mathbf{y}_{i,cen})$. The observed data likelihood can be written as

$$L_o(\boldsymbol{\theta}) = \prod_{i=1}^{n} \left[\int \int \int \left\{ \prod_{j=1}^{n_i} f(y_{ij}|\mathbf{z}_{ij}, \mathbf{b}_i, \boldsymbol{\theta})^{1-c_{ij}} F(d|\mathbf{z}_{ij}, \mathbf{b}_i, \boldsymbol{\theta})^{c_{ij}} \right\} \right.$$
$$\left. \times f(\mathbf{b}_i|\boldsymbol{\theta}) f(\mathbf{r}_i|\mathbf{y}_i, \mathbf{z}_i, \mathbf{c}_i, \boldsymbol{\phi}) \, d\mathbf{y}_{i,mis} \, d\mathbf{y}_{i,cen} \, d\mathbf{b}_i \right],$$

where

$$F(d \mid z_{ij}, \mathbf{b}_i, \boldsymbol{\theta}) = \int_{-\infty}^{d} f(y_{ij}|\mathbf{z}_{ij}, \mathbf{b}_i, \boldsymbol{\theta}) \, dy_{ij}.$$

Since the observed data likelihood is intractable, we can again consider a Monte Carlo EM algorithm for likelihood inference. Let $\boldsymbol{\theta}^{(k)}$ be the parameter estimates from the k-th EM iteration. The E-step for individual i at the $(k+1)$st EM iteration can be written as

$$Q_i(\boldsymbol{\theta}|\boldsymbol{\theta}^{(k)}) = \int \int \int \left[\left\{ \log f(\mathbf{y}_i|\mathbf{z}_i, \mathbf{b}_i, \boldsymbol{\theta}) + \log f(\mathbf{b}_i|\boldsymbol{\theta}) \right. \right.$$
$$\left. + \log f(\mathbf{r}_i|\mathbf{y}_i, \mathbf{z}_i, \mathbf{c}_i, \boldsymbol{\phi}) \right\}$$
$$\left. \times f(\mathbf{y}_{cen,i}, \mathbf{y}_{mis,i}, \mathbf{b}_i|\mathbf{y}_{obs,i}, \mathbf{z}_i, \mathbf{c}_i, \mathbf{r}_i, \boldsymbol{\theta}^{(k)}) \right]$$
$$d\mathbf{b}_i \, d\mathbf{y}_{cen,i} \, d\mathbf{y}_{mis,i}.$$

To approximate $Q_i(\boldsymbol{\theta}|\boldsymbol{\theta}^{(k)})$ using a Monte Carlo method, we may again use the Gibbs sampler to generate samples of the "missing data" $(\mathbf{y}_{cen,i}, \mathbf{y}_{mis,i}, \mathbf{b}_i)$ from the distribution $f(\mathbf{y}_{cen,i}, \mathbf{y}_{mis,i}, \mathbf{b}_i|\mathbf{y}_{obs,i}, \mathbf{z}_i, \mathbf{c}_i, \mathbf{r}_i, \boldsymbol{\theta}^{(t)})$ by iteratively sampling from the full conditionals $f(\mathbf{y}_{cen,i}|\mathbf{y}_{obs,i}, \mathbf{y}_{mis,i}, \mathbf{z}_i, \mathbf{b}_i, \mathbf{c}_i, \mathbf{r}_i, \boldsymbol{\theta}^{(k)})$, $f(\mathbf{y}_{mis,i} |\mathbf{y}_{obs,i}, \mathbf{y}_{cen,i}, \mathbf{z}_i, \mathbf{b}_i, \mathbf{c}_i, \boldsymbol{\theta}^{(k)})$, and $f(\mathbf{b}_i|\mathbf{y}_i, \mathbf{z}_i, \mathbf{c}_i, \mathbf{r}_i, \boldsymbol{\theta}^{(k)})$. The method is similar to that in Chapter 4 so the detail is omitted here.

When a time-dependent covariate $\mathbf{z}_i$ is also measured with errors, a similar likelihood method can be considered. For example, suppose that we consider a classical measurement error model for covariate $\mathbf{z}_i$, as in previous sections, and the missingness of the response is assumed to depend on the random effects in the models. Then, we have the following observed-data likelihood

$$L_o(\boldsymbol{\theta}) = \prod_{i=1}^{n} \left[\int \int \int \left\{ \prod_{j=1}^{n_i} f(y_{ij}|\mathbf{a}_i, \mathbf{b}_i, \boldsymbol{\theta})^{1-c_{ij}} F(d|\mathbf{a}_i, \mathbf{b}_i, \boldsymbol{\theta})^{c_{ij}} \right\} \right.$$

$$\left. \times f(\mathbf{z}_i|\mathbf{a}_i, \boldsymbol{\theta}) f(\mathbf{a}_i|\boldsymbol{\theta}) f(\mathbf{b}_i|\boldsymbol{\theta}) f(\mathbf{r}_i|\mathbf{a}_i, \mathbf{b}_i, \boldsymbol{\phi}) \, d\mathbf{y}_{i,cen} \, d\mathbf{a}_i \, d\mathbf{b}_i \right].$$

An EM algorithm or computationally more efficient approximate methods, similar to those in the previous sections, can be then developed. The details are omitted here.

Censoring and Time-to-Event

In some applications, when modeling the longitudinal process, we also need to model a time-to-event process, such as time to dropout and time to death. In these situations, *joint modeling* of the longitudinal process and the time-to-event process is often required since the two processes are associated. For example, inference for the longitudinal model must incorporate the time-to-dropout model if the dropouts are informative. In this section, we briefly discuss joint inference for a mixed effects model with censored response and a time-to-event model. Survival models or time-to-event models will be described in details in Chapter 7. In Chapter 8, we will provide a more comprehensive discussion of joint models for longitudinal data and survival data, so we skip the details here.

Note that here both the longitudinal response and the event time may be censored, so we have two censored processes which must be taken into account simultaneously. We focus on longitudinal responses with left censoring and survival data with right censoring. Let $\mathbf{s}_i = (s_{i1}, \ldots, s_{im})^T$ be a vector of censored response indicators for the longitudinal process such that $s_{ij} = 1$ if y_{ij} is censored and 0 otherwise. Let $\mathbf{r}_i = (r_{i1}, \ldots, r_{im})^T$ be a vector of event indicators such that $r_{ij} = 1$ if the event has happened by time t_{ij} for individual i and $r_{ij} = 0$ otherwise.

Let T_i be the time to an event for individual i. In many cases in practice, the distribution of event time T_i may depend on the random effects $\mathbf{b}_i$ in the longitudinal model. For example, in a longitudinal study, time to dropout of an individual in the study may be related to the initial intercept and slope of this individual's longitudinal trajectory, such as the initial viral load level and decay rate in an AIDS study. Thus we may assume the following frailty model for modeling the time to the event of interest:

$$\lambda(t_i|\mathbf{z}_i, \mathbf{b}_i) = \lambda_0(t_i) \exp(\boldsymbol{\gamma}_1^T \mathbf{z}_i + \boldsymbol{\gamma}_2^T \mathbf{b}_i), \qquad (6.21)$$

where $\lambda(t_i|\mathbf{z}_i, \mathbf{b}_i)$ is a hazard function, $\lambda_0(t_i)$ is an unspecified baseline hazard function, and γ_1 and γ_2 are vectors of unknown parameters (see Chapter 7 for more details of this survival model). We assume that the longitudinal response $\mathbf{y}_i$ and the event process $\mathbf{r}_i$ are conditionally independent given the random effects $\mathbf{b}_i$ in the longitudinal model.

The observed data are

$$\{(\mathbf{y}_{obs,i}, \mathbf{r}_i, \mathbf{z}_i, \mathbf{s}_i), \ i = 1, 2, \cdots, n\}.$$

The joint likelihood for the observed data can be written as

$$
L_o(\boldsymbol{\theta}) \ = \ \prod_{i=1}^{n} \left[\int \int \int f(\mathbf{y}_i | \mathbf{z}_i, \mathbf{b}_i, \boldsymbol{\beta}, \boldsymbol{\sigma}) f(\mathbf{b}_i | D) \right.
$$

$$
\left. \times f(\mathbf{r}_i | \mathbf{z}_i, \mathbf{b}_i, \boldsymbol{\gamma}) f(\mathbf{s}_i | \mathbf{y}_i, \boldsymbol{\phi}) \ d\mathbf{y}_{i,cen} \ d\mathbf{b}_i \right],
$$

where

$$
f(\mathbf{r}_i | \mathbf{z}_i, \mathbf{b}_i, \boldsymbol{\gamma}) = \prod_{k=1}^{m} f(r_{ik} | r_{i0}, \cdots, r_{i,k-1}; \mathbf{z}_i, \mathbf{b}_i, \boldsymbol{\gamma}),
$$

$$
f(r_{ik} | r_{i0}, \cdots, r_{i,k-1}; \mathbf{z}_i, \mathbf{b}_i, \boldsymbol{\gamma}) = p_{ik}^{r_{ik}} (1 - p_{ik})^{1 - r_{ik}}.
$$

$$
p_{ik} = 1 - \exp\left[-\exp(\gamma_{0k} + \boldsymbol{\gamma}_1^T \mathbf{z}_i + \boldsymbol{\gamma}_2^T \mathbf{b}_i) \right].
$$

We can again use a Monte Carlo EM algorithm to obtain the MLE of $\boldsymbol{\theta}$, but the computation is highly intensive. We will present some details in Chapter 8 for similar joint models.

For likelihood methods addressing censoring, missing data, and measurement errors simultaneously, the major difficulties are computational challenges and potential identifiability problems. Monte Carlo EM algorithms for joint likelihood estimation can be computationally very intensive, so computationally much more efficient approximate methods based on linearization procedures or Laplace approximations are extremely valuable for these problems. In fact, sometimes these approximate methods may be the only choice since exact likelihood estimation may be computationally infeasible. For avoid parameter or model non-identifiability, one should avoid building too complicated models or impose some restrictions on parameter space such as diagonal covariance matrices.

The foregoing methods may also be extended to mixed effects models with censored covariates. Such an extension is conceptually straightforward, but the implementations may be very tedious due to intensive computation.

6.5 Appendix

In this Appendix, we provide detailed derivations of some results for the EM algorithms in Section 6.3, following Wu (2002, 2004). To simplify presentation, we focus on the following quadratic LME model for the response

$$
\begin{aligned}
\mathbf{y}_i &= \tilde{X}_i\boldsymbol{\beta} + \tilde{Z}_i\mathbf{b}_i + \mathbf{e}_i \\
&= \mathbf{x}_{ij}^T\boldsymbol{\beta}^{(1)} + \tilde{\mathbf{x}}_{ij}\boldsymbol{\beta}^{(2)}\mathbf{x}_{ij}^T\boldsymbol{\alpha}_i + \mathbf{x}_{ij}^T\mathbf{b}_i + e_{ij},
\end{aligned}
$$

where $\tilde{\mathbf{x}}_{ij} = (t_{ij}, t_{ij}^2)$, $\boldsymbol{\beta}^{(1)} = (\beta_1, \beta_2, \beta_3)^T$, and $\boldsymbol{\beta}^{(2)} = (\beta_4, \beta_5)^T$.

We first consider the M-step. The updated estimates in the M-step of the EM algorithm can be obtained as follows. To facilitate the maximization procedure, we can write $Q_i(\boldsymbol{\theta}|\boldsymbol{\theta}^{(k)})$, the conditional expectation of the complete-data log-likelihood given the observed data and current estimates at k-th EM iteration, as follows

$$
\begin{aligned}
Q_i(\boldsymbol{\theta}|\boldsymbol{\theta}^{(k)}) &= E(l_c(\boldsymbol{\theta})|\mathbf{q}_i, \mathbf{c}_i, \mathbf{z}_i; \boldsymbol{\theta}^{(k)}) \\
&= h_1(\boldsymbol{\beta}, \sigma^2) + h_2(\delta^2) + h_3(\boldsymbol{\alpha}, A) + h_4(B).
\end{aligned}
$$

The first part contains the parameters $\boldsymbol{\beta}$ and σ^2:

$$
\begin{aligned}
h_1(\boldsymbol{\beta}, \sigma^2) &= -\frac{1}{2}\sum_{i=1}^n n_i\log(2\pi\sigma^2) - \frac{1}{2\sigma^2}\sum_{i=1}^n E\Big[(\mathbf{y}_i - \tilde{X}_i\boldsymbol{\beta} - \tilde{Z}_i\mathbf{b}_i)^T \\
&\qquad\times(\mathbf{y}_i - \tilde{X}_i\boldsymbol{\beta} - \tilde{Z}_i\mathbf{b}_i) \mid \mathbf{q}_i, \mathbf{c}_i, , \mathbf{z}_i, \hat{\boldsymbol{\theta}}^{(k)}\Big].
\end{aligned}
$$

The updated estimates $\hat{\boldsymbol{\beta}}^{(k+1)}$ and $\hat{\sigma}^{2(k+1)}$ can be obtained from the equations

$$
\partial h_1(\boldsymbol{\beta}, \sigma^2)/\partial\boldsymbol{\beta} = 0, \qquad \partial h_1(\boldsymbol{\beta}, \sigma^2)/\partial\sigma^2 = 0.
$$

The second part contains the parameter δ^2:

$$
\begin{aligned}
h_2(\delta^2) &= -\frac{1}{2}\sum_{i=1}^n n_i\log(2\pi\delta^2) \\
&\quad -\frac{1}{2\delta^2}\sum_{i=1}^n\sum_{k=1}^{n_i} E\left((z_{ik} - \mathbf{u}_{ik}^T\boldsymbol{\alpha}_i)^2|\mathbf{q}_i, \mathbf{c}_i, \mathbf{z}_i, \hat{\boldsymbol{\theta}}^{(k)}\right),
\end{aligned}
$$

and the updated estimate $\hat{\delta}^{2(k+1)}$ can be obtained from the equation

$$
\partial h_2(\delta^2)/\partial\delta^2 = 0.
$$

The third part contains the parameters $\boldsymbol{\alpha}$ and A:

$$
\begin{aligned}
h_3(\boldsymbol{\alpha}, A) &= -\frac{p}{2}\log(2\pi|A|) \\
&\quad -\frac{1}{2}E\left((\boldsymbol{\alpha}_i - \boldsymbol{\alpha})^T A^{-1}(\boldsymbol{\alpha}_i - \boldsymbol{\alpha})|\mathbf{q}_i, \mathbf{c}_i, \mathbf{z}_i, \hat{\boldsymbol{\theta}}^{(k)}\right),
\end{aligned}
$$

and the updated estimates $\hat{\alpha}^{(k+1)}$ and $\hat{A}^{(k+1)}$ can be obtained from the equations

$$\partial h_3(\alpha, A)/\partial \alpha = 0, \qquad \partial h_3(\alpha, A)/\partial A = 0.$$

The fourth part contains parameters B:

$$h_4(B) = -\frac{p}{2}\log(2\pi|B|) - \frac{1}{2}E(\mathbf{b}_i^T B^{-1}\mathbf{b}_i^T|\mathbf{q}_i, \mathbf{c}_i, \mathbf{z}_i, \hat{\boldsymbol{\theta}}^{(k)}),$$

and the updated estimate $\hat{B}^{(k+1)}$ can be obtained from the equation

$$\partial h_4(B)/\partial B = 0.$$

Note that, in the foregoing equations, we need to compute some conditional expectations, which are given below.

To compute the conditional expectations in the foregoing equations, note that

$$(\boldsymbol{\alpha}_i, \mathbf{y}_i, \mathbf{z}_i) \sim N(\mu_{\boldsymbol{\alpha}_i}, \Sigma_{\boldsymbol{\alpha}_i}),$$

where

$$\mu_{\boldsymbol{\alpha}_i} = \begin{pmatrix} \alpha \\ \mu_2 \\ U_i\alpha \end{pmatrix}, \quad \Sigma_{\boldsymbol{\alpha}_i} = \begin{pmatrix} A & \Sigma_{21}^T & A^T U_i^T \\ \Sigma_{21} & Q_i & \Sigma_{32}^T \\ U_i A & \Sigma_{32} & U_i A U_i^T + \sigma^2 I \end{pmatrix},$$

$U_i = (\mathbf{u}_{i1}, \cdots, \mathbf{u}_{in_i})^T, \quad \mu_2 = (\mu_2(1), \cdots, \mu_2(n_i))^T, \quad \Sigma_{21} = (\Sigma_{21}(j,k)), \Sigma_{32} = (\Sigma_{32}(j,k))$, with

$$\begin{aligned} \mu_2(k) &= \mathbf{x}_{ik}^T \boldsymbol{\beta}^{(1)} + \tilde{\mathbf{x}}_{ik}^T \boldsymbol{\beta}^{(2)} \mathbf{x}_{ik}^T \alpha, \\ \Sigma_{21}(j,k) &= \tilde{\mathbf{x}}_{ik}^T \boldsymbol{\beta}^{(2)} \mathbf{x}_{ik}^T A, \\ Q_i &= \tilde{\mathbf{x}}_{ij}^T \boldsymbol{\beta}^{(2)} \mathbf{x}_{ij}^T A \mathbf{x}_{ik} \boldsymbol{\beta}^{(2)} \tilde{\mathbf{x}}_{ik}^T + \mathbf{x}_{ij}^T B \mathbf{x}_{ik}, \\ \Sigma_{32}(j,k) &= \tilde{\mathbf{x}}_{ij}^T \boldsymbol{\beta}^{(2)} \mathbf{x}_{ij}^T A \mathbf{x}_{ik}. \end{aligned}$$

We also have

$$(\mathbf{b}_i, \mathbf{y}_i, \mathbf{z}_i) \sim N(\mu_{\mathbf{b}_i}, \Sigma_{\mathbf{b}_i}),$$

where

$$\mu_{\mathbf{b}_i} = \begin{pmatrix} \mathbf{0} \\ \mu_2 \\ U_i\alpha \end{pmatrix}, \quad \Sigma_{\mathbf{b}_i} = \begin{pmatrix} B & B^T \tilde{Z}_i^T & 0 \\ \tilde{Z}_i B & Q_i & \Sigma_{32}^T \\ 0 & \Sigma_{32} & U_i A U_i^T + \sigma^2 I \end{pmatrix}.$$

Based on the above results, the conditional expectations and conditional covariances $E(\boldsymbol{\alpha}_i|\mathbf{y}_i, \mathbf{z}_i, \boldsymbol{\theta})$, $E(\mathbf{b}_i|\mathbf{y}_i, \mathbf{z}_i, \boldsymbol{\theta})$, $\text{Cov}(\boldsymbol{\alpha}_i|\mathbf{y}_i, \mathbf{z}_i, \boldsymbol{\theta})$, and $\text{Cov}(\mathbf{b}_i|\mathbf{y}_i, \mathbf{z}_i, \boldsymbol{\theta})$ can be computed using well-known properties of multivariate normal distributions.

To compute $E(\boldsymbol{\alpha}_i \boldsymbol{\alpha}_i^T|\mathbf{y}_i, \mathbf{z}_i, \boldsymbol{\theta})$ and $E(\mathbf{b}_i \mathbf{b}_i^T|\mathbf{y}_i, \mathbf{z}_i, \boldsymbol{\theta})$, note that

$$E(\boldsymbol{\alpha}_i \boldsymbol{\alpha}_i^T|\mathbf{y}_i, \mathbf{z}_i, \boldsymbol{\theta}) = E(\boldsymbol{\alpha}_i|\mathbf{y}_i, \mathbf{z}_i, \boldsymbol{\theta}) E(\boldsymbol{\alpha}_i|\mathbf{y}_i, \mathbf{z}_i, \boldsymbol{\theta})^T$$

$$+ \text{Cov}(\boldsymbol{\alpha}_i | \mathbf{y}_i, \mathbf{z}_i, \boldsymbol{\theta}),$$

$$E(\mathbf{b}_i \mathbf{b}_i^T | \mathbf{y}_i, \mathbf{z}_i, \boldsymbol{\theta}) = E(\mathbf{b}_i | \mathbf{y}_i, \mathbf{z}_i, \boldsymbol{\theta}) E(\mathbf{b}_i | \mathbf{y}_i, \mathbf{z}_i, \boldsymbol{\theta})^T$$
$$+ \text{Cov}(\mathbf{b}_i | \mathbf{y}_i, \mathbf{z}_i, \boldsymbol{\theta}).$$

To compute the observed information matrix $I(\mathbf{q}_i, \mathbf{c}_i, \mathbf{z}_i; \hat{\boldsymbol{\theta}})$, note that

$$\partial l_c^{(i)}(\boldsymbol{\theta}) / \partial \boldsymbol{\alpha} = -A^{-1}(\boldsymbol{\alpha}_i - \boldsymbol{\alpha}),$$

$$\partial l_c^{(i)}(\boldsymbol{\theta}) / \partial \boldsymbol{\beta}^{(1)} = -\frac{1}{\sigma^2} \sum_{j=1}^{n_i} e_{ij} \mathbf{x}_{ij},$$

$$\partial l_c^{(i)}(\boldsymbol{\theta}) / \partial \boldsymbol{\beta}^{(2)} = -\frac{1}{\sigma^2} \sum_{j=1}^{n_i} e_{ij} \tilde{\mathbf{x}}_{ij} \boldsymbol{\alpha}_i^T \mathbf{x}_{ij},$$

$$\partial^2 l_c^{(i)}(\boldsymbol{\theta}) / \partial \boldsymbol{\alpha} \partial \boldsymbol{\alpha}^T = A^{-1},$$

$$\partial^2 l_c^{(i)}(\boldsymbol{\theta}) / \partial \boldsymbol{\beta}^{(1)} \partial \boldsymbol{\beta}^{(1)T} = -\frac{1}{\sigma^2} \sum_{j=1}^{n_i} \mathbf{x}_{ij} \mathbf{x}_{ij}^T,$$

$$\partial^2 l_c^{(i)}(\boldsymbol{\theta}) / \partial \boldsymbol{\alpha} \partial \boldsymbol{\alpha}^T = \frac{1}{\sigma^2} \sum_{j=1}^{n_i} (\tilde{\mathbf{x}}_{ij} \boldsymbol{\alpha}_i^T \mathbf{x}_{ij}) \mathbf{x}_{ij}^T \boldsymbol{\alpha}_i \tilde{\mathbf{x}}_{ij}^T,$$

$$\partial^2 l_c^{(i)}(\boldsymbol{\theta}) / \partial \boldsymbol{\beta}^{(1)} \partial \boldsymbol{\beta}^{(2)T} = \frac{1}{\sigma^2} \sum_{j=1}^{n_i} (\tilde{\mathbf{x}}_{ij} \boldsymbol{\alpha}_i^T \mathbf{x}_{ij}) \mathbf{x}_{ij}^T,$$

$$\partial^2 l_c^{(i)}(\boldsymbol{\theta}) / \partial \boldsymbol{\beta}^{(2)} \partial \boldsymbol{\beta}^{(1)T} = \frac{1}{\sigma^2} \sum_{j=1}^{n_i} \mathbf{x}_{ij} \mathbf{x}_{ij}^T \boldsymbol{\alpha}_i \tilde{\mathbf{x}}_{ij}^T.$$

Therefore, the observed information matrix can be computed using the conditional expectations given above.

For the EM algorithm in Section 6.3.1, the expressions of the M-step are similar as above. For the E-step, note that

$$(\boldsymbol{\alpha}_i, \mathbf{y}_i, \mathbf{z}_i) \sim N(\mu_{\boldsymbol{\alpha}_i}, \Sigma_{\boldsymbol{\alpha}_i}),$$

where

$$\mu_{\boldsymbol{\alpha}_i} = \begin{pmatrix} \boldsymbol{\alpha} \\ \mathbf{g}_i(\boldsymbol{\beta}, \mathbf{b}_i) \\ U_i \boldsymbol{\alpha} \end{pmatrix},$$

$$\Sigma_{\boldsymbol{\alpha}_i} = \begin{pmatrix} A & \beta_r A^T \mathbf{u}_{ij} \tilde{X}_{i(2)}^T & A^T U_i^T \\ \beta_r \tilde{X}_{i(2)} \mathbf{u}_{ij}^T A & Q_i & U_i A^T \mathbf{u}_{ij} \tilde{X}_{i(2)} \beta_r \\ U_i A & \tilde{X}_{i(2)}^T \mathbf{u}_{ij}^T A U_i^T \beta_r & U_i A U_i^T + \sigma^2 I \end{pmatrix},$$

$$Q_i = \tilde{X}_{i(2)} \mathbf{u}_{ij}^T A \mathbf{u}_{ij} \tilde{X}_{i(2)}^T \beta_r^2 + \tilde{Z}_i D \tilde{Z}_i^T + \delta^2 I,$$

and $U_i = (\mathbf{u}_{i1}, \cdots, \mathbf{u}_{in_i})^T$. We also have

$$(\mathbf{b}_i, \mathbf{y}_i, \mathbf{z}_i) \sim N(\mu_{\mathbf{b}_i}, \Sigma_{\mathbf{b}_i}),$$

where

$$\mu_{\mathbf{b}_i} = \begin{pmatrix} \mathbf{0} \\ \mathbf{g}_i(\boldsymbol{\beta}, \mathbf{b}_i) \\ U_i \boldsymbol{\alpha} \end{pmatrix},$$

$$\Sigma_{\mathbf{b}_i} = \begin{pmatrix} D & D^T \tilde{Z}_i^T & 0 \\ \tilde{Z}_i D & Q_i & U_i A^T \mathbf{u}_{ij} \tilde{X}_{i(2)} \beta_r \\ 0 & \tilde{X}_{i(2)}^T \mathbf{u}_{ij}^T A U_i^T \beta_r & U_i A U_i^T + \sigma^2 I \end{pmatrix}.$$

Then, the conditional expectations and the conditional covariance matrices $E(\boldsymbol{\alpha}_i | \mathbf{y}_i, \mathbf{z}_i, \boldsymbol{\theta})$, $E(\mathbf{b}_i | \mathbf{y}_i, \mathbf{z}_i, \boldsymbol{\theta})$, $\text{Cov}(\boldsymbol{\alpha}_i | \mathbf{y}_i, \mathbf{z}_i, \boldsymbol{\theta})$, and $\text{Cov}(\mathbf{b}_i | \mathbf{y}_i, \mathbf{z}_i, \boldsymbol{\theta})$ can be computed using well-known properties of multivariate normal distributions.

CHAPTER 7

Survival Mixed Effects (Frailty) Models

7.1 Introduction

In practice we are often interested in modeling the time to an event of interest. For example, in a longitudinal study some subjects may drop out before the end of the study, so one may be interested in finding any possible relationship between dropout times and covariates such as age and gender (e.g., are younger subjects more likely to drop out earlier?). Other common events of interest include time to death, time to infection of a disease, time to a car accident, time to completion of a task, etc. These types of data are called *event-time data* or *survival data*. The analysis of event-time data or survival data is called *survival analysis*. Survival data or event time data are very common in practice. For simplicity, in this chapter we will often treat the event as "death", but the event can be defined in a much broader sense (e.g., dropout, infection, accident, etc.).

There are some special characteristics of survival data:

- survival data may be *censored*. For example, the event of interest may not be observed for some subjects in the study, possibly due to subjects' dropouts, loss of follow-up, or early termination of the study.

- survival data are often *skewed*, and many data are skewed to the right. So survival data may not follow symmetric distributions such as the normal distributions.

- survival data may have *unequal* follow-up times. For example, subjects may enter the study at different times.

In the analysis of survival data or event-time data, these special features must be incorporated, so special statistical methods are often required. Figure 7.1 shows typical survival data for four subjects and how the data may be censored: one event time is censored due to early termination of the study, the other

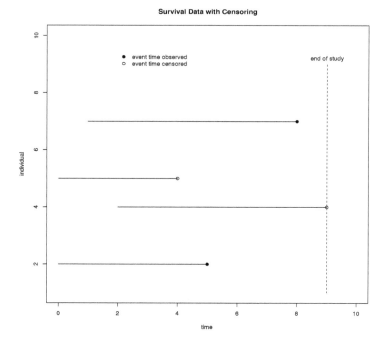

Figure 7.1 *Survival data for four individuals. The event times for two individuals are observed, while the event times for the other two individuals are censored (one censoring is due to the termination of the study while the other censoring is possibly due to dropout).*

event time is censored possibly due to dropout or loss of follow-up, while the other two event times are observed. So the observed survival data for these four subjects consist of two observed event times and two censoring times.

In survival regression models, our goal is to find any possible relationship between survival times and important covariates. A common approach is to model the *hazard* (or risk) of the event (see definition of hazard in Section 7.2.1), rather than the mean of the response as in classical regression models, incorporating censoring in the data. Thus, statistical inference for survival regression models typically requires special considerations. Semiparametric regression models, such as the Cox proportional hazards models, are particularly popular in survival analysis.

In some applications, survival data may be *clustered*. For example, in a multi-center study survival data from the same center may be more similar than data from different centers; people living in the same neighborhood may have more similar records of car accidents than people living in different neighborhoods; patients' survival rates may differ substantially across different hospitals but may be similar within the same hospitals, and so on. In these cases, each center or neighborhood or hospital may be viewed as a cluster. While survival data between different clusters may be independent, survival data within the same cluster may be more similar so they may be *correlated*. In the analysis of clustered survival data, one approach is to introduce random effects to represent cluster effects or variations between clusters and to incorporate correlation within the same clusters since the same cluster shares the same random effects. This leads to survival models with random effects, or *survival mixed effects models*, or *frailty models*.

In a survival mixed effects regression model, covariates may be missing or may be measured with errors. In these cases, special statistical methods are often required due to special features of survival data and models. In this chapter, we first give a brief review of survival models and methods, and then we discuss frailty models with missing data and measurement errors. In Chapter 8, we will consider joint inference of survival models and longitudinal models.

7.2 Survival Models

There have been extensive developments of statistical models and methods for the analysis of survival data or event time data. Due to the nature of survival data, nonparametric and semiparametric models are widely used, since these models do not make distributional assumptions for the survival data. Parametric models are also used in survival analysis, and these models may be more efficient than nonparametric or semiparametric models *if* the distributional assumptions hold. In the following sections, we give a brief review of commonly used nonparametric, semiparametric, and parametric survival models and methods. More comprehensive discussions of survival models and methods can be found in Fleming and Harrington (1991), Andersen et al. (1993), Collett (2003), Lawless (2003), and Cook and Lawless (2007).

Let T be the time to an event of interest, called *survival time*, such as the time to dropout or time to death. Suppose that there are n individuals in the sample with independently observed survival times $t_1, t_2, \cdots, t_n$. We may arrange the distinct survival times in ascending order: $t_{(1)} < t_{(2)} < \cdots < t_{(r)}$, where $r \leq n$ since there may be ties. In practice, some of the survival times t_j's may be censored. We first focus on *right censored* data, i.e., for an individual with censored data the observed time t_j is the censoring time, with the unobserved true survival time known to be larger than the censoring time. We assume that the censoring is *random* or *non-informative* in the sense that the (true) survival time is independent of the censoring time, i.e., individuals with censored data may be viewed as a random sub-sample.

For individual i, let s_i be the true survival time, which may not be observed for some individuals due to censoring, and let c_i be the censoring time if the survival time is censored. Then, the observed survival times are

$$t_i = \min(s_i, c_i), \qquad i = 1, 2, \cdots, n. \tag{7.1}$$

Let

$$\delta_i = I(s_i \leq c_i) = \begin{cases} 1 & \text{if an event is observed for individual } i \\ 0 & \text{if event time is right censored for individual } i \end{cases}$$

be the censoring indicator. Then, the *observed data* can be written as

$$\{(t_i, \delta_i), \ i = 1, 2, \cdots, n\}, \tag{7.2}$$

i.e., for each individual we either observe a survival time or observe a censoring time and the censoring status.

7.2.1 Nonparametric Methods

In survival analysis, the survival function and the hazard functions play important roles. In this section, we briefly introduce these concepts and related estimates and properties. The *survival function* is the probability that an individual survives to some time beyond time t, i.e.,

$$S(t) = P(T \geq t) = 1 - F(t), \qquad t > 0,$$

where $F(t) = P(T < t)$ is the usual cumulative distribution function (cdf). The survival function summarizes the survival experience of the event-time process.

Given observed data $t_{(1)} < t_{(2)} < \cdots < t_{(r)}$, let n_j be the number of individuals who have not experienced any events before time $t_{(j)}$ (i.e., the number of individuals who are *at risk* at time $t_{(j)}$), and let d_j be the number of individuals who experience the event at time $t_{(j)}$. Then, the survival function $S(t)$ can be estimated by the following nonparametric estimator $\hat{S}(t)$, called *Kaplan-Meier estimator*,

$$\hat{S}(t) = \prod_{j=1}^{k} \left(\frac{n_j - d_j}{n_j} \right), \quad \text{for } t_{(k)} \leq t < t_{(k+1)}, \quad k = 1, 2, \cdots, r, \tag{7.3}$$

with $\hat{S}(t) = 1$ for $t < t_{(1)}$ and the assumption that $t_{(r+1)} = \infty$. The standard error of the estimated survival function is given by

$$SE(\hat{S}(t)) = \hat{S}(t) \left(\sum_{i=1}^{k} \frac{d_j}{n_j(n_j - d_j)} \right)^{1/2}.$$

Note that, when there is no censoring, the Kaplan-Meier estimate reduces to the empirical survival function $\hat{S}(t) = \tilde{S}(t) = 1 - \tilde{F}(t)$ where $\tilde{F}(t)$ is the empirical cdf.

In survival regression models, a popular approach is to model the hazard function, rather than the mean functions as in classical regression models. The *hazard function* is defined as

$$h(t) = \text{limit}_{\Delta t \to \infty} \frac{P(t \le T \le t + \Delta t \mid T \ge t)}{\Delta t}, \qquad t > 0,$$

which is the *risk* or *hazard* of death (or event) at time t, i.e., the probability that an individual dies at time t given that he/she has survived to time t. In other words, the hazard function $h(t)$ may be interpreted as the approximate probability that an individual, who is alive on day t, dies in the following day. The *cumulative hazard function* is defined as

$$H(t) = \int_0^t h(u)du, \qquad t > 0.$$

It is easy to see that the relationships between survival function, hazard function, and cumulative hazard function are

$$f(t) = h(t)S(t), \quad h(t) = -\frac{d}{dt}\left(\log S(t)\right), \quad H(t) = -\log S(t), \quad (7.4)$$

where $f(t)$ is the probability density function of T. Thus, a model based on one of the three functions may also be expressed in terms of another function. The choice may be based on simplicity of the formulation and easy interpretation.

The probability distribution for the observed data (t_i, δ_i) is given by

$$f(t_i, \delta_i) \propto f(t_i)^{\delta_i} S(t_i)^{1-\delta_i} = h(t_i)^{\delta_i} S(t_i). \qquad (7.5)$$

So the log-likelihood for individual i is given by

$$l_i = \delta_i \log f(t_i) + (1 - \delta_i) \log S(t_i), \qquad (7.6)$$

which can be used for likelihood inference.

Often we wish to compare the survival experiences of two groups, such as a treatment group and a control group. In this case, a commonly used test is the so-called (weighted) *log-rank test*. There are different versions of this test. The test statistics of the log-rank tests can be expressed as weighted (and scaled) sums of the differences between the observed number of death and the expected number of death. One can choose appropriate weights to emphasize early or late survival differences between the two groups (Fleming and Harrington 1991; Wu and Gilbert 2002). The test statistics asymptotically follow χ^2-distributions under the null hypothesis of no survival difference between the two groups.

Both the Kaplan-Meier estimate of the survival function and the log-rank test for survival differences are *nonparametric* since they do not require any distributional assumptions for the survival data.

7.2.2 Semiparametric Models

In survival regression models, we wish to determine if the variation in subjects' survival experiences may be partially explained by covariates. In survival analysis, a popular approach is to model the hazard function rather than the mean of the survival times as in classical regression models. Since a hazard function may be complicated, we can avoid a parametric assumption and allow the hazard function to be nonparametric. Then, one may link the hazard function to covariates $\mathbf{x}_i$ through the usual (parametric) linear predictor $\mathbf{x}_i^T \boldsymbol{\beta}$, which is sometimes called a *risk score* or *prognostic index* in survival analysis. This leads to a *semiparametric* regression model.

A widely used semiparametric survival regression model is the following *Cox proportional hazards model* (Cox 1972)

$$h_i(t) = h_0(t) \exp(\mathbf{x}_i^T \boldsymbol{\beta}), \qquad \text{or} \qquad S_i(t) = S_0(t)^{\exp(\mathbf{x}_i^T \boldsymbol{\beta})}, \qquad (7.7)$$

where $h_0(t)$ is an unspecified baseline hazard function, $\boldsymbol{\beta}$ is a vector of unknown regression parameters, and $\mathbf{x}_i = (x_{i1}, \cdots, x_{ip})^T$ is a vector of covariates, $i = 1, 2, \cdots, n$. The baseline hazard function $h_0(t)$ may be interpreted as the hazard when all the covariate values are zero (i.e., $\mathbf{x}_i = \mathbf{0}$). Note that in the Cox proportional hazards model (7.7), no distributional assumption is made for the survival data, so it is very flexible. The assumption in the model is that the hazards ratio $h_i(t)/h_0(t)$ does not change over time (i.e., proportional hazards), which should be checked in a particular application.

Statistical inference for the regression parameters $\boldsymbol{\beta}$ can be based on the likelihood method. The log-likelihood for individual i is given by

$$\begin{aligned} l_i(\boldsymbol{\beta}) &= \delta_i \log(f(t_i|\mathbf{x}_i, \boldsymbol{\beta})) + (1 - \delta_i) \log(S(t_i|\mathbf{x}_i, \boldsymbol{\beta})) \\ &= \delta_i \left[\log h_0(t_i) + \mathbf{x}_i^T \boldsymbol{\beta} \right] - \exp(\mathbf{x}_i^T \boldsymbol{\beta}) H_0(t_i), \qquad (7.8) \end{aligned}$$

where $h_0(t)$ and $H_0(t)$ are nonparametric functions, and

$$H_0(t) = \int_0^t h_0(u) du$$

is the baseline cumulative hazard function. For inference on $\boldsymbol{\beta}$, the nonparametric functions $h_0(t)$ and $H_0(t)$ may be viewed as nuisance parameters. To avoid estimating these nuisance parameters, inference can be based on the following *partial likelihood* (Cox, 1972), assuming no ties in the data,

$$L(\boldsymbol{\beta}) = \prod_{i=1}^{n} \left\{ \frac{\exp(\mathbf{x}_i^T \boldsymbol{\beta})}{\sum_{j \in R(t_i)} \exp(\mathbf{x}_j^T \boldsymbol{\beta})} \right\}^{\delta_i}$$

$$= \prod_{i=1}^{n} \left\{ \frac{\exp(\mathbf{x}_i^T \boldsymbol{\beta})}{\sum_{j=1}^{n} I(t_j \geq t_i) \exp(\mathbf{x}_j^T \boldsymbol{\beta})} \right\}^{\delta_i},$$

where $R(t_i)$ is the set of individuals who are at risk at time t_i, i.e., the set of individuals who are still alive at a time just before t_i, and $I(t_j \geq t_i)$ is the usual indicator function. An estimate of $\boldsymbol{\beta}$ can then be obtained by maximizing the partial likelihood, e.g., solving the following partial likelihood score equation

$$\mathbf{u}(\boldsymbol{\beta}) = \partial \log L(\boldsymbol{\beta})/\partial \boldsymbol{\beta} = 0$$

to obtain an estimator $\hat{\boldsymbol{\beta}}$, using (say) the Newton-Raphson method.

Standard results for maximum likelihood estimation carry over without modification to maximum partial likelihood estimation. For example, under some regularity conditions, we have

$$\sqrt{n}(\hat{\boldsymbol{\beta}} - \boldsymbol{\beta}) \xrightarrow{d} N(0, nI^{-1}(\boldsymbol{\beta})), \qquad \text{as } n \to \infty,$$

where the information matrix is given by

$$I(\boldsymbol{\beta}) = -E \left(\frac{\partial^2 \log L(\boldsymbol{\beta})}{\partial \boldsymbol{\beta} \partial \boldsymbol{\beta}^T} \right).$$

See Fleming and Harrington (1991) and Andersen et al. (1993) for details of asymptotic results. Approximate variance of $\hat{\boldsymbol{\beta}}$ can be obtained based on the diagonal elements of matrix $I^{-1}(\hat{\boldsymbol{\beta}})$.

Model selection or variable selection can be based on the likelihood ratio test and AIC criteria, similar to classical regression models. However, model diagnostics for Cox proportional hazards models are more complicated than that for classical regression models because of the censoring (usual residuals may not be easily defined due to censoring). Model diagnostics can still be based on "residuals", as in classical regression models, but there are various ways to define residuals, including the Cox-Snell residuals, the martingale residuals, the deviance residuals, the Schoenfeld residuals, and score residuals. Collett (2003) provided a discussion and overview.

In longitudinal studies, some covariates may be measured over time, i.e., some covariates are time dependent. Let $x_{ij}(t)$ be the j-th covariate for individual i at time t. In this case, the Cox model may be modified as follows

$$h_i(t) = h_0(t) \exp \left(\sum_{j=1}^{p} x_{ij}(t)\beta_j \right), \qquad i = 1, 2, \cdots, n.$$

However, in this case the above model is no longer a proportional hazards model since the hazards ratio $h_i(t)/h_0(t)$ changes over time. Nevertheless, statistical inference can still be based on partial likelihoods, similar to that for the Cox proportional hazards model.

7.2.3 Parametric Models

Although the semiparametric Cox proportional hazards model (7.7) is widely used in the analysis of survival data, parametric regression models for survival data have also been developed. These parametric models assume that the survival data follow some parametric distributions, and they may be preferred if the distributional assumptions hold. A major advantage of parametric regression models is that, if the parametric distributional assumption holds for the survival data, statistical inference based on the parametric model will be more efficient than a semiparametric model which makes no distributional assumptions. On the other hand, a major advantage of semiparametric models such as the Cox proportional hazards models is that they are robust against distributional assumptions.

Weibull Distribution and Model

For modeling survival data, the (parametric) *Weibull distribution* plays an important role, similar to the normal distribution in linear regression models in some sense. The Weibull distribution is popular in survival analysis because

- its hazard function can take a variety of forms, which offers much flexibility in modeling survival data, and it includes the *exponential distribution* as a special case;
- summary statistics such as the median and percentiles can be easily obtained (in survival analysis the median and percentiles are perhaps more useful in summarizing data than the mean and standard deviation since survival data are often skewed);
- it has both the proportional hazards property and the accelerated failure time property (to be described later).

For these reasons, the Weibull distribution is widely used in survival analysis.

The Weibull distribution, denoted by $W(\lambda, \gamma)$, has the following probability density function (pdf):

$$f(t) = \lambda \gamma t^{\gamma-1} \exp(-\lambda t^{\gamma}), \qquad 0 \le t < \infty,$$

where $\lambda > 0$ is the *scale parameter* and $\gamma > 0$ is the *shape parameter*. When the shaper parameter $\gamma = 1$, the Weibull distribution reduces to an *exponential distribution*. The survival function and the hazard function of the Weibull distribution $W(\lambda, \gamma)$ are given by

$$S(t) = \exp(-\lambda t^{\gamma}), \qquad h(t) = \lambda \gamma t^{\gamma-1}, \qquad 0 \le t < \infty.$$

Note that the hazard function $h(t)$ is monotonically increasing for $\gamma > 1$ and is monotonically decreasing for $\gamma < 1$. Figure 7.2 shows the probability density functions and the hazard functions for the Weibull distribution for selected

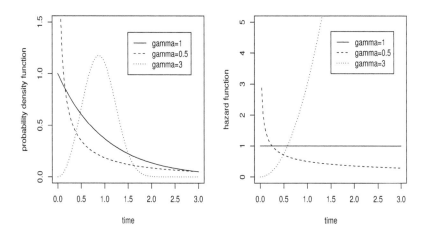

Figure 7.2 *The probability density function (left figure) and hazard function (right figure) for Weibull distributions with $\lambda = 1$ and $\gamma = 1, 0.5$, and 3 respectively.*

parameter values. We see that the Weibull distribution can take a variety of forms and is thus very flexible in modeling survival data.

The *Weibull proportional hazards model*, which is a parametric survival regression model, can be written as

$$h_i(t) = h_0(t) \exp(\mathbf{x}_i^T \boldsymbol{\beta}), \qquad i = 1, 2, \cdots, n, \tag{7.9}$$

where $\boldsymbol{\beta} = (\beta_1, \cdots, \beta_p)^T$, $\mathbf{x}_i = (x_{1i}, \cdots, x_{pi})^T$, and $h_0(t)$ is the hazard function of a Weibull distribution $W(\lambda, \gamma)$, i.e.,

$$h_0(t) = \lambda \gamma t^{\gamma-1},$$

which is a parametric function.

Note that, based on the Weibull proportional hazards model (7.9), we have

$$h_i(t) = \lambda \gamma t^{\gamma-1} \exp(\mathbf{x}_i^T \boldsymbol{\beta}) = \left[\lambda \exp(\mathbf{x}_i^T \boldsymbol{\beta}) \right] \gamma t^{\gamma-1},$$

which is just the hazard function of the Weibull distribution $W(\lambda e^{\mathbf{x}_i^T \boldsymbol{\beta}}, \gamma)$. Thus, the effect of covariate $\mathbf{x}_i$ is to alter the scale parameter from λ to $\lambda \exp(\mathbf{x}_i^T \boldsymbol{\beta})$, while the shape parameter γ remains unchanged. Note also that, in the Weibull proportional hazards model (7.9), the hazard functions $h_i(t)$ and $h_0(t)$ are parametric, while in the Cox proportional hazards model (7.7), the corresponding hazard functions are nonparametric, although these two types of proportional hazards models have the same form.

For parametric survival regression models, statistical inference can be based on standard likelihood methods. For example, the likelihood for the Weibull proportional hazards model is given

$$L(\boldsymbol{\beta}, \lambda, \gamma) = \prod_{i=1}^{n} (h_i(t_i))^{\delta_i} S_i(t_i),$$

where $h_i(t)$ and $S_i(t)$ are the hazard function and survival function of the Weibull distribution respectively. Numerical methods such as the Newton-Raphson method can then be used to obtain the MLEs of the model parameters.

Accelerated failure time (AFT) models

For both the Cox and Weibull proportional hazards models, we assume that the hazards ratio $h_i(t)/h_0(t)$ is constant over time. In practice, however, the proportional hazards assumption may not be valid, so the Cox and/or Weibull proportional hazards models may not be appropriate in these situations. An alternative and popular survival regression model is the so-called *accelerated failure time (AFT) model*, which does not require the proportional hazards assumption.

An AFT model may be interpreted as the speed of disease progression. For example, the effect of a treatment may be assumed to "speed up" or "slow down" the passage of time, which can be expressed as

$$S_T(t) = S_C \left(\frac{t}{\phi} \right), \qquad \text{or} \qquad h_T(t) = \frac{1}{\phi} h_C \left(\frac{t}{\phi} \right),$$

where T denotes the treatment group, C denotes the control group, $S_T(t)$ and $h_T(t)$ ($S_C(t)$ and $h_C(t)$) are the survival function and hazard function in the treatment (control) group respectively, and $\phi > 0$ reflects the impact of the treatment. We call $1/\phi$ the *acceleration factor*. Thus, if the "event" is death, "$\phi < 1$" suggests an acceleration in the time to death in the treatment group compared with the control group. Such an interpretation is very appealing in practice.

A general AFT model can be written as

$$h_i(t) = \exp(-\mathbf{x}_i^T \boldsymbol{\beta}) \, h_0 \left(t \exp(-\mathbf{x}_i^T \boldsymbol{\beta}) \right), \qquad i = 1, 2, \cdots, n. \quad (7.10)$$

For a parametric AFT model, the following alternative log-linear representation is widely used

$$\log(T_i) = \mathbf{x}_i^T \boldsymbol{\beta} + \sigma \epsilon_i, \qquad i = 1, 2, \cdots, n, \quad (7.11)$$

where σ is a scale parameter and ϵ_i's are random errors. If we assume that ϵ_i follows a parametric distribution, we have a parametric AFT model.

Different choices of the distributions for ϵ_i lead to different AFT models. The following are three commonly used parametric AFT models. One common choice for the distribution of ϵ_i is the *Gumbel distribution*, which is an extreme value distribution with survival function and hazard function given by

$$S(t) = \exp(-e^t), \qquad h(t) = e^t, \qquad -\infty < t < \infty.$$

If ϵ_i follows the Gumbel distribution, the survival time T_i follows a Weibull distribution. Thus, the Weibull distribution has both the proportional hazards property and the accelerated failure time property, which is very appealing and makes the Weibull survival model very attractive.

Another common choice for the distribution of ϵ_i is the standard normal distribution $N(0, 1)$. If ϵ_i follows $N(0, 1)$, the survival time T_i follows a *log-normal distribution*, whose survival functions are given by

$$S_0(t) = 1 - \Phi\left(\frac{\log(t) - \beta_0}{\sigma}\right), \qquad S_i(t) = 1 - \Phi\left(\frac{\log(t) - \mathbf{x}_i^T\boldsymbol{\beta}}{\sigma}\right).$$

The third common choice for the distribution of ϵ_i is the *logistic distribution*, such as a logistic distribution with mean zero and variance $\pi^2/3$ which has the survival function and hazard function given respectively by

$$S(t) = \frac{1}{1 + \exp(t)}, \qquad h(t) = \frac{1}{1 + \exp(-t)}.$$

If ϵ_i follows a logistic distribution, the survival time T_i follows a *log-logistic distribution*, whose survival function is given by

$$S_i(t) = \frac{1}{1 + \exp[(\log(t) - \mathbf{x}_i^T\boldsymbol{\beta})/\sigma]}.$$

The foregoing three parametric AFT models are perhaps most commonly used in practice.

Inference for parametric AFT model (7.10) or (7.11) can be based on the likelihood method. The likelihood is given by

$$L(\boldsymbol{\beta}, \sigma) = \prod_{i=1}^n (f_i(t_i))^{\delta_i} (S_i(t_i))^{1-\delta_i},$$

where $f_i(t)$ and $S_i(t)$ are the parametric density function and survival function for individual i respectively. The MLE of $(\boldsymbol{\beta}, \sigma)$ satisfies the following equation

$$\frac{\partial \log L(\boldsymbol{\beta}, \sigma)}{\partial(\boldsymbol{\beta}, \sigma)} = 0,$$

which may be solved using the Newton-Raphson method. Likelihood inference then proceeds in the usual way, and standard asymptotic properties for MLE also hold under some regularity conditions. Note that, when the distributional assumptions hold, likelihood inference based on parametric models leads to asymptotically most efficient estimates.

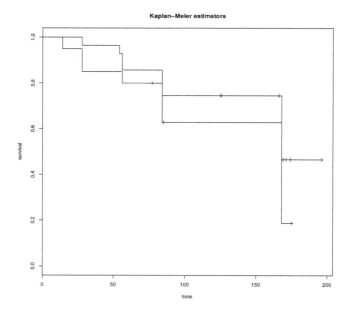

Figure 7.3 *Kaplan-Meier estimators for dropout times of subjects with CD4 less than 200 and subjects with CD4 greater or equal to 200.*

Example 7.1 *Modeling time to dropout*

In the AIDS study described in Chapter 1, some patients dropped out of the study early for various reasons such as drug side-effects. It would be of interest to explore the characteristics of dropout patients which may be predictive for times to dropout. For example, CD4 cell count is an important measure of disease progression, so we may be interested in testing whether baseline CD4 values are predictive for patients' dropout times. For AIDS patients, a threshold CD4 value of 200 is commonly used in many studies. Therefore, we wish to compare the dropout times of patients with CD4 values of 200 or higher to those with CD4 values less than 200. The dropout times for patients who never dropped out may be considered as right censored, with the study end time as the censoring time.

Figure 7.3 shows the Kaplan-Meier estimates of the survival functions of the dropout times (in days) for the two groups of patients. From the figure, it is not clear if there is a significant difference between these two groups. In fact, the log-rank test for testing the survival differences between the two groups produces a p-value of 0.103, indicating some marginal evidence of survival differences between the two groups.

Table 7.1 *Parameter estimates of the three survival regression models*

Model	Par.	Est.	S.E.	p-value	Par.	Est.	S.E.	p-value
Cox PH		–	–	–	β_1	0.042	0.128	0.74
Weibull	β_0	5.228	0.108	0.000	β_1	–0.161	0.093	0.083
Log-Normal	β_0	5.033	0.149	0.000	β_1	–0.241	0.148	0.103

Par.: parameter, Est.: estimate, S.E.: standard error.

In the following, we consider several survival regression models for time to dropout, with the original baseline CD4 value as a covariate. Each type of regression models has its advantages and disadvantages. For example, a Cox proportional hazards model is robust to distributional assumptions but it has the proportional hazards assumption. A Weibull model may produce more efficient parameter estimates than a Cox model but it has the distributional assumption. An parametric AFT model does not require the proportional hazards assumption but it also has the distributional assumption. For comparison, here we consider the following three models for the times T_i to dropout: a semiparametric Cox proportional hazards model, a parametric Weibull model, and a parametric log-normal AFT model. We standardize CD4 values to avoid very small/large estimates, which may be unstable.

Here the Cox proportional hazards model can be written as

$$h_i(t) = h_0(t) \exp(\beta_1 z_i), \qquad i = 1, 2, \cdots, n, \tag{7.12}$$

where $h_0(t)$ is a unspecified nonparametric baseline hazard function and z_i is the baseline CD4 value for individual i. The Weibull regression model can be written in a log-linear form as follows:

$$\log(T_i) = \beta_0 + \beta_1 z_i + \sigma \epsilon_i, \qquad i = 1, 2, \cdots, n, \tag{7.13}$$

where the random error ϵ_i follows the Gumbel distribution so the event time T_i follows a Weibull distribution. The log-normal AFT model has the same form as the model (7.13), but with the error ϵ_i following a standard normal distribution $N(0, 1)$. Both the Weibull model and the log-normal AFT model are parametric regression models, and the Weibull model has both the proportional hazards and accelerated failure time properties.

Table 7.1 shows the parameter estimates of the regression coefficients in the three models. Note that the parameters in the semiparametric Cox model and the Weibull/log-normal models have different interpretations, due to different model formulations, so they are not directly comparable. Moreover, the Cox model does not have an intercept term since the intercept is absorbed into the baseline hazard. However, parameter β_1 in all three models measures the effects of baseline CD4 values on event times, so its estimates and the associated

p-values under the three models can be used to check if baseline CD4 values are predictive for times to dropout.

From Table 7.1, there seems a big difference between the p-values of $\hat{\beta}_1$ under the Cox model and the two parametric models. This is probably because the two parametric models are more efficient (e.g., smaller standard errors) than the semiparametric Cox model. The semiparametric Cox model shows no significant effect of CD4 value on dropout time, while the two parametric models (Weibull and log-normal) show some evidence of CD4 effects on times to dropout: patients with higher initial CD4 values appear to have later dropout times. The estimates of the scale parameter σ are 0.52 (Weibull model) and 0.86 (log-normal model). Model diagnostics, such as residual plots, are needed to check the assumptions of the models. In conclusion, baseline CD4 values may be related to times to dropout, but the evidence is not strong.

7.2.4 Interval Censoring and Informative Censoring

Interval Censored Data

In previous sections we assume that the exact survival times and censoring times can be observed and the censoring is right censoring. In practice, however, we sometimes do not observe the *exact* survival times. Instead, we only know that the event has occurred during a particular time interval. For example, we may know that a subject in a study dropped out last month but we do not know the exact dropout day, or we may know that a patient got an infection last week but we do not know the exact time of infection. In both cases, the exact event times are not observed, but the event times are known to be in some time intervals. Such survival data are called *interval-censored*. In practice, perhaps most survival data may be viewed as interval-censored since the event times are often recorded as the nearest day or week or month.

Suppose that, for each individual, information is recorded on whether or not an event has occurred at scheduled times $t_1 < t_2 < \cdots < t_k$, with $t_0 = 0$ and $t_{k+1} = \infty$. Then, an individual who has an event detected at time t_j has an actual or exact event time t between t_{j-1} and t_j, i.e., $t_{j-1} \leq t < t_j$. Let $p_{ij} = P(t_{j-1} \leq T_i < t_j)$ be the probability that individual i experiences an event in the j-th time interval $[t_{j-1}, t_j)$. Let

$$\pi_{ij} = P(t_{j-1} \leq T_i < t_j \mid T_i > t_{j-1})$$

be the probability that individual i is free of the event at time t_{j-1} but has an event between t_{j-1} and t_j. Then we have

$$p_{ij} = (1 - \pi_{i1})(1 - \pi_{i2}) \cdots (1 - \pi_{i,j-1})\pi_{ij}, \qquad j = 2, 3, \cdots, k+1.$$

Let r_{ij} be an event indicator such that $r_{ij} = 1$ if subject i has an event in

$[t_{j-1}, t_j)$ and $r_{ij} = 0$ otherwise. The likelihood for the observed data can be written as

$$L = \prod_{i=1}^{n} \prod_{j=1}^{k+1} p_{ij}^{r_{ij}} = \prod_{i=1}^{n} \prod_{j=1}^{k+1} [(1 - \pi_{i1})(1 - \pi_{i2}) \cdots (1 - \pi_{i,j-1}) \pi_{ij}]^{r_{ij}}.$$

Note that for a proportional hazards model, we have

$$\pi_{ij} = 1 - P(T_i \geq t_j | T_i \geq t_{j-1}) = \left(\frac{S_i(t_j)}{S_i(t_{j-1})} \right)^{\exp(\mathbf{x}_i^T \boldsymbol{\beta})}.$$

Likelihood estimation and inference can then proceed in the usual way, based on the partial likelihood or full likelihood methods.

Informative Censoring

The models and methods for survival analysis in previous sections are based on the assumption that the censoring is non-informative or random, i.e., the censoring is not related to any factors associated with the survival times. In practice, however, sometimes the censoring may be *informative*. For example, an individual may drop out from a study because of drug side effects or other factors associated with the survival times, so individuals with censored survival times may offer valuable information about the study, and individuals without censoring may not be representative of the sample or the population. Thus, in statistical analysis we must incorporate such information in order to avoid biased results. We may, for example, assume a possible model for the informative censoring to check if the probability of censoring is related to covariates. For example, a logistic regression model may be used to model the censoring indicator. The procedure is similar to non-ignorable missing data problems.

7.3 Frailty Models

7.3.1 Clustered Survival Data

In Section 7.2, the event times or the sample members are assumed to be *independent*, as in cross-sectional studies. In some applications, however, the event times or the sample members may be *correlated*. The correlation often arises because there are *clusters* in the data, so the event time data in the same clusters may be correlated since they may be more similar than data from different centers, although event time data from different clusters are still assumed to be independent.

The following examples show how clustered data may arise in practice: (i) in a multi-center study data from the same center may be correlated but data from different centers may be independent (here each center may be viewed as a cluster); (ii) in a sample survey individuals from the same family may be correlated but individuals from different families may be independent (here each family may be viewed as a cluster); (iii) in a large-scale medical study patients from the same hospital may have similar survival rates (here each hospital may be viewed as a cluster); (iv) in a study with repeated measurements, such as multiple car accidents caused by the same individuals in a given year, the event data on the same individual may be correlated (here each individual may be viewed as a cluster). In these examples, the survival data from the same clusters may be correlated because each cluster shares the same characteristics, but survival data from different clusters may be independent. These survival data are called *clustered survival data*.

For clustered survival data, although the data within the same cluster may be similar, data from different clusters are typically more different, i.e., there is often a large variation between different clusters. In analysis of clustered survival data, we can use random effects to represent *cluster-effects* or variations between clusters. These random effects also incorporate correlation within the same clusters because the random effects are common to all data within the same cluster, which induces some dependence among the data in the same cluster. A survival model with random effects is often called a *frailty model* (Clayton and Cuzick 1985; Oakes 1989; Liang, Self, Bandeen-Roche, and Zeger 1995; Li and Thompson 1997). Here a random effect is called a *frailty*, because individuals with a larger (smaller) value of the effect will have a larger (smaller) value of the hazard function, i.e., they are more (less) likely to die or experience an event sooner so are considered to be more (less) frail.

As in GLMM or NLME models, the random effects or frailties in a survival model for clustered data play two roles:

- they incorporate the *correlation* between the survival data within the same clusters because the data in the same cluster share the same traits;
- they incorporate the *variation* in the survival data between different clusters since survival data may vary substantially across clusters.

Thus, random effects in a frailty model have similar interpretations as that in GLMM and NLME models. Moreover, frailty models are also nonlinear models, similar to GLMM and NLME models. The unique characteristics of survival or frailty models are that the data are often censored and the assumed distributions in parametric models are not necessarily in the exponential family. Moreover, semiparametric models are perhaps more commonly used in survival or frailty models in which one models the hazard functions rather than the mean functions of the responses, without distributional assumptions for the response data.

Statistical inference for frailty models is often based on the likelihood or partial likelihood methods. The computation can be more tedious for frailty models than for GLMM and NLME models, especially for semiparametric Cox proportional hazards models where the baseline hazard functions are unspecified. In the next section, we provide more details of frailty models and methods for inference.

7.3.2 Models and Inference

Consider clustered survival data. For individual j in cluster i, let s_{ij} be the true survival time, some of which may not be observed due to censoring, and let c_{ij} be the censoring time, $j = 1, 2, \cdots, n_i$; $i = 1, 2, \cdots, n$, where n_i is the number of individuals in the i-th cluster. The observed survival times are $t_{ij} = \min(s_{ij}, c_{ij})$, which are either true survival times or censoring times. Let $\delta_{ij} = I(s_{ij} \leq c_{ij})$ be the censoring indicator function. The observed data are

$$\{(t_{ij}, \delta_{ij}), \ j = 1, 2, \cdots, n_i; \ i = 1, 2, \cdots, n\}.$$

Let $h_{ij}(t), \mathbf{x}_{ij}, \mathbf{z}_{ij}$ be respectively the hazard function and covariate vectors for individual j in cluster i. To model clustered survival data, we can extend the Cox proportional hazards model to a frailty model by introducing random effects as follows

$$\begin{aligned}
h_{ij}(t_{ij}|\mathbf{b}_i, \boldsymbol{\beta}) &= h_0(t_{ij}) \exp(\mathbf{x}_{ij}^T \boldsymbol{\beta} + \mathbf{z}_{ij}^T \mathbf{b}_i), & (7.14) \\
\mathbf{b}_i &\sim N(0, D), & i = 1, 2, \cdots, n, \ j = 1, 2, \cdots, n_i,
\end{aligned}$$

where $\mathbf{b}_i$ contains random effects for cluster i. Note that the distribution of the random effects $\mathbf{b}_i$ is not restricted to a normal distribution. We assume that, conditional on the random effects $\mathbf{b}_i$, the event times T_{ij} are independent. A commonly used class of frailty models is the multiplicative frailty model with $\mathbf{z}_{ij} = 1$ and $\exp(b_i) \sim \text{gamma}(\lambda, \lambda)$.

Statistical inference for frailty model (7.14) can be based on the likelihood methods. Assuming non-informative censoring, the probability distribution for the observed data (t_{ij}, δ_{ij}) of individual j in cluster i, given the random effects $\mathbf{b}_i$, can be written as

$$\begin{aligned}
f(t_{ij}, \delta_{ij}|\mathbf{b}_i, \boldsymbol{\beta}) &\propto f(t_{ij}|\mathbf{b}_i, \boldsymbol{\beta})^{\delta_{ij}} S(t_{ij}|\mathbf{b}_i, \boldsymbol{\beta})^{1-\delta_{ij}} \\
&= h_{ij}(t_{ij}|\mathbf{b}_i, \boldsymbol{\beta})^{\delta_{ij}} S(t_{ij}|\mathbf{b}_i, \boldsymbol{\beta}), & (7.15)
\end{aligned}$$

suppressing the covariates. So the corresponding log-likelihood can be written as

$$\begin{aligned}
l_{ij}(\boldsymbol{\beta}) &= \delta_{ij} \log(f(t_{ij}|\mathbf{b}_i, \boldsymbol{\beta})) + (1 - \delta_{ij}) \log(S(t_{ij}|\mathbf{b}_i, \boldsymbol{\beta})) & (7.16) \\
&= \delta_{ij} \left[\log h_0(t_{ij}) + \mathbf{x}_i^T \boldsymbol{\beta} + \mathbf{z}_{ij}^T \mathbf{b}_i \right] - \exp(\mathbf{x}_i^T \boldsymbol{\beta} + \mathbf{z}_{ij}^T \mathbf{b}_i) H_0(t_{ij}).
\end{aligned}$$

Inference can again be based on the partial likelihood (Klein 1992), but the computation can be challenging due to the unobservable random effects.

To model clustered survival data based on an accelerated failure time model, the corresponding frailty model can be written in the following log-linear form

$$\log(T_{ij}) = \mathbf{x}_i^T \boldsymbol{\beta} + \mathbf{z}_{ij}^T \mathbf{b}_i + \epsilon_{ij}, \tag{7.17}$$
$$\mathbf{b}_i \sim N(0, D), \qquad i = 1, 2, \cdots, n, \ j = 1, 2, \cdots, n_i.$$

If ϵ_{ij} is assumed to follow a parametric distribution, inference can be based on the following likelihood

$$L(\boldsymbol{\beta}, D) = \prod_{i=1}^{n} \int \left\{ \prod_{j=1}^{n_i} h_{ij}(t_{ij})^{\delta_{ij}} S_{ij}(t_{ij}) f(\mathbf{b}_i | D) \right\} d\mathbf{b}_i. \tag{7.18}$$

MLE of $(\boldsymbol{\beta}, D)$ can be obtained by maximizing $L(\boldsymbol{\beta}, D)$. Usual asymptotic results hold for the MLE under some regularity conditions.

For a parametric frailty model, such as that based on the Weibull proportional hazards model or a parametric accelerated failure time model, likelihood inference can be based on the *full* likelihood such as (7.18). For a semiparametric frailty model, such as that based on the Cox proportional hazards model, likelihood inference can be based on the partial likelihood. For either semiparametric or parametric frailty models, the EM algorithm can be used for estimation where the unobservable random effects are treated as missing data (Nielsen et al. 1992). Likelihood computation for frailty models can be computationally very intensive due to the unobservable random effects and the nonparametric hazard functions.

7.4 Survival and Frailty Models with Missing Covariates

7.4.1 Survival Models with Missing Covariates

In survival regression models, covariates may contain missing data, as in other regression models. The complete-case method, which discards all incomplete observations, may lead to biased or inefficient estimates. So the missing data must be addressed appropriately. Statistical methods for missing covariates in survival regression models are similar to those for nonlinear or generalized linear models, but the computation can become more challenging due to the natures of survival models. In particular, for Cox proportional hazards models, one models the hazard function rather than the mean function, leading to more computational challenges. In this section, we focus on Cox proportional hazards models with missing covariates, and consider likelihood methods for inference, following Herring and Ibrahim (2001). We assume that the missing data are missing at random or ignorable and that the censoring distribution is independent of missing data.

Consider the following Cox proportional hazards model

$$h_i(t) = h_0(t) \exp(\mathbf{x}_i^T \boldsymbol{\beta}), \qquad i = 1, 2, \cdots, n, \tag{7.19}$$

where the baseline hazard function $h_0(t)$ is unspecified. Suppose that covariates $\mathbf{x}_i$ contain missing values. We write $\mathbf{x}_i = (\mathbf{x}_{mis,i}, \mathbf{x}_{obs,i})$ as in previous chapters. To address missing data in covariates $\mathbf{x}_i$, we assume a covariate distribution with density $f(\mathbf{x}_i | \boldsymbol{\alpha})$. The likelihood for the observed data $(\mathbf{x}_{obs,i}, t_i, \delta_i)$ for individual i can be written as

$$
\begin{aligned}
L_i(\boldsymbol{\theta}) \quad \propto \quad & \int h(t_i | \mathbf{x}_{mis,i}, \mathbf{x}_{obs,i}, \boldsymbol{\theta})^{\delta_i} S(t_i | \mathbf{x}_{mis,i}, \mathbf{x}_{obs,i}, \boldsymbol{\theta}) \\
& \times f(\mathbf{x}_{mis,i}, \mathbf{x}_{obs,i} | \boldsymbol{\alpha}) \, d\, \mathbf{x}_{mis,i},
\end{aligned}
\tag{7.20}
$$

where $\boldsymbol{\theta} = (\boldsymbol{\beta}, h_0(t), \boldsymbol{\alpha})$, and the completely observed covariates are suppressed. The observed-data likelihood $L_i(\boldsymbol{\theta})$ involves an intractable integral.

When covariates $\mathbf{x}_i$ are categorical, the integral in (7.20) reduces to a summation, which simplifies the computation. When covariates $\mathbf{x}_i$ are continuous, we can use a Monte Carlo EM algorithm for likelihood computation, similar to that in Chapter 4. As shown in Herring and Ibrahim (2001), the computation of E-step in this case can be quite tedious, so they proposed an approximation based on a first-order Taylor expansion which simplifies computation in the E-step. The M-step is like a complete-data maximization, so usual optimization procedures can be used.

As in Chapter 4, the Monte Carlo EM algorithm approximates the expectation in the E-step by a Monte Carlo mean, where the missing data are substituted by many simulated values from the predictive distribution of the missing covariates given the observed values and current parameter estimates. Specifically, at the k-th iteration of the EM algorithm, one simulates large samples of $\mathbf{x}_{mis,i}$ from the conditional distribution $f(\mathbf{x}_{mis,i} | \mathbf{x}_{obs,i}, t_i, \delta_i, \boldsymbol{\theta}^{(k)})$. To simulate the desired values, note that

$$f(\mathbf{x}_{mis,i} | \mathbf{x}_{obs,i}, t_i, \delta_i, \boldsymbol{\theta}^{(k)}) \propto f(t_i, \delta_i | \mathbf{x}_i, \boldsymbol{\theta}^{(k)}) f(\mathbf{x}_i | \boldsymbol{\alpha}^{(k)}),$$

where

$$f(t_i, \delta_i | \mathbf{x}_i, \boldsymbol{\theta}^{(k)}) \propto h(t_i | \mathbf{x}_i, \boldsymbol{\theta}^{(k)})^{\delta_i} S(t_i | \mathbf{x}_i, \boldsymbol{\theta}^{(k)}).$$

Thus, we can use the Gibbs sampler along with rejection sampling methods to simulate samples from the conditional distribution $f(\mathbf{x}_{mis,i} | \mathbf{x}_{obs,i}, t_i, \delta_i, \boldsymbol{\theta}^{(k)})$. Details can be found in Herring and Ibrahim (2001).

For parametric survival models with missing covariates, such as parametric accelerated failure time models and Weibull proportional hazards models, the approach is similar and computation may be simpler since there are no non-parametric components in the models.

7.4.2 Frailty Models with Missing Covariates

The methods in the previous section can be extended to frailty models with missing covariates. In a frailty model, we may view the unobservable random effects as additional "missing covariates", and then we may apply the methods in the previous section in a relatively straightforward way. The idea is similar to the methods for missing covariates in GLMM and NLME models in Chapter 4. In the following, we focus on Cox proportional hazards models with random effects and missing covariates, and we follow the approach in Herring, Ibrahim, and Lipsitz (2002). We assume that the missing data are ignorable and censoring is non-informative.

Let $t_{ij}, \mathbf{x}_{ij}, \mathbf{z}_{ij}$ be respectively the event time and covariates from individual j in cluster i, $i = 1, 2, \cdots, n$, $j = 1, 2, \cdots, n_i$. Consider the following frailty model for clustered survival data

$$
\begin{aligned}
h_{ij}(t_{ij}|\mathbf{b}_i, \boldsymbol{\beta}) &= h_0(t_{ij}) \exp(\mathbf{x}_{ij}^T\boldsymbol{\beta} + \mathbf{z}_{ij}^T\mathbf{b}_i), \qquad (7.21)\\
\mathbf{b}_i &\sim N(0, D), \qquad i = 1, 2, \cdots, n, \ j = 1, 2, \cdots, n_i,
\end{aligned}
$$

where the event times are assumed to be conditionally independent given the random effects. Suppose that covariates $\mathbf{x}_{ij}$ contain missing values. Let $\mathbf{x}_{ij} = (\mathbf{x}_{mis,ij}, \mathbf{x}_{obs,ij})$. The observed data are

$$
\{(t_{ij}, \delta_{ij}, \mathbf{x}_{obs,ij}, \mathbf{z}_{ij}), \ i = 1, 2, \cdots, n, \ j = 1, 2, \cdots, n_i\},
$$

and the observed-data likelihood for cluster i can be written as

$$
\begin{aligned}
L_i(\boldsymbol{\theta}) = \int \prod_{j=1}^{n_i} \Big[\int & h_{ij}(t_{ij}|\mathbf{x}_{mis,ij}, \mathbf{x}_{obs,ij}, \mathbf{b}_i, \boldsymbol{\beta})^{\delta_{ij}} \\
& \times S_{ij}(t_{ij}|\mathbf{x}_{mis,ij}, \mathbf{x}_{obs,ij}, \mathbf{b}_i, \boldsymbol{\beta}) \\
& \times f(\mathbf{x}_{mis,ij}, \mathbf{x}_{obs,ij}|\boldsymbol{\alpha}) \ d\mathbf{x}_{mis,ij} \Big] f(\mathbf{b}_i|D) \ d\mathbf{b}_i, \quad (7.22)
\end{aligned}
$$

where $\boldsymbol{\theta}$ contains all parameters and covariates $\mathbf{z}_{ij}$ are suppressed. In the following we briefly describe a Monte Carlo EM algorithm for likelihood computation.

By treating the random effects as additional "missing data", in the E-step of the k-th EM iteration, we need to simulate large samples of the "missing data" from the conditional distribution of these "missing data" given the observed data and current parameter estimates, i.e., $f(\mathbf{x}_{mis,ij}, \mathbf{b}_i|\mathbf{x}_{obs,ij}, t_{ij}, \delta_{ij}, \boldsymbol{\theta}^{(k)})$. This Monte Carlo sampling may be accomplished by noting that

$$
\begin{aligned}
f(\mathbf{x}_{mis,ij}, \mathbf{b}_i|\mathbf{x}_{obs,ij}, t_{ij}, \delta_{ij}, \boldsymbol{\theta}^{(k)}) \ \propto \ & f(t_{ij}, \delta_{ij}|\mathbf{x}_{ij}, \mathbf{b}_i, \boldsymbol{\beta}^{(k)}) \\
& \times f(\mathbf{x}_{ij}|\boldsymbol{\alpha}^{(k)}) f(\mathbf{b}_i|D^{(k)}).
\end{aligned}
$$

Therefore, we can use the Gibbs sampler as follows, at the l-th iteration of the Gibbs sampler,

- generate a sample $\mathbf{x}_{mis,ij}^{(l)}$ from

$$f(\mathbf{x}_{mis,ij}|\mathbf{x}_{obs,ij},t_{ij},\delta_{ij},\mathbf{b}_i^{(l-1)},\boldsymbol{\theta}^{(k)})$$
$$\propto f(t_{ij},\delta_{ij}|\mathbf{x}_i,\mathbf{b}_i^{(k-1)},\boldsymbol{\beta}^{(k)})f(\mathbf{x}_{mis,ij}|\mathbf{x}_{obs,ij},\boldsymbol{\alpha}^{(k)})$$

- generate a sample $\mathbf{b}_i^{(l)}$ from

$$f(\mathbf{b}_i|\mathbf{x}_{mis,ij}^{(l)},\mathbf{x}_{obs,ij},t_{ij},\delta_{ij},\boldsymbol{\theta}^{(k)})$$
$$\propto f(\mathbf{b}_i|D^{(k)})\prod_{j=1}^{n_i} f(t_{ij},\delta_{ij}|\mathbf{x}_{mis,ij}^{(k)},\mathbf{x}_{obs,ij},\mathbf{b}_i^{(l-1)},\boldsymbol{\theta}^{(k)}).$$

After a burn-in period for the Gibbs sampling, we obtain a desired sample of $(\mathbf{x}_{mis,ij},\mathbf{b}_i)$ from the conditional distribution $f(\mathbf{x}_{mis,ij},\mathbf{b}_i|\mathbf{x}_{obs,ij},t_{ij},\delta_{ij},\boldsymbol{\theta}^{(k)})$ in the k-th EM iteration. Repeating the procedure m_k times, we obtain m_k independent samples. Then, we can approximate the expectation in the E-step by an empirical mean using the simulated values. The M-step is then like a complete-data maximization, so standard optimization procedures can be used.

When the missing data are non-ignorable, one only needs to specify a non-ignorable missing data model and then incorporate it in the likelihood. The approach is similar to that in Chapter 4.

Multiple imputation methods can also be used to address missing covariates in survival models and frailty models. A major advantage of a multiple imputation method is that standard complete-data methods can be used, once each missing value is imputed by several plausible values. To generate proper multiple imputations, one can use a Bayesian framework and generate $\mathbf{x}_{mis,ij}$ from the following predictive distribution of the missing data given the observed data

$$f(\mathbf{x}_{mis,ij}|\mathbf{x}_{obs,ij},t_{ij},\delta_{ij}) \quad \propto \quad \int\int\int f(t_{ij},\delta_{ij}|\mathbf{x}_{ij},\mathbf{b}_i,\boldsymbol{\beta})f(\mathbf{x}_{ij}|\boldsymbol{\alpha})$$
$$\times f(\mathbf{b}_i|D)f(\boldsymbol{\theta}|\boldsymbol{\eta})\,d\,\mathbf{x}_{mis,ij}\,d\,\mathbf{b}_i\,d\,\boldsymbol{\theta},$$

where $\boldsymbol{\theta}$ contains all parameters and $\boldsymbol{\eta}$ contains hyper-parameters for the prior distribution $f(\boldsymbol{\theta}|\boldsymbol{\eta})$. The sampling can be accomplished using the Gibbs sampler or data augmentation methods, as described in Chapter 4.

7.5 Frailty Models with Measurement Errors

The Models

Covariates are often introduced in frailty models to partially explain the variation in the event time data, but some of these covariates may be measured with errors. As in GLMM and NLME models, statistical analysis ignoring covariate measurement errors in frailty models may lead to biased results (Li and

Lin 2000), so measurement errors in covariates must be addressed. In this section, we focus on covariate measurement errors in Cox models with random effects, following Li and Lin (2000). We assume that the survival data are right censored and the censoring is non-informative.

For individual j in cluster i, let s_{ij} be the true survival time, which may not be observed due to censoring, and let c_{ij} be the censoring time, $j = 1, 2, \cdots, n_i$; $i = 1, 2, \cdots, n$. The observed survival times are $t_{ij} = \min(s_{ij}, c_{ij})$. Let $\delta_{ij} = I(s_{ij} \leq c_{ij})$ be the censoring indicator function. For simplicity, we focus on a single univariate error-prone covariate x_{ij}. Let x_{ij}^* be the true but unobserved covariate value for individual j in cluster i, whose observed but mis-measured value is x_{ij}. We assume that x_{ij} is a surrogate of x_{ij}^* or that the measurement error is nondifferential. Let $\mathbf{t}_i = (t_{i1}, t_{i2}, \cdots, t_{in_i})^T$, and define $\boldsymbol{\delta}_i$, $\mathbf{x}_i$, $\mathbf{x}_i^*$, and $\mathbf{z}_i$ similarly, where $\mathbf{z}_{ij}$ contains covariates without measurement errors. Conditional on the random effects $\mathbf{b}_i$, we assume that the observed data $\{(t_{ij}, \delta_{ij}), \ j = 1, 2, \cdots, n_i; i = 1, 2, \cdots, n\}$ are independent and follow the following proportional hazards model

$$\lambda_{ij}(t|x_{ij}^*, \mathbf{b}_i) = \lambda_0(t) \exp(x_{ij}^* \beta + \mathbf{z}_{ij}^T \mathbf{b}_i), \qquad (7.23)$$
$$\mathbf{b}_i \sim N(0, D). \qquad (7.24)$$

where the hazard function $\lambda_{ij}(t|x_{ij}^*, \mathbf{b}_i)$ is related to the unobserved true covariate value $\mathbf{x}_i^*$ rather than the observed but mis-measured value $\mathbf{x}_i$.

We consider the following classical measurement error model

$$x_{ij} = x_{ij}^* + \epsilon_{ij}, \qquad (7.25)$$

where ϵ_{ij} i.i.d. $\sim N(0, \sigma_e^2)$. For likelihood inference, we consider a structural approach for measurement errors and assume that $\mathbf{x}_i$ follows a LME model

$$\mathbf{x}_i = U_i \, \boldsymbol{\alpha} + V_i \, \mathbf{a}_i + \boldsymbol{\epsilon}_i \equiv \mathbf{x}_i^* + \boldsymbol{\epsilon}_i, \qquad (7.26)$$

where U_i and V_i are design matrices, $\boldsymbol{\alpha}$ contains unknown population parameters (fixed effects) and $\mathbf{a}_i \sim N(0, A)$ are random effects, and $\boldsymbol{\epsilon}_i \sim N(0, \delta^2 I)$ are the measurement errors.

Likelihood Inference

The observed data are

$$\{(\mathbf{t}_i, \boldsymbol{\delta}_i, \mathbf{x}_i, \mathbf{z}_i), \ i = 1, 2, \cdots, n\},$$

and the observed-data likelihood is given by

$$L_o(\boldsymbol{\theta}) = \prod_{i=1}^n \int f(\mathbf{t}_i, \boldsymbol{\delta}_i | \mathbf{x}_i^*, \mathbf{b}_i, \boldsymbol{\theta}) f(\mathbf{x}_i | \mathbf{a}_i, \boldsymbol{\theta}) f(\mathbf{b}_i | \boldsymbol{\theta}) f(\mathbf{a}_i | \boldsymbol{\theta}) \, d\mathbf{a}_i \, d\mathbf{b}_i,$$

where θ denotes the collection of all unknown parameters,

$$
\begin{aligned}
f(\mathbf{t}_i, \boldsymbol{\delta}_i | \mathbf{x}_i^*, \mathbf{b}_i, \boldsymbol{\theta}) \;=\; &\prod_{j=1}^{n_i} \Big\{ \big[\lambda_0(t_{ij}) \exp(x_{ij}^* \beta + \mathbf{z}_{ij}^T \mathbf{b}_i) \big]^{\delta_{ij}} \\
&\times \exp\big(- \Lambda_0(t_{ij}) \exp(x_{ij}^* \beta + \mathbf{z}_{ij}^T \mathbf{b}_i) \big) \Big\},
\end{aligned}
$$

and $\Lambda_0(t)$ is the cumulative baseline hazard. Here the observed-data likelihood $L_o(\boldsymbol{\theta})$ is highly complicated and does not have a closed form expression. Li and Lin (2000) proposed a Monte Carlo EM algorithm for likelihood estimation, in a way similar to the ones in previous sections (but more complicated). For computing standard errors, Li and Lin (2000) proposed to use the profile likelihood method (Hu et al. 1998; Murphy 1995), since there are too many parameters.

Due to the nonparametric hazards functions and censoring, there are many unknown parameters in $\boldsymbol{\theta}$, so the above Monte Carlo EM algorithm can be computationally very intensive and may exhibit convergence problems. Therefore, computationally more efficient approximate methods are highly valuable to frailty models with measurement errors. These approximate methods can be developed in a way similar to the ones for GLMM and NLME models with measurement errors, as described in previous chapters.

Li and Lin (2000) showed that, after adjusted for measurement errors, the original frailty model leads to a new (but more complicated) frailty model, which suggests that ignoring measurement errors in frailty models may lead to misspecification of both the fixed effects and the frailty structures. This also allows some specific bias analysis, in a way similar to GLMMs.

CHAPTER 8

Joint Modeling Longitudinal Data and Survival Data

8.1 Introduction

As shown in Chapter 7, survival data often arise in longitudinal studies. In practice, we may need to model a time-to-event process and a longitudinal process *jointly*. A common situation is in survival models with time-dependent covariates, in which covariate data may be missing at failure times, so a longitudinal model for the covariates is required to address missing covariates or covariate measurement errors. As another example, in modeling longitudinal data with dropouts, we may also need to model the times to dropout or times to death, with the objectives of avoiding possible biases in the estimation of the longitudinal model, as well as studying the association between the time to event and characteristics of the longitudinal trajectories such as initial slopes or intercepts. In both cases, *joint modeling* of the longitudinal data and the survival data is required.

In joint modeling of longitudinal data and survival data, the main focus may be either the longitudinal model, or the survival model, or both models, depending on the objectives of the studies. When the main focus is on one model, the other model is then secondary so its parameters may be viewed as nuisance parameters. In this case, one should focus on correct specification of the main model and simplify the secondary model to reduce the number of nuisance parameters and avoid potential parameter non-identifiability. If both models are of primary interest, we may reduce the number of other secondary parameters such as the association parameters.

In joint modeling of longitudinal data and survival data, we often have the following three situations:

- the main focus is on modeling survival data, with modeling longitudinal data being secondary. This situation may arise, for example, in survival models with measurement errors or missing data in time-dependent covariates, where modeling the longitudinal covariate processes is only used to address measurement errors or missing data (covariate data may be missing at failure times);

- the main focus is on modeling longitudinal data, with modeling survival data being secondary. This situation may arise, for example, in the analysis of longitudinal data with informative dropouts, where a survival model is only used to address the informative dropouts and is not of primary interest.

- the main focus is on modeling both the longitudinal data and the survival data, with a goal of understanding the *association* between the two processes. This situation may arise, for example, in AIDS studies where we are interested in modeling both the viral load process and time to death, and our goal is to understand the association between the survival times and characteristics of the viral load process. In this case, the survival model and the longitudinal model are usually linked (e.g., share some parameters).

In all three cases, we must model the longitudinal process and the time-to-event process *jointly*, in order to avoid potential bias.

In some cases, the longitudinal and survival models may be linked through shared parameters or shared variables, leading to so-called *shared parameter models*. For example, when a mixed effects model is used to model the longitudinal process and a frailty model is used to model the event time process (e.g., the dropout process in the longitudinal study), the two models may share the same random effects since these random effects characterize the individual-specific longitudinal process. Alternatively, the two models may be governed by the same underlying latent process (shared variables). In these cases, the association between the longitudinal and survival models is made clear so joint inference is required.

Sometimes we may also need to model two or more longitudinal processes simultaneously. This situation may arise, for example, when the longitudinal processes are correlated. For example, in many AIDS studies both viral load and CD4 are measured repeatedly over time, and these two processes are usually correlated. We may treat both variables as responses and model them jointly, with a goal of understanding their association over time, or we may treat one variable as the response and the other variable as a time-dependent covariate.

A Simple Two-Step Method

For joint inference of several models with shared parameters or shared unobserved variables, a commonly used simple approach is the so-called *two-step method*:

- Step 1: estimate the shared variables or parameters in one model based on the observed data;

- Step 2: estimate parameters in the other model *separately*, with the shared variables or parameters substituted by their estimates from the first step as if they were observed values.

Such a simple two-step method is closely related to the *regression calibration* method in measurement error literature.

The naive two-step method is simple, and standard statistical software can be readily used. However, the naive two-step method may lead to biased estimation. Consider the two-step method for joint modeling of longitudinal data and survival data. In the first step we can estimate the parameters in the longitudinal model separately, and then in the second step we substitute the estimated shared parameters or variables in the survival model and then proceed with standard inference. This simple approach may lead to the following problems: (i) estimation may be *biased*, especially when the longitudinal process and the survival process are strongly associated; and (ii) the *uncertainty* of estimation in the first step is not incorporated in the second step, so standard errors of the parameter estimates in the survival model may be under-estimated. Therefore, a third step is needed to correct the bias and incorporate the estimation uncertainty in the first step, which may be difficult for some complex problems. Often, a *joint* inference is preferred, such as a joint likelihood inference.

Joint Likelihood Method

Statistical inference for joint models can be based on the *joint likelihood* of all the observed data. Such an approach is appealing because it provides valid and reliable inference and the resulting MLEs are asymptotically efficient and asymptotically normal under the usual regularity conditions. MLEs of all model parameters can be obtained *simultaneously* by maximizing the joint likelihood. Since the likelihood method is a standard approach for inference in mixed effects models, the joint likelihood method is a natural choice for joint mixed effects models. So we focus on the joint likelihood method in this chapter.

There are two potential problems for joint likelihood approaches: (i) model or parameter *identifiability* may be a potential problem since there are often many unknown parameters in joint models so the models or parameters may become non-identifiable in the sense that two sets of different parameters may lead to the same likelihood; and (ii) computation can be quite challenging, since joint likelihoods typically involve high-dimensional and intractable integrals. We will discuss parameter identifiability and computational issues in this chapter.

Missing Data and Measurement Errors

Since missing data and measurement errors are very common in longitudinal studies, in jointly modeling of longitudinal data and survival data, we often need to address these incomplete data problems. For the joint likelihood approach, this is relatively straightforward since the joint likelihood methods with incomplete data are similar to the ones in previous chapters. A major difficulty is the computational challenge, since joint models are more complex, especially joint inferences for GLMM or NLME models and Cox proportional hazards models. Therefore, computationally more efficient approximate methods are highly valuable for joint models.

In this chapter, we will mainly focus on joint likelihood methods for joint mixed effects models. We will also discuss missing data, measurement errors, and computational issues. Due to space limitation, we will focus on selected topics rather than a comprehensive overview.

8.2 Joint Modeling for Longitudinal Data and Survival Data

There is an extensive literature on joint modeling for longitudinal data and survival data. Some recent work includes Tseng, Hsieh, and Wang (2005), Nathoo and Dean (2008), Ye, Lin, and Taylor (2008), Song and Wang (2008), and Wu et al. (2008), among others. Tsiatis and Davidian (2004) provided a comprehensive review of earlier work. If the longitudinal and survival processes are unrelated, we can simply model them separately. However, in practice the longitudinal process and the survival process in the same study are often associated or linked, so a joint inference is required to avoid biased results.

Sometimes a longitudinal model and a survival model may be from different studies, but one wishes to borrow information across studies. In this case, a joint modeling approach may also be desirable. For example, when addressing measurement errors in time-dependent covariates of a survival model, we may use external data or data from a similar study to estimate the measurement errors in the covariates. Then a joint modeling approach would be appealing. In some cases, we may also use the joint model framework to make more efficient inference on the survival model by incorporating the longitudinal data as auxiliary information (e.g., Hogan and Laird 1997; Xu and Zeger 2001; Faucett et al. 2002).

In the following sections, we consider mixed effects models for modeling the longitudinal processes. For modeling the survival process, we consider data with either right censoring or interval censoring.

8.2.1 Joint Models with Right Censored Survival Data

In this section, we consider joint modeling for longitudinal data and survival data when the survival data are right censored. We focus on the case where the main interest is in survival modeling, with modeling longitudinal data being secondary. Thus, parameters in the survival model are of primary interest, while parameters in the longitudinal model are usually treated as nuisance parameters. This situation often arises in modeling survival data with measurement errors or missing data in time-dependent covariates. Here a longitudinal model is assumed for the time-dependent covariates in order to address measurement error and missing data, so the longitudinal model is secondary.

For individual i, let s_i be the survival time or event time, subject to right censoring, $i = 1, 2, \cdots, n$. We assume that the censoring is random or non-informative. Let c_i be the censoring time. Due to censoring, we only observe

$$t_i = \min\{s_i, c_i\},$$

i.e., we only observe either the true survival time s_i or the censored time c_i. In the survival times are right censored, we know that the true but unobserved survival times are greater than the censoring times, i.e., $s_i > c_i$. Let

$$\delta_i = I(s_i \leq c_i)$$

be the censoring indicator function such that $\delta_i = 0$ if the survival time for individual i is right censored and $\delta_i = 1$ otherwise.

The objective is to model the survival data, with covariates being introduced in the model to partially explain the variation in the survival data. Consider a time-dependent covariate with possible measurement errors in the observed data. Denote the observed covariate value for individual i at time u_{ij} by $z_{ij} = z_i(u_{ij})$ and denote the corresponding unobserved true covariate value by $z_{ij}^*, j = 1, 2, \cdots, m_i$. Let $\mathbf{z}_i = (z_{i1}, z_{i2}, \cdots, z_{im_i})^T$, and let $\mathbf{x}_i$ be other covariates without measurement errors.

Survival Model

For individual i, let T_i be the time to an event of interest and let t_i be its observed value, which may be right censored. For modeling survival data, we consider the following model:

$$\lambda_i(t) = \lambda_0(t) \exp(z_i^*(t)\beta_1 + \mathbf{x}_i^T \boldsymbol{\beta}_2), \qquad i = 1, \cdots, n, \tag{8.1}$$

where $\boldsymbol{\beta} = (\beta_1, \boldsymbol{\beta}_2)$ are unknown regression parameters. In model (8.1) we link the hazard function to the unobserved true covariate $z_i^*(t)$ rather than the observed but mis-measured covariate $z_i(t)$. We assume that the measurement errors are *non-differential* so that

$$\lambda_i(t \mid z_i^*(t), z_i(t), \mathbf{x}_i) = \lambda_i(t \mid z_i^*(t), \mathbf{x}_i).$$

Statistical inference can be based on the following partial likelihood (Cox

1972, 1975; Kalbfleisch and Prentice 2002)

$$L_p^* = \prod_{i=1}^{n} \left[\frac{\exp\{z_i^*(t_i)\beta_1 + \mathbf{x}_i^T\beta_2\}}{\sum_{j=1}^{n} \exp\{z_j^*(t_i)\beta_1 + \mathbf{x}_i^T\beta_2\}I(t_j \geq t_i)} \right]^{\delta_i}, \qquad (8.2)$$

where the survival times are assumed to be independent without ties.

For statistical inference based on the partial likelihood (8.2), the time-dependent covariate value $z_i(t)$ must be available at *each* event time t_j for *all* individuals. In practice, however, this is usually not the case since covariate measurement times usually do not coincide with event times. This leads to missing data in time-dependent covariates in the survival model. Such missingness may be assumed to be ignorable or MAR since the missingness may not be related to the missing values. The last-value-carried-forward (LVCF) method for missing longitudinal data has been widely used to impute the missing covariates, but it may lead to biased estimation (Prentice 1982). Moreover, the observed covariates $\mathbf{z}_i$ may be measured with errors, so the true covariate values $\mathbf{z}_i^*$ are unobserved. Therefore, we must address both measurement error and missing data in the time-dependent covariate $z_i(t)$ in survival model (8.1).

Longitudinal Model

For time-dependent covariate $z_i(t)$ with measurement errors, we can viewed the repeated measurements on each individual as replicates, which allows us to partially address measurement error in the covariate. Assuming the covariate values change smoothly over time, we can empirically model the covariate process in order to address both measurement error and missing data. Consider the following additive classical measurement error model

$$\mathbf{z}_i = U_i \boldsymbol{\alpha} + V_i \mathbf{a}_i + \boldsymbol{\epsilon}_i \equiv \mathbf{z}_i^* + \boldsymbol{\epsilon}_i, \qquad i = 1,\ldots,n, \qquad (8.3)$$

where U_i and V_i are known design matrices, $\boldsymbol{\alpha}$ is a vector of fixed-effects parameters, $\mathbf{a}_i$ is a vector of random effects, and $\boldsymbol{\epsilon}_i$ is a vector of measurement errors for individual i. Assume that $\mathbf{a}_i \sim N(0, A)$ and $\boldsymbol{\epsilon}_i \sim N(0, \delta^2 I)$ and $\mathbf{a}_i$ and $\boldsymbol{\epsilon}_i$ are independent. We also assume that the true covariates are

$$\mathbf{z}_i^* = U_i \boldsymbol{\alpha} + V_i \mathbf{a}_i,$$

which can be chosen empirically, such as a low-order polynomial with random coefficients, to mimic the "true" covariate process. Ye, Lin, and Taylor (2008) considered a more flexible semiparametric mixed effects model for the covariate process to address measurement errors. Note that a semiparametric mixed effects model may be approximated by a parametric mixed effects model.

To address missing data in the covariates at any event times, we can recast covariate model (8.3) in continuous time:

$$z_i(t) = \mathbf{u}_i^T(t) \boldsymbol{\alpha} + \mathbf{v}_i^T(t) \mathbf{a}_i + \epsilon_i(t), \qquad i = 1,\ldots,n.$$

Thus, at any event time t_i, the unobserved true covariate value can be taken as

$$z_i^*(t_i) = \mathbf{u}_i^T(t_i)\boldsymbol{\alpha} + \mathbf{v}_i^T(t_i)\mathbf{a}_i.$$

Ye, Lin, and Taylor (2008) showed in a simulation study that this approach is better than the naive last-value-carried-forward method.

Joint Likelihood

The *joint likelihood* of the survival model (8.1) and the longitudinal model (8.3) based on all observed data is given by

$$
L(\boldsymbol{\theta}) = \prod_{i=1}^{n} \int \Bigg\{ \big[\lambda_0(t_i)\exp\{z_i^*(t_i)\beta_1 + \mathbf{x}_i^T\boldsymbol{\beta}_2\}\big]^{\delta_i}
$$
$$
\times \exp\left[-\int_0^{t_i}\lambda_0(u)\exp\{z_i^*(u)\beta_1 + \mathbf{x}_i^T\boldsymbol{\beta}_2\}du\right]
$$
$$
\times f(\mathbf{z}_i|\mathbf{a}_i,\boldsymbol{\alpha})f(\mathbf{a}_i|A) \Bigg\} d\mathbf{a}_i.
$$

The computation for likelihood inference based on the above joint likelihood can be highly intensive. In Section 8.4.1 we will consider a Monte Carlo EM algorithm, and in Section 8.4.2 we will consider a computationally more efficient approximate method based on Laplace approximations. Alternatively, Tsiatis and Davidian (2001) proposed a conditional score approach which is also computationally more efficient than a Monte Carlo EM algorithm and yields consistent and asymptotically normal estimators.

Clustered Survival Data

Sometimes survival data may be clustered, such as data from a multi-center study. As discussed in Chapter 7, for clustered survival data we can consider a frailty model, in which random effects are introduced to represent cluster effects and to incorporate correlation within clusters. Frailty models with covariate measurement error are discussed in Chapter 7. We briefly review the models and methods here, using the same notation as in Chapter 7.

For individual j in cluster i, let s_{ij} be the survival time, subject to right censoring, let c_{ij} be the censoring time, and let

$$t_{ij} = \min(s_{ij}, c_{ij}), \qquad j = 1, 2, \cdots, n_i; \ i = 1, 2, \cdots, n$$

be the observed time. Let

$$\delta_{ij} = I(s_{ij} \leq c_{ij})$$

be the censoring indicator. We again consider the classical measurement error model (8.3). Assume that the data

$$\{(t_{ij}, \delta_{ij}), \ j = 1, 2, \cdots, n_i; i = 1, 2, \cdots, n\}$$

are conditionally independent given the random effects $\mathbf{b}_i$. To model clustered survival data, we consider the following frailty model

$$\lambda_{ij}(t|z_{ij}^*, \mathbf{b}_i) = \lambda_0(t) \exp(z_{ij}^* \beta + \mathbf{w}_{ij}^T \mathbf{b}_i),$$

where z_{ij}^* is the true but unobserved covariate value for individual j in cluster i, and $\mathbf{w}_{ij}$ are covariates without measurement errors. We assume that the random effects $\mathbf{b}_i \sim N(0, D)$. The *joint* likelihood for all the observed data is then given by

$$L_o(\boldsymbol{\theta}) = \prod_{i=1}^{n} \int \int f(\mathbf{t}_i, \boldsymbol{\delta}_i | \mathbf{z}_i^*, \mathbf{b}_i, \boldsymbol{\theta}) f(\mathbf{z}_i | \mathbf{a}_i, \boldsymbol{\theta}) f(\mathbf{b}_i | \boldsymbol{\theta}) f(\mathbf{a}_i | \boldsymbol{\theta}) \, d\mathbf{a}_i \, d\mathbf{b}_i,$$

where

$$f(\mathbf{t}_i, \boldsymbol{\delta}_i | \mathbf{z}_i^*, \mathbf{b}_i, \boldsymbol{\theta}) = \prod_{j=1}^{n_i} \left\{ \left[\lambda_0(t_{ij}) \exp(z_{ij}^* \beta + \mathbf{w}_{ij}^T \mathbf{b}_i) \right]^{\delta_{ij}} \right.$$
$$\left. \times \exp\left(- \Lambda_0(t_{ij}) \exp(z_{ij}^* \beta + \mathbf{w}_{ij}^T \mathbf{b}_i) \right) \right\},$$

and $\Lambda_0(t)$ is the cumulative baseline hazard. Computation for joint likelihood inference in this case can be very tedious due to additional random effects (frailties) in the survival models. We will discuss joint likelihood estimation in Section 8.4.

8.2.2 *Joint Models with Interval Censored Survival Data*

In Section 8.2.1 we consider right censored survival data and assume that the *exact* survival times and censoring times are observed. In practice, however, we often cannot observe the exact survival or censoring times, but we only know that some events have occurred during some particular time intervals. The resulting survival data are called *interval censored*. In this section we consider joint modeling for longitudinal data and interval-censored survival data.

To illustrate the points, we focus on the case where the main objective is to model the longitudinal process, with modeling survival data being secondary. For example, we may mainly focus on inference for a longitudinal model, but we also consider a time-to-event model to incorporate informative dropouts in the longitudinal model. Usually we do not know the exact dropout time of an individual, but we do know that the individual dropped out during a particular time period, so the observed event-time data is interval censored. In this case, the longitudinal model and the survival model are clearly associated, so joint modeling is required to avoid bias inference for the longitudinal model. Here the longitudinal model is the main focus while the time-to-dropout model is secondary, so the parameters in the survival model can be treated as nuisance parameters. We do not consider covariate measurement errors in this section.

Let y_{ij} be the longitudinal response measurement for individual i at time t_{ij}, $i = 1, \ldots, n$; $j = 1, \ldots, m$, and let $\mathbf{y}_i = (y_{i1}, \ldots, y_{im})^T$. Let $\mathbf{z}_i$ be time-independent covariates for individual i, assuming no measurement errors. For simplicity, we first assume that the measurement schedules are the same for all individuals without missing data. Extensions to more general cases will be discussed in later sections. Let T_i be the time to an event for individual i. Let $\mathbf{r}_i = (r_{i1}, \ldots, r_{im})^T$ be the vector of event indicators such that $r_{ij} = 1$ if the event has happened by time t_{ij} for individual i and $r_{ij} = 0$ otherwise. We assume that $r_{i1} = 0$ for all i.

For modeling the longitudinal data, we consider a mixed-effects model with the conditional density given the random effects being $f(\mathbf{y}_i | \mathbf{z}_i, \mathbf{b}_i, \boldsymbol{\beta}, \boldsymbol{\sigma})$, such as a GLMM or NLME model, where $\mathbf{b}_i \sim N(0, D)$ are random effects and $\boldsymbol{\beta}$ and $\boldsymbol{\sigma}$ are mean parameters and variance-covariance parameters respectively. For the time-to-event process, we assume that the event times T_i may depend on the random effects $\mathbf{b}_i$ in the longitudinal model, since these random effects characterize the individual-specific longitudinal trajectories. This assumption may also be reasonable when the event times depend on a latent process which governs the longitudinal trajectories. For example, in a longitudinal study the time to drop out may depend on an individual's initial intercept and slope of his/her longitudinal trajectory, which may be represented by the corresponding random effects in the longitudinal model (DeGruttola and Tu 1994).

Therefore, for modeling event times T_i, we consider a Cox model in which the random effects $\mathbf{b}_i$ in the longitudinal model serve as additional "covariates" in the Cox model:

$$\lambda(t_i | \mathbf{z}_i, \mathbf{b}_i) = \lambda_0(t_i) \exp(\boldsymbol{\gamma}_1^T \mathbf{z}_i + \boldsymbol{\gamma}_2^T \mathbf{b}_i), \qquad i = 1, 2, \cdots, n, \qquad (8.4)$$

where $\lambda_0(t_i)$ is the baseline hazard function and $\boldsymbol{\gamma}_1$ and $\boldsymbol{\gamma}_2$ are unknown parameters linking covariates $\mathbf{z}_i$ and random effects $\mathbf{b}_i$ to the conditional hazard rate respectively. In the above setting, the longitudinal model and the survival model are linked through the shared random effects $\mathbf{b}_i$. Such models are also called *shared parameter models* (Wu and Carroll 1988; Little 1995). Wu et al. (2008) also considered such a joint model, which is briefly described as follows.

Let

$$
\begin{aligned}
p_{ik} &= P(r_{ik} = 1 | r_{i0} = \cdots = r_{i,k-1} = 0, \mathbf{z}_i, \mathbf{b}_i) \\
&= 1 - P(T_i \geq t_{ik} | T_i \geq t_{i,k-1}, \mathbf{z}_i, \mathbf{b}_i), \quad k = 1, 2, \cdots, m, \quad (8.5)
\end{aligned}
$$

be the probability that an event has occurred during time interval $[t_{i,k-1}, t_{ik})$ given that no event was observed before time $t_{i,k-1}$. Then, based on the assumed survival model, we have

$$p_{ik} = 1 - \exp\left[-\exp(\gamma_{0k} + \boldsymbol{\gamma}_1^T \mathbf{z}_i + \boldsymbol{\gamma}_2^T \mathbf{b}_i)\right], \qquad (8.6)$$

where

$$\gamma_{0k} = \log \int_{t_{i,k-1}}^{t_{ik}} \lambda_0(u)du, \qquad k = 1, \cdots, m.$$

Let $\boldsymbol{\gamma}_0 = (\gamma_{01}, \cdots, \gamma_{0m})^T$ and $\boldsymbol{\gamma} = (\boldsymbol{\gamma}_0^T, \boldsymbol{\gamma}_1^T, \boldsymbol{\gamma}_2^T)^T$. The density function for the event indicator vector $\mathbf{r}_i$ can be written as

$$f(\mathbf{r}_i | \mathbf{z}_i, \mathbf{b}_i, \boldsymbol{\gamma}) = \prod_{k=1}^{m} f(r_{ik} | r_{i0} = \cdots = r_{i,k-1} = 0, \mathbf{z}_i, \mathbf{b}_i, \boldsymbol{\gamma}), \qquad (8.7)$$

where

$$f(r_{ik} | r_{i0} = \cdots = r_{i,k-1} = 0; \mathbf{z}_i, \mathbf{b}_i, \boldsymbol{\gamma}) = p_{ik}^{r_{ik}} (1 - p_{ik})^{1-r_{ik}},$$

is the conditional density for r_{ik} given previous values.

Simultaneous inference for all parameters in the longitudinal model and the survival model can be based on the joint likelihood of the observed data. Specifically, let $\boldsymbol{\theta} = (\boldsymbol{\beta}, \sigma, \boldsymbol{\gamma}, D)$ denote the collection of all unknown parameters. We assume that $\mathbf{y}_i$ and $\mathbf{r}_i$ are conditionally independent given the random effects $\mathbf{b}_i$, i.e., $\mathbf{r}_i$ depends on $\mathbf{y}_i$ through the random effects $\mathbf{b}_i$. Then the joint conditional density of $(\mathbf{y}_i, \mathbf{r}_i)$ given the random effects can be written as

$$f(\mathbf{y}_i, \mathbf{r}_i | \mathbf{z}_i, \mathbf{b}_i, \boldsymbol{\theta}) = f(\mathbf{y}_i | \mathbf{z}_i, \mathbf{b}_i, \boldsymbol{\beta}, \sigma) f(\mathbf{r}_i | \mathbf{z}_i, \mathbf{b}_i, \boldsymbol{\gamma}).$$

The *joint likelihood* for all the observed data can then be written as

$$L_o(\boldsymbol{\theta}) = \prod_{i=1}^{n} \left[\int f(\mathbf{y}_i | \mathbf{z}_i, \mathbf{b}_i, \boldsymbol{\beta}, \sigma) f(\mathbf{r}_i | \mathbf{z}_i, \mathbf{b}_i, \boldsymbol{\gamma}) f(\mathbf{b}_i | D) \, d\mathbf{b}_i \right]. \quad (8.8)$$

MLE of parameters $\boldsymbol{\theta}$ can be obtained by maximizing the observed data likelihood $L_o(\boldsymbol{\theta})$. The observed-data likelihood $L_o(\boldsymbol{\theta})$ may be difficult to evaluate because it involves an intractable and possibly high dimensional integral. We will describe an Monte Carlo EM algorithm and a computationally more efficient approximate method in Section 8.4.

Comments

For joint inference of a longitudinal model and a survival model, the joint likelihood approach is appealing if the assumed models hold, since it not only provides valid inference but also retains the asymptotic normality and optimality of the resulting MLEs under usual regularity conditions. A major difficulty with the joint likelihood approach is that it can be computationally very challenging, due to highly intractable joint likelihoods. Thus in Sections 8.4 we will consider computationally much more efficient approximate methods.

Alternatively, a simple and widely used method is the two-step method or regression calibration methods. These two-step methods can use available software for implementation, so they are often computationally much simpler.

However, as discussed earlier, sometimes such an approach may lead to biased results, and one must adjust standard errors of the resulting estimates to incorporate the uncertain in the estimation in the first step. We will provide more details of two-step methods in Section 8.3.

In previous sections, we consider the cases where either the longitudinal model or the survival model is of primary interest, with the other model being secondary. In some studies, however, the main objective is to model both the longitudinal process and the survival process, with a goal of understanding the *association* between the two processes. In this case, the parameters in both the longitudinal model and the survival model may be of primary interest, especially the association parameters which link the two processes. For example, during an anti-HIV treatment in an AIDS study, we may be interested in modeling both the viral load trajectories and the times to viral rebound or the times to viral suppression, and we may be interested in the relationship between viral load trajectories and time to viral rebound or viral suppression. In this case, both the longitudinal model and the survival model are of interest and a main objective is to estimate the association between the two processes.

8.3 Two-Step Methods

8.3.1 Simple Two-Step Methods

In joint modeling of longitudinal data and survival data, the longitudinal model and the survival model are usually linked through some shared parameters or shared variables. For example, the following two cases arise frequently in practice:

- the response of the longitudinal model is a time-dependent covariate in the survival model. For example, this situation arises in survival analysis with measurement error or missing data in time-dependent covariates;
- the longitudinal model and the survival model share the same parameters or random effects. For example, this situation arises in longitudinal analysis with dropouts or there is a latent process which governs both the longitudinal and event time processes.

In both cases, the link between the two models is made clear.

In the above cases, a simple two-step approach is to first fit one model (often the secondary model) to the observed data *separately*, ignoring the other model, and then in the second step the shared parameters or variables in the other model are substituted by their estimates from the first step. Then, one proceeds with inference in the usual way as if the estimated parameters or variables were observed data. This simple two-step method is related to the regression calibration method in measurement error literature. A major advantage of the two-step method is that it is simple and standard software is usually available. However, as noted earlier, such a simple or naive two-step method may lead to biased estimates and standard errors. In the following we discuss the two-step method in more details.

We consider the following semiparametric survival model for right-censored survival data, with measurement errors or (ignorable) missing data in a time-dependent covariate z_i (see (8.1) on page 257)

$$\lambda_i(t) = \lambda_0(t) \exp(z_i^*(t)\beta_1 + \mathbf{x}_i^T\boldsymbol{\beta}_2), \tag{8.9}$$

where $\mathbf{x}_i^T$ contains covariates without measurement errors, and $z_i^*(t)$ is the unobserved true covariate value and $z_i(t)$ is the observed error-prone covariate value for individual i at time t. For the error prone time-dependent covariate $z_i(t)$, we consider the following additive classical measurement error model (see (8.3) on page 258)

$$\mathbf{z}_i = U_i\,\boldsymbol{\alpha} + V_i\,\mathbf{a}_i + \boldsymbol{\epsilon}_i \equiv \mathbf{z}_i^* + \boldsymbol{\epsilon}_i, \qquad i = 1, \ldots, n, \tag{8.10}$$

where U_i and V_i are known design matrices, $\boldsymbol{\alpha}$ contains population parameters, $\mathbf{a}_i$ are random effects, and $\boldsymbol{\epsilon}_i$ represent measurement errors.

Given the observed data $\{(\mathbf{z}_i, \mathbf{x}_i, t_i, \delta_i), i = 1, 2, \cdots, n\}$, with $t_i = \min\{s_i, c_i\}$ being either the event time or the (right) censoring time and δ_i being the censoring indicator, we can write the survival model (8.9) as

$$\lambda_i(t;\, z_i(t), \mathbf{x}_i) = \lambda_0(t) E\left[\exp(z_i^*(t)\beta_1 + \mathbf{x}_i^T\boldsymbol{\beta}_2) \mid z_i(t), \mathbf{x}_i, t_i > t\right]. \tag{8.11}$$

The conditional expectation in (8.11) is quite intractable. However, following Dafni and Tsiatis (1998) and Ye, Lin, and Taylor (2008), we can approximate this expectation by

$$E\left[\exp(z_i^*(t)\beta_1 + \mathbf{x}_i^T\boldsymbol{\beta}_2) \mid z_i(t), \mathbf{x}_i, t_i > t\right]$$
$$\approx \exp\left[E\left(z_i^*(t)\beta_1 + \mathbf{x}_i^T\boldsymbol{\beta}_2 \mid z_i(t), \mathbf{x}_i, t_i > t)\right)\right],$$

where is often called a regression calibration approximation.

The two-step method proceeds as follows. In the first step, we estimate the conditional expectation $E\left(z_i^*(t)\beta_1 + \mathbf{x}_i^T\boldsymbol{\beta}_2 \mid z_i(t), \mathbf{x}_i, t_i > t\right)$ by fitting the covariate model (8.10) to the observed covariate data, ignoring the survival model, and in the second step we substitute the conditional expectation in (8.11) by its estimate from the first step and then proceed inference on the Cox model in a usual way. Ye, Lin, and Taylor (2008) proposed two approaches for the first step, called *risk set regression calibration (RRC)* method and *ordinary regression calibration (ORC)* method respectively. The idea is either to fit the LME covariate model (8.10) to the observed covariate data in the risk set or to fit the LME covariate model (8.10) to all observed covariate data.

8.3.2 Modified Two-Step Methods

As pointed out by Ye, Lin, and Taylor (2008) and Albert and Shih (2009), the two-step method in Section 8.3.1 may lead to bias in two ways: (i) the covariate trajectories of subjects who experience an event (e.g., die or drop out) may be different from those who do not experience any event, so estimation of the covariate model in the first step based on all covariate data may be biased; and (ii) inference in the second step that ignores the estimation uncertainty in the first step may lead to biased results (e.g., under-estimating standard errors). The bias in case (i), called bias from informative dropouts, may depend on the *strength of the association* between the longitudinal process and the survival process, and the bias in case (ii) may depend on the *magnitude of measurement errors* in covariates. In the following, we consider some modified two-step methods to address these biases.

The bias from informative dropout is due to the fact that the covariate trajectory is related to the length of follow-up (e.g., subjects who drop out early or die may have different trajectories), so much of the bias may be removed if we can recapture these missing covariate measurements due to dropouts by incorporating the event time information. Albert and Shih (2009) proposed to recapture the missing measurements by generating data from the conditional distribution of the covariate given the event time $f(\mathbf{z}_i|T_i)$:

$$f(\mathbf{z}_i|T_i;\boldsymbol{\theta}) = \int f(\mathbf{z}_i|\mathbf{a}_i, T_i;\boldsymbol{\theta}) f(\mathbf{a}_i|T_i;\boldsymbol{\theta}) \, d\mathbf{a}_i$$

$$= \int f(\mathbf{z}_i|\mathbf{a}_i;\boldsymbol{\theta}) f(\mathbf{a}_i|T_i;\boldsymbol{\theta}) \, d\mathbf{a}_i,$$

where $\mathbf{z}_i$ and T_i are assumed to be conditionally independent given the random effects $\mathbf{a}_i$.

Albert and Shih (2009) approximate $f(\mathbf{z}_i|T_i;\boldsymbol{\theta})$ using a LME model, as in Wu and Bailey (1989) and Wu and Follmann (1999), and then use standard software to simulate missing data from $f(\mathbf{z}_i|T_i;\boldsymbol{\theta})$. Once the missing measurements are simulated, the covariate model is fitted to the "complete data" to get less biased estimates, which are used in the second step. The procedure is iterated several times to incorporate the missing data uncertainty. Thus, the idea is similar to a multiple imputation method with non-ignorable missing data.

In order to adjust the standard errors of the parameter estimates in the survival model to incorporate the estimation uncertainty in the first step, we can consider a *parametric bootstrap method* as follows:

STEP 1. generate covariate values based on the assumed covariate model, with the unknown parameters substituted by their estimates;

STEP 2. generate survival times from the fitted survival model;

STEP 3. for each generated bootstrap dataset from steps 1 and 2, fit the models using the two-step method and obtain new parameter estimates.

Repeating the procedure B times (say, $B = 500$), we can obtain the estimated standard errors for the fixed parameters from the sample covariance matrix across the B bootstrap datasets. This Bootstrap method produces more reliable estimates of the standard errors than the naive two-step method, if the assumed models are correct.

The foregoing two-step methods rely on certain assumptions and approximations and may not completely remove biases of the naive two-step method. In Section 8.4, we consider a unified approach based on the *joint likelihood* of all the observed data.

8.4 Joint Likelihood Inference

In this section we consider a unified approach based on the joint likelihood of all observed longitudinal data and event-time data. We can estimate all parameters in the longitudinal model and the survival model *simultaneously* based on the joint likelihood. This approach avoids much of the bias in the naive two-step method noted in Section 8.3. Such a joint likelihood method provides the most efficient estimation if the assumed models are correct. The joint likelihood approach is quite general and can be extended to joint inference for more than two models which are linked in some ways, either explicitly or implicitly. Likelihood estimation is usually based on the EM algorithms.

In Section 8.2, we see that the joint likelihood for a survival model and a longitudinal model is often highly complicated, due to the unobservable random effects, censoring, and the semiparametric or nonlinear structures of the models. Therefore, a major difficulty in joint likelihood inference is computational challenges. In the following sections, we consider two approaches: a Monte Carlo EM algorithm for exact likelihood inference, which can be computationally very intensive, and an approximate method for likelihood inference, which is computationally much more efficient. We focus the longitudinal models and survival models presented in Section 8.2.2.

8.4.1 Exact Likelihood Inference

In this section, we consider a Monte Carlo EM algorithm to simultaneously obtain "exact" MLEs of all parameters in the survival and longitudinal models presented in Section 8.2.2 (page 260). Recall that in Section 8.2.2 we considered a joint model based on a mixed-effects longitudinal model and a survival model for *interval-censored* survival data. In the following, we consider joint likelihood inference for that joint model, with time-independent covariates for simplicity.

Let y_{ij} be the longitudinal response measurements, $\mathbf{z}_i$ be time-independent covariates, and $\mathbf{r}_i = (r_{i1}, \ldots, r_{im})^T$ be the vector of event indicators such that

$$r_{ij} = \begin{cases} 1 & \text{an event has happened by time } t_{ij} \text{ for individual } i \\ 0 & \text{otherwise,} \quad\quad i = 1, \ldots, n; \; j = 1, \ldots, m. \end{cases}$$

The observed data are $\{(\mathbf{y}_i, \mathbf{z}_i, \mathbf{r}_i), \; i = 1, 2, \cdots, n\}$. The joint observed-data likelihood $L_o(\boldsymbol{\theta})$, as given in (8.8) (page 262) in Section 8.2.2, is quite intractable. In the following, we consider an Monte Carlo EM algorithm for estimation.

The "complete data" are given by

$$\{(\mathbf{y}_i, \mathbf{z}_i, \mathbf{r}_i, \mathbf{b}_i), \; i = 1, 2, \cdots, n\}.$$

The complete-data log-likelihood for individual i can be written as

$$l_c^{(i)}(\boldsymbol{\theta}) = \log f(\mathbf{y}_i|\mathbf{z}_i, \mathbf{b}_i, \boldsymbol{\beta}, \boldsymbol{\sigma}) + \log f(\mathbf{b}_i|D) + \log f(\mathbf{r}_i|\mathbf{z}_i, \mathbf{b}_i, \boldsymbol{\gamma}).$$

The E-step of the Monte Carlo EM algorithm computes the conditional expectation of the complete-data loglikelihood given the observed data and current parameter estimates. Thus, at the t-th EM iteration, the E-step for individual i can be written as

$$Q_i(\boldsymbol{\theta}|\boldsymbol{\theta}^{(t)}) = \int \left\{ \log f(\mathbf{y}_i|\mathbf{z}_i, \mathbf{b}_i, \boldsymbol{\beta}, \boldsymbol{\sigma}) + \log f(\mathbf{b}_i|D) \right.$$
$$\left. + \log f(\mathbf{r}_i|\mathbf{z}_i, \mathbf{b}_i, \boldsymbol{\gamma}) \right\} f(\mathbf{b}_i|\mathbf{y}_i, \mathbf{z}_i, \mathbf{r}_i, \boldsymbol{\theta}^{(t)}) \, d\mathbf{b}_i.$$

It is difficult to evaluate the integral $Q_i(\boldsymbol{\theta}|\boldsymbol{\theta}^{(t)})$ analytically. However, we can approximate the integral using Monte Carlo methods by simulating many samples of the unobservable random effects $\mathbf{b}_i$ from its conditional distribution given the observed data $f(\mathbf{b}_i|\mathbf{y}_i, \mathbf{z}_i, \mathbf{r}_i, \boldsymbol{\theta}^{(t)})$, which are then used to approximate the integral with an empirical mean, similar to the methods in Chapter 4.

To generate independent samples of the unobservable random effects $\mathbf{b}_i$ from the conditional distribution $f(\mathbf{b}_i|\mathbf{y}_i, \mathbf{z}_i, \mathbf{r}_i, \boldsymbol{\theta}^{(t)})$, we note that

$$f(\mathbf{b}_i|\mathbf{y}_i, \mathbf{z}_i, \mathbf{r}_i, \boldsymbol{\theta}^{(t)}) \propto f(\mathbf{y}_i|\mathbf{z}_i, \mathbf{b}_i, \boldsymbol{\beta}^{(t)}, \boldsymbol{\sigma}^{(t)}) f(\mathbf{r}_i|\mathbf{z}_i, \mathbf{b}_i, \boldsymbol{\gamma}^{(t)}) f(\mathbf{b}_i|D^{(t)}),$$

where the density functions on the right-hand side are all known. Thus, we may use the Gibbs sampler along with rejection sampling methods to generate Monte Carlo samples of the random effects $\mathbf{b}_i$, similar to the methods in Chapter 4.

Let $(\tilde{\mathbf{b}}_i^{(1)}, \tilde{\mathbf{b}}_i^{(2)}, \cdots, \tilde{\mathbf{b}}_i^{(m_t)})$ be a random sample of size m_t simulated from $f(\mathbf{b}_i|\mathbf{y}_i, \mathbf{z}_i, \mathbf{r}_i, \boldsymbol{\theta}^{(t)})$ at t-th EM iteration. The E-step of the Monte Carlo EM algorithm at the $(t+1)$st EM iteration can be approximated as follows

$$Q(\boldsymbol{\theta}|\boldsymbol{\theta}^{(t)}) \;\; = \;\; \sum_{i=1}^{n} Q_i(\boldsymbol{\theta}|\boldsymbol{\theta}^{(t)}) \approx \sum_{i=1}^{n} \left\{ \frac{1}{m_t} \sum_{j=1}^{m_t} \left[\log f\left(\mathbf{y}_i|\mathbf{z}_i, \widetilde{\mathbf{b}}_i^{(j)}, \boldsymbol{\beta}, \boldsymbol{\sigma}\right) \right.\right.$$

$$\left.\left. + \log f\left(\widetilde{\mathbf{b}}_i^{(j)}|D\right) + \log f\left(\mathbf{r}_i|\mathbf{z}_i, \widetilde{\mathbf{b}}_i^{(j)}, \boldsymbol{\gamma}\right)\right]\right\}. \qquad (8.12)$$

The above approximation can be made arbitrary accurate by increasing the number of Monte Carlo sample m_t.

The M-step of the Monte Carlo EM algorithm is then to maximize $Q(\boldsymbol{\theta}|\boldsymbol{\theta}^{(t)})$, which is just like a complete-data maximization, so standard optimization procedures for complete-data mixed effects models and event-time models can be used to obtain the updated parameter estimates $\boldsymbol{\theta}^{(t+1)}$. If we assume that the parameters in the models are distinct, we can maximize each term of $Q(\boldsymbol{\theta}|\boldsymbol{\theta}^{(t)})$ separately using standard methods. Iterating the E-step and M-step until convergence, we obtain the MLE (or a local maximizer) $\hat{\boldsymbol{\theta}}$ of $\boldsymbol{\theta}$.

The variance-covariance matrix of $\hat{\boldsymbol{\theta}}$ can be approximated as follows. At the convergence of the Monte Carlo EM algorithm, let

$$S_{ij}(\hat{\boldsymbol{\theta}}) = \frac{\partial l_c^{(ij)}(\boldsymbol{\theta})}{\partial \boldsymbol{\theta}} = \frac{\partial l_c^{(i)}(\boldsymbol{\theta}|\mathbf{y}_i, \mathbf{z}_i, \widetilde{\mathbf{b}}_i^{(j)}, \mathbf{r}_i)}{\partial \boldsymbol{\theta}},$$

evaluated at $\boldsymbol{\theta} = \hat{\boldsymbol{\theta}}$. We have

$$I(\hat{\boldsymbol{\theta}}) \approx \sum_{i=1}^{n} \sum_{j=1}^{m_t} \frac{1}{m_t} S_{ij}(\hat{\boldsymbol{\theta}}) S_{ij}^T(\hat{\boldsymbol{\theta}}).$$

The approximate asymptotic covariance matrix of $\hat{\boldsymbol{\theta}}$ is $I^{-1}(\hat{\boldsymbol{\theta}})$.

Zeng and Cai (2005) derived some asymptotic results for maximum likelihood estimators in joint analysis of longitudinal data and survival data. They showed the consistency of the maximum likelihood estimators, derived their asymptotic distributions, and showed that the maximum likelihood estimators in joint analysis are semiparametrically efficient.

8.4.2 Approximate Inference

A major difficulty of the likelihood method for joint inference of longitudinal and survival models is the intensive computation (Tsiatis and Davidian 2004), due to evaluations of highly intractable integrals. The Monte Carlo EM algorithm for exact likelihood estimation in the previous section can be computationally very intensive and may exhibit convergence problems. Therefore, in this section we consider an approximate likelihood method based on a Laplace approximation to the observed-data joint loglikelihood, which approximates the intractable integral in the observed-data joint likelihood by an analytic expression, similar to the ones described in previous chapters. This approximate method is computationally much more efficient than the Monte Carlo EM algorithm, so it is very appealing for joint likelihood inference of longitudinal models and survival models.

We consider again the joint models in Section 8.2.2 or Section 8.4.1. For approximate inference, we consider a first-order Laplace approximation to the observed-data joint likelihood $L_o(\boldsymbol{\theta})$ in (8.8) (page 262). Specifically, let

$$
\begin{aligned}
l_c(\boldsymbol{\theta}, \mathbf{b}) &\equiv \sum_{i=1}^{n} l_c^{(i)}(\boldsymbol{\theta}, \mathbf{b}_i) \\
&= \sum_{i=1}^{n} \left[\log f(\mathbf{y}_i|\mathbf{z}_i, \mathbf{b}_i, \boldsymbol{\beta}, \boldsymbol{\sigma}) + \log f(\mathbf{b}_i|D) + \log f(\mathbf{r}_i|\mathbf{z}_i, \mathbf{b}_i, \boldsymbol{\gamma}) \right]
\end{aligned}
$$

be the "complete-data" log-likelihood (or h-likelihood as in Lee et al. 2006). Let $\tilde{\mathbf{b}} = \{\tilde{\mathbf{b}}_i, i = 1, 2, \cdots, n\}$ solve the following equations

$$
\partial l_c^{(i)}(\boldsymbol{\theta}, \mathbf{b}_i)/\partial \mathbf{b}_i = \mathbf{0}, \qquad i = 1, 2, \cdots, n.
$$

It can be shown that, following Lee et al. (2006), the first-order Laplacian approximation to the observed-data joint log-likelihood $l_o(\boldsymbol{\theta}) = \log L_o(\boldsymbol{\theta})$ is given by

$$
\tilde{l}_o(\boldsymbol{\theta}, \tilde{\mathbf{b}}) = l_c(\boldsymbol{\theta}, \tilde{\mathbf{b}}) - \frac{1}{2} \log \left| -\frac{1}{2\pi} \frac{\partial^2 l_c(\boldsymbol{\theta}, \mathbf{b})}{\partial \mathbf{b}^2} \right|_{\mathbf{b}=\tilde{\mathbf{b}}}, \tag{8.13}
$$

i.e., $\tilde{l}_o(\boldsymbol{\theta}, \tilde{\mathbf{b}})$ is a first-order Laplacian approximation to $l_o(\boldsymbol{\theta})$ by integrating out the unobservable random effects $\mathbf{b}$ in $l_o(\boldsymbol{\theta})$. Thus, an approximate MLE of $\boldsymbol{\theta}$ can be obtained by solving the following equation

$$
\partial \tilde{l}_o(\boldsymbol{\theta}, \tilde{\mathbf{b}})/\partial \boldsymbol{\theta} = \mathbf{0}.
$$

Given starting values $\mathbf{b}^{(0)}$ and $\boldsymbol{\theta}^{(0)}$, iterating the above procedure leads to an approximate MLE $\tilde{\boldsymbol{\theta}}$. Similarly, REMLs of the variance-covariance parameters can be obtained by integrating out the mean parameters and the random effects in $l_o(\boldsymbol{\theta})$ in a similar way (see Lee et al. 2006).

The above approximate method consists of the following steps:

STEP 1. Obtain initial parameter estimate $\boldsymbol{\theta}^{(0)}$ using a naive method, e.g., fit the survival model and the longitudinal model separately to obtain initial parameter estimates;

STEP 2. Given the current parameter estimate $\boldsymbol{\theta}^{(t)}$ at t-th iteration ($t =$

$0, 1, 2, \cdots$), obtain a updated random-effects estimate $\mathbf{b}^{(t+1)}$ by solving

$$\frac{\partial l_c^{(i)}(\boldsymbol{\theta}^{(t)}, \mathbf{b}_i)}{\partial \mathbf{b}_i} = \mathbf{0}, \qquad i = 1, 2, \cdots, n;$$

STEP 3. Given the random-effects estimates $\mathbf{b}^{(t+1)}$, obtain a updated parameter estimate $\boldsymbol{\theta}^{(t+1)}$ by solving

$$\frac{\partial \tilde{l}_o(\boldsymbol{\theta}, \mathbf{b}^{(t+1)})}{\partial \boldsymbol{\theta}} = \mathbf{0};$$

STEP 4. Iterate between Step 2 and Step 3 until convergence.

Note that the above approximate method does not involve any intractable integration, so it offers a major computational advantage in joint estimation of longitudinal models and survival models, which is highly desirable in joint model inference.

The estimates of the random effects obtained in the above procedure can be interpreted as empirical Bayes estimates. Approximate standard errors of the approximate estimates can be obtained based on the following formula

$$\mathrm{Cov}(\tilde{\boldsymbol{\theta}}) \approx \left[-\frac{\partial^2 \tilde{l}_o(\boldsymbol{\theta}, \tilde{\mathbf{b}})}{\partial \boldsymbol{\theta} \partial \boldsymbol{\theta}^T} \right]^{-1}_{\boldsymbol{\theta} = \tilde{\boldsymbol{\theta}}},$$

where the estimates $\tilde{\mathbf{b}}$ and $\tilde{\boldsymbol{\theta}}$ are from the last iteration.

8.5 Joint Models with Incomplete Data

In jointly modeling of longitudinal data and survival data, we often need to address missing data and measurement error problems since these problems are very common in longitudinal studies. In this section, we discuss missing data and measurement error problems in joint models. Statistical methods for joint models with incomplete data are conceptually similar to the ones discussed in earlier chapters, but the computation typically becomes more tedious. In fact, in likelihood inference computation may be the major challenge, due to complexities of the models, censoring, missing data, and measurement errors. Thus one should use computationally more efficient approximate methods for estimation and inference.

Another major issue for joint models with incomplete data is parameter identifiability. In the presence of missing data and measurement errors, we need to assume additional models to address the incomplete data. This leads to a large number of unknown parameters to be estimated, so parameters may be non-identifiable in the sense that two sets of unknown parameters may lead to the same joint likelihood. If the sample size is not large enough, we may be unable to estimate all unknown parameters even if the parameters are identifiable. In these cases, a reasonable strategy is to simplify the models, especially secondary models such as models for missing data or measurement errors, and to impose certain restrictions on secondary parameters such as assuming diagonal covariance matrices.

In the following sections, we consider likelihood methods for joint models with missing data and measurement errors. Since the likelihood methods are conceptually similar to the ones in earlier chapters, we omit some of the details.

8.5.1 Joint Models with Missing Data

In joint models, missing data may arise in either longitudinal models or survival models, and missing data may occur in either covariates or responses or both in any models. In practice, dropouts in a longitudinal study may lead to missing data in any variables that are measured over time, including the longitudinal response variable and time-dependent covariates, and the missingness may be nonignorable in the sense that the dropouts or missingness may be related to the longitudinal trajectories or the missing values. In this section, we focus on the case where there are missing data in the response of the longitudinal model and in the covariates of both the longitudinal model and the survival model, following Wu et al. (2008). We consider the joint model in Section 8.2 where the survival data may be interval-censored, and we focus on time-independent covariates for simplicity.

Let y_{ij} be the response value in the longitudinal model for individual i at time t_{ij}, $i = 1, \ldots, n$; $j = 1, \ldots, m$, and let $\mathbf{y}_i = (y_{i1}, \ldots, y_{im})^T$ be the repeated measurements on individual i. For simplicity, here we assume that the measurement schedules are fixed and are the same to all individuals. A response value is assumed missing if its value is not observed at or near the scheduled measurement time. Let $\mathbf{z}_i$ be the collection of time-independent (baseline) covariates for individual i. We write $\mathbf{y}_i = (\mathbf{y}_{i,mis}, \mathbf{y}_{i,obs})$, where $\mathbf{y}_{i,mis}$ is a collection of missing responses and $\mathbf{y}_{i,obs}$ is a collection of observed responses. Similarly, we write $\mathbf{z}_i = (\mathbf{z}_{i,mis}, \mathbf{z}_{i,obs})$. Let $\mathbf{d}_i = (d_{i1}, \ldots, d_{im})^T$ be a vector of missing response indicators such that

$$d_{ij} = \begin{cases} 1 & \text{if } y_{ij} \text{ is missing} \\ 0 & \text{if } y_{ij} \text{ is observed} \end{cases} .$$

If subject i drops out between time t_{ik} and $t_{i,k+1}$ and does not return to study later, then $d_{i1} = \cdots = d_{ik} = 0, d_{ij} = 1$ for $j > k$, but we allow subjects to possibly return to the study at a later time after he/she drops out.

We also consider a time-to-event process. Let $\mathbf{r}_i = (r_{i1}, \ldots, r_{im})^T$ be the vector of event indicators such that $r_{ij} = 1$ if the event has happened on individual

i by time t_{ij} and $r_{ij} = 0$ otherwise. We assume that $r_{i1} = 0$ for all i. Let T_i be the time to an event for individual i, or the duration time until an event occurs. In practice, the exact event time T_i usually cannot be directly observed. However, if we observe no events at times $t_{i1}, \cdots, t_{i,k-1}$ but know an event has occurred by time t_{ik}, we know that the actual event time is between $t_{i,k-1}$ and t_{ik}, i.e., $t_{i,k-1} < T_i \leq t_{ik}$. Note that if no event has been observed by the last time t_{im} for individual i, it indicates that $T_i > t_{im}$. This type of event time data structure is referred to as interval censored event times (Lawless 2003).

The Models

For modeling the longitudinal data, we consider a mixed effects model with conditional density function $f(\mathbf{y}_i|\mathbf{z}_i, \mathbf{b}_i, \beta, \sigma)$ given the random effects $\mathbf{b}_i$, such as a LME or a GLMM or a NLME model, where β contains mean parameters and σ contains variance-covariance parameters. For the survival data, we consider a Cox proportional hazards model which links to the longitudinal mixed effects model through the shared random effects $\mathbf{b}_i$. That is, the event times are assumed to be associated with individual-specific longitudinal trajectories such as individual slopes. For example, patients with faster increase in viral load may be more likely to drop out earlier in an AIDS study. Thus, the time-to-event model can be written as

$$\lambda(t_i|\mathbf{z}_i, \mathbf{b}_i) = \lambda_0(t_i) \exp(\gamma_1^T \mathbf{z}_i + \gamma_2^T \mathbf{b}_i), \qquad (8.14)$$

where $\lambda_0(t_i)$ is a unspecified baseline hazard function and γ_1 and γ_2 are unknown parameters linking baseline covariates $\mathbf{z}_i$ and random effects $\mathbf{b}_i$ to the conditional hazard rate respectively.

We follow the setup in Section 8.2 and repeat some notation below. Let

$$p_{ik} = P(r_{ik} = 1|r_{il} = 0, 0 \leq l < k; \mathbf{z}_i, \mathbf{b}_i).$$

Then we have

$$p_{ik} = 1 - \exp\left[-\exp(\gamma_{0k} + \gamma_1^T \mathbf{z}_i + \gamma_2^T \mathbf{b}_i)\right], \qquad (8.15)$$

where

$$\gamma_{0k} = \log \int_{t_{i,k-1}}^{t_{ik}} \lambda_0(u)du, \qquad k = 1, \cdots, m.$$

Note that, given the current observation mechanism, we only need to deal with the finite number of parameters $\{\gamma_{0k}, \ k = 1, 2, \cdots, m\}$ instead of the unknown nonparametric function $\lambda_0(t)$ in the likelihood estimation, which simplifies computation. We assume that the event indicators $r_{i1}, \cdots, r_{i,m}$ are conditionally independent given the random effects $\mathbf{b}_i$.

Let $\gamma_0 = (\gamma_{01}, \cdots, \gamma_{0m})^T$ and $\gamma = (\gamma_0^T, \gamma_1^T, \gamma_2^T)^T$, and let $\gamma =$

$(\boldsymbol{\gamma}_0^T, \boldsymbol{\gamma}_1^T, \boldsymbol{\gamma}_2^T)^T$. Note that

$$f(\mathbf{r}_i | \mathbf{z}_i, \mathbf{b}_i, \boldsymbol{\gamma}) = \prod_{k=1}^{m} f(r_{ik} | r_{il}, 0 \leq l < k; \mathbf{z}_i, \mathbf{b}_i, \boldsymbol{\gamma}),$$

where

$$f(r_{ik} | r_{il} = 0, 0 \leq l < k; \mathbf{z}_i, \mathbf{b}_i, \boldsymbol{\gamma}) = p_{ik}^{r_{ik}} (1 - p_{ik})^{1 - r_{ik}},$$

and r_{ik} equals 0 before an event and equals 1 after an event. We have

$$P(l_i < T_i \leq u_i | \mathbf{z}_i, \mathbf{b}_i) = \exp\left[-\int_0^{u_i} \lambda_0(t) dt \cdot \exp(\boldsymbol{\gamma}_1^T \mathbf{z}_i + \boldsymbol{\gamma}_2^T \mathbf{b}_i) \right]$$
$$-\exp\left[-\int_0^{l_i} \lambda_0(t) dt \cdot \exp(\boldsymbol{\gamma}_1^T \mathbf{z}_i + \boldsymbol{\gamma}_2^T \mathbf{b}_i) \right],$$

where $l_i = \max(t_{ij} : r_{ij} = 0)$ and $u_i = \min(t_{il} : r_{il} = 1)$, which may reduce some computing and simplify the presentation. Here $u_i = \infty$ if $r_{il} = 0$ for $l = 1, \ldots, m$.

When there are informative dropouts or nonignorable missing longitudinal responses, the missing data mechanism must be taken into account for valid likelihood inference. Since the random effects $\mathbf{b}_i$ characterize individual-specific longitudinal trajectories, we can assume a missing data model which allows the probability of missing a value to possibly depend on the random effects $\mathbf{b}_i$, i.e., we can assume a shared-parameter model or a random-effect-based dropout model (Wu and Carroll 1988; Follmann and Wu 1995; Ten Have et al. 1998). In other words, the missingness depends on both the missing values $\mathbf{y}_{mis,i}$ and the observed values $\mathbf{y}_{obs,i}$ through the random effects $\mathbf{b}_i$. The probability of missing a response value at time t_{ij} may also depend on the missing status at the previous time point $t_{i,j-1}$.

We assume that the missing baseline covariates are missing at random (or ignorable), so we do not need to specify a missing covariate mechanism. However, we need to make a distributional assumption for the incompletely observed covariates for likelihood inference. No distributional assumption is needed for covariates without missing data.

Based on the above arguments, for example, we may consider the following model for the missing responses:

$$\text{logit}(P(d_{ij} = 1 | d_{i,j-1}, \mathbf{b}_i, \boldsymbol{\phi})) = \phi_0 + \phi_1 d_{i,j-1} + \boldsymbol{\phi}_2^T \mathbf{b}_i,$$
$$f(\mathbf{d}_i | \mathbf{b}_i, \boldsymbol{\phi}) = f(d_{i1} | \mathbf{b}_i, \boldsymbol{\phi}) \prod_{j=2}^{m} f(d_{ij} | d_{i,j-1}, \mathbf{b}_i, \boldsymbol{\phi}),$$

where the parameters $\boldsymbol{\phi}$ may be viewed as nuisance parameters and are usually not of inferential interest.

Joint Likelihood Inference

We consider simultaneous estimation for all parameters based on the joint likelihood of all the observed data $\{(\mathbf{y}_{i,obs}, \mathbf{z}_{i,obs}, \mathbf{r}_i, \mathbf{d}_i), i = 1, 2, \cdots, n\}$. Let $\boldsymbol{\theta} = (\boldsymbol{\beta}, \boldsymbol{\sigma}, \boldsymbol{\gamma}, \boldsymbol{\phi}, D)$ denote the collection of all unknown parameters. We assume that the longitudinal responses $\mathbf{y}_i$ and the event indicators $\mathbf{r}_i$ are conditionally independent given the random effects $\mathbf{b}_i$, i.e., $\mathbf{r}_i$ depends on $\mathbf{y}_i$ through the random effects $\mathbf{b}_i$, so the association between the longitudinal process and the survival process is incorporated through the shared random effects $\mathbf{b}_i$. Based on earlier discussion, we also assume that the missing data model $f(\mathbf{d}_i|\mathbf{y}_i, \mathbf{b}_i, \boldsymbol{\phi}) = f(\mathbf{d}_i|\mathbf{b}_i, \boldsymbol{\phi})$. Thus we have

$$f(\mathbf{y}_i, \mathbf{r}_i, \mathbf{d}_i|\mathbf{z}_i, \mathbf{b}_i, \boldsymbol{\theta}) = f(\mathbf{y}_i|\mathbf{z}_i, \mathbf{b}_i, \boldsymbol{\beta}, \boldsymbol{\sigma})f(\mathbf{r}_i|\mathbf{z}_i, \mathbf{b}_i, \boldsymbol{\gamma})f(\mathbf{d}_i|\mathbf{b}_i, \boldsymbol{\phi}).$$

The joint likelihood for all the observed data can then be written as

$$L_o(\boldsymbol{\theta}) = \prod_{i=1}^{n} \left[\int \int \int f(\mathbf{y}_i|\mathbf{z}_i, \mathbf{b}_i, \boldsymbol{\beta}, \boldsymbol{\sigma})f(\mathbf{r}_i|\mathbf{z}_i, \mathbf{b}_i, \boldsymbol{\gamma})f(\mathbf{d}_i|\mathbf{b}_i, \boldsymbol{\phi}) \right.$$
$$\left. \times f(\mathbf{z}_i|\boldsymbol{\alpha})f(\mathbf{b}_i|D) \, d\mathbf{y}_{i,mis} \, d\mathbf{z}_{i,mis} \, d\mathbf{b}_i \right],$$

where $f(\mathbf{z}_i|\boldsymbol{\alpha})$ is the assumed distribution for the incompletely observed covariates $\mathbf{z}_i$ with unknown parameters $\boldsymbol{\alpha}$.

Wu et al. (2008) used a Monte Carlo EM (or ECM) algorithm to obtain the MLE of $\boldsymbol{\theta}$. If we treat the unobservable random effects $\mathbf{b}_i$ as additional "missing data", we can write the "complete data" as

$$\{(\mathbf{y}_i, \mathbf{z}_i, \mathbf{r}_i, \mathbf{d}_i, \mathbf{b}_i), \ i = 1, 2, \cdots, n\}.$$

The E-step at the t-th iteration of the EM algorithm for individual i can then be written as

$$Q_i(\boldsymbol{\theta}|\boldsymbol{\theta}^{(t)}) = E\left(l_c^{(i)}(\mathbf{y}_i, \mathbf{z}_i, \mathbf{r}_i, \mathbf{d}_i, \mathbf{b}_i) \mid \mathbf{y}_{i,obs}, \mathbf{z}_{i,obs}, \mathbf{d}_i, \mathbf{r}_i, \boldsymbol{\theta}^{(t)} \right)$$
$$= \int \int \int \left\{ \log f(\mathbf{y}_i|\mathbf{z}_i, \mathbf{b}_i, \boldsymbol{\beta}, \boldsymbol{\sigma}) + \log f(\mathbf{z}_i|\boldsymbol{\alpha}) \right.$$
$$\left. + \log f(\mathbf{b}_i|D) + \log f(\mathbf{r}_i|\mathbf{b}_i, \boldsymbol{\gamma}) + \log f(\mathbf{d}_i|\mathbf{b}_i, \boldsymbol{\phi}) \right\}$$
$$\times f(\mathbf{y}_{i,mis}, \mathbf{z}_{i,mis}, \mathbf{b}_i|\mathbf{y}_{i,obs}, \mathbf{z}_{i,obs}, \mathbf{d}_i, \mathbf{r}_i, \boldsymbol{\theta}^{(t)})$$
$$\times d\mathbf{y}_{i,mis} \, d\mathbf{z}_{i,mis} \, d\mathbf{b}_i,$$

which generally does not have an analytic expression. As in Wu et al. (2008), we can approximate the integral $Q_i(\boldsymbol{\theta}|\boldsymbol{\theta}^{(t)})$ by Monte Carlo methods combined with Gibbs sampler and rejection sampling methods.

Specifically, we can simulate large samples of $(\mathbf{y}_{i,mis}, \mathbf{z}_{i,mis}, \mathbf{b}_i)$ from the conditional distribution $f(\mathbf{y}_{i,mis}, \mathbf{z}_{i,mis}, \mathbf{b}_i|\mathbf{y}_{i,obs}, \mathbf{z}_{i,obs}, \mathbf{d}_i, \mathbf{r}_i, \boldsymbol{\theta}^{(t)})$, and then approximate $Q_i(\boldsymbol{\theta}|\boldsymbol{\theta}^{(t)})$ by an empirical mean. To generate random samples from the conditional density $f(\mathbf{y}_{i,mis}, \mathbf{z}_{i,mis}, \mathbf{b}_i | \mathbf{y}_{i,obs}, \mathbf{z}_{i,obs}, \mathbf{d}_i, \mathbf{r}_i, \boldsymbol{\theta}^{(t)})$, one approach is to use the Gibbs sampler by iteratively sampling from the full conditionals $f(\mathbf{y}_{i,mis}|\mathbf{y}_{i,obs}, \mathbf{z}_i, \mathbf{b}_i, \mathbf{d}_i, \mathbf{r}_i, \boldsymbol{\theta}^{(t)})$, $f(\mathbf{z}_{i,mis}|\mathbf{z}_{i,obs}, \mathbf{y}_i, \mathbf{b}_i, \mathbf{d}_i, \mathbf{r}_i, \boldsymbol{\theta}^{(t)})$, and $f(\mathbf{b}_i|\mathbf{y}_i, \mathbf{z}_i, \mathbf{d}_i, \mathbf{r}_i, \boldsymbol{\theta}^{(t)})$. To sample these full conditionals, note that

$$
\begin{aligned}
f(\mathbf{y}_{i,mis}|\mathbf{y}_{i,obs}, \mathbf{z}_i, \mathbf{b}_i, \mathbf{d}_i, \mathbf{r}_i, \boldsymbol{\theta}^{(t)}) &\propto f(\mathbf{y}_i|\mathbf{z}_i, \mathbf{b}_i, \boldsymbol{\beta}^{(t)}, \boldsymbol{\sigma}^{(t)}), \\
f(\mathbf{z}_{i,mis}|\mathbf{z}_{i,obs}, \mathbf{y}_i, \mathbf{b}_i, \mathbf{d}_i, \mathbf{r}_i, \boldsymbol{\theta}^{(t)}) &\propto f(\mathbf{y}_i|\mathbf{z}_i, \mathbf{b}_i, \boldsymbol{\beta}^{(t)}, \boldsymbol{\sigma}^{(t)}) f(\mathbf{z}_i|\boldsymbol{\alpha}^{(t)}) \\
&\quad \times f(\mathbf{r}_i|\mathbf{z}_i, \mathbf{b}_i, \boldsymbol{\gamma}^{(t)}), \\
f(\mathbf{b}_i|\mathbf{y}_i, \mathbf{z}_i, \mathbf{d}_i, \mathbf{r}_i, \boldsymbol{\theta}^{(t)}) &\propto f(\mathbf{b}_i|D^{(t)}) f(\mathbf{y}_i|\mathbf{z}_i, \mathbf{b}_i, \boldsymbol{\beta}^{(t)}, \boldsymbol{\sigma}^{(t)}) \\
&\quad \times f(\mathbf{r}_i|\mathbf{z}_i, \mathbf{b}_i, \boldsymbol{\gamma}^{(t)}) f(\mathbf{d}_i|\mathbf{b}_i, \boldsymbol{\phi}^{(t)}).
\end{aligned}
$$

Since the density functions on the right-hand sides of the above expressions are known, rejection sampling methods such as the adaptive rejection sampling method or a multivariate rejection method may be used to sample from each of the full conditionals (Wu et al. 2008). Iteratively sampling from each of the full conditionals in turn for a burn-in period until the Markov chain converges, we obtain a random sample of $(\mathbf{y}_{i,mis}, \mathbf{z}_{i,mis}, \mathbf{b}_i)$ from the conditional distribution $f(\mathbf{y}_{i,mis}, \mathbf{z}_{i,mis}, \mathbf{b}_i | \mathbf{y}_{i,obs}, \mathbf{z}_{i,obs}, \mathbf{d}_i, \mathbf{r}_i, \boldsymbol{\theta}^{(t)})$. Repeating this procedure m_t times, we obtain m_t independent random samples from $f(\mathbf{y}_{i,mis}, \mathbf{z}_{i,mis}, \mathbf{b}_i | \mathbf{y}_{i,obs}, \mathbf{z}_{i,obs}, \mathbf{d}_i, \mathbf{r}_i, \boldsymbol{\theta}^{(t)})$. Then, in the E-step $Q_i(\boldsymbol{\theta}|\boldsymbol{\theta}^{(t)})$ is approximated by the sample average.

The M-step of the Monte Carlo EM algorithm is just like a complete-data maximization, so standard complete-data optimization procedures can be used. Due to the large number of unknown parameters, it is preferred to use an Monte Carlo ECM algorithm, in which the M-step is replaced by a sequence of maximizations over a subset of the parameters holding the remaining parameters fixed. To choose the subsets of parameters in the M-step of an ECM algorithm, we can consider the parameters in each model separately and consider the mean parameters and variance-covariance parameters in each model separately.

The variance-covariance matrix of $\hat{\boldsymbol{\theta}}$ can be approximated as follows. At the convergence of the ECM algorithm, let $(\tilde{\mathbf{y}}_{mis,i}^{(j)}, \tilde{\mathbf{z}}_{mis,i}^{(j)}, \tilde{\mathbf{b}}_i^{(j)})$ be j-th simulated sample from the last ECM iteration, and

$$
S_{ij}(\hat{\boldsymbol{\theta}}) = \frac{\partial}{\partial \boldsymbol{\theta}} l_c^{(i)}(\boldsymbol{\theta}|\mathbf{y}_{obs,i}, \tilde{\mathbf{y}}_{mis,i}^{(j)}, \mathbf{z}_{obs,i}, \tilde{\mathbf{z}}_{mis,i}^{(j)}, \tilde{\mathbf{b}}_i^{(j)}, \mathbf{d}_i, \mathbf{r}_i),
$$

evaluated at $\boldsymbol{\theta} = \hat{\boldsymbol{\theta}}$. We have

$$
I(\hat{\boldsymbol{\theta}}) \approx \sum_{i=1}^{n} \sum_{j=1}^{m_t} \frac{1}{m_t} S_{ij}(\hat{\boldsymbol{\theta}}) S_{ij}^{T}(\hat{\boldsymbol{\theta}}),
$$

where t is the last ECM iteration number. The approximate asymptotic covariance matrix of $\hat{\theta}$ is $I^{-1}(\hat{\theta})$.

The foregoing likelihood method based on a Monte Carlo ECM algorithm can be computationally very intensive and may exhibit convergence problems. If the ECM algorithm diverges quickly, it is possible that the joint models are non-identifiable. If the ECM algorithm converges, it may converge very slowly. To reduce the huge computational burden, we may consider approximation methods based on Laplace or Taylor approximations, in a way similar to that in previous chapters. We will show such an approximate method in Section 8.5.2 for joint models with measurement errors and missing data in time-dependent covariates.

Example 8.1 *Joint inference for an NLME model and a survival model with missing data*

In Example 4.2 of Chapter 4, we considered an NLME model for modeling HIV viral dynamics in the early period during an anti-HIV treatment, where viral loads typically decline in the initial period and then some viral loads may rebound later in the study. We may be interested in the relationship between individual-specific characteristics of the viral dynamics in the early period, such as individual initial viral decay rates, and long-term antiviral responses such as times to viral rebound later in the study (or times to death). For example, an important question is whether patients with faster initial viral decay rates have earlier viral rebounds later in the study. Thus, we need to *jointly model* HIV viral dynamics and times to viral rebound, i.e., a joint model of longitudinal data and survival data. Since missing data are very common in these studies, we also need to address missing data problems in the joint model.

Following Wu et al. (2008), we consider the following NLME model for modeling HIV viral dynamics in the early period during an anti-HIV treatment

$$y_{ij} = \log_{10}(P_{1i}e^{-\lambda_{1i}t_{ij}} + P_{2i}e^{-\lambda_{2i}t_{ij}}) + e_{ij}, \qquad (8.16)$$

$$\log(P_{1i}) = \beta_1 + b_{1i}, \qquad \lambda_{1i} = \beta_2 + \beta_3 z_i + b_{2i},$$

$$\log(P_{2i}) = \beta_4 + b_{3i}, \qquad \lambda_{2i} = \beta_5 + b_{4i}, \qquad (8.17)$$

$$i = 1, 2, \cdots, n; \ j = 1, 2, \cdots, n_i,$$

where y_{ij} is the $\log_{10}$-transformation of the viral load measurement for the i-th patient at j-th time point, z_i is the baseline CD4 value for patient i, λ_{1i} and λ_{2i} represent individual-specific first and second phases of viral decay rates respectively, P_{1i} and P_{2i} are individual-specific baseline values, $\beta = (\beta_1, \cdots, \beta_5)^T$ are fixed effects, e_{ij} represents the within individual random error, and b_{ki}'s are random effects. We assume that e_{ij} i.i.d. $\sim N(0, \sigma^2 I)$ and $\mathbf{b}_i \sim N(\mathbf{0}, D)$, and e_{ij} and $\mathbf{b}_i$ are independent.

One of the objectives is to test if the time to viral rebound depends on baseline CD4 values and the random effects which characterize individual-specific viral load trajectories in the early period. For the time T_i to viral rebound, we consider the following Cox proportional hazards model, which links the hazard function of T_i to baseline CD4 values and the random effects,

$$\lambda(t_i|\mathbf{z}_i, \mathbf{b}_i) = \lambda_0(t_i) \exp(\gamma_1 z_i + \gamma_2 b_{i1} + \gamma_3 b_{i2} + \gamma_4 b_{i3}), \qquad (8.18)$$

where $\lambda_0(t_i)$ is a unspecified baseline hazard function and γ_j's are unknown parameters. Thus, the viral dynamic (NLME) model (8.16) – (8.17) for the longitudinal viral load data and the survival model (8.18) for the time to viral rebound data are linked together through the shared random effects and covariates. Note that the NLME model (8.16) – (8.17) is fitted to viral load data in the *early* period while the survival model (8.18) is fitted to viral rebound data in the *late* period of the study.

Since the baseline CD4 values z_i contain missing data, we need to make a distributional assumption for z_i in likelihood inference: we assume that $z_i \sim N(\alpha_1, \alpha_2)$. We also assume that the missing CD4 values are missing at random for simplicity. There are also some dropouts in the dataset, which lead to missing values in the longitudinal responses $\mathbf{y}_i$. The missingness due to dropouts may be informative or non-ignorable, so a missing data model for the missing responses needs to be assumed and incorporated in likelihood inference to avoid possible biased results. Let d_{ij} be a missing data indicator for the missing responses y_{ij}. We assume the following simple missing data model for possibly non-ignorable missing data in the responses

$$f(\mathbf{d}_i|\mathbf{y}_i, \mathbf{b}_i, \mathbf{z}_i, \boldsymbol{\phi}) = \prod_{j=1}^{m} P(d_{ij} = 1|\boldsymbol{\phi}, \mathbf{b}_i)^{d_{ij}}$$

$$\times (1 - P(d_{ij} = 1|\boldsymbol{\phi}, \mathbf{b}_i))^{1-d_{ij}},$$

$$\log\left(\frac{P(d_{ij} = 1|\boldsymbol{\phi}, \mathbf{b}_i)}{1 - P(d_{ij} = 1|\boldsymbol{\phi}, \mathbf{b}_i)}\right) = \phi_0 + \phi_1 b_{1i} + \phi_2 b_{2i} + \phi_3 b_{3i} + \phi_4 b_{4i},$$

where the parameters ϕ_j's may be viewed as nuisance parameters. In the above missing data model, we assume that the missingness of the response depends on the unobservable random effects which characterize individual-specific viral load trajectories.

We use the joint likelihood method described in this section, implemented by a Monte Carlo EM algorithm, to obtain the parameter estimates. Table 8.1 shows the resulting parameter estimates and the associated standard errors (from Wu et al. 2008). The p-values are computed based on Wald-type tests. Note that the estimate of the parameter β_3 indicates that higher initial CD4 values are associated with faster initial viral decay (p-value = 0.001). The estimate of parameter β_5 reflects viral rebounds at later stage of the study. Time to viral rebound depends on initial CD4 values (p-value associated with $\hat{\gamma}_1$ is smaller

Table 8.1 *Estimates of parameters in the NLME model and the survival model based on the joint likelihood method*

	Estimates of the NLME model parameters									
	β_1		β_2		β_3		β_4		β_5	
	Est.	S.E.	Est.	S.E.	Est.	S.E.	Est.	S.E.	Est.	S.E.
	10.32	0.15	53.98	9.17	14.01	4.37	6.02	0.22	−0.55	0.34
P-value	< 0.001		< 0.001		0.001		< 0.001		0.10	

	Estimates of the survival model parameters							
	γ_1		γ_2		γ_3		γ_4	
	Est.	S.E.	Est.	S.E.	Est.	S.E.	Est.	S.E.
	−1.47	0.06	0.041	0.046	−0.012	4.20	0.001	3.81
P-value	< 0.001		0.37		0.99		0.99	

Est.: estimate, S.E.: standard error.

than 0.001): smaller baseline CD4 values are associated with earlier viral rebounds. However, the viral rebound time does not appear to be significantly associated with initial viral decay rate.

8.5.2 Joint Models with Measurement Errors

In longitudinal studies many covariates are measured over time, along with the response measurements. Some of these covariates may be measured with errors and may be missing at the response measurement times. In this section, we consider joint modeling of longitudinal data and survival data with measurement errors and missing data in time-dependent covariates, following Wu et al. (2009). We consider a mixed effects model for the longitudinal process. As discussed in previous sections, the random effects in the longitudinal model may serve as "error-free covariates" for the survival model, i.e., the longitudinal model and the survival model are linked through shared random effects. So joint inference is desirable.

Let z_{ik} be the observed value of an error-prone covariate and z_{ik}^* be the corresponding unobserved "true" covariate value for individual i at time u_{ik}, $i = 1, \ldots, n$, $k = 1, \ldots, m_i$. Here we allow the covariate measurement times u_{ik} to possibly differ from the response measurement times t_{ij}, so we allow missing data in the time-dependent covariate where the missingness is assumed to be missing at random. We suppress the accurately observed covariates in the models. Let $\mathbf{z}_i = (z_{i1}, \ldots, z_{im_i})^T$ and $\mathbf{z}_i^* = (z_{i1}^*, \ldots, z_{im_i}^*)^T$.

As in Section 8.5.1, we consider a mixed effects model with density function $f(\mathbf{y}_i|\mathbf{z}_i^*, \mathbf{b}_i, \boldsymbol{\beta}, \boldsymbol{\sigma})$ given the random effects $\mathbf{b}_i$, such as a LME or GLMM or NLME model. The longitudinal response $\mathbf{y}_i$ is assumed to depend on the un-observed true covariate $\mathbf{z}_i^*$ rather than the observed but mis-measured covariate $\mathbf{z}_i$.

Let T_i be the time to an event of interest for individual i, and let t_i be the observed version, $i = 1, 2, \cdots, n$. In this section we consider event times with *right censoring*. For individual i with censoring, let c_i be the censoring time (so the unobserved true event time $s_i > c_i$). The observed data are $\{(t_i, \delta_i), i = 1, \cdots, n\}$, where

$$t_i = \min(s_i, c_i), \qquad \delta_i = I(s_i \leq c_i),$$

and $I(\cdot)$ is an indicator function. For the survival model, here we consider a parametric accelerated failure time (AFT) model as described in Chapter 7. We assume that the event time T_i depends on the unobserved true covariate value $\mathbf{z}_i^*$ and the random effects $\mathbf{b}_i$ in the longitudinal model. Specifically, we consider the following mixed-effects model for the time-to-event process T_i

$$\log(T_i) = \gamma_0 + \boldsymbol{\gamma}_1^T \mathbf{z}_i^* + \boldsymbol{\gamma}_2^T \mathbf{b}_i + \epsilon_i, \qquad i = 1, 2, \cdots, n, \qquad (8.19)$$

where $\boldsymbol{\gamma} = (\gamma_0, \boldsymbol{\gamma}_1^T, \boldsymbol{\gamma}_2^T)^T$ are unknown parameters and ϵ_i are i.i.d. and follow a parametric distribution with mean 0 and other parameters λ such as a normal distribution.

The parametric AFT model (8.19) may be a good alternative to a Cox pro-portional hazards model in that (i) the interpretation of model parameters in model (8.19) may be easier (Klein et al. 1999), (ii) the fixed-effect estimates in model (8.19) are robust against neglected covariates (Hougaard 1999), (iii) model (8.19) is computationally more manageable; and finally (iv) AFT model (8.19) does not require the proportional hazards assumption.

To address measurement errors and missing data in the covariates, we need to model the covariate process. Since the covariate trajectories are often very complex in many longitudinal studies, as an illustration, here we consider a flexible nonparametric mixed effects model to empirically model the covariate process $z_{ik} = z_i(u_{ik})$ (see Section 2.5.2 in Chapter 2, page 74)

$$z_i(t) = r(t) + h_i(t) + \xi_i(t) \equiv z_i^*(t) + \xi_i(t), \qquad i = 1, \ldots, n, \qquad (8.20)$$

where $z_i^*(t) = r(t) + h_i(t)$ is the true but unobserved covariate value at time t, $r(t)$ and $h_i(t)$ are unknown nonparametric smooth fixed-effects and random-effects functions respectively, and $\xi_i(t) \sim N(0, \sigma^2)$ is the measurement er-ror. The random smooth function $h_i(t)$ is introduced to incorporate the inter-individual variation in the covariate process, while the fixed smooth function $r(t)$ represents population average of the covariate process. We assume that $h_i(t)$ is the realization of a zero-mean stochastic process.

As discussed in Chapter 2 (Section 2.5.2), we can approximate the non-parametric functions $r(t)$ and $h_i(t)$ by linear combinations of some basis functions. Let $\mathbf{\Psi}_p(t) = [\psi_0(t), \psi_1(t), \ldots, \psi_{p-1}(t)]^T$ and $\mathbf{\Phi}_q(t) = [\phi_0(t), \phi_1(t), \ldots, \phi_{q-1}(t)]^T$ be the corresponding basis functions. Then, we can approximate the nonparametric mixed effects covariate model (8.20) by the following parametric LME model

$$z_i(t) \approx \mathbf{\Psi}_p(t)^T \boldsymbol{\alpha} + \mathbf{\Phi}_q(t)^T \mathbf{a}_i + \xi_i(t) = z_i^*(t) + \xi_i(t), \qquad (8.21)$$

where $\boldsymbol{\alpha} = (\alpha_0, \cdots, \alpha_{p-1})^T$ are fixed effects and $\mathbf{a}_i = (a_{i0}, \cdots, a_{i,q-1})^T$ are random effects with $\mathbf{a}_i \sim N(0, A)$. We assume that $\mathbf{a}_i, \mathbf{b}_i, \mathbf{e}_i, \epsilon_i, \xi_i$ are all independent of each other. Ye, Lin, and Taylor (2008) considered a flexible semiparametric mixed effects model for the covariate process, which can also be approximated by a parametric LME model.

When there are more than one error-prone covariates, we can model each covariate separately, or consider a multivariate version of the above model, such as a multivariate LME model (Shah et al. 1997).

Likelihood Inference

We consider simultaneous likelihood inference for all model parameters based on the joint likelihood of the observed data. Let $\boldsymbol{\theta} = (\boldsymbol{\alpha}, \boldsymbol{\beta}, \sigma, \nu, A, B, \boldsymbol{\gamma}, \lambda)$ be the collection of all model parameters. The joint likelihood for all the observed data can be written as

$$
\begin{aligned}
L_o(\boldsymbol{\theta}|\mathbf{y}, \mathbf{z}, \mathbf{t}, \boldsymbol{\delta}) &\equiv \prod_{i=1}^{n} L_o(\boldsymbol{\theta}|\mathbf{y}_i, \mathbf{z}_i, t_i, \delta_i) \\
&= \prod_{i=1}^{n} \left[\int \int f(\mathbf{y}_i|\mathbf{a}_i, \mathbf{b}_i; \boldsymbol{\theta}) f(\mathbf{b}_i; B) f(\mathbf{z}_i|\mathbf{a}_i; \boldsymbol{\theta}) \right. \\
&\qquad\qquad \left. \times f(\mathbf{a}_i; A) f^*(t_i, \delta_i|\mathbf{a}_i, \mathbf{b}_i; \boldsymbol{\gamma}, \lambda) \, d\mathbf{a}_i \, d\mathbf{b}_i \right],
\end{aligned}
$$

where

$$f^*(t_i, \delta_i|\mathbf{a}_i, \mathbf{b}_i; \boldsymbol{\gamma}, \lambda) = f(t_i|\mathbf{a}_i, \mathbf{b}_i; \boldsymbol{\gamma}, \lambda)^{\delta_i} \left[1 - F(t_i|\mathbf{a}_i, \mathbf{b}_i; \boldsymbol{\gamma}, \lambda)\right]^{1-\delta_i}$$

and $F(t_i|\mathbf{a}_i, \mathbf{b}_i; \boldsymbol{\gamma}, \lambda)$ is the cumulative distribution function corresponding to the density function $f(t_i|\mathbf{a}_i, \mathbf{b}_i; \boldsymbol{\gamma}, \lambda)$. If in the AFT model (8.19) we assume that the error $\epsilon_i \sim N(0, \lambda^2)$, then we have

$$
\begin{aligned}
\log f^*(t_i, \delta_i|\mathbf{a}_i, \mathbf{b}_i; \boldsymbol{\gamma}, \lambda) = -\frac{1}{2}\delta_i \Bigg\{ & \log(2\pi\lambda^2) \\
& + \left[\frac{\log(t_i) - (\gamma_0 + \boldsymbol{\gamma}_1^T \mathbf{z}_i^* + \boldsymbol{\gamma}_2^T \mathbf{b}_i)}{\lambda}\right]^2 + 2\log(t_i) \Bigg\}
\end{aligned}
$$

$$+(1 - \delta_i) \log \left[1 - \Phi \left(\frac{\log(t_i) - (\gamma_0 + \boldsymbol{\gamma}_1^T \mathbf{z}_i^* + \boldsymbol{\gamma}_2^T \mathbf{b}_i)}{\lambda}\right)\right],$$

where $\Phi(\cdot)$ is the cumulative distribution function for the standard normal distribution $N(0,1)$.

To get the MLE of $\boldsymbol{\theta}$, we may again consider an Monte Carlo EM algorithm, similar to that in Section 8.5.1. However, such a method can be computationally very demanding. The convergence of such a Monte Carlo EM algorithm depends on the dimension of the random effects $(\mathbf{a}_i, \mathbf{b}_i)$. When the dimension of the random effects $(\mathbf{a}_i, \mathbf{b}_i)$ is not small, the Monte Carlo EM method can be computationally extremely intensive. Thus, in the following we approximate the observed-data joint log-likelihood $\log L_o(\boldsymbol{\theta}|\mathbf{y}, \mathbf{z}, \mathbf{w}, \boldsymbol{\delta})$ using a first-order Laplacian approximation, as in Wu et al. (2009). This approximate method is computationally much more efficient than the Monte Carlo EM algorithm.

An Approximate Method

The idea of the approximate method is similar to the ones described in previous chapters, but applied to the joint model of a longitudinal model and a survival model. Specifically, let

$$
\begin{aligned}
l_c(\boldsymbol{\theta}, \mathbf{a}, \mathbf{b}) &\equiv \sum_{i=1}^{n} l_c^{(i)}(\boldsymbol{\theta}, \mathbf{a}_i, \mathbf{b}_i) \\
&= \sum_{i=1}^{n} \Big[\log f(\mathbf{y}_i|\mathbf{a}_i, \mathbf{b}_i; \boldsymbol{\theta}) + \log f(\mathbf{b}_i; B) \\
&\qquad + \log f(\mathbf{z}_i|\mathbf{a}_i; \boldsymbol{\theta}) + \log f(\mathbf{a}_i; A) \\
&\qquad + \log f^*(t_i, \delta_i|\mathbf{a}_i, \mathbf{b}_i; \boldsymbol{\gamma}, \lambda) \Big]
\end{aligned}
$$

be the "complete-data" log-likelihood or the h-likelihood as in Lee et al. (2006). Let $(\tilde{\mathbf{a}}, \tilde{\mathbf{b}}) = \{(\tilde{\mathbf{a}}_i, \tilde{\mathbf{b}}_i), i = 1, 2, \cdots, n\}$ solve the following equations

$$\frac{\partial l_c^{(i)}(\boldsymbol{\theta}, \mathbf{a}_i, \mathbf{b}_i)}{\partial(\mathbf{a}_i, \mathbf{b}_i)} = \mathbf{0}, \qquad i = 1, 2, \cdots, n.$$

The first-order Laplacian approximation to the observed-data log-likelihood $l_o(\boldsymbol{\theta}|\mathbf{y}, \mathbf{z}, \mathbf{w}, \boldsymbol{\delta}) = \log L_o(\boldsymbol{\theta}|\mathbf{y}, \mathbf{z}, \mathbf{w}, \boldsymbol{\delta})$ is given by

$$\tilde{l}_o(\boldsymbol{\theta}, \tilde{\mathbf{a}}, \tilde{\mathbf{b}}) = l_c(\boldsymbol{\theta}, \tilde{\mathbf{a}}, \tilde{\mathbf{b}}) - \frac{1}{2} \log \left| -\frac{1}{2\pi} \frac{\partial^2 l_c(\boldsymbol{\theta}, \mathbf{a}, \mathbf{b})}{\partial(\mathbf{a}, \mathbf{b})^2} \right|_{(\mathbf{a},\mathbf{b})=(\tilde{\mathbf{a}},\tilde{\mathbf{b}})}, \quad (8.22)$$

i.e., $\tilde{l}_o(\boldsymbol{\theta}, \tilde{\mathbf{a}}, \tilde{\mathbf{b}})$ is a first-order Laplacian approximation to $l_o(\boldsymbol{\theta}|\mathbf{y}, \mathbf{z}, \mathbf{w}, \boldsymbol{\delta})$ by integrating out the random effects $(\mathbf{a}, \mathbf{b})$. Thus, an approximate estimate of $\boldsymbol{\theta}$ can be obtained by solving the following equation

$$\frac{\partial \tilde{l}_o(\boldsymbol{\theta}, \tilde{\mathbf{a}}, \tilde{\mathbf{b}})}{\partial \boldsymbol{\theta}} = 0.$$

Given starting values $(\mathbf{a}^{(0)}, \mathbf{b}^{(0)})$ and $\boldsymbol{\theta}^{(0)}$, iterating the above procedure until convergence leads to an approximate estimate $\tilde{\boldsymbol{\theta}}$. REMLs of the variance-covariance parameters can be obtained in a similar way. The procedure is similar to that in Section 8.4.2.

8.6 Joint Modeling of Several Longitudinal Processes

In the previous sections, we have discussed joint inference of a longitudinal model and a survival model. Sometimes, we may also want to jointly model two or more longitudinal processes, since these processes may be correlated. The following two situations may arise in practice: (i) two or more longitudinal processes may be associated, so joint inference may be more efficient than separate analyses; and (ii) in a longitudinal model one or more time-dependent covariates may be measured with errors or may have missing data, so joint modeling of the response and covariate processes is required to address measurement errors and missing data. For example, for the mental distress data in Chapter 1, the repeated measurements of the GSI scores, depression scores, anxiety scores, and some other variables are highly correlated, so a joint model would be desirable, which can borrow information across different processes.

In this section, we briefly discuss statistical methods for joint modeling of several longitudinal processes.

8.6.1 Multivariate Mixed Effects Models with Incomplete Data

When several longitudinal processes are of similar types, say all normally distributed, we may consider a multivariate mixed-effects model, which incorporates both the correlation between the longitudinal processes and the correlation among the repeated measurements within each process. Such a multivariate mixed effects model may lead to more efficient inference than separate analysis of each longitudinal process. Shah, Laird, and Schoenfeld (1997) considered a multivariate LME model and focus on bivariate case. In the following, we follow Shah et al. (1997) and describe a multivariate LME model in the general form.

For modeling several longitudinal processes, it is likely that the measurement times and the measurement frequencies are different for different longitudinal processes, which leads to missing data in the responses of a standard multivariate LME model. Sometimes the missing data may be non-ignorable or informative. In the following, we first consider a complete-data multivariate LME model and then consider missing data problems.

Suppose that there are n independent individuals, with individual i having

n_i measurements on each of K response variables. Let y_{ijk} be the k-th response value for individual i at time t_{ij}, and let e_{ijk} be the corresponding random error, $i = 1, \ldots, n$, $j = 1, \ldots, n_i$, $k = 1, \ldots, K$. Let $\mathbf{y}_{ik} = (y_{i1k}, \ldots, y_{in_ik})^T$ be the repeated measurements for individual i on response k, and let $\mathbf{y}_i = (\mathbf{y}_{i1}^T, \ldots, \mathbf{y}_{iK}^T)^T$ be all the response measurements for individual i. Define $\mathbf{e}_{ik}$ and $\mathbf{e}_i$ in a similar way. Let $\mathbf{z}_{il} = (z_{i1l}, \ldots, z_{in_il})^T$ be the covariate repeated measurements of the l-th covariate for individual i, $l = 1, \ldots, p$, and let $\mathbf{z}_i = (\mathbf{z}_{i1}^T, \ldots, \mathbf{z}_{ip}^T)^T$. Let $N = \sum_{i=1}^n n_i$ be the total number of measurements. For the k-th response $\mathbf{y}_{ik}$, we may consider the following *univariate* LME model

$$\mathbf{y}_{ik} = \mathbf{X}_{ik}\boldsymbol{\beta}_k + \mathbf{T}_{ik}\mathbf{b}_{ik} + \mathbf{e}_{ik}, \quad i = 1, 2, \cdots, n, \; k = 1, 2, \cdots, K, \quad (8.23)$$

where $\boldsymbol{\beta}_k$ are fixed effects, $\mathbf{b}_{ik}$ are random effects, and $\mathbf{X}_{ik}$ and $\mathbf{T}_{ik}$ are known design matrices. We assume that the within-individual measurements are conditionally independent given the random effects, i.e., we assume $\mathbf{e}_{ik} \sim \sigma_k^2 I$. The univariate LME model (8.23) may be used if we model each response process separately, but it fails to incorporate the *correlation* or association between different processes.

To incorporate the correlation between different processes or to borrow information across different processes, we consider a multivariate version of LME model (8.23), following Shah et al. (1997). Let Σ be the $K \times K$ covariance matrix for the K response variables, i.e.,

$$\text{Cov}(y_{ij1}, \ldots, y_{ijK}) = \Sigma = (\sigma_{ij})_{K \times K}.$$

Let $X_i = \text{diag}(\mathbf{X}_{i1}, \ldots, \mathbf{X}_{iK})$ be a block diagonal matrix with the k-th block being matrix $\mathbf{X}_{ik}$, and let $T_i = \text{diag}(\mathbf{T}_{i1}, \ldots, \mathbf{T}_{iK})$ be a block diagonal matrix with the k-th block being matrix $\mathbf{T}_{ik}$. Then, we can combine the K univariate LME models in (8.23) and obtain the following *multivariate linear mixed effects (LME) model*

$$\mathbf{y}_i = X_i\boldsymbol{\beta} + T_i\mathbf{b}_i + \mathbf{e}_i, \quad i = 1, 2, \cdots, n, \quad (8.24)$$

where $\boldsymbol{\beta} = (\boldsymbol{\beta}_1^T, \ldots, \boldsymbol{\beta}_K^T)^T$ are fixed effects, $\mathbf{b}_i = (\mathbf{b}_{i1}^T, \ldots, \mathbf{b}_{iK}^T)^T$ are random effects, $\mathbf{b}_i \sim N(\mathbf{0}, D)$, and $\mathbf{e}_i \sim N(\mathbf{0}, \Sigma \otimes I_i)$.

The multivariate LME model (8.24) is a joint model for all the K longitudinal processes. It incorporates the between-process correlation via the covariance matrix Σ and incorporates the within-process correlation via random effects $\mathbf{b}_i$. The multivariate LME model (8.24) borrows information both across processes and across individuals.

Missing Data Problems

For several longitudinal processes, the observed data may be available at different time points across processes. This leads to missing data. Sometimes the missing data may be non-ignorable in the sense that the missingness may be related to the missing values. Covariates may also be missing. For simplicity we assume that the missing covariates are missing at random. In the following we consider likelihood inference for the multivariate LME model (8.24) in the presence of missing data. The method is similar to that for a univariate LME model.

Let r_{ij} be a missing response indicator such that $r_{ij} = 1$ if the j-th component of $\mathbf{y}_i$ is missing and 0 otherwise, and let $\mathbf{r}_i = (r_{i1}, \ldots, r_{i,Kn_i})^T$. We write $\mathbf{y}_i = (\mathbf{y}_{mis,i}, \mathbf{y}_{obs,i})$, where $\mathbf{y}_{mis,i}$ contains the missing components of $\mathbf{y}_i$ and $\mathbf{y}_{obs,i}$ contains the observed components of $\mathbf{y}_i$, and write $\mathbf{z}_i = (\mathbf{z}_{mis,i}, \mathbf{z}_{obs,i})$ similarly.

By treating the unobservable random effects $\mathbf{b}_i$ as additional "missing data", we have "missing data" $\{(\mathbf{y}_{mis,i}, \mathbf{z}_{mis,i}, \mathbf{b}_i)\}$. The "complete data" are $\{(\mathbf{y}_i, \mathbf{z}_i, \mathbf{b}_i, \mathbf{r}_i), \ i = 1, \ldots, n\}$, and the "complete-data" log-likelihood can be written as

$$
\begin{aligned}
l_c(\psi) &= \sum_{i=1}^{n} l_c(\psi|\mathbf{y}_i, \mathbf{z}_i, \mathbf{r}_i, \mathbf{b}_i) \\
&= \sum_{i=1}^{n} \Big\{ \log f(\mathbf{y}_i|\mathbf{z}_i, \mathbf{b}_i; \boldsymbol{\beta}, \Sigma) + \log f(\mathbf{z}_i; \boldsymbol{\alpha}) \\
&\qquad\qquad + \log f(\mathbf{b}_i; D) + \log f(\mathbf{r}_i|\mathbf{y}_i, \mathbf{z}_i; \boldsymbol{\phi}) \Big\},
\end{aligned}
$$

where $\psi = (\boldsymbol{\alpha}, \boldsymbol{\beta}, \Sigma, D, \boldsymbol{\phi})$ denotes the collection of all parameters. We consider an EM algorithm for likelihood estimation. Let $\psi^{(t)}$ be the parameter estimates from the t-th EM iteration. The E-step at the $(t+1)$-th EM iteration can be written as

$$
\begin{aligned}
Q(\psi|\psi^{(t)}) &= \sum_{i=1}^{n} Q_i(\psi|\psi^{(t)}) \\
&= \sum_{i=1}^{n} E\left[l_c(\psi|\mathbf{y}_i, \mathbf{z}_i, \mathbf{r}_i, \mathbf{b}_i)|\mathbf{y}_{obs,i}, \mathbf{z}_{obs,i}, \mathbf{r}_i; \psi^{(t)} \right] \\
&= \sum_{i=1}^{n} \int\int\int \Big\{ \log f(\mathbf{y}_i|\mathbf{z}_i, \mathbf{b}_i; \boldsymbol{\beta}, \Sigma) + \log f(\mathbf{z}_i; \boldsymbol{\alpha}) \\
&\qquad\qquad + \log f(\mathbf{b}_i; D) + \log f(\mathbf{r}_i|\mathbf{y}_i, \mathbf{z}_i; \boldsymbol{\phi}) \Big\} \\
&\qquad\qquad \times f(\mathbf{y}_{mis,i}, \mathbf{z}_{mis,i}, \mathbf{b}_i|\mathbf{y}_{obs,i}, \mathbf{z}_{obs,i}, \mathbf{r}_i; \psi^{(t)}) \\
&\qquad\qquad d\,\mathbf{b}_i\, d\,\mathbf{y}_{mis,i} d\,\mathbf{z}_{mis,i}.
\end{aligned}
$$

Thus $Q(\psi|\psi^{(t)})$ involves a high dimensional integral. For multivariate LME models, however, we can integrate out the random effects $\mathbf{b}_i$ from $Q(\psi|\psi^{(t)})$

in the E-step. This greatly simplifies the computation. We briefly outline the derivation below.

Note that

$$
f(\mathbf{y}_{mis,i}, \mathbf{z}_{mis,i}, \mathbf{b}_i | \mathbf{y}_{obs,i}, \mathbf{z}_{obs,i}, \mathbf{r}_i; \boldsymbol{\psi}^{(t)})
$$
$$
= f(\mathbf{b}_i | \mathbf{y}_i, \mathbf{z}_i, \mathbf{r}_i; \boldsymbol{\psi}^{(t)}) f(\mathbf{y}_{mis,i}, \mathbf{z}_{mis,i} | \mathbf{y}_{obs,i}, \mathbf{z}_{obs,i}, \mathbf{r}_i; \boldsymbol{\psi}^{(t)}).
$$

It can be shown that

$$
[\mathbf{b}_i | \mathbf{y}_i, \mathbf{z}_i, \mathbf{r}_i; \boldsymbol{\psi}^{(t)}] \sim N(\widetilde{\mathbf{b}}_i, \widetilde{D}_i),
$$

where

$$
\begin{aligned}
\widetilde{\mathbf{b}}_i &= D^{(t)} T_i^T \left(T_i D^{(t)} T_i^T + \Sigma^{(t)} \otimes I_i \right)^{-1} (\mathbf{y}_i - X_i \boldsymbol{\beta}^{(t)}), \\
\widetilde{D}_i &= D^{(t)} - D^{(t)} T_i^T \left(T_i D^{(t)} T_i^T + \Sigma^{(t)} \otimes I_i \right)^{-1} T_i D^{(t)}.
\end{aligned}
$$

Then, after some algebra, we can obtain the following result

$$
\begin{aligned}
Q_i(\boldsymbol{\psi} | \boldsymbol{\psi}^{(t)}) = \Bigg\{ & -\frac{1}{2} \log |2\pi \Sigma \otimes I_i| - \frac{1}{2} \int \int \Big[\mathrm{tr}(T_i^T (\Sigma \otimes I_i)^{-1} T_i \widetilde{D}_i) \\
& + \left(\mathbf{y}_i - X_i \boldsymbol{\beta} - T_i \widetilde{\mathbf{b}}_i \right)^T (\Sigma \otimes I_i)^{-1} \left(\mathbf{y}_i - X_i \boldsymbol{\beta} - T_i \widetilde{\mathbf{b}}_i \right) \Big] \\
& \times f(\mathbf{y}_{mis,i}, \mathbf{z}_{mis,i} | \mathbf{y}_{obs,i}, \mathbf{z}_{obs,i}, \mathbf{r}_i; \boldsymbol{\psi}^{(t)}) \, d\mathbf{y}_{mis,i} d\mathbf{z}_{mis,i} \Bigg\} \\
& + \int \int \log f(\mathbf{z}_i; \boldsymbol{\alpha}) f(\mathbf{w}_{mis,i} | \mathbf{w}_{obs,i}, \mathbf{r}_i; \boldsymbol{\psi}^{(t)}) \, d\mathbf{y}_{mis,i} d\mathbf{z}_{mis,i} \\
& - \Bigg\{ \frac{1}{2} \log |2\pi D| + \frac{1}{2} \int \int \Big[\mathrm{tr}(D^{-1} \widetilde{D}_i) + \left(\widetilde{\mathbf{b}}_i' D^{-1} \widetilde{\mathbf{b}}_i \right) \Big] \\
& \times f(\mathbf{y}_{mis,i}, \mathbf{z}_{mis,i} | \mathbf{y}_{obs,i}, \mathbf{z}_{obs,i}, \mathbf{r}_i; \boldsymbol{\psi}^{(t)}) \, d\mathbf{y}_{mis,i} d\mathbf{z}_{mis,i} \Bigg\} \\
& + \int \int \log f(\mathbf{r}_i | \mathbf{y}_i, \mathbf{z}_i; \boldsymbol{\phi}) f(\mathbf{w}_{mis,i} | \mathbf{w}_{obs,i}, \mathbf{r}_i; \boldsymbol{\psi}^{(t)}) \, d\mathbf{y}_{mis,i} d\mathbf{z}_{mis,i}.
\end{aligned}
$$

where $\mathbf{w}_{mis,i} = (\mathbf{y}_{mis,i}, \mathbf{z}_{mis,i})$ and $\mathbf{w}_{obs,i} = (\mathbf{y}_{obs,i}, \mathbf{z}_{obs,i})$.

Since the above integrals do not involve the random effects $\mathbf{b}_i$, in the E-step we only need to simulate samples from $f(\mathbf{y}_{mis,i}, \mathbf{z}_{mis,i} | \mathbf{y}_{obs,i}, \mathbf{z}_{obs,i}, \mathbf{r}_i; \boldsymbol{\psi}^{(t)})$, which has a lower dimension so is easier to sample from. The Monte Carlo sampling in the E-step can again be accomplished by Gibbs sampler combined with rejection sampling methods, in a way similar to that in earlier chapters. The computational advantage of integrating out the random effects in the E-step can be substantial, especially when the dimension of the random effects is high, since it is often the sampling of random effects in the E-step that causes much of the convergence problems.

When the covariates $\mathbf{z}_i$ are categorical or discrete, the above integration with respect to $\mathbf{z}_{mis,i}$ reduces to a summation, which further simplifies the computation.

For multivariate longitudinal categorical data, Zeng and Cook (2007) proposed a *joint transitional model* using generalized estimating equations that allow modeling of covariate effects on marginal transition probabilities as well as the association parameters. Their model simultaneously incorporates a Markov structure in each longitudinal process and an association structure between different processes. They demonstrated via simulation that a substantial efficiency can be gained using a multivariate model rather than separate univariate models. Similarly, Wu, Liu, and Liu (2009) considered a multivariate transitional mixed effects model for multivariate continuous responses.

8.6.2 Other Joint Modeling Approaches

There are other approaches for joint modeling of several longitudinal processes. In the following, we briefly describe two of them: one is based on a factorization of the joint distribution and the other one uses random effects to incorporate the correlation between processes.

Factorization of Joint Distributions

The multivariate LME model in Section 8.6.1 is useful when the responses are all normally distributed. In practice, however, the longitudinal responses may be of different types. For example, one response may be a continuous variable while the other response may be a binary variable. In this case, a multivariate LME model is clearly inappropriate, and it may even be hard to define the correlation between the two processes. In the following, we consider a joint modeling approach which factors the joint distribution of the two responses into a product of univariate regression models.

Suppose that $\mathbf{y}_i^{(1)} = (y_{i1}^{(1)}, \cdots, y_{in_i}^{(1)})^T$ and $\mathbf{y}_i^{(2)} = (y_{i1}^{(2)}, \cdots, y_{in_i}^{(2)})^T$ are two different longitudinal response processes, say $y_{ij}^{(1)}$ is a continuous variable and $y_{ij}^{(2)}$ is a binary variable. Let $\boldsymbol{\theta}$ contains all parameters and let $\mathbf{z}_i$ be covariates. Then we can write the joint likelihood as follows

$$
\begin{aligned}
L(\boldsymbol{\theta}) &= \prod_{i=1}^{n} f(\mathbf{y}_i^{(1)}, \mathbf{y}_i^{(2)} | \mathbf{z}_i, \boldsymbol{\theta}) \\
&= \prod_{i=1}^{n} \int \int f(\mathbf{y}_i^{(1)} | \mathbf{z}_i, \mathbf{b}_i^{(1)}, \boldsymbol{\theta}) f(\mathbf{y}_i^{(2)} | \mathbf{y}_i^{(1)}, \mathbf{z}_i, \mathbf{b}_i^{(2)}, \boldsymbol{\theta}) \, d\mathbf{b}_i^{(1)} \, d\mathbf{b}_i^{(2)},
\end{aligned}
$$

where $\mathbf{b}_i^{(1)}$ and $\mathbf{b}_i^{(2)}$ are random effects in each model respectively. We can then assume standard mixed effects models for $f(\mathbf{y}_i^{(1)}|\mathbf{z}_i, \mathbf{b}_i^{(1)}, \boldsymbol{\theta})$ and $f(\mathbf{y}_i^{(2)}|\mathbf{y}_i^{(1)}, \mathbf{z}_i, \mathbf{b}_i^{(2)}, \boldsymbol{\theta})$ respectively, such as a LME model for $f(\mathbf{y}_i^{(1)}|\mathbf{z}_i, \mathbf{b}_i^{(1)}, \boldsymbol{\theta})$ and a logistic mixed model for $f(\mathbf{y}_i^{(2)}|\mathbf{y}_i^{(1)}, \mathbf{z}_i, \mathbf{b}_i^{(2)}, \boldsymbol{\theta})$.

This factorization approach can be extended to more than two processes in a straightforward way. However, one should perform sensitivity analysis to check if the final results are sensitive to the *order* of the factorization.

Random Effects Approach

In a mixed effects model, we use random effects to incorporate the correlation among the repeated measurements within each individual. This idea can be extended to modeling several longitudinal processes by using random effects to incorporate the correlation between different processes. That is, we can use random effects to combine several correlated processes. We describe such an approach as follows.

Let y_{ijk} be the response measurement for individual i, process j, at time t_{ijk}, $i = 1, \cdots, n$; $j = 1, \cdots, J$; $k = 1, \cdots, n_i$. The response y_{ijk} can either be a continuous or discrete variable. We can incorporate the correlation among the repeated measurements within each individual via a transitional (Markov) structure, and incorporate the correlation between different processes via random effects. Specifically, we may consider the following mixed effects transitional joint model

$$
\begin{aligned}
E(y_{ijk}) &= g(y_{ij,k-1}, \mathbf{z}_{ijk}, \boldsymbol{\beta}_j, \boldsymbol{\sigma}_j, \mathbf{b}_{ik}), & (8.25)\\
\mathbf{b}_{ik} &\sim N(0, D), \quad i = 1, \cdots, n; \ j = 1, \cdots, J; \ k = 1, \cdots, n_i,
\end{aligned}
$$

where $g(\cdot)$ is a known function such as a link function or a nonlinear function, $\mathbf{z}_{ijk}$ are covariates, $\mathbf{b}_{ik}$ are random effects, $\boldsymbol{\beta}_j$ is a vector of fixed parameters for process j, $\boldsymbol{\sigma}_j$ contains variance-covariance parameters for within individual measurements of process j, and D is a unknown variance-covariance matrix.

In the mixed effects transitional model (8.25), the between-process correlation is incorporated by the random effects $\mathbf{b}_{ik}$ and the within-process correlation is incorporated by a first-order transition (Markov) model. The random effects $\mathbf{b}_{ik}$ also account for individual-time-specific deviations from the population trajectories of the longitudinal processes. The first-order Markov dependence in model (8.25) can easily be extended to more general Markov structures, such as a second- or third-order Markov structure. We may assume that y_{ijk}'s are conditionally independent given the random effects $\mathbf{b}_{ik}$, i.e., at each time point the processes are assumed to be conditionally independent given the random effects.

Let $\boldsymbol{\theta}$ be the collection of all unknown parameters. The observed-data likelihood function can be written as

$$
L_o(\boldsymbol{\theta}) \;=\; \prod_{i=1}^{n} \prod_{k=1}^{n_i} \int \left[\prod_{j=1}^{J} f(y_{ijk}|y_{ij,k-1}, \mathbf{z}_{ijk}, \mathbf{b}_{ik}, \boldsymbol{\beta}, \sigma) \right] f(\mathbf{b}_{ik}|D) \, d\, \mathbf{b}_{ik}.
$$

The EM algorithm can be used to obtain MLE of $\boldsymbol{\theta}$. The "complete-data" log-likelihood function is given by

$$
l_c(\boldsymbol{\theta}) \;=\; \sum_{i=1}^{n} \sum_{k=1}^{n_i} \left\{ \sum_{j=1}^{J} \log f(y_{ijk}|y_{ij,k-1}, \mathbf{z}_{ijk}, \mathbf{b}_{ik}, \boldsymbol{\beta}, \sigma) + \log f(\mathbf{b}_{ik}|D) \right\}.
$$

The E-step at the t-th EM iteration can then be written as

$$
\begin{aligned}
Q(\boldsymbol{\theta}|\boldsymbol{\theta}^{(t)}) \;=\;& E(l_c(\boldsymbol{\theta})|\mathbf{y}_{ik}, \mathbf{z}_{ik}, \mathbf{r}_{ik}; \boldsymbol{\theta}^{(t)}) \\
\;=\;& \sum_{i=1}^{n} \sum_{k=1}^{n_i} \int \left[\sum_{j=1}^{J} \log f(y_{ijk}|y_{ij,k-1}, \mathbf{z}_{ijk}, \mathbf{b}_{ik}, \boldsymbol{\beta}, \sigma) \right. \\
& \left. + \log f(\mathbf{b}_{ik}|D) \right] f(\mathbf{b}_{ik}|\mathbf{y}_{ik}, \mathbf{z}_{ik}, \mathbf{r}_{ik}; \boldsymbol{\theta}^{(t)}) \, d\mathbf{b}_i.
\end{aligned}
$$

The M-step is to maximize $Q(\boldsymbol{\theta}|\boldsymbol{\theta}^{(t)})$ to produce an updated estimate of $\boldsymbol{\theta}$.

Cook et al. (2004a) proposed a conditional Markov model for clustered progressive multi-state processes where multiplicative random effects for each transition intensity were used to address the clustering in processes within each subject. They considered a multi-state Markov model and applied their method to clustered progressive chronic disease processes.

Carroll, Midthune, Freedman, and Kipnis (2006) proposed a multivariate measurement error model which combines linear mixed measurement error models and linear seemingly unrelated regression models. They considered separate marginal mixed measurement error models for different nutrients which seemingly unrelated but aspects of each model are highly correlated. They showed a substantial efficiency gain from joint modeling over separate modeling.

Example 8.2 *Joint transitional models*

Suppose that we observe J continuous longitudinal processes where each process may be modeled by a linear or nonlinear model. We may then consider the following joint transitional model in which the within process longitudinal correlation is modeled by a transitional structure and the between process correlation is incorporated by individual-time specific random effects:

$$
\begin{aligned}
y_{ijk} \;&=\; g(y_{ij,k-1}, t_{ijk}, \boldsymbol{\beta}_{ijk}) + e_{ijk}, \\
\boldsymbol{\beta}_{ijk} \;&=\; A_{ijk}\, \boldsymbol{\beta}_j + \mathbf{b}_{ik}, \\
e_{ijk} \quad \text{i.i.d.} \quad &\sim N(0, \sigma_j^2), \qquad \mathbf{b}_{ik} \sim N(0, D),
\end{aligned}
$$

$$i = 1, \cdots, n; \; j = 1, \cdots, J; \; k = 1, \cdots, n_i,$$

where $g(\cdot)$ is a known linear or nonlinear function, $\boldsymbol{\beta}_j$ is a vector of fixed parameters for process j, and A_{ijk} is a design matrix containing covariates. Different processes are linked through the shared random effects in the model.

If we observe J binary longitudinal processes, we may consider the following joint transitional model

$$
\begin{aligned}
\text{logit}(P(y_{ijk} = 1)) &= \beta_{1ijk} + \beta_{2ijk} t_{ijk} + \beta_{3ijk} y_{ij,k-1} + \boldsymbol{\beta}_{4ijk}^T \mathbf{z}_{ijk} \\
\beta_{1ijk} &= \beta_{1j} + b_{1ik}, \quad \beta_{2ijk} = \beta_{2j} + b_{2ik}, \\
\beta_{3ijk} &= \beta_{3j} + b_{3ik}, \quad \beta_{4ijk} = \beta_{4j} + b_{4ik}, \\
\mathbf{b}_{ik} &= (b_{1ik}, b_{2ik}, b_{3ik}, b_{4ik})^T \sim N(0, D), \\
& \quad i = 1, \cdots, n; \; j = 1, \cdots, J; \; k = 1, \cdots, n_i,
\end{aligned}
$$

where covariates may also be used to partially explain variations in the individual-specific parameters β_{lijk}'s. Different processes are again linked through shared random effects.

8.6.3 Joint Longitudinal Models with Incomplete Data: A Summary

In modeling longitudinal data, some time-dependent covariates may be measured with errors or may be missing. In this case, additional longitudinal models for the time-dependent covariates are required to address measurement errors or missing data in likelihood inference. This leads to joint models of two (or more) longitudinal processes in which one is a covariate process and the other is a response process. When missing data in either the response or the covariates are non-ignorable, the missing data indicators are longitudinal binary processes, and these missing data processes must also be incorporated in the joint likelihood. This is an example of joint models of several longitudinal processes.

In the above example, the covariate models and the missing data models are secondary, so the parameters in these models may be viewed as nuisance parameters and are usually not of primary interest. To increase the precision or efficiency of the main parameter estimates, and to avoid potential identifiability problems, we should simplify the secondary models to reduce the number of nuisance parameters. Examples of joint modeling for longitudinal response process, longitudinal covariate process, and missing data process can be found in Chapters 4 – 6. In the following we give a brief summary of the basic approaches.

Let $\mathbf{y}_i = (y_{i1}, \cdots, y_{in_i})^T$ be the response process and $\mathbf{z}_i = (z_{i1}, \cdots, z_{in_i})^T$ be the covariate process. Suppose that we model the response process by a mixed effects model with density function $f(\mathbf{y}_i|\mathbf{z}_i, \mathbf{b}_i, \boldsymbol{\beta}, \boldsymbol{\sigma})$ given the random effects $\mathbf{b}_i$, such as a GLMM or NLME model. If the covariates $\mathbf{z}_i$ are measured with errors or have missing values, we can model the covariate process by another mixed effects model $f(\mathbf{z}_i|\mathbf{a}_i, \boldsymbol{\alpha})$ to address the measurement errors or missing data:

$$\mathbf{z}_i = \mathbf{z}_i^* + \boldsymbol{\epsilon}_i = h(\mathbf{t}_i, \boldsymbol{\alpha}, \mathbf{a}_i) + \boldsymbol{\epsilon}_i, \qquad i = 1, 2, \cdots, n,$$

where $\mathbf{t}_i = (t_{i1}, \cdots, t_{in_i})$ are measurement times for individual i, $\mathbf{z}_i$'s are observed but mis-measured covariate values and $\mathbf{z}_i^* = h(\mathbf{t}_i, \boldsymbol{\alpha}, \mathbf{a}_i)$ are the corresponding unobserved true covariate values, $h(\cdot)$ is a known function (often linear), $\boldsymbol{\alpha}$ contains fixed parameters, $\mathbf{a}_i$ contains random effects, and $\boldsymbol{\epsilon}_i = (\epsilon_{i1}, \cdots, \epsilon_{in_i})^T$ represents measurement errors in the data for individual i. We assume that $\mathbf{z}_i$ is a surrogate of $\mathbf{z}_i^*$ and that $f(\mathbf{y}_i|\mathbf{z}_i, \mathbf{z}_i^*, \mathbf{b}_i, \boldsymbol{\beta}, \boldsymbol{\sigma}) = f(\mathbf{y}_i|\mathbf{z}_i^*, \mathbf{b}_i, \boldsymbol{\beta}, \boldsymbol{\sigma})$.

The *joint likelihood* of the two longitudinal processes for the observed data can be written as

$$L_o(\boldsymbol{\theta}) = \prod_{i=1}^{n} \int \int f(\mathbf{y}_i|\mathbf{z}_i^*, \mathbf{b}_i, \boldsymbol{\beta}, \boldsymbol{\sigma}) f(\mathbf{z}_i|\mathbf{a}_i, \boldsymbol{\alpha}) \, d\mathbf{a}_i \, d\mathbf{b}_i.$$

Maximizing the joint likelihood $L_o(\boldsymbol{\theta})$ gives the MLEs of all model parameters simultaneously. Standard errors of the MLEs can be obtained from the joint observed information matrix. The joint likelihood estimation incorporates the uncertainty in estimating all parameters. For two-step methods or regression calibration methods, a third step is required to address the uncertainty of the estimation in the first step.

If both the response and the covariate processes are LME models, we can integrate out the random effects $(\mathbf{a}_i, \mathbf{b}_i)$ in the joint likelihood $L_o(\boldsymbol{\theta})$ and derive an analytic expression for $L_o(\boldsymbol{\theta})$. Wang, Wang, and Wang (2000) discussed such an approach. For LME models, however, parameter or model identifiability is more likely an issue than GLMM or NLME models. Wang and Heckman (2009) discussed parameter identifiability in LME models. For GLMM or NLME or frailty models, however, the integral in the joint likelihood $L_o(\boldsymbol{\theta})$ can be highly intractable, so Monte Carlo EM algorithms or its variations are often used but these algorithms are often computationally very demanding. Therefore, computationally more efficient approximate methods based on Taylor or Laplace approximations are highly valuable for joint model inference.

When missing data are non-ignorable, a model for the missing data process is typically assumed and incorporated in joint likelihood inference. In other words, one jointly models the response process, the covariate process, and the missing data process for joint likelihood inference. In this case, the association between these models is clear, and a joint inference is necessary for valid analysis. Specifically, let r_{ij} be a missing data indicator such that $r_{ij} = 1$ if z_{ij} is missing and $r_{ij} = 0$ if z_{ij} is observed. One then assumes a model for the longitudinal binary process $\mathbf{r}_i = (r_{i1}, \cdots, r_{in_i})^T$ to mimic a possible non-ignorable missing data mechanism, i.e., how the missingness may be related to the missing and observed data. Two commonly assumed non-ignorable missing data models are:

- missing probabilities depend on the observed and missing values, i.e., a missing data model $f(\mathbf{r}_i|\mathbf{y}_i, \mathbf{z}_i, \boldsymbol{\phi})$,
- missing probabilities depend on the unobserved random effects in the models, i.e., a missing data model $f(\mathbf{r}_i|\mathbf{a}_i, \mathbf{b}_i, \boldsymbol{\phi})$.

In both models, the missing probabilities depend on unobserved quantities, so the missing data mechanism is non-ignorable.

The choice of missing data models can be based on scientific considerations or subject-area knowledge. For example, the above second missing data model may be more reasonable if the responses and covariates are measured with errors and one believes that the missingness depends on the true but unobserved response and covariate values or some underlying latent processes that govern the response and covariate processes. Since a non-ignorable missing data model is not testable based on the observed data, it is a good strategy to consider different missing data models and then perform sensitivity analysis.

For missing data model $f(\mathbf{r}_i|\mathbf{y}_i, \mathbf{z}_i, \boldsymbol{\phi})$, the joint likelihood for all three longitudinal models can be written as

$$L_o^{(1)}(\boldsymbol{\theta}) = \prod_{i=1}^{n} \int \int \int f(\mathbf{y}_i|\mathbf{z}_i^*, \mathbf{b}_i, \boldsymbol{\beta}, \boldsymbol{\sigma}) f(\mathbf{z}_i|\mathbf{a}_i, \boldsymbol{\alpha})$$
$$\times f(\mathbf{r}_i|\mathbf{y}_i, \mathbf{z}_i, \boldsymbol{\phi})\, d\mathbf{z}_{mis,i}\, d\mathbf{a}_i\, d\mathbf{b}_i.$$

For missing data model $f(\mathbf{r}_i|\mathbf{a}_i, \mathbf{b}_i, \boldsymbol{\phi})$, assuming that the within individual covariate measurements are conditionally independent given the random effects, the joint likelihood for all three longitudinal models can be written as

$$L_o^{(2)}(\boldsymbol{\theta}) = \prod_{i=1}^{n} \int \int f(\mathbf{y}_i|\mathbf{z}_i^*, \mathbf{b}_i, \boldsymbol{\beta}, \boldsymbol{\sigma}) f(\mathbf{z}_{obs,i}|\mathbf{a}_i, \boldsymbol{\alpha}) f(\mathbf{r}_i|\mathbf{a}_i, \mathbf{b}_i, \boldsymbol{\phi})\, d\mathbf{a}_i\, d\mathbf{b}_i.$$

The joint likelihood $L_o^{(2)}(\boldsymbol{\theta})$ has a lower dimension and is computationally easier to manage. In both cases, the main parameters of interest are usually $\boldsymbol{\beta}$, and the parameters $\boldsymbol{\alpha}$ and $\boldsymbol{\phi}$ are often viewed as nuisance parameters. MLEs of all the parameters are then obtained simultaneously by maximizing the joint likelihoods. When the responses also have nonignorable missing data, the joint likelihood inference is similar.

In the context of missing data and measurement errors described in this section, the associations among the longitudinal models are clear and joint inference is desirable. Likelihood methods provide a unified framework for such joint inference, with desirable asymptotic properties of the resulting estimates. However, likelihood methods are based on distributional assumptions, which may be restrictive in some cases, and the asymptotic optimality and normality of the resulting MLEs rely on certain regularity conditions. Alternative joint modeling approaches are available, such as the two-step methods or regression calibration methods discussed earlier and GEE methods (see Chapter 10).

CHAPTER 9

Robust Mixed Effects Models

9.1 Introduction

In previous chapters, we have mostly focused on parametric mixed effects models. In these models, the random effects are typically assumed to follow (multivariate) normal distributions and the within individual (cluster) random errors are assumed to follow parametric distributions in the exponential family. For example, in LME and NLME models we assume that both the random effects and the responses (or the within individual errors) follow normal distributions, and in GLMMs we assume that the random effects follow normal distributions and the responses (or the within individual errors) follow parametric distributions in the exponential family. Likelihood inference is then based on the assumed distributions.

In practice, however, the assumed parametric distributions may not hold. Moreover, outliers may be present. Sometimes, outliers may not be easily detected, especially for multi-dimensional data. Likelihood inference for standard (parametric) mixed effects models are typically sensitive to the assumed parametric distributions and to outliers. For example, a few outliers in a dataset can have a large influence on the parameter estimation and inference. Therefore, likelihood inference ignoring outliers is unreliable and misleading.

In a broad sense, outliers may be viewed as incompletely observed data. Sometimes outliers may be data measured with errors. If the outliers can be detected or screened manually in advance, one may remove these outliers in the analysis or check their sources since sometimes outliers may contain valuable information. For multi-dimensional (or multivariate) data, however, it may be difficult to detect outliers in advance. Therefore, robust methods which are less sensitive to outliers are highly valuable. Robust methods usually outperform classical methods in the presence of outliers. Good robust estimates can be reasonably efficient when the assumed distributions hold and more efficient than standard estimates when the assumed distributions do not hold.

An *outlier* can be defined as an observation which appears to be inconsistent with the remainder of the data. For longitudinal data, there are two types of outliers:

- outliers can occur at *individual level*, sometimes called *e-outliers*, which arise among the repeated measurements within a given individual, and

- outliers can also occur at *population level*, sometimes called *b-outliers*, which are unusual individuals or clusters in the sample.

The terms *e-outliers* and *b-outliers* have appeared in the literature and are based on the standard notation for a LME model. Specifically, for LME model (9.4) (page 302), the within individual errors are usually denoted by $\mathbf{e}_i$ and the random effects are denoted by $\mathbf{b}_i$, so outliers in the repeated measurements within each individual are called e-outliers while outlying clusters or individuals (corresponding large random effects) are called b-outliers. Since different notation may be used for a LME model, it may be a good idea to avoid using the terms e-outliers and b-outliers. Figure 9.1 shows these two types of outliers in a longitudinal AIDS study. Sometimes it may not be possible to distinguish between the two cases. In either cases, likelihood inference based on standard mixed effects models are sensitive to these outliers and thus may lead to misleading results if the outliers are not addressed.

In this chapter, we consider some commonly used robust methods for mixed effects models. We focus on the following two common approaches for robust inference:

- one approach is to replace standard distributions assumed for a mixed effects model by more robust distributions. For example, the usual normal distributions assumed in mixed effects models can be replaced by heavier tail *t-distributions*;

- the other approach is to bound or downweight the influence of outliers in estimating equations, i.e., estimation and inference are based on estimating equations which bound or downweight the influence of outliers. A well-known example is the so-called *M-estimators*.

Note that a t-distribution with small degrees of freedom has heavier tails than a corresponding normal distribution with the same mean, so a t-distribution can accommodate some outliers. A t-distribution approaches to a normal distribution as its degrees of freedom increase, so one can choose the degree of freedom in a t-distribution to either increase or decrease its robustness to outliers. For M-estimators, their robustness can be adjusted by choosing appropriate turning points. Both approaches provide robust inference for mixed effects models.

For a robust method, we need some measures of *robustness*. There are two

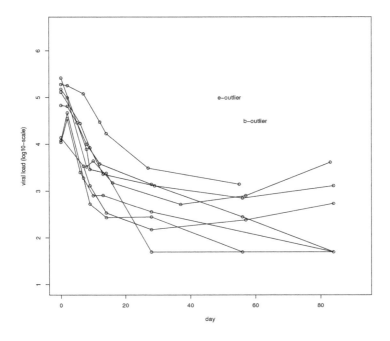

Figure 9.1 *Two types of possible outliers in a longitudinal AIDS study.*

commonly used measures of robustness: breakdown point and influence function. The *breakdown point* of an estimator is the proportion of outliers an estimator can handle before giving an arbitrarily large or small result. For example, the sample *mean* has a breakdown point of 0 since the mean can become arbitrarily large by increasing any one value in the sample. On the other hand, the *median* has a breakdown point of 0.5 (the maximum breakdown point) since the median remains unchanged even if we change near half of the values in the sample. So the median is a robust statistic while the mean is not, although both are commonly used to summarize the data in a sample. An *influence function* shows what happens to an estimator when we change the assumed distribution of the data slightly.

Detailed discussions of basic ideas and theories for robust methods can be found in Huber (1981), Rousseeuw and Leroy (1987), and Maronna et al. (2006), among others.

9.2 Robust Methods

There is an extensive literature for robust methods. Maronna, Martin, and Yohai (2006) provided a recent review. In this section, we briefly review the basic ideas of the two commonly used robust methods: robust distributions and M-estimators.

9.2.1 Robust Distributions

In a standard mixed effects model, typically the random effects are assumed to follow a (multivariate) normal distribution and the response is assumed to follow a distribution in the exponential family, including the normal distribution. It is well known that normal distributions do not accommodate outliers well, i.e., they are sensitive to outliers, and a t-distribution is more robust to outliers. A t-distribution is similar to a corresponding normal distribution (with the same mean) in shape, i.e., they are both bell shaped and symmetric, but a t-distribution has *heavier tails* than the corresponding normal distribution, especially when the degree of freedom of the t-distribution is small. Thus, a t-distribution can better accommodate outliers. Therefore, we can use a t distribution to replace the corresponding normal distribution assumed in the model for robust inference.

As the number of degrees of freedom increases, the standard t-distribution approaches to the standard normal distribution. The number of degrees of freedom in a t-distribution controls how heavy the tails of the distribution are. For robust inference, we may fix the degrees of freedom of a t-distribution and choose a value between 3 to 9, but the degrees of freedom can also be estimated from the data.

The probability density function of the standard t-*distribution* with k degrees of freedom, denoted by $t(k)$, can be written as

$$f(x) = \frac{\Gamma\left(\frac{k+1}{2}\right)}{\sqrt{k\pi}\,\Gamma\left(\frac{k}{2}\right)} \left(1 + \frac{x^2}{k}\right)^{-\frac{k+1}{2}}, \qquad -\infty < x < \infty,$$

where $\Gamma(x)$ is the Gamma function

$$\Gamma(x) = \int_0^\infty u^{x-1} e^{-u} du.$$

For the standard t-distribution with k degrees of freedom, the mean is 0 and variance is $k/(k-2)$ for $k > 2$. Since a t-distribution is symmetric, its mean and its median are the same. When the degree of freedom $k = 1$, the corresponding standard t-distribution becomes the Cauchy distribution. Figure 9.2

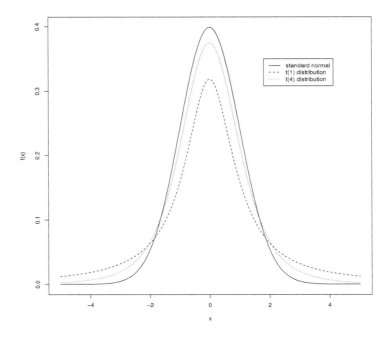

Figure 9.2 *Density functions for the standard normal distribution and the t distributions with degrees of freedom being 1 and 4 respectively.*

shows the density functions of standard t-distributions, with degrees of freedom being 1 and 4 respectively, and the standard normal distribution $N(0, 1)$. We see that the three distributions are similar in shape but the t-distributions have heavier tails than the normal distribution.

A multivariate t-distribution is a multivariate generalization of the univariate t-distribution. The probability density function of a p-dimension random vector $\mathbf{x} = (x_1, x_2, \cdots, x_p)^T$ following a general *multivariate t-distribution with k degrees of freedom and parameters* $\boldsymbol{\mu}$ *and* Σ, denoted by $t_p(\boldsymbol{\mu}, \Sigma, k)$, is given by

$$f(\mathbf{x}) = \frac{\Gamma\left(\frac{k+p}{2}\right)}{\Gamma(k/2)k^{p/2}\pi^{p/2}|\Sigma|^{1/2}\left[1 + \frac{1}{k}(\mathbf{x} - \boldsymbol{\mu})^T\Sigma^{-1}(\mathbf{x} - \boldsymbol{\mu})\right]^{(k+p)/2}},$$
$$\mathbf{x} \in R^p,$$

where $\boldsymbol{\mu}$ is a $p \times 1$ vector and Σ is a $p \times p$ matrix. The mean vector and the variance-covariance matrix of $\mathbf{x}$ are given respectively by

$$E(\mathbf{x}) = \boldsymbol{\mu}, \qquad Cov(\mathbf{x}) = \left(\frac{k}{k-2} \right) \Sigma \text{ for } k > 2.$$

Note that the variance-covariance matrix of $\mathbf{x}$ exists only for $k > 2$.

For robust inference, we can replace the multivariate normal distributions typically assumed in a standard regression model by the corresponding multivariate t-distributions. For example, for a LME model, we can assume that the random effects or the within-individual random errors or both follow t-distributions. Such robust approaches can be found in Lange et al. (1989), Pinheiro et al. (2001), and Song et al. (2007).

There are other robust methods. For example, one may replace a normal distribution assumed in a model by a *mixture* of two or more normal distributions. One may also replace a parametric distribution assumed in a model by a nonparametric distribution. Lai and Shih (2003) considered such a nonparametric method for NLME models.

9.2.2 M-Estimators

A widely used class of robust methods is to bound or downweight the influence of outlying observations in a set of estimating equations for parameter estimation. The set of estimating equations is typically motivated from likelihood equations for maximum likelihood estimation. Specifically, consider a distribution with density function $f(y, \boldsymbol{\theta})$. The MLEs of parameters in $\boldsymbol{\theta}$ based on data $\{y_1, y_2, \cdots, y_n\}$ can be obtained by solving the following estimating equation

$$\sum_{i=1}^{n} \psi(y_i, \boldsymbol{\theta}) = 0, \tag{9.1}$$

where

$$\psi(y_i, \boldsymbol{\theta}) = \frac{\partial \log f(y_i, \boldsymbol{\theta})}{\partial \boldsymbol{\theta}}.$$

This is equivalent to *minimize* the quantity $\sum_{i=1}^{n} \rho(y_i, \boldsymbol{\theta})$, where

$$\rho(y_i, \boldsymbol{\theta}) = -\log f(y_i, \boldsymbol{\theta}), \qquad \psi(y_i, \boldsymbol{\theta}) = -\frac{\partial \rho(y_i, \boldsymbol{\theta})}{\partial \boldsymbol{\theta}}.$$

For example, if $y_1, y_2, \cdots, y_n$ i.i.d. $\sim N(\mu, 1)$, then $\psi(y_i, \mu) = y_i - \mu$ and $\rho(y_i, \mu) = (y_i - \mu)^2 / 2$ (see Example 9.1 for details). This motivates the following M-estimators.

M-estimators, or maximum likelihood type estimators, are generalizations of MLEs in which the functions $\rho(y_i, \boldsymbol{\theta})$ or $\psi(y_i, \boldsymbol{\theta})$ in (9.1) are not necessarily

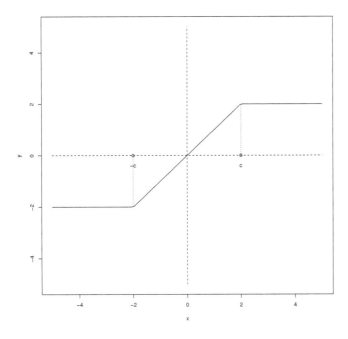

Figure 9.3 *Huber's ψ-function with turning point $c = 2$.*

related to a probability density function. Instead, function $\psi(y_i, \boldsymbol{\theta})$ can be cho-
sen to *downweight* outliers in such a way that $\psi(x)$ is close to $|x|$ when $|x|$ is
small and $\psi(x)$ remains small when $|x|$ is large. For example, Huber suggested
the following well-known function (Huber 1981)

$$\psi(x) \;=\; \begin{cases} x & \text{if } |x| \leq c \\ c & \text{if } x > c \\ -c & \text{if } x < -c \end{cases} \tag{9.2}$$

where the constant $c > 0$ is a *turning point*. Figure 9.3 shows the ψ-function
with turning point $c = 2$. The corresponding *Huber's ρ-function* is given by

$$\rho(x) \;=\; \begin{cases} x^2/2, & \text{for } |x| \leq c, \\ c|x| - x^2/2, & \text{for } |x| > c. \end{cases} \tag{9.3}$$

So Huber replaces the quadratic function in the normal likelihood by a function
with a bounded derivative.

When the turning point c goes to ∞, the M-estimator approaches to the sample
mean. When the turning point c goes to 0, the M-estimator approaches to the
sample median. Thus, the value of the turning point c is important: the larger

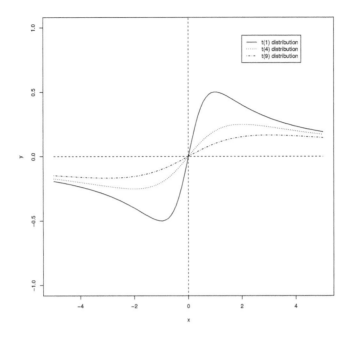

Figure 9.4 *The ψ-functions of standard t distributions with degrees of freedom being 1, 4, and 9 respectively.*

the value of c, the closer the M-estimator to MLE. In other words, the smaller the value of c, the more robust the M-estimator. In practice, we can choose the turning point c which achieves a *balance* between robustness and efficiency, such as $c = 2$.

Note that, for a standard t-distribution with k degrees of freedom, it can be shown that

$$\psi(x) = \frac{x}{x^2 + k}, \qquad -\infty < x < \infty.$$

Figure 9.4 shows the ψ-functions for t distributions with degrees of freedom being 1, 4, and 9 respectively. We see that large or small values of x are downweighted in a way similar to the Huber's ϕ-function.

It can be shown that M-estimators are asymptotically normally distributed. M-estimators are popular in robust inference due to their generality, efficiency, and high breakdown point. Note that, however, M-estimators are not necessarily unique, i.e., there may be more than one solution to the equations (9.1). In practice, the choice of the ψ function may not be critical, and many choices may lead to similar results. M-estimators have good large-sample properties. For small samples, an alternative approach is the bootstrap method.

There are many other robust methods such as R-estimators and L-estimators, but M-estimators may be the most popular (Richardson 1997; Preisser and Qaqish 1999; Cantoni and Ronchetti 2001; Yeap and Davidian, 2001; Sinha 2004; Copt and Victoria-Feser 2006; Noh and Lee 2007).

Example 9.1 *Robust estimation for normal distributions.*

Suppose that $\{y_1, y_2, \cdots, y_n\}$ is an i.i.d. sample from normal distribution $N(\mu, 1)$, where the mean parameter μ needs to be estimated. The log-likelihood function for estimating parameter μ is given by

$$l(\mu) = -\frac{n}{2}\log(2\pi) - \frac{1}{2}\sum_{i=1}^{n}(y_i - \mu)^2.$$

So the likelihood equation is given by

$$\sum_{i=1}^{n}\psi(y_i, \mu) = \sum_{i=1}^{n}(y_i - \mu) = 0,$$

where $\psi(y_i, \mu) = y_i - \mu$. The solution (MLE) of the above estimating equation has an *unbounded* influence function, i.e., the influence of an outlier on the MLE is unbounded. In other words, an outlier can have a large influence on the resulting MLE $\hat{\mu}$. Note that maximizing $l(\mu)$ is equivalent to minimizing $\sum_{i=1}^{n}\rho(y_i, \mu)$, where $\rho(y_i, \mu) = (y_i - \mu)^2/2$.

For a robust estimator of μ, we can replace the above $\psi(y_i, \mu)$ by the Huber's function $\psi_c(y_i, \mu)$ given in (9.2). Then, one solves the following robust estimating equation

$$\sum_{i=1}^{n}\psi_c(y_i, \mu) = 0,$$

which bounds or downweights influences of large or small values of y_i's. The solution of the above robust estimating equation, $\tilde{\mu}_c$, is a robust estimator of μ. With an appropriate choice of the turning point c, the robust estimator $\tilde{\mu}_c$ will be less sensitive to outliers.

9.3 Mixed Effects Models with Robust Distributions

9.3.1 LME Models with Multivariate t-Distributions

In a standard LME model, the random effects and the within-individual errors are assumed to follow multivariate normal distributions. Thus, likelihood inference for standard LME model is sensitive to outliers. For robust inference, one approach is to replace the multivariate normal distributions in the LME models by the corresponding multivariate t-distributions with the same means and variance-covariance matrices. Since t-distributions have heavier tails than normal distributions, they can accommodate some outliers. Thus, LME models with t-distributions for the random effects and/or the within-individual errors are more robust to outliers than standard LME models with normal distributions.

As noted in Section 9.2, in longitudinal studies outliers may occur at *population level*, suggesting a t-distribution for the random effects to accommodate these outliers, and/or at the *individual level*, suggesting a t-distribution for the within-individual error to accommodate these outliers. In other words, for robust LME models, we may consider t-distributions for the random effects or for the within-individual errors or for both. In the following, we focus on LME models where t-distributions are assumed for both the random effects and the within-individual errors.

LME models with t-distributions

Let $\mathbf{y}_i = (y_{i1}, y_{i2}, \cdots, y_{in_i})^T$ be the n_i response measurements on individual i, $i = 1, 2, \cdots, n$. A standard LME model is given by

$$
\begin{aligned}
\mathbf{y}_i &= X_i\boldsymbol{\beta} + Z_i\mathbf{b}_i + \mathbf{e}_i, \qquad i = 1, 2, \cdots, n, \\
\mathbf{b}_i &\sim N(0, D), \quad \mathbf{e}_i \sim N(0, R_i),
\end{aligned}
\tag{9.4}
$$

where $\boldsymbol{\beta} = (\beta_0, \beta_1, \cdots, \beta_p)^T$ are fixed effects, $\mathbf{b}_i = (b_{i1}, \cdots, b_{iq})^T$ are random effects, X_i and Z_i are known design matrices, $\mathbf{e}_i = (e_{i1}, e_{i2}, \cdots, e_{in_i})^T$ are within-individual errors, R_i is the covariance matrix for the within-individual errors, and D is the covariance matrix of the random effects.

Since normal distributions are sensitive to outliers, we can assume multivariate t-distributions for the random effects $\mathbf{b}_i$ and the within-individual errors $\mathbf{e}_i$ in the LME model (9.4). This leads to the following *LME model with t-distributions* (Lange et al. 1989; Pinheiro et al. 2001):

$$
\begin{aligned}
\mathbf{y}_i &= X_i\boldsymbol{\beta} + Z_i\mathbf{b}_i + \mathbf{e}_i, \qquad i = 1, 2, \cdots, n, \\
\mathbf{b}_i &\sim t_q(0, D, \nu), \quad \mathbf{e}_i \sim t_{n_i}(0, R_i, \nu),
\end{aligned}
\tag{9.5}
$$

where ν is the degrees of freedom of the multivariate t distributions. See Section 9.2.1 for definition and properties of multivariate t distributions.

For the multivariate t LME model (9.5), it can be shown that the marginal distribution of the response $\mathbf{y}_i$ is given by (Johnson and Kotz 1972)

$$
\mathbf{y}_i \sim t_{n_i}(X_i\boldsymbol{\beta}, \ Z_iDZ_i^T + R_i, \ \nu),
$$

i.e., the marginal distribution of the response $\mathbf{y}_i$ is a multivariate t-distribution with mean $X_i\boldsymbol{\beta}$ and variance-covariance $(\nu/(\nu - 2))(Z_iDZ_i^T + R_i)$. The variance-covariance matrices of the random effects and the within-individual errors in the multivariate t LME model (9.5) are given respectively by

$$Var(\mathbf{b}_i) = \frac{\nu}{\nu - 2}D, \quad Var(\mathbf{e}_i) = \frac{\nu}{\nu - 2}R_i, \quad \text{for } \nu > 2, \; i = 1, 2, \cdots, n.$$

In both the standard LME model (9.4) and the robust LME model (9.5), the marginal means of the response are the same:

$$E(\mathbf{y}_i) = X_i\boldsymbol{\beta}.$$

Thus, in both models the fixed effects $\boldsymbol{\beta}$ have the same interpretation and can be interpreted as the population averages, as in standard linear regression models. However, the interpretations of the variance-covariance matrices D and R_i are different in the two models. Moreover, the marginal variance-covariance matrix of the response in the robust LME model (9.5) is given by

$$Var(\mathbf{y}_i) = \frac{\nu}{\nu - 2}(Z_i D Z_i^T + R_i),$$

while marginal variance-covariance matrix of the response in the standard LME model (9.4) is given by

$$Var(\mathbf{y}_i) = Z_i D Z_i^T + R_i.$$

So the variance-covariance matrices in the two models differ by a factor of $\nu/(\nu - 2)$.

To facilitate likelihood inference, we can re-write the multivariate t LME model (9.5) in a *gamma-normal hierarchical form* as follows (Kotz and Nadarajah 2004)

$$
\begin{aligned}
\mathbf{y}_i \mid \mathbf{b}_i, \tau_i &\sim N(X_i\boldsymbol{\beta} + Z_i\mathbf{b}_i, \; \tau_i^{-1}R_i), &\qquad (9.6)\\
\mathbf{b}_i \mid \tau_i &\sim N(0, \; \tau_i^{-1}D),\\
\tau_i &\sim Gamma\left(\frac{\nu}{2}, \frac{\nu}{2}\right), &\qquad i = 1, 2, \cdots, n,
\end{aligned}
$$

where the gamma distribution $Gamma(\alpha, \beta)$ has the following probability density function:

$$f(x) = \frac{\beta^\alpha x^{\alpha - 1}}{\Gamma(\alpha)} \exp(-\beta x), \qquad x > 0, \alpha > 0, \beta > 0.$$

This gamma-normal hierarchical representation is an alternative formulation of the multivariate t LME model (9.5) and is convenient for maximum likelihood estimation of the parameters using the EM algorithm, as described next.

Likelihood Inference

For maximum likelihood estimation of the parameters in the multivariate t LME model (9.5), we consider the EM algorithm based on the gamma-normal hierarchical model (9.6), following Pinheiro et al. (2001). Let $R_i = \sigma^2 I_{n_i}$ for simplicity. For the gamma-normal hierarchical model (9.6), we can treat the random variables $\mathbf{b}_i$ and τ_i as "missing data", so the "complete data" are

$$\{(\mathbf{y}_i, \mathbf{b}_i, \tau_i), \quad i = 1, 2, \cdots, n\}.$$

Let $\boldsymbol{\theta} = (\boldsymbol{\beta}, D, \sigma^2)$ denote all parameters. Let $\mathbf{y} = (\mathbf{y}_i^T, \cdots, \mathbf{y}_n^T)^T$ and $\mathbf{b} = (\mathbf{b}_i^T, \cdots, \mathbf{b}_n^T)^T$.

The "complete data" log-likelihood based on the gamma-normal hierarchical model (9.6) is given by

$$
\begin{aligned}
l_c(\boldsymbol{\theta}) &= \sum_{i=1}^{n} \left[-\frac{n_i}{2} \log(\sigma^2) - \frac{\tau_i}{2\sigma^2} (\mathbf{y}_i - X_i\boldsymbol{\beta} - Z_i\mathbf{b}_i)^T (\,\cdot\,) \right] \\
&\quad + \left[-\frac{n}{2} \log |D| - \frac{1}{2} \mathrm{tr} \left(D^{-1} \sum_{i=1}^{n} \tau_i \mathbf{b}_i \mathbf{b}_i^T \right) \right] \\
&\quad + \sum_{i=1}^{n} \left[\frac{\nu}{2} \left\{ \log \left(\frac{\nu}{2} \right) + \log(\tau_i) - \tau_i \right\} - \log(\tau_i) - \log \left(\Gamma \left(\frac{\nu}{2} \right) \right) \right] \\
&\quad + c \\
&= l_c^{(1)}(\boldsymbol{\beta}, \sigma^2) + l_c^{(2)}(D) + l_c^{(3)}(\nu) + c.
\end{aligned}
$$

The E-step of the EM algorithm computes the conditional expectation of the complete-data log-likelihood given the observed data and current parameter estimate $\hat{\boldsymbol{\theta}}$, i.e.,

$$Q(\boldsymbol{\theta}|\hat{\boldsymbol{\theta}}) = E(l_c(\boldsymbol{\theta})\,|\mathbf{y}, \hat{\boldsymbol{\theta}}).$$

It can be shown that the E-step reduces to the computation of the following quantities:

$$
\begin{aligned}
\hat{\Omega}_i &= \tau_i \,\mathrm{cov}(\mathbf{b}_i|\hat{\boldsymbol{\theta}}, \mathbf{y}) = \left(\hat{D}^{-1} + \frac{1}{\hat{\sigma}^2} Z_i^T Z_i \right)^{-1}, \\
\hat{\mathbf{b}}_i &= E(\mathbf{b}_i|\hat{\boldsymbol{\theta}}, \mathbf{y}) = \hat{D} Z_i^T (Z_i \hat{D} Z_i^T + \sigma^2 I)^{-1} (\mathbf{y}_i - X_i\hat{\boldsymbol{\beta}}), \\
\hat{\tau}_i &= E(\tau_i|\hat{\boldsymbol{\theta}}, \mathbf{y}) = \frac{\hat{\nu} + n_i}{\hat{\nu} + \delta_i^2(\hat{\boldsymbol{\beta}}, \hat{D}, \hat{\sigma}^2)},
\end{aligned}
$$

where

$$\delta_i^2(\boldsymbol{\beta}, D, \sigma^2) = (\mathbf{y}_i - X_i\boldsymbol{\beta})^T (Z_i D Z_i^T + \sigma^2 I)^{-1} (\mathbf{y}_i - X_i\boldsymbol{\beta}).$$

We consider the ECM algorithm (Meng and Rubin 1993) in which the M-step is replaced by a sequence of constrained maximization (CM) steps, with each step being a maximization over a subset of parameters while holding others fixed. Specifically, the CM-step consists of the following sequence:

- updates $\hat{\boldsymbol{\beta}}$ by maximizing $E(l_c^{(1)}(\boldsymbol{\beta}, \hat{\sigma}^2)\,|\,\mathbf{y}, \hat{\boldsymbol{\theta}})$ over $\boldsymbol{\beta}$, holding $\hat{\sigma}^2$ fixed;

- updates $\hat{\sigma}^2$ by maximizing $E(l_c^{(1)}(\beta, \sigma^2) \mid \mathbf{y}, \hat{\boldsymbol{\theta}})$ over σ^2, holding $\hat{\beta}$ fixed;
- updates $\hat{D}$ by maximizing $E(l_c^{(2)}(D) \mid \mathbf{y}, \hat{\boldsymbol{\theta}})$ over D;
- updates $\hat{\nu}$ by maximizing $E(l_c^{(3)}(\nu) \mid \mathbf{y}, \hat{\boldsymbol{\theta}})$ over ν.

These lead to the following updated parameter estimates:

$$\hat{\beta} = \left(\sum_{i=1}^n \frac{\hat{\tau}_i}{\hat{\sigma}^2} X_i^T X_i \right)^{-1} \sum_{i=1}^n \frac{\hat{\tau}_i}{\hat{\sigma}^2} X_i^T (\mathbf{y}_i - Z_i \hat{\mathbf{b}}_i),$$

$$\hat{\sigma}^2 = \frac{\sum_{i=1}^n \left[\hat{\tau}_i (\mathbf{y}_i - X_i \hat{\beta} - Z_i \hat{\mathbf{b}}_i)^T (\mathbf{y}_i - X_i \hat{\beta} - Z_i \hat{\mathbf{b}}_i) + \mathrm{tr}(\hat{\Omega}_i Z_i^T Z_i) \right]}{\sum_{i=1}^n n_i},$$

$$\hat{D} = \frac{1}{n} \sum_{i=1}^n (\hat{\tau}_i \hat{\mathbf{b}}_i \hat{\mathbf{b}}_i^T + \hat{\Omega}_i),$$

$$\hat{\nu} = \arg\max_\nu \sum_{i=1}^n \left\{ \frac{\nu}{2} \left[\log\left(\frac{\nu}{2}\right) + E(\log(\tau_i) \mid \mathbf{y}, \hat{\boldsymbol{\theta}}) - \hat{\tau}_i \right] \right.$$
$$\left. - \log\left(\Gamma\left(\frac{\nu}{2}\right)\right) \right\}.$$

Iterating between the above E- and CM-steps until convergence, we obtain the MLEs (or local maximums) of the parameters in the multivariate t LME model (9.5).

The MLEs from the multivariate t LME model (9.5) are more robust against outliers than those based on the standard LME model (9.4). In a simulation study, Pinheiro et al. (2001) showed that the multivariate t LME model substantially outperforms the normal or standard LME model when outliers are present in the data. The gains in efficiency in estimating the parameters is particularly high for the variance-covariance parameters (Pinheiro et al. 2001; Song et al. 2007).

9.3.2 GLMM and NLME Models with Multivariate t-Distributions

The multivariate t LME model in Section 9.3.1 can be extended to other mixed effects models such as GLMM, NLME, and frailty models for robust inference. The idea is to replace the normal distributions typically assumed in the models by the corresponding t-distributions to accommodate outliers in the data. For LME and NLME models, we may consider t-distributions for both the random effects and the within-individual errors, which accommodate outliers at both the individual levels and population levels. For GLMM and frailty models, we may consider a t-distribution only for the random effects, which accommodate outliers at the population level.

The ideas of multivariate t GLMM and NLME models are similar to the multivariate t LME models in Section 9.3.1. However, when the models are *nonlinear* in the random effects, we are unable to obtain closed-form or analytic expressions similar to those in Section 9.3.1. Therefore, numerical or Monte Carlo methods may be needed for likelihood estimation for multivariate t GLMM and NLME models. In the following, we focus on NLME models with multivariate t-distributions. The idea also applies to GLMM and frailty models.

Suppose that we only consider outliers in the population level, not in the individual level, for illustration. Then we can assume a multivariate t-distribution for the random effects, with the within individual errors still assumed to follow normal distributions. Thus, a robust NLME model can be written as

$$
\begin{aligned}
\mathbf{y}_i &= \mathbf{g}(\mathbf{x}_i, \mathbf{b}_i, \boldsymbol{\beta}) + \mathbf{e}_i, \quad i = 1, 2, \cdots, n, \\
\mathbf{b}_i &\sim t_q(\mathbf{0}, D, \nu), \qquad \mathbf{e}_i \sim N(\mathbf{0}, R_i),
\end{aligned}
\tag{9.7}
$$

where $\mathbf{g}(\cdot)$ is a known vector-value nonlinear function. For simplicity, assume that $R_i = \sigma^2 I_{n_i}$. Let $\boldsymbol{\theta}$ denote a collection of all unknown parameters. The likelihood function can be written as

$$
L(\boldsymbol{\theta}) = \prod_{i=1}^{n} L_i(\boldsymbol{\theta}) = \prod_{i=1}^{n} \int f(\mathbf{y}_i | \mathbf{b}_i, \boldsymbol{\beta}, \sigma^2) f(\mathbf{b}_i | D, \nu) \, d\mathbf{b}_i,
\tag{9.8}
$$

where

$$
f(\mathbf{y}_i | \mathbf{b}_i, \boldsymbol{\beta}, \sigma^2) = (2\pi\sigma^2)^{-n_i} \exp\left[-\frac{(\mathbf{y}_i - \mathbf{g}(\mathbf{x}_i, \mathbf{b}_i, \boldsymbol{\beta}))^T (\,\cdot\,)}{2\sigma^2} \right],
$$

$$
f(\mathbf{b}_i | D, \nu) = \frac{\Gamma\left(\frac{\nu+q}{2}\right)}{\Gamma(\nu/2)\nu^{q/2}\pi^{q/2}|D|^{1/2}\left[1 + \frac{1}{\nu}\mathbf{b}_i^T D^{-1}\mathbf{b}_i\right]^{(\nu+q)/2}}.
$$

Similarly, for a GLMM the likelihood has the same form as (9.8) but with $\mathbf{y}_i | \mathbf{b}_i$ follows a distribution in the exponential family.

If there are outliers in both the population level and the individual level, we can consider multivariate t-distributions for both the random effects and the within-individual errors in the NLME model. The likelihood has the same form as (9.8) but with $\mathbf{y}_i | \mathbf{b}_i$ follows a multivariate t-distribution. If outliers only occur at the individual level but not the population level, we may consider a multivariate t-distribution for the within individual error $\mathbf{e}_i$ but not for the random effects in the NLME model.

When the dimension of the random effects $\mathbf{b}_i$ is small, we may consider the Gauss-Hermit quadrature method to numerically evaluate the integral in the likelihood $L(\boldsymbol{\theta})$. When the dimension of the random effects $\mathbf{b}_i$ is not small, we may consider a Monte Carlo EM algorithm for likelihood estimation, as in previous chapters.

9.3.3 Robust Models with Incomplete Data

In the presence of missing data and/or measurement errors, for likelihood inference we need to assume models for the incompletely observed variables and, in the case of nonignorable missing data, we also need to assume models for the missing data mechanisms, as in previous chapters. For robust inference, we can replace the normal distributions assumed in the models by multivariate t-distributions with the same means and variance-covariances, which accommodate some outliers in the data. Likelihood inference then proceeds in the usual way. As an illustration, in the following we consider robust methods for missing covariates in a mixed effects model.

Consider a mixed effects model with conditional density $f(\mathbf{y}|\mathbf{z}_i, \mathbf{b}_i, \boldsymbol{\beta}, \boldsymbol{\sigma})$ given the random effects $\mathbf{b}_i$, where $\mathbf{z}_i$ contains p covariates with missing data or measurement errors. If the covariates $\mathbf{z}_i$ are time-independent and continuous, we can assume a multivariate t-distribution $t_p(\boldsymbol{\alpha}, A, k)$ to accommodate outliers in the covariates. For random effects $\mathbf{b}_i$, we can also assume a multivariate t-distribution $t_q(0, B, k)$ to accommodate outliers in the response (population level). The likelihood function can be written as

$$L(\boldsymbol{\theta}) = \prod_{i=1}^{n} \int \int f(\mathbf{y}_i|\mathbf{z}_i, \mathbf{b}_i, \boldsymbol{\beta}, \boldsymbol{\sigma}) f(\mathbf{b}_i|B, k) f(\mathbf{z}_i|\boldsymbol{\alpha}, A, k) \, d\mathbf{b}_i \, d\mathbf{z}_{i,mis},$$

where $f(\mathbf{y}_i|\mathbf{z}_i, \mathbf{b}_i, \boldsymbol{\beta}, \boldsymbol{\sigma})$ may also be assumed to follow a multivariate t distribution $t_{n_i}(g(\mathbf{z}_i, \mathbf{b}_i, \boldsymbol{\beta}, \boldsymbol{\sigma}), R_i, k)$ for LME or NLME models.

If the covariates $\mathbf{z}_i$ are time-dependent and continuous, we can assume a multivariate t LME model to model the covariate process and to accommodate outliers in the covariates. The likelihood function can then be written as

$$L(\boldsymbol{\theta}) = \prod_{i=1}^{n} \int \int \int f(\mathbf{y}_i|\mathbf{z}_i, \mathbf{b}_i, \boldsymbol{\beta}, \boldsymbol{\sigma}) f(\mathbf{b}_i|B, k)$$
$$\times f(\mathbf{z}_i|\boldsymbol{\alpha}, \mathbf{a}_i, k) f(\mathbf{a}_i|A, k) \, d\mathbf{a}_i \, d\mathbf{b}_i \, d\mathbf{z}_{i,mis},$$

where a multivariate t distribution $t_d(0, A, k)$ is assumed for the random effects $\mathbf{a}_i$ in the covariate LME model to accommodate outliers in the covariate population level.

If the missing data are nonignorable, we only need to add a model for the missing data mechanism in the above likelihoods. Likelihood inference then proceeds as in previous chapters, but the computation may become more tedious. The above procedures can be extended to missing responses in a straightforward way.

9.4 M-Estimators for Mixed Effects Models

In the previous sections we used multivariate t-distributions to replace multivariate normal distributions in mixed effects models for robust inference, since t-distributions have heavier tails than normal distributions so they can accommodate some outliers in the data. Inference is then based on the usual likelihood methods, so standard likelihood theories, such as asymptotic normality and asymptotic optimality, apply to the resulting MLEs if usual regularity conditions hold. However, a t-distribution is a parametric distribution, so it is still not very flexible in modeling real data in some cases, i.e., the parametric distributional assumption is still restrictive and may not hold in some cases.

In this section, we consider the M-estimator approach which bounds or down-weights outliers in a set of estimating equations motivated from likelihood equations. The resulting M-estimators are robust against outliers. The M-estimators are also asymptotically normally distributed. This approach is quite general, and the M-estimators are often reasonably efficient and have high breakdown points. In the following, we first consider M-estimators for GLMs to illustrate the basic ideas, and then we extend the methods to mixed effects models.

9.4.1 M-Estimators for GLM

In this section we consider robust estimating equations for generalized linear models (GLMs) for cross-sectional data to illustrate the ideas, and then in next section we extend the idea to mixed effects models for longitudinal data. Note that a GLM may be viewed as a special nonlinear model, so the idea in this section also applies to nonlinear regression models. Note also that an outlier may occur in either the response or the covariates or both, so a robust method for regression models should address these cases simultaneously.

Let $\{y_1, y_2, \cdots, y_n\}$ be an i.i.d. sample from a distribution in the exponential family, with mean $E(y_i) = \mu_i$ and variance $Var(y_i) = V(\mu_i)$. Consider the following GLM

$$g(\mu_i) = \mathbf{x}_i^T \boldsymbol{\beta} \equiv \eta_i, \tag{9.9}$$

where $\mathbf{x}_i$ contains covariates, $\boldsymbol{\beta}$ contains unknown parameters, $g(\cdot)$ is a link function, and $\eta_i = \mathbf{x}_i^T \boldsymbol{\beta}$ is the linear predictor. The MLE for $\boldsymbol{\beta}$ is a solution of the following estimating equation (score equation):

$$\sum_{i=1}^{n} \frac{\partial \log f(y_i|\mathbf{x}_i, \boldsymbol{\beta})}{\partial \boldsymbol{\beta}} = \sum_{i=1}^{n} \left[\left(\frac{\partial \mu_i}{\partial \eta_i} \right) V(\mu_i)^{-1} (y_i - \mu_i) \mathbf{x}_i \right] = 0, \tag{9.10}$$

where $f(y_i|\mathbf{x}_i, \boldsymbol{\beta})$ is the probability density function of the response y_i.

Note that the solution of the estimating equation (9.10) has an *unbounded* influence function. This can be seen from the score function for β:

$$\frac{\partial \log f(y_i|\mathbf{x}_i, \boldsymbol{\beta})}{\partial \boldsymbol{\beta}} = \left(\frac{\partial \mu_i}{\partial \eta_i}\right) V(\mu_i)^{-1}(y_i - \mu_i)\mathbf{x}_i.$$

Thus, the influence of an outlier on the MLE is unbounded, i.e., an outlier in the response or covariates can have a large influence on the resulting estimators. In other words, the MLE from estimating equation (9.10) is very sensitive to outliers so is not robust.

To address outliers in the response y_i and covariates $\mathbf{x}_i$ simultaneously, we consider a general class of M-estimators of Mallows's type (Mallow 1975), where the influence of outliers in the response and covariates are bounded separately (e.g., Cantoni and Ronchetti 2001). Specifically, let

$$r_i = V(\mu_i)^{-1/2}(y_i - \mu_i), \qquad \psi_c(r_i) = \max(-c, \, \min(r_i, c)),$$

where $\psi_c(\cdot)$ is the Huber function (see (9.2) on page 299) which controls deviations in the response y_i. For covariates $\mathbf{x}_i$, we may consider a weight function $w(\mathbf{x}_i)$ to down-weight the leverage points, e.g., we may choose

$$w(\mathbf{x}_i) = \sqrt{1 - h_i}$$

where h_i the i-th diagonal element of the usual hat matrix $H = X(X^TX)^{-1}X^T$ in regression models. Then, a robust M-estimator can be obtained from the following estimating equation (Hampel et al. 1986; Cantoni and Ronchetti 2001)

$$\sum_{i=1}^{n} \left[\left(\frac{\partial \mu_i}{\partial \eta_i}\right) w(\mathbf{x}_i)V(\mu_i)^{-1/2}\psi_c(r_i) - a(\boldsymbol{\beta})\right] = 0, \qquad (9.11)$$

where

$$a(\boldsymbol{\beta}) = \frac{1}{n}\sum_{i=1}^{n} \left(\frac{\partial \mu_i}{\partial \eta_i}\right) w(\mathbf{x}_i)V(\mu_i)^{-1/2}E(\psi_c(r_i))$$

is a bias correction term to ensure the Fisher consistency of the resulting estimator and the expectation is taken with respect to the conditional distribution of y_i given $\mathbf{x}_i$. The turning point c in $\psi_c(r_i)$ can be chosen to ensure a given level of asymptotic efficiency. Thus, Mallows's method downweights outliers in all the covariates by the same amount, regardless of the response values.

When $\psi_c(r_i) = r_i$ (i.e., when $c \to \infty$) and $w(\mathbf{x}_i) = 1$ (i.e., no weighting) for all i, the M-estimator from the estimating equation (9.11) is the usual MLE, which is fully efficient (asymptotically) but is sensitive to outliers.

9.4.2 M-Estimators for Mixed Effects Models

The M-estimators for GLM for cross-sectional data based on robust estimating equation (9.11) in Section 9.4.1 can be extended to mixed effects models for longitudinal data or clustered data. Such extensions have been considered in the literature (Richardson 1997; Sinha 2004; Noh and Lee 2007). In this section, we consider a class of M-estimators of Mallows's type for general mixed effects models, including LME, GLMM, and NLME models.

Let y_{ij} and $\mathbf{x}_{ij}$ be the response and covariates for individual (or cluster) i at measurement j respectively, $i = 1, 2, \cdots, n$, $j = 1, 2, \cdots, n_i$. We write $\mathbf{y}_i = (y_{i1}, y_{i2}, \cdots, y_{in_i})^T$ and write $\mathbf{x}_i$ similarly. Suppose that there may be outliers in both the response and the covariates. We consider a mixed effects model with conditional density function $f(\mathbf{y}_i|\mathbf{x}_i, \mathbf{b}_i, \boldsymbol{\beta}, \boldsymbol{\sigma})$ given the random effects $\mathbf{b}_i$, where $\boldsymbol{\beta}$ contains fixed mean parameters and $\boldsymbol{\sigma}$ contains variance-covariance parameters (for some GLMMs, parameters $\boldsymbol{\sigma}$ may not be distinct from parameters $\boldsymbol{\beta}$). Assume that the random effects $\mathbf{b}_i$ are i.i.d. $\sim N(0, B)$, where B is a covariance matrix. Let $\boldsymbol{\theta} = (\boldsymbol{\beta}, B, \boldsymbol{\sigma})$ be the collection of all unknown parameters.

The likelihood for the observed data $\{(\mathbf{y}_i, \mathbf{x}_i), i = 1, \ldots, n\}$ is

$$L(\boldsymbol{\theta}) = \prod_{i=1}^{n} \int f(\mathbf{y}_i|\mathbf{x}_i, \mathbf{b}_i, \boldsymbol{\beta}, \boldsymbol{\sigma}) f(\mathbf{b}_i|B) \, d\mathbf{b}_i.$$

Usual MLE of $\boldsymbol{\theta}$ may be obtained by solving equation

$$\frac{\partial \log L(\boldsymbol{\theta})}{\partial \boldsymbol{\theta}} = 0.$$

Thus, for example, the ML estimating equation for the mean parameters $\boldsymbol{\beta}$ can be written as

$$\sum_{i=1}^{n} E\left(\frac{\partial \log f(\mathbf{y}_i|\mathbf{x}_i, \mathbf{b}_i, \boldsymbol{\beta}, \boldsymbol{\sigma})}{\partial \boldsymbol{\beta}} \Big| \mathbf{y}_i, \mathbf{x}_i, \boldsymbol{\theta}\right) = 0,$$

where the conditional expectation is taken with respect to the conditional distribution $f(\mathbf{b}_i|\mathbf{y}_i, \mathbf{x}_i, \boldsymbol{\theta})$. Estimating equations for other parameters in $\boldsymbol{\theta}$ can be written similarly. For standard maximum likelihood methods, it is known that statistical inference can be very sensitive to outliers in either the responses or the covariates, as noted in Section 9.4.1. Similar to the robust approach described in Section 9.4.1, in the following we can consider a robust version of the above estimating equations.

Let

$$\begin{aligned}
\boldsymbol{\mu}_i &= \mathbf{g}_i(\boldsymbol{\beta}, \mathbf{b}_i) = E(\mathbf{y}_i|\mathbf{x}_i, \mathbf{b}_i, \boldsymbol{\beta}), \\
R_i &= Cov(\mathbf{y}_i|\mathbf{x}_i, \mathbf{b}_i, \boldsymbol{\sigma}),
\end{aligned}$$

$$\mathbf{r}_i = \mathbf{r}_i(\boldsymbol{\beta}, \mathbf{b}_i) = R_i^{-\frac{1}{2}} (\mathbf{y}_i - \boldsymbol{\mu}_i),$$

$$D_i = D_i(\boldsymbol{\beta}, \mathbf{b}_i) = \frac{\partial \mathbf{g}_i(\boldsymbol{\beta}, \mathbf{b}_i)}{\partial \boldsymbol{\beta}}.$$

For a robust estimator of the mean parameters $\boldsymbol{\beta}$, we can bound the influence of any outlying response values and down-weight leverage points in the covariates separately by the following robust estimating equation

$$\sum_{i=1}^{n} E\left[\left\{ D_i^T W_i(\mathbf{x}_i) R_i^{-\frac{1}{2}} \psi_i(\mathbf{r}_i) - a(\boldsymbol{\beta}) \right\} \Big| \mathbf{y}_i, \mathbf{x}_i, \boldsymbol{\theta} \right] = 0, \qquad (9.12)$$

where

$$a(\boldsymbol{\beta}) = \frac{1}{n} \sum_{i=1}^{n} E\left(D_i^T W_i(\mathbf{x}_i) R_i^{-1/2} \psi_i(\mathbf{r}_i) \Big| \mathbf{b}_i \right)$$

is a bias correction term to ensure the Fisher consistency of the robust estimator, the function $\psi_i : R^{n_i} \to R^{n_i}$ is a monotone function which bounds the influence of outlying observations in the response $\mathbf{y}_i$ such as the Huber's function $\psi_c(r_i) = \max(-c, \min(r_i, c))$, the function $W_i(\cdot)$ is a weight function which downweights leverage points in covariates $\mathbf{x}_i$, and the conditional expectation is taken with respect to the conditional distribution $f(\mathbf{b}_i | \mathbf{y}_i, \mathbf{z}_i)$. Estimating equations for other parameters can be written similarly.

A simple choice of the weight function is $W_i(\mathbf{x}_i) = \sqrt{1 - h_i}$ where h_i the i-th diagonal element of the usual hat matrix H in regression models. Sometimes it may be more appropriate to choose a more complex weight function, such as a function of the Mahalanobis distance (e.g., in Sinha 2004)

$$W_i(\mathbf{x}_i) = \min\left[1, \left(\frac{d_0}{(\mathbf{x}_i - \mathbf{m}_i)^T S_i^{-1}(\mathbf{x}_i - \mathbf{m}_i)} \right)^{\gamma/2} \right],$$

where $\gamma \geq 1$, d_0 may be chosen as the 95th percentile of the $\chi^2(p)$ distribution with $p = dim(\mathbf{x}_i)$, and $\mathbf{m}_i$ and S_i are respectively robust estimates of $E(\mathbf{x}_i)$ and $Cov(\mathbf{x}_i)$ such as the minimum volume ellipsoid (MVE) estimates of Rousseeuw and van Zomeren (Maronna et al. 2006).

Note that, when the weight function $W_i(\mathbf{x}_i) = 1$ and $\psi_i(\mathbf{r}_i) = \mathbf{r}_i$ (or $\rho(u) = u^2/2$), robust estimating equation (9.12) reduces to the standard likelihood equation, and the resulting estimate reduces to the standard MLE.

Robust estimating equation (9.12) for $\boldsymbol{\beta}$ can be solved using the Newton-Raphson method as follows. Taking a first-order Taylor expansion about $\boldsymbol{\beta}$, after some algebra, the k-th iteration of the Newton-Raphson method for the parameter $\boldsymbol{\beta}$ is given by

$$\boldsymbol{\beta}_n^{(k+1)} = \boldsymbol{\beta}_n^{(k)} - \left\{ \sum_{i=1}^{n} E\left[\left(D_i^{(k)T} W_i(\mathbf{x}_i) R_i^{-\frac{1}{2}} \frac{\partial \Delta_i^{(k)}}{\partial \boldsymbol{\beta}} \right) \Big| \mathbf{y}_i, \mathbf{x}_i, \boldsymbol{\theta} \right] \right\}^{-1}$$

$$\times \sum_{i=1}^{n} E\left[\left(D_i^{(k)T} W_i(\mathbf{x}_i) R_i^{-1/2} \Delta_i^{(k)}\right) \middle| \mathbf{y}_i, \mathbf{x}_i, \boldsymbol{\theta}\right], \qquad (9.13)$$

$$k = 1, 2, 3, \cdots,$$

where

$$D_i^{(k)} = D_i(\boldsymbol{\beta}_n^{(k)}, \mathbf{b}_i) = \frac{\partial \mathbf{g}_i(\boldsymbol{\beta}_n^{(k)}, \mathbf{b}_i)}{\partial \boldsymbol{\beta}},$$

$$\Delta_i^{(k)} = \Delta_i(\boldsymbol{\beta}_n^{(k)}, \mathbf{b}_i) = \psi_i\left(\mathbf{r}_i(\boldsymbol{\beta}_n^{(k)}, \mathbf{b}_i)\right) - E\left[\psi_i(\mathbf{r}_i(\boldsymbol{\beta}_n^{(k)}, \mathbf{b}_i)) | \mathbf{b}_i\right].$$

At convergence, we have the robust estimator $\hat{\boldsymbol{\beta}}_n$. We will discuss how to evaluate the conditional expectations in the above expressions in the next section. In the following, we show some asymptotic results of the robust estimators.

Asymptotics

Under the regularity conditions A1 – A7 given below, we can show that the above robust estimators are consistent and asymptotically follow a multivariate normal distribution when the sample size $n \to \infty$ (Wu, 2009). Let

$$\phi(\boldsymbol{\beta}, \mathbf{y}_i, \mathbf{x}_i) = E\left(D_i^T(\boldsymbol{\beta}, \mathbf{b}_i) R_i^{-1/2} W_i(\mathbf{x}_i) \Delta_i(\boldsymbol{\beta}, \mathbf{b}_i) \middle| \mathbf{y}_i, \mathbf{x}_i, \boldsymbol{\theta}\right),$$

and let $\Phi(\boldsymbol{\beta}, \mathbf{y}_i, \mathbf{x}_i)$ be a function such that its derivative with respect to $\boldsymbol{\beta}$ is $\phi(\boldsymbol{\beta}, \mathbf{y}_i, \mathbf{x}_i)$. Let $Q_n(\boldsymbol{\beta}) = \sum_{i=1}^{n} \Phi(\boldsymbol{\beta}, \mathbf{y}_i, \mathbf{x}_i)$ and

$$\hat{\boldsymbol{\beta}}_n = \text{argmax}_{\boldsymbol{\beta} \in \Gamma} Q_n(\boldsymbol{\beta}).$$

We denote convergence in probability and convergence in distribution by $\xrightarrow{p}$ and $\xrightarrow{d}$ respectively.

The following regularity conditions are similar to those appeared in the literature (e.g., Amemiya 1985; Domowitz and White 1982; Vonesh et al. 2002). They are mostly used to ensure the compactness of the parameter space, sufficient smoothness of appropriate functions, and parameter identifiability. In most applications, these conditions are satisfied. Note that in a real data application the parameter space may be viewed as compact. The regularity conditions are as follows:

R1. The parameter space Γ of $\boldsymbol{\beta}$ is compact. The true parameter $\boldsymbol{\beta}_0$ is in the interior of Γ.

R2. The function $\Phi(\boldsymbol{\beta}, \mathbf{y}_i, \mathbf{x}_i)$ is a continuous function of $\boldsymbol{\beta}$ and a measurable function of $(\mathbf{y}_i, \mathbf{x}_i)$, and is twice continuously differentiable with respect to $\boldsymbol{\beta}$ and is dominated by a uniformly integrable function.

R3. The appropriate functions appeared in the proofs (in the Appendix) satisfy the necessary regularity conditions needed for change of the order of integration and differentiation.

R4. The function $\phi^2(\boldsymbol{\beta}, \mathbf{y}_i, \mathbf{x}_i)$ is dominated by a uniformly integrable function.

R5. The true parameter value $\boldsymbol{\beta}_0$ is the unique minimizer of $Q_n(\boldsymbol{\beta})$.

R6. The function $\partial\phi(\boldsymbol{\beta}, \mathbf{y}_i, \mathbf{x}_i)/\partial\boldsymbol{\beta}$ is dominated by a uniformly integrable function.

R7. The following limits exist and the matrices are positive definite

$$\Omega(\boldsymbol{\beta}) = \text{limit}_{n\to\infty} \frac{1}{n} \sum_{i=1}^{n} E\left(\phi(\boldsymbol{\beta}, \mathbf{y}_i, \mathbf{x}_i)\phi(\boldsymbol{\beta}, \mathbf{y}_i, \mathbf{x}_i)^T\right),$$

$$\Lambda(\boldsymbol{\beta}) = \text{limit}_{n\to\infty} \frac{1}{n} \sum_{i=1}^{n} E\left(\frac{\partial\phi(\boldsymbol{\beta}, \mathbf{y}_i, \mathbf{x}_i)}{\partial\boldsymbol{\beta}}\right).$$

Theorem 9.1 *Under the regularity conditions R1 – R7, we have the following results:*
(a) the robust estimator $\hat{\boldsymbol{\beta}}_n$ is consistent, i.e.,

$$\hat{\boldsymbol{\beta}}_n \xrightarrow{p} \boldsymbol{\beta}_0, \quad as\ n \to \infty.$$

(b) the robust estimator $\hat{\boldsymbol{\beta}}_n$ asymptotically follows a normal distribution, i.e.,

$$\sqrt{n}(\hat{\boldsymbol{\beta}}_n - \boldsymbol{\beta}_0) \xrightarrow{d} N(0, \Lambda^{-1}(\boldsymbol{\beta}_0)\Omega(\boldsymbol{\beta}_0)\Lambda^{-1}(\boldsymbol{\beta}_0)), \quad as\ n \to \infty,$$

where $\boldsymbol{\beta}_0$ is the true parameter value.

The proof is given in the Appendix (Section 9.4.5).

Therefore, the robust estimator $\hat{\boldsymbol{\beta}}_n$ is consistent and asymptotically normally distributed. This asymptotic result can be used for approximate inference. For example, the asymptotic variance-covariance matrix of $\hat{\boldsymbol{\beta}}_n$ can be used to compute approximate standard error of the robust estimator $\hat{\boldsymbol{\beta}}_n$, i.e., we have the following estimate for the variance-covariance of $\hat{\boldsymbol{\beta}}_n$

$$\widehat{Var}(\hat{\boldsymbol{\beta}}_n) = \frac{1}{n}\Lambda^{-1}(\hat{\boldsymbol{\beta}}_n)\Omega(\hat{\boldsymbol{\beta}}_n)\Lambda^{-1}(\hat{\boldsymbol{\beta}}_n).$$

Then the approximate standard errors of the components of $\hat{\boldsymbol{\beta}}_n$ are the square roots of the diagonal elements of matrix $\widehat{Var}(\hat{\boldsymbol{\beta}}_n)$. These approximate standard errors can then be used to construct approximate confidence intervals of $\boldsymbol{\beta}$ and conduct hypothesis testing about $\boldsymbol{\beta}$, such as Wald-type tests.

9.4.3 A Monte Carlo Newton-Raphson Method

For the robust method in the previous section, we need to evaluate conditional expectations of some nonlinear functions with respect to the distribution $f(\mathbf{b}_i | \mathbf{y}_i, \mathbf{z}_i, \boldsymbol{\theta})$, as can be seen in (9.13) (page 312). Since the dimension of the random effects $\mathbf{b}_i$ may be large, evaluations of these conditional expectations can be computationally challenging. Here we consider approximating these conditional expectations using Monte Carlo methods, as in previous chapters. That is, we can approximate the conditional expectations by their empirical means based on large samples simulated from the conditional distribution $f(\mathbf{b}_i | \mathbf{y}_i, \mathbf{x}_i, \boldsymbol{\theta})$. Random samples from the conditional distribution $f(\mathbf{b}_i | \mathbf{y}_i, \mathbf{x}_i, \boldsymbol{\theta})$ can be generated using the Gibbs sampler along with rejection sampling methods by noting that

$$f(\mathbf{b}_i | \mathbf{y}_i, \mathbf{x}_i, \boldsymbol{\theta}) \quad \propto \quad f(\mathbf{y}_i | \mathbf{x}_i, \mathbf{b}_i, \boldsymbol{\theta}) f(\mathbf{b}_i | B).$$

The procedure is similar to those in previous chapters, so we omit the details here.

Suppose that $\{\mathbf{b}_i^{(1)}, \mathbf{b}_i^{(2)}, \cdots, \mathbf{b}_i^{(m_k)}\}$ is an i.i.d. sample of size m_k (large) simulated from distribution $f(\mathbf{b}_i | \mathbf{y}_i, \mathbf{x}_i, \boldsymbol{\theta})$ at the k-th iteration of the Newton-Raphson method. Then the robust estimating equation (9.12) can be approximated as

$$\sum_{i=1}^{n} \frac{1}{m_k} \sum_{j=1}^{m_k} \left[D_i^T(\boldsymbol{\beta}, \mathbf{b}_i^{(j)}) R_i^{-1/2} W_i(\mathbf{x}_i) \psi_i(\mathbf{r}_i(\boldsymbol{\beta}, \mathbf{b}_i^{(j)})) - a(\boldsymbol{\beta}) \right] = 0. \quad (9.14)$$

Thus, a robust estimate for $\boldsymbol{\beta}$ can be obtained by solving equation (9.14) using the Newton-Raphson method, without evaluations of intractable integrals. For example, the k-th iteration of the Newton-Raphson method for parameters $\boldsymbol{\beta}$ can be written as

$$\begin{aligned}
\boldsymbol{\beta}_n^{(k+1)} &= \boldsymbol{\beta}_n^{(k)} - \left\{ \sum_{i=1}^{n} \frac{1}{m_k} \sum_{j=1}^{m_k} \left[\left(D_i^{(k,j)T} W_i(\mathbf{x}_i) R_i^{-\frac{1}{2}} \frac{\partial \Delta_i^{(k,j)}}{\partial \boldsymbol{\beta}} \right) \right] \right\}^{-1} \\
&\quad \times \sum_{i=1}^{n} \frac{1}{m_k} \sum_{j=1}^{m_k} \left[\left(D_i^{(k,j)T} W_i(\mathbf{x}_i) R_i^{-1/2} \Delta_i^{(k,j)} \right) \right],
\end{aligned}$$

where

$$\begin{aligned}
D_i^{(k,j)} &= D_i(\boldsymbol{\beta}_n^{(k)}, \mathbf{b}_i^{(j)}) = \frac{\partial \mathbf{g}_i(\boldsymbol{\beta}_n^{(k)}, \mathbf{b}_i^{(j)})}{\partial \boldsymbol{\beta}}, \\
\Delta_i^{(k,j)} &= \Delta_i(\boldsymbol{\beta}_n^{(k)}, \mathbf{b}_i^{(j)}) \\
&= \psi_i \left(\mathbf{r}_i(\boldsymbol{\beta}_n^{(k)}, \mathbf{b}_i^{(j)}) \right) - \frac{1}{m_k} \sum_{j=1}^{m_k} \left[\psi_i(\mathbf{r}_i(\boldsymbol{\beta}_n^{(k)}, \mathbf{b}_i^{(j)})) \right].
\end{aligned}$$

We can increase the number of Monte Carlo samples m_k as the number of iteration k increases. The accuracy of the Monte Carlo approximation increases as m_k increases. We can show that equation (9.14) approaches to equation

(9.12) as the number of Monte Carlo samples m_k increases. A major advantage of the Monte Carlo method is to avoid intractable integrations. Such a Monte Carlo approach is especially useful in the presence of missing data or covariate measurement errors, as will be discussed later.

The following theorem shows that the approximate robust estimating equations will converge to the exact robust estimating equations as the number of Monte Carlo samples goes to infinite.

Theorem 9.2 *As $m \to \infty$, we have*

$$\sum_{i=1}^{n} \frac{1}{m} \sum_{j=1}^{m} \left[D_i^T(\boldsymbol{\beta}, \mathbf{b}_i^{(j)}) R_i^{-1/2} W_i(\mathbf{x}_i)\psi_i(\mathbf{r}_i(\boldsymbol{\beta}, \mathbf{b}_i^{(j)})) - a(\boldsymbol{\beta}) \right]$$

$$\xrightarrow{p} \sum_{i=1}^{n} E\left\{ \left[D_i^T(\boldsymbol{\beta}, \mathbf{b}_i) R_i^{-1/2} W_i(\mathbf{x}_i)\psi_i(\mathbf{r}_i(\boldsymbol{\beta}, \mathbf{b}_i)) - a(\boldsymbol{\beta}) \right] \middle| \mathbf{y}_i, \mathbf{x}_i, \boldsymbol{\theta} \right\}.$$

The proof is given in the Appendix (Section 9.4.5).

The following theorem shows that the Monte Carlo robust estimators are consistent and asymptotically normal as the sample size n goes to infinite.

Theorem 9.3 *Let*

$$\phi_m(\boldsymbol{\beta}, \mathbf{y}_i, \mathbf{x}_i) = \frac{1}{m} \sum_{j=1}^{m} \left[D_i^T(\boldsymbol{\beta}, \mathbf{b}_i^{(j)}) R_i^{-1/2} W_i \Delta_i(\boldsymbol{\beta}, \mathbf{b}_i^{(j)}) \right].$$

Let $\Phi_m(\boldsymbol{\beta}, \mathbf{y}_i, \mathbf{x}_i)$ be a function such that its derivative with respect to $\boldsymbol{\beta}$ is $\phi_m(\boldsymbol{\beta}, \mathbf{y}_i, \mathbf{x}_i)$, $Q_{n,m}(\boldsymbol{\beta}) = \sum_{i=1}^{n} \Phi_m(\boldsymbol{\beta}, \mathbf{y}_i, \mathbf{x}_i)$, and

$$\hat{\boldsymbol{\beta}}_{n,m} = argmax_{\boldsymbol{\beta} \in \Gamma} Q_{n,m}(\boldsymbol{\beta}).$$

Under regularity conditions R1 – R7, we have
(a) Consistency:

$$\hat{\boldsymbol{\beta}}_{n,m} \xrightarrow{p} \boldsymbol{\beta}_0, \qquad as \ n \to \infty.$$

(b) Normality:

$$\sqrt{n}(\hat{\boldsymbol{\beta}}_{n,m} - \boldsymbol{\beta}_0) \xrightarrow{d} N(0, \Lambda_m^{-1}(\boldsymbol{\beta}_0)\Omega_m(\boldsymbol{\beta}_0)\Lambda_m^{-1}(\boldsymbol{\beta}_0)), \quad as \ n \to \infty,$$

where

$$\Omega_m(\boldsymbol{\beta}) = limit_{n \to \infty} \frac{1}{n} \sum_{i=1}^{n} E\left(\phi_m(\boldsymbol{\beta}, \mathbf{y}_i, \mathbf{x}_i)\phi_m(\boldsymbol{\beta}, \mathbf{y}_i, \mathbf{x}_i)^T \right),$$

$$\Lambda_m(\boldsymbol{\beta}) = limit_{n \to \infty} \frac{1}{n} \sum_{i=1}^{n} E\left(\frac{\partial \phi_m(\boldsymbol{\beta}, \mathbf{y}_i, \mathbf{x}_i)}{\partial \boldsymbol{\beta}} \right).$$

Matrix $\Lambda_m(\beta)$ can be consistently estimated by

$$\hat{\Lambda}_m(\hat{\beta}_{n,m}) = n^{-1} \sum_{i=1}^{n} E(\partial\phi_m(\hat{\beta}_{n,m}, \mathbf{y}_i, \mathbf{x}_i)/\partial\beta \mid \mathbf{y}_i, \mathbf{x}_i, \hat{\theta}),$$

matrix $\Omega_m(\beta)$ can be consistently estimated by

$$\hat{\Omega}_m(\hat{\beta}_{n,m}) = n^{-1} \sum_{i=1}^{n} E(\phi_m(\hat{\beta}_{n,m}, \mathbf{y}_i, \mathbf{x}_i)\phi_m(\hat{\beta}_{n,m}, \mathbf{y}_i, \mathbf{x}_i)^T \mid \mathbf{y}_i, \mathbf{x}_i, \hat{\theta}),$$

and the conditional expectations can also be approximated by Monte Carlo methods.

The proof is given in the Appendix (Section 9.4.5).

9.4.4 A Robust Approximate Method

The Monte Carlo Newton-Raphson method in the previous section can be computationally very intensive if the dimension of the random effects $\mathbf{b}_i$ is not small. Following the ideas of the approximate methods in previous chapters, in this section we follow Wu (2009) and consider an approximate method based on a first-order Taylor approximation, which avoids sampling the random effects so it is computationally more efficient than the Monte Carlo method. Here is the idea: at each iteration, take a first-order Taylor expansion of $\mathbf{g}_i(\beta, \mathbf{b}_i) = E(\mathbf{y}_i|\mathbf{x}_i, \mathbf{b}_i, \beta)$ about the current estimates of the random effects $\mathbf{b}_i$ and the parameters β to linearize the model, and then we iteratively solve the resulting "working" LME model and estimate the parameters in the LME model using a robust method. At convergence, approximate robust estimates are obtained from the LME model at the last iteration.

Specifically, at each iteration, let the current estimate of $(\theta, \mathbf{b}_i)$ be $(\tilde{\theta}, \tilde{\mathbf{b}}_i)$, suppressing the iteration number. Taking a first-order Taylor expansion of $\mathbf{g}_i = (g_{i1}, \cdots, g_{in_i})^T$ around the current parameter estimate $\tilde{\beta}$ and current random effects estimate $\tilde{\mathbf{b}}_i$, we obtain the following "working" LME model

$$\tilde{\mathbf{y}}_i \;=\; P_i\beta + H_i\,\mathbf{b}_i + \mathbf{e}_i, \qquad i = 1, 2, \cdots, n, \tag{9.15}$$

where $P_i = (\mathbf{p}_{i1}, \ldots, \mathbf{p}_{in_i})^T$, $\mathbf{p}_{ij} = \partial g_{ij}(\beta, \mathbf{b}_i)/\partial\beta$, $H_i = (\mathbf{h}_{i1}, \ldots, \mathbf{h}_{in_i})^T$, $\mathbf{h}_{ij} = \partial g_{ij}(\beta, \mathbf{b}_i)/\partial\mathbf{b}_i$,

$$\tilde{\mathbf{y}}_i \;=\; \mathbf{y}_i - \mathbf{g}_i(\tilde{\beta}, \tilde{\mathbf{b}}_i) + P_i\tilde{\beta} + H_i\tilde{\mathbf{b}}_i,$$
$$\tilde{\mathbf{b}}_i \;=\; E(\mathbf{b}_i|\mathbf{y}_i, \mathbf{x}_i, \tilde{\theta}) = \tilde{B}H_i^T L_i^{-1}(\tilde{\mathbf{y}}_i - P_i\tilde{\beta}),$$
$$L_i \;=\; \tilde{\delta}^2 I + H_i\tilde{B}H_i^T,$$

and all the partial derivatives in the foregoing expressions are evaluated at the current estimates $(\tilde{\alpha}, \tilde{\beta}, \tilde{\mathbf{b}}_i)$. Let $\eta = (\delta^2, B)$ be the variance-covariance parameters. We can *iteratively* solve LME model (9.15). At each iteration, estimates of the parameters in model (9.15) can be obtained by solving the following estimating equations (Richardson 1997):

$$\sum_{i=1}^{n} P_i^T L_i^{-1}(\tilde{\mathbf{y}}_i - P_i\beta) = 0, \tag{9.16}$$

$$\sum_{i=1}^{n} \left[(\tilde{\mathbf{y}}_i - P_i\beta)^T A_i(\tilde{\mathbf{y}}_i - P_i\beta) - \operatorname{tr}\left(L_i^{-1}\frac{\partial L_i}{\partial \eta} \right) \right] = 0, \tag{9.17}$$

where $A_i = L_i^{-1}(\partial L_i/\partial \eta)L_i^{-1}$.

Estimates from estimating equations (9.16) and (9.17) are sensitive to outliers in either the response or the covariates. So we consider a robust version of these estimating equations. Let $A_i^* = L_i^{-1/2}(\partial L_i/\partial \eta)L_i^{-1/2}$. We consider the following robust versions of estimating equations (9.16) and (9.17):

$$\sum_{i=1}^{n} P_i^T W_i(\mathbf{x}_i)L_i^{-1/2}\psi_i\left(L_i^{-1/2}(\tilde{\mathbf{y}}_i - P_i\beta) \right) = 0, \tag{9.18}$$

$$\sum_{i=1}^{n} \left[\psi_i\left(L_i^{-1/2}(\tilde{\mathbf{y}}_i - P_i\beta) \right)^T W_i^{-1/2}(\mathbf{x}_i)A_i^* W_i^{-1/2}(\mathbf{x}_i) \right.$$

$$\left. \times\psi_i\left(L_i^{-1/2}(\tilde{\mathbf{y}}_i - P_i\beta) \right) - \operatorname{tr}\left(L_i^{-1}\frac{\partial L_i}{\partial \eta} \right) \right] = 0. \tag{9.19}$$

Estimating equations (9.18) and (9.19) can again be solved using the Newton-Raphson method. For example, the k-th iteration for parameter β can be written as

$$\beta^{(k+1)} = \beta^{(k)} - \left[\sum_{i=1}^{n} P_i^T W_i(\mathbf{x}_i)L_i^{-1/2}\frac{\partial \psi_i(\tilde{\mathbf{r}}_i(\beta^{(k)}))}{\partial \beta} \right]^{-1}$$

$$\times \left[\sum_{i=1}^{n} P_i^T W_i(\mathbf{x}_i)L_i^{-1/2}\psi_i(\tilde{\mathbf{r}}_i(\beta^{(k)})) \right], \tag{9.20}$$

where $\tilde{\mathbf{r}}_i(\beta) = L_i^{-1/2}(\tilde{\mathbf{y}}_i - P_i\beta)$. Equations for other parameters can be written similarly. Thus, we obtain robust estimates of the parameters by solving equations (9.18) and (9.19) at each iteration. Then, at convergence a final robust estimate is obtained at the last iteration. In equations (9.18) – (9.19), we do not need to evaluate intractable integrals, so the approximate robust method is computationally more efficient than the Monte Carlo method in the previous section.

Next we study the asymptotic properties of the above robust approximate estimates. Let

$$\phi_i^*(\boldsymbol{\beta}) = E(P_i^T W_i(\mathbf{x}_i) L_i^{-1/2} \psi_i(L_i^{-1/2}(\tilde{\mathbf{y}}_i - P_i\boldsymbol{\beta})),$$

$$\Omega^*(\boldsymbol{\beta}) = \lim_{n\to\infty} \frac{1}{n} \sum_{i=1}^{n} \phi_i^*(\boldsymbol{\beta})\phi_i^*(\boldsymbol{\beta})^T,$$

$$\Lambda^*(\boldsymbol{\beta}) = \lim_{n\to\infty} \frac{1}{n} \sum_{i=1}^{n} \frac{\partial \phi_i^*(\boldsymbol{\beta})}{\partial \boldsymbol{\beta}}.$$

For the regularity conditions R1 – R7 in section 9.4.2, we replace $\phi^2(\boldsymbol{\beta}, \mathbf{y}_i, \mathbf{x}_i)$ in R4 by $\phi_i^{*2}(\boldsymbol{\beta})$, replace $\partial\phi(\boldsymbol{\beta}, \mathbf{y}_i, \mathbf{x}_i)/\partial\boldsymbol{\beta}$ in R6 by $\partial\phi^*(\boldsymbol{\beta})/\partial\boldsymbol{\beta}$, and replace $\Omega(\boldsymbol{\beta})$ and $\Lambda(\boldsymbol{\beta})$ in R7 by $\Omega^*(\boldsymbol{\beta})$ and $\Lambda^*(\boldsymbol{\beta})$ respectively, and call the resulting regularity conditions R1* – R7*. Moreover, we add the following two additional regularity conditions

R8*. The numbers of within individual measurements $n_i = O(\min_i(n_i))$ uniformly for $i = 1, 2, \cdots, n$,

R9*. The density function $f(y_{ij}|\mathbf{x}_i, \mathbf{b}_i, \boldsymbol{\theta})$ satisfies the necessary regularity conditions such that, for fixed $\boldsymbol{\theta}$, the MLE of $\mathbf{b}_i$ is $\sqrt{n_i}$ – consistent for $\mathbf{b}_i$ as $n_i \to \infty$.

The following theorem shows that the approximate robust estimates are consistent and asymptotically normal when both the sample size and the number of measurements per individual go to infinite, i.e., when both $n \to \infty$ and $\min_i(n_i) \to \infty$.

Theorem 9.4 *Let $\tilde{\beta}_n$ be the approximate robust estimate of the true parameter value β_0 obtained by solving (9.20). Under regularity conditions R1* – R9*, we have*
(a) the approximate robust estimate is consistent when both the sample size and the number of within-individual measurements go to infinite, i.e.,

$$\tilde{\beta}_n \xrightarrow{p} \beta_0, \quad as\ n \to \infty, \ \min_i(n_i) \to \infty.$$

(b) the approximate robust estimate is asymptotically normal, i.e.,

$$\sqrt{n}(\tilde{\beta}_n - \beta_0) \xrightarrow{d} N(0, \Lambda^*(\beta_0)^{-1}\Omega^*(\beta_0)\Lambda^*(\beta_0)^{-1}),$$
$$as\ n \to \infty, \ \min_i(n_i) \to \infty.$$

We defer the proof to Section 9.5 where a similar result holds.

9.4.5 Appendix

A0: Some Lemmas

The following three lemmas will be needed in the proofs of the Theorems (see, e.g., Amemiya 1985; Vonesh and Chinchilli 1997). Here we let $f(\cdot)$ denote a generic function and let $X, X_i, i = 1, 2, \cdots$, be random variables.

Lemma 1. Let X_n be a random variable satisfying $X_n = c + O_p(a_n)$ where $a_n = o(1)$ and c is a constant. If $f(x)$ is a continuously differentiable function at $x = c$, then
$$f(X_n) = f(c) + O_p(a_n).$$
The result holds when $O_p(\cdot)$ is replaced by $o_p(\cdot)$ or when X_n and c are replaced by vectors.

Lemma 2. Suppose that function $f(x, \theta)$ is dominated by a uniformly integrable function. Then $\frac{1}{n} \sum_{i=1}^{n} f(X_i, \theta)$ converges in probability to $E f(X, \theta)$ uniformly in θ.

Lemma 3. Suppose that $f_n(X, \theta) \xrightarrow{p} f(\theta)$ and $f(\theta)$ attains a unique minimum at θ_0. Then
$$\hat{\theta}_n \equiv \mathrm{argmin}_\theta f_n(X, \theta) \xrightarrow{p} \theta_0, \quad \text{as } n \to \infty.$$

Lemma 4. Suppose that $f(\cdot)$ is a continuous function and $X_n \xrightarrow{p} c$. Then $f(X_n) \xrightarrow{p} f(c)$.

A1. Proof of Theorem 9.1

Let $Q(\beta) = E(\Phi(\beta, \mathbf{y}_i, \mathbf{x}_i))$. By Lemma 2 and condition R2, we have
$$\frac{1}{n} Q_n(\beta) \xrightarrow{p} Q(\beta), \quad \text{as } n \to \infty,$$
where $Q_n(\beta)$ is defined in Section 9.4. Then, by Lemma 3, we have
$$\hat{\beta}_n \xrightarrow{p} \beta_0, \quad \text{as } n \to \infty.$$
This proves part (a) of Theorem 9.1.

To prove part (b), let
$$\tilde{\phi}(\beta, \mathbf{y}, \mathbf{x}) = \sum_{i=1}^{n} \phi(\beta, \mathbf{y}_i, \mathbf{x}_i).$$

Since $\tilde{\phi}(\hat{\beta}_n, \mathbf{y}, \mathbf{x}) = 0$ and $\hat{\beta}_n \xrightarrow{p} \beta_0$, by first-order Taylor expansion of $\tilde{\phi}(\beta, \mathbf{y}, \mathbf{x})$ around β_0 and the mean-value theorem, we have
$$0 = \tilde{\phi}(\beta_0, \mathbf{y}, \mathbf{x}) + \frac{\partial \tilde{\phi}(\beta^*, \mathbf{y}, \mathbf{x})}{\partial \beta} (\hat{\beta}_n - \beta_0),$$

where $||\boldsymbol{\beta}^* - \boldsymbol{\beta}_0|| \le ||\hat{\boldsymbol{\beta}}_n - \boldsymbol{\beta}_0||$, and $\partial \tilde{\phi}(\boldsymbol{\beta}^*, \mathbf{y}, \mathbf{x})/\partial \boldsymbol{\beta} = \partial \tilde{\phi}(\boldsymbol{\beta}, \mathbf{y}, \mathbf{x})/\partial \boldsymbol{\beta}|_{\boldsymbol{\beta}=\boldsymbol{\beta}^*}$.
So

$$\sqrt{n}(\hat{\boldsymbol{\beta}}_n - \boldsymbol{\beta}_0) = \left[-\frac{1}{n} \frac{\partial \tilde{\phi}(\boldsymbol{\beta}^*, \mathbf{y}, \mathbf{x})}{\partial \boldsymbol{\beta}} \right]^{-1} \left(\frac{1}{\sqrt{n}} \tilde{\phi}(\boldsymbol{\beta}_0, \mathbf{y}, \mathbf{x}) \right).$$

By the Lindeberg's Central Limit Theorem, we have

$$\frac{1}{\sqrt{n}} \tilde{\phi}(\boldsymbol{\beta}_0, \mathbf{y}, \mathbf{x}) = \frac{1}{\sqrt{n}} \sum_{i=1}^{n} \phi(\boldsymbol{\beta}_0, \mathbf{y}_i, \mathbf{x}_i) \xrightarrow{d} N(0, \Omega(\boldsymbol{\beta}_0)), \qquad \text{as } n \to \infty.$$

By Lemma 2, result in part (a), and conditions R4 and R7, we have

$$\frac{1}{n} \frac{\partial \tilde{\phi}(\boldsymbol{\beta}^*, \mathbf{y}, \mathbf{x}))}{\partial \boldsymbol{\beta}} \quad = \quad \frac{1}{n} \sum_{i=1}^{n} \frac{\partial \phi(\boldsymbol{\beta}^*, \mathbf{y}_i, \mathbf{x}_i)}{\partial \boldsymbol{\beta}}$$

$$\xrightarrow{p} \quad \lim_{n \to \infty} \frac{1}{n} \sum_{i=1}^{n} E\left(\frac{\partial \phi(\boldsymbol{\beta}_0, \mathbf{y}_i, \mathbf{x}_i)}{\partial \boldsymbol{\beta}} \right) = \Lambda(\boldsymbol{\beta}_0),$$

$$\text{as } n \to \infty.$$

Therefore, by Slutsky's Theorem, we have

$$\sqrt{n}(\hat{\boldsymbol{\beta}}_n - \boldsymbol{\beta}_0) \xrightarrow{d} N(0, \Lambda^{-1}(\boldsymbol{\beta}_0)\Omega(\boldsymbol{\beta}_0)\Lambda^{-1}(\boldsymbol{\beta}_0)), \quad \text{as } n \to \infty.$$

This proves Theorem 9.1.

A2. Proof of Theorem 9.2

By Lemma 2, we have

$$\sum_{i=1}^{n} \frac{1}{m} \sum_{j=1}^{m} \left[D_i^T(\boldsymbol{\beta}, \mathbf{d}_i^{(j)}) R_i^{-1/2} W_i(\mathbf{x}_i) \psi_i(\mathbf{r}_i(\boldsymbol{\beta}, \mathbf{d}_i^{(j)})) - a(\boldsymbol{\beta}) \right]$$

$$\xrightarrow{p} \sum_{i=1}^{n} E\left[D_i^T(\boldsymbol{\beta}, \mathbf{d}_i) R_i^{-1/2} W_i(\mathbf{x}_i) \psi_i(\mathbf{r}_i(\boldsymbol{\beta}, \mathbf{d}_i)) - a(\boldsymbol{\beta}) \right]$$

$$= \sum_{i=1}^{n} E\left\{ E\left[D_i^T(\boldsymbol{\beta}, \mathbf{d}_i) R_i^{-1/2} W_i(\mathbf{x}_i) \psi_i(\mathbf{r}_i(\boldsymbol{\beta}, \mathbf{d})) - a(\boldsymbol{\beta}) \right] \Big| \mathbf{y}_i, \mathbf{x}_i, \boldsymbol{\theta} \right\}$$

$$= \sum_{i=1}^{n} E\left\{ \left[D_i^T(\boldsymbol{\beta}, \mathbf{d}_i) R_i^{-1/2} W_i(\mathbf{x}_i) \psi_i(\mathbf{r}_i(\boldsymbol{\beta}, \mathbf{d}_i)) - a(\boldsymbol{\beta}) \right] \Big| \mathbf{y}_i, \mathbf{x}_i, \boldsymbol{\theta} \right\}.$$

This proves Theorem 9.2.

A3. Proof of Theorem 9.3

This proof is analogous to that for Theorem 9.1. By Lemma 2 and R2, we have

$$
\begin{aligned}
\frac{1}{n} Q_{n,m}(\boldsymbol{\beta}) \quad &= \quad \frac{1}{n} \sum_{i=1}^{n} \Phi_m(\boldsymbol{\beta}, \mathbf{y}_i, \mathbf{x}_i) \\
&\xrightarrow{p} \quad E(\Phi_m(\boldsymbol{\beta}, \mathbf{y}_1, \mathbf{x}_1)) \\
&= \quad E\left(E(\Phi_m(\boldsymbol{\beta}, \mathbf{y}_1, \mathbf{x}_1)|\mathbf{y}_1, \mathbf{x}_1)\right) \\
&= \quad E\left(\Phi(\boldsymbol{\beta}, \mathbf{y}_1, \mathbf{x}_1)\right) = Q(\boldsymbol{\beta}), \qquad \text{as } n \to \infty.
\end{aligned}
$$

Thus, by Lemma 3, we have $\hat{\boldsymbol{\beta}}_{n,m} \xrightarrow{p} \boldsymbol{\beta}_0$. This completes the proof of part (a).

Next, we prove part (b). Let $\tilde{\phi}_m(\boldsymbol{\beta}, \mathbf{y}, \mathbf{x}) = \sum_{i=1}^{n} \phi_m(\boldsymbol{\beta}, \mathbf{y}_i, \mathbf{x}_i)$. Then

$$
\tilde{\phi}_m(\hat{\boldsymbol{\beta}}_{n,m}, \mathbf{y}, \mathbf{x}) = 0.
$$

By the result in (a), we have $\hat{\boldsymbol{\beta}}_{n,m} \xrightarrow{p} \boldsymbol{\beta}_0$. Taking a first-order Taylor expansion of $\tilde{\phi}_m(\boldsymbol{\beta}, \mathbf{y}, \mathbf{x})$ around $\boldsymbol{\beta}_0$ and the mean-value theorem, we have

$$
0 = \tilde{\phi}_m(\boldsymbol{\beta}_0, \mathbf{y}, \mathbf{x}) + \frac{\partial \tilde{\phi}_m(\boldsymbol{\beta}_m^*, \mathbf{y}, \mathbf{x})}{\partial \boldsymbol{\beta}}(\hat{\boldsymbol{\beta}}_{n,m} - \boldsymbol{\beta}_0),
$$

where $\|\boldsymbol{\beta}_m^* - \boldsymbol{\beta}_0\| \le \|\hat{\boldsymbol{\beta}}_{n,m} - \boldsymbol{\beta}_0\|$. So

$$
\sqrt{n}(\hat{\boldsymbol{\beta}}_{n,m} - \boldsymbol{\beta}_0) = \left[-\frac{1}{n} \frac{\partial \tilde{\phi}(\boldsymbol{\beta}_m^*, \mathbf{y}, \mathbf{x})}{\partial \boldsymbol{\beta}} \right]^{-1} \left(\frac{1}{\sqrt{n}} \tilde{\phi}_m(\boldsymbol{\beta}_0, \mathbf{y}, \mathbf{x}) \right).
$$

By the Central Limit Theorem, we have

$$
\frac{1}{\sqrt{n}} \tilde{\phi}_m(\boldsymbol{\beta}_0, \mathbf{y}, \mathbf{x}) \quad = \quad \frac{1}{\sqrt{n}} \sum_{i=1}^{n} \phi_m(\boldsymbol{\beta}_0, \mathbf{y}_i, \mathbf{x}_i) \xrightarrow{d} N(0, \Omega_m(\boldsymbol{\beta}_0))
$$

$$
\text{as } n \to \infty,
$$

where

$$
\Omega_m(\boldsymbol{\beta}_0) = \text{limit}_{n \to \infty} \frac{1}{n} \sum_{i=1}^{n} E\left(\phi_m(\boldsymbol{\beta}_0, \mathbf{y}_i, \mathbf{x}_i) \phi_m(\boldsymbol{\beta}_0, \mathbf{y}_i, \mathbf{x}_i)^T \right).
$$

By Lemma 2, we have

$$
\frac{1}{n} \frac{\partial \tilde{\phi}_m(\boldsymbol{\beta}_m^*, \mathbf{y}, \mathbf{x}))}{\partial \boldsymbol{\beta}} \xrightarrow{p} \lim_{n \to \infty} \frac{1}{n} \sum_{i=1}^{n} E\left(\frac{\partial \phi_m(\boldsymbol{\beta}_0, \mathbf{y}_i, \mathbf{x}_i)}{\partial \boldsymbol{\beta}} \right)
$$

$$
= \lim_{n \to \infty} \frac{1}{n} \sum_{i=1}^{n} \left\{ \frac{1}{m} \sum_{j=1}^{m} E\left[E(D_{ij}^T R_i^{-1/2} W_i \Delta_{ij} | \mathbf{y}_i, \mathbf{x}_i) \right] \right\}
$$

$$= \lim_{n \to \infty} \frac{1}{n} \sum_{i=1}^{n} E \left(\frac{\partial \phi(\boldsymbol{\beta}_0, \mathbf{y}_i, \mathbf{x}_i)}{\partial \boldsymbol{\beta}} \right) = \Lambda(\boldsymbol{\beta}_0),$$

$$\text{as } n \to \infty,$$

where $D_{ij} = D_i^T(\boldsymbol{\beta}, \mathbf{d}_i^{(j)})$ and $\Delta_{ij} = \Delta_i(\boldsymbol{\beta}, \mathbf{d}_i^{(j)})$. Therefore, by Slutsky's Theorem, we have

$$\sqrt{n}(\hat{\boldsymbol{\beta}}_{n,m} - \boldsymbol{\beta}_0) \xrightarrow{d} N(0, \Lambda^{-1}(\boldsymbol{\beta}_0)\Omega_m(\boldsymbol{\beta}_0)\Lambda^{-1}(\boldsymbol{\beta}_0)), \qquad \text{as } n \to \infty.$$

By result (a) and Lemma 4, we have

$$\hat{\Lambda}(\hat{\boldsymbol{\beta}}_{n,m}) \xrightarrow{p} \Lambda(\boldsymbol{\beta}_0) \quad \text{and} \quad \hat{\Omega}_m(\hat{\boldsymbol{\beta}}_{n,m}) \xrightarrow{p} \Omega_m(\boldsymbol{\beta}_0), \qquad \text{as } n \to \infty.$$

This proves Theorem 9.3.

9.5 Robust Inference for Mixed Effects Models with Incomplete Data

In many studies, in addition to possible outliers, there may also be measurement errors and missing values in the data. Ignoring any one of these problems may lead to biased analyses. Thus, it is important to addressed outliers, measurement errors, and missing data *simultaneously*. In this section, we consider robust inference for mixed effects models in the presence of outliers, measurement errors, and missing data.

9.5.1 Robust Inference with Covariate Measurement Errors

We first consider mixed effects models in the presence of outliers in the response and covariates, as well as measurement errors in time-dependent covariates. Let z_{ij} be an error-prone time-varying covariate for individual i at measurement time t_{ij}, $i = 1, 2, \cdots, n$, $j = 1, 2, \cdots, n_i$. For simplicity of presentation, we focus on one error-prone covariate z_{ij}, but the method can be easily extended to more than one error-prone covariates. Let z_{ij}^* be the true but unobserved covariate value, whose observed but possibly mis-measured value is z_{ij}. We write $\mathbf{z}_i = (z_{i1}, z_{i2}, \cdots, z_{in_i})^T$ and write $\mathbf{z}_i^*$ similarly. We assume that the response depends on z_{ij}^* rather than z_{ij}, as in previous chapters. That is, we assume a response model with density $f(\mathbf{y}_i|\mathbf{z}_i^*, \mathbf{x}_i, \mathbf{b}_i, \boldsymbol{\beta}, \boldsymbol{\sigma})$ given the random effects $\mathbf{b}_i$, where $\mathbf{x}_i$ contains covariates without measurement errors and $\mathbf{b}_i \sim N(0, B)$.

We consider the following classical measurement error model for the error-prone covariate $\mathbf{z}_i$:

$$\begin{aligned} \mathbf{z}_i &= U_i \boldsymbol{\alpha} + V_i \mathbf{a}_i + \boldsymbol{\epsilon}_i = \mathbf{z}_i^* + \boldsymbol{\epsilon}_i, \\ \mathbf{a}_i &\sim N(0, A), \qquad \boldsymbol{\epsilon}_i \sim N(0, \delta^2 I), \quad i = 1, 2, \cdots, n, \end{aligned} \qquad (9.21)$$

where U_i and V_i are design matrices, and A is a unknown covariance matrix. The true but unobserved covariate values for individual i are

$$\mathbf{z}_i^* = U_i\boldsymbol{\alpha} + V_i\mathbf{a}_i, \qquad i = 1, 2, \cdots, n.$$

The covariate model (9.21) can also be used to incorporate missing covariates when the missing data mechanism is ignorable.

Let $\boldsymbol{\gamma} = (\boldsymbol{\alpha}, \boldsymbol{\beta})$ and $\mathbf{d}_i = (\mathbf{a}_i, \mathbf{b}_i)$ be the fixed mean parameters and the random effects in the response and covariate models respectively. Let $\boldsymbol{\theta} = (\boldsymbol{\alpha}, \boldsymbol{\beta}, A, B, \delta, \sigma)$ be the collection of all unknown parameters in the response and covariate models. Let

$$\mathbf{g}_i(\boldsymbol{\gamma}, \mathbf{d}_i) \quad = \quad E(\mathbf{y}_i|\mathbf{z}_i^*, \mathbf{x}_i, \mathbf{b}_i, \boldsymbol{\theta}) = \mathbf{g}_i(\boldsymbol{\alpha}, \boldsymbol{\beta}, \mathbf{a}_i, \mathbf{b}_i),$$

where $\mathbf{g}_i = (g_{i1}, \cdots, g_{in_i})^T$. The joint likelihood for the observed data $\{(\mathbf{y}_i, \mathbf{z}_i, \mathbf{x}_i), i = 1, \ldots, n\}$ is given by

$$L(\boldsymbol{\theta}) = \prod_{i=1}^{n} \int \int f(\mathbf{y}_i|\mathbf{x}_i, \mathbf{d}_i, \boldsymbol{\gamma}, \boldsymbol{\sigma}) f(\mathbf{z}_i|\mathbf{a}_i, \boldsymbol{\alpha}, \delta) f(\mathbf{a}_i|A) f(\mathbf{b}_i|B) \ d\mathbf{a}_i \ d\mathbf{b}_i.$$

Let $\mathbf{s}_i = (\mathbf{y}_i^T, \mathbf{z}_i^T)^T$, $R_i = \mathrm{diag}(Var(\mathbf{y}_i|\mathbf{b}_i), \ Var(\mathbf{z}_i|\mathbf{a}_i))$, and let

$$\begin{aligned}
\mathbf{r}_i(\boldsymbol{\gamma}, \mathbf{d}_i) &= R_i^{-1/2}(\mathbf{s}_i - \mathbf{g}_i^*(\boldsymbol{\gamma}, \mathbf{d}_i)), & \mathbf{e}_i^* &= (\mathbf{e}_i^T, \boldsymbol{\epsilon}_i^T)^T, \\
\mathbf{g}_i^*(\boldsymbol{\gamma}, \mathbf{d}_i) &= E(\mathbf{y}_i|\mathbf{d}_i), \\
E(\mathbf{z}_i|\mathbf{a}_i) &= (\mathbf{g}_i(\boldsymbol{\gamma}, \mathbf{d}_i))^T, \ (U_i\boldsymbol{\alpha} + V_i\mathbf{a}_i)^T)^T = (\mathbf{g}_i^T, \mathbf{z}_i^{*T})^T.
\end{aligned}$$

For robust inference, we can bound the influence of any outlying responses and error-prone covariates and down-weight leverage points in the error-free covariates separately. Thus, we consider the following estimating equation for robust estimate of the mean parameters in $\boldsymbol{\gamma}$:

$$\sum_{i=1}^{n} E\left\{\left[D_i^T R_i^{-\frac{1}{2}} W_i(\mathbf{x}_i)\psi_i(\mathbf{r}_i(\boldsymbol{\gamma}, \mathbf{d}_i)) - a(\boldsymbol{\gamma})\right]\Big|\mathbf{y}_i, \mathbf{z}_i, \mathbf{x}_i, \boldsymbol{\theta}\right\} = 0, \quad (9.22)$$

where

$$\begin{aligned}
D_i &= D_i(\boldsymbol{\gamma}, \mathbf{d}_i) = \partial \mathbf{g}_i^*(\boldsymbol{\gamma}, \mathbf{d}_i)/\partial\boldsymbol{\gamma}, \\
a(\boldsymbol{\gamma}) &= \frac{1}{n}\sum_{i=1}^{n} E\left(D_i^T(\boldsymbol{\gamma}, \mathbf{d}_i)R_i^{-1/2}W_i(\mathbf{x}_i)\psi_i(\mathbf{r}_i(\boldsymbol{\gamma}, \mathbf{d}_i))\Big|\mathbf{d}_i, \mathbf{y}_i, \mathbf{z}_i, \mathbf{x}_i\right),
\end{aligned}$$

$\psi_i : R^{n_i} \to R^{n_i}$ is a monotone function which bounds the influence of outlying data in the response $\mathbf{y}_i$ and in the error-prone covariate $\mathbf{z}_i$ such as the Huber's function, $W_i(\cdot)$ is a weight function which downweights leverage points in the error-free covariates $\mathbf{x}_i$. When $W_i(\mathbf{x}_i) = 1$ and $\psi_i(\mathbf{r}_i) = \mathbf{r}_i$, the resulting estimate from equation (9.22) reduces to the standard MLE.

Robust estimating equation (9.22) for estimating the mean parameters $\boldsymbol{\gamma}$ can be solved using the Newton-Raphson method as follows. Taking a first-order Taylor expansion about γ, after some algebra, the k-th iteration of the Newton-Raphson method for the parameters γ is given by

$$\gamma_n^{(k+1)} = \gamma_n^{(k)} - \left\{ \sum_{i=1}^{n} E\left[\left(D_{ik}^T R_i^{-1/2} W_i(\mathbf{x}_i) \frac{\partial \Delta_{ik}}{\partial \gamma}\right) \Big| \mathbf{y}_i, \mathbf{z}_i, \mathbf{x}_i, \boldsymbol{\theta}\right]\right\}^{-1}$$

$$\times \sum_{i=1}^{n} E\left[\left(D_{ik}^T R_i^{-1/2} W_i(\mathbf{x}_i) \Delta_{ik}\right) \Big| \mathbf{y}_i, \mathbf{z}_i, \mathbf{x}_i, \boldsymbol{\theta}\right],$$

where

$$D_{ik} = D_i(\gamma_n^{(k)}, \mathbf{d}_i),$$
$$\Delta_{ik} = \Delta_i(\gamma_n^{(k)}, \mathbf{d}_i) = \psi_i(\mathbf{r}_i(\gamma_n^{(k)}, \mathbf{d}_i)) - E(\psi_i(\mathbf{r}_i(\gamma_n^{(k)}, \mathbf{d}_i))|\mathbf{d}_i).$$

After the Newton-Raphson iterations converge, we obtain a robust estimate $\hat{\gamma}_n$ for parameters γ.

Asymptotic results similar to that in Section 9.4 can also be obtained. Specifically, under some regularity conditions similar to R1 – R7, we can prove the consistency of the robust estimates:

$$\hat{\gamma}_n \xrightarrow{p} \gamma_0, \qquad \text{as } n \to \infty,$$

and the asymptotic normality of the robust estimates:

$$\sqrt{n}(\hat{\gamma}_n - \gamma_0) \xrightarrow{d} N(0, \Lambda^{-1}(\gamma_0)\Omega(\gamma_0)\Lambda^{-1}(\gamma_0)), \qquad \text{as } n \to \infty,$$

where

$$\Omega(\gamma) = \lim_{n\to\infty} \frac{1}{n}\sum_{i=1}^{n} E\left(\phi(\gamma, \mathbf{s}_i, \mathbf{x}_i)\phi(\gamma, \mathbf{s}_i, \mathbf{x}_i)^T\right),$$

$$\Lambda(\gamma) = \lim_{n\to\infty} \frac{1}{n}\sum_{i=1}^{n} E\left(\frac{\partial \phi(\gamma, \mathbf{s}_i, \mathbf{x}_i)}{\partial \gamma}\right).$$

The proof is similar to that for Theorem 9.1.

To evaluate the conditional expectations in the above expressions, we can again use a Monte Carlo approach. That is, we generate random samples from the conditional distribution $f(\mathbf{d}_i|\mathbf{y}_i, \mathbf{z}_i, \mathbf{x}_i, \boldsymbol{\theta}) \equiv f(\mathbf{a}_i, \mathbf{b}_i|\mathbf{y}_i, \mathbf{z}_i, \mathbf{x}_i, \boldsymbol{\theta})$ using the Gibbs sampler along with rejection sampling methods by iteratively sampling from the full conditionals $f(\mathbf{a}_i|\mathbf{b}_i, \mathbf{y}_i, \mathbf{z}_i, \mathbf{x}_i, \boldsymbol{\theta})$ and $f(\mathbf{b}_i|\mathbf{a}_i, \mathbf{y}_i, \mathbf{z}_i, \mathbf{x}_i, \boldsymbol{\theta})$ respectively until convergence. Suppose that $\{\mathbf{d}_i^{(1)}, \mathbf{d}_i^{(2)}, \cdots, \mathbf{d}_i^{(m)}\}$ is a large independent sample generated from $f(\mathbf{d}_i|\mathbf{y}_i, \mathbf{z}_i, \mathbf{x}_i, \boldsymbol{\theta})$. Then, estimating equation (9.22) can be approximated as follows:

$$\sum_{i=1}^{n} \frac{1}{m}\sum_{j=1}^{m} \left[D_i^T(\gamma, \mathbf{d}_i^{(j)})R_i^{-1/2}W_i(\mathbf{x}_i)\psi_i(\mathbf{r}_i(\gamma, \mathbf{d}_i^{(j)})) - a(\gamma)\right] = 0.$$

Similar to Theorem 9.2, we can show that the above Monte Carlo approximation can be made arbitrary accurate by increasing m. That is, we have

$$\sum_{i=1}^{n} \frac{1}{m} \sum_{j=1}^{m} \left[D_i^T(\boldsymbol{\gamma}, \mathbf{d}_i^{(j)}) R_i^{-1/2} W_i(\mathbf{x}_i) \psi_i(\mathbf{r}_i(\boldsymbol{\gamma}, \mathbf{d}_i^{(j)})) - a(\boldsymbol{\gamma}) \right]$$

$$\xrightarrow{p} \sum_{i=1}^{n} E \left\{ \left[D_i^T(\boldsymbol{\gamma}, \mathbf{d}_i) R_i^{-1/2} W_i(\mathbf{x}_i) \psi_i(\mathbf{r}_i(\boldsymbol{\gamma}, \mathbf{d}_i)) - a(\boldsymbol{\gamma}) \right] \bigg| \mathbf{u}_i, \boldsymbol{\theta} \right\},$$

as $m \rightarrow \infty$, where $\mathbf{u}_i = (\mathbf{y}_i, \mathbf{z}_i, \mathbf{x}_i)$.

Let

$$\phi_m(\boldsymbol{\gamma}, \mathbf{y}_i, \mathbf{z}_i, \mathbf{x}_i) = \frac{1}{m} \sum_{j=1}^{m} \left[D_i^T(\boldsymbol{\gamma}, \mathbf{d}_i^{(j)}) R_i^{-1/2} W_i \Delta_i(\boldsymbol{\gamma}, \mathbf{d}_i^{(j)}) \right],$$

and let $\Phi_m(\boldsymbol{\gamma}, \mathbf{y}_i, \mathbf{z}_i, \mathbf{x}_i)$ be a function such that its derivative with respect to
$\boldsymbol{\gamma}$ is $\phi_m(\boldsymbol{\gamma}, \mathbf{y}_i, \mathbf{z}_i, \mathbf{x}_i)$, $Q_{n,m}(\boldsymbol{\gamma}) = \sum_{i=1}^{n} \Phi_m(\boldsymbol{\gamma}, \mathbf{y}_i, \mathbf{z}_i, \mathbf{x}_i)$, and

$$\hat{\boldsymbol{\gamma}}_{n,m} = \text{argmax}_{\boldsymbol{\gamma} \in \Gamma} Q_{n,m}(\boldsymbol{\gamma}).$$

Under regularity conditions similar to R1 – R7 in section 9.4.2, we have
(a) consistency:

$$\hat{\boldsymbol{\gamma}}_{n,m} \xrightarrow{p} \boldsymbol{\gamma}_0, \qquad \text{as } n \rightarrow \infty,$$

(b) asymptotic normality:

$$\sqrt{n}(\hat{\boldsymbol{\gamma}}_{n,m} - \boldsymbol{\gamma}_0) \xrightarrow{d} N(0, \Lambda_m^{-1}(\boldsymbol{\gamma}_0) \Omega_m(\boldsymbol{\gamma}_0) \Lambda_m^{-1}(\boldsymbol{\gamma}_0)), \qquad \text{as } n \rightarrow \infty,$$

where

$$\Omega_m(\boldsymbol{\gamma}) = \text{limit}_{n \rightarrow \infty} \frac{1}{n} \sum_{i=1}^{n} E \left(\phi_m(\boldsymbol{\gamma}, \mathbf{y}_i, \mathbf{z}_i, \mathbf{x}_i) \phi_m(\boldsymbol{\gamma}, \mathbf{y}_i, \mathbf{z}_i, \mathbf{x}_i)^T \right),$$

$$\Lambda_m(\boldsymbol{\gamma}) = \text{limit}_{n \rightarrow \infty} \frac{1}{n} \sum_{i=1}^{n} E \left(\frac{\partial \phi_m(\boldsymbol{\gamma}, \mathbf{y}_i, \mathbf{z}_i, \mathbf{x}_i)}{\partial \boldsymbol{\gamma}} \right).$$

Matrix $\Lambda_m(\boldsymbol{\gamma})$ can be consistently estimated by

$$\hat{\Lambda}_m(\hat{\boldsymbol{\gamma}}_{n,m}) = \frac{1}{n} \sum_{i=1}^{n} E \left(\frac{\partial \phi_m(\hat{\boldsymbol{\gamma}}_{n,m}, \mathbf{y}_i, \mathbf{z}_i, \mathbf{x}_i)}{\partial \boldsymbol{\gamma}} \bigg| \mathbf{y}_i, \mathbf{z}_i, \mathbf{x}_i, \hat{\boldsymbol{\theta}} \right),$$

matrix $\Omega_m(\boldsymbol{\gamma})$ can be consistently estimated by

$$\hat{\Omega}_m(\hat{\boldsymbol{\gamma}}_{n,m}) = \frac{1}{n} \sum_{i=1}^{n} E(\phi_m(\hat{\boldsymbol{\gamma}}_{n,m}, \mathbf{y}_i, \mathbf{z}_i, \mathbf{x}_i) \phi_m(\,\cdot\,)^T | \mathbf{y}_i, \mathbf{z}_i, \mathbf{x}_i, \hat{\boldsymbol{\theta}}),$$

and the conditional expectations can again be approximated by Monte Carlo
methods.

9.5.2 A Robust Approximate Method

The Monte Carlo Newton-Raphson method in the previous section can be computationally very intensive if the dimension of the random effects $\mathbf{d}_i \equiv (\mathbf{a}_i, \mathbf{b}_i)$ is not small. In this section, we follow Wu (2009) and briefly describe an approximate method similar to that in Section 9.4.4. The method is computationally more efficient than the Monte Carlo Newton-Raphson method.

We iterate the following procedure. At each iteration, let the current estimate of $(\boldsymbol{\theta}, \mathbf{a}_i, \mathbf{b}_i)$ be $(\tilde{\boldsymbol{\theta}}, \tilde{\mathbf{a}}_i, \tilde{\mathbf{b}}_i)$, where $(\tilde{\mathbf{a}}_i, \tilde{\mathbf{b}}_i)$ are empirical Bayes estimates of the random effects. We suppress the iteration number to simplify the notation. Taking a first-order Taylor expansion of $\mathbf{g}_i(\boldsymbol{\gamma}, \mathbf{d}_i) = \mathbf{g}_i(\boldsymbol{\alpha}, \boldsymbol{\beta}, \mathbf{a}_i, \mathbf{b}_i)$ around the current estimate of the mean parameters $\boldsymbol{\gamma}$, $\tilde{\boldsymbol{\gamma}} = (\tilde{\boldsymbol{\alpha}}, \tilde{\boldsymbol{\beta}})$, and the current random effects estimate $\tilde{\mathbf{d}}_i = (\tilde{\mathbf{a}}_i, \tilde{\mathbf{b}}_i)$, we obtain the following "working" LME model

$$\tilde{\mathbf{y}}_i = P_{1i}\,\boldsymbol{\alpha} + P_{2i}\,\boldsymbol{\beta} + H_{1i}\,\mathbf{a}_i + H_{2i}\,\mathbf{b}_i + \mathbf{e}_i, \tag{9.23}$$

where $P_{1i} = (\mathbf{p}_{1i1}, \ldots, \mathbf{p}_{1in_i})^T$, $P_{2i} = (\mathbf{p}_{2i1}, \ldots, \mathbf{p}_{2in_i})^T$, $H_{1i} = (\mathbf{h}_{1i1}, \ldots, \mathbf{h}_{1in_i})^T$, $H_{2i} = (\mathbf{h}_{2i1}, \ldots, \mathbf{h}_{2in_i})^T$,

$$
\begin{aligned}
\mathbf{p}_{1ij} &= \partial g_{ij}(\boldsymbol{\gamma}, \mathbf{d}_i)/\partial\boldsymbol{\alpha}, & \mathbf{p}_{2ij} &= \partial g_{ij}(\boldsymbol{\gamma}, \mathbf{d}_i)/\partial\boldsymbol{\beta}, \\
\mathbf{h}_{1ij} &= \partial g_{ij}(\boldsymbol{\gamma}, \mathbf{d}_i)/\partial\mathbf{a}_i, & \mathbf{h}_{2ij} &= \partial g_{ij}(\boldsymbol{\gamma}, \mathbf{d}_i)/\partial\mathbf{b}_i, \\
\tilde{\mathbf{y}}_i &= \mathbf{y}_i - \mathbf{g}_i(\tilde{\boldsymbol{\alpha}}, \tilde{\boldsymbol{\beta}}, \tilde{\mathbf{a}}_i, \tilde{\mathbf{b}}_i) + P_{1i}\tilde{\boldsymbol{\alpha}} + P_{2i}\tilde{\boldsymbol{\beta}} + H_{1i}\tilde{\mathbf{a}}_i + H_{2i}\tilde{\mathbf{b}}_i, \\
\tilde{\mathbf{a}}_i &= E(\mathbf{a}_i|\mathbf{z}_i, \mathbf{y}_i, \tilde{\boldsymbol{\theta}}) = \tilde{A}V_i^T L_{2i}^{-1}(\mathbf{z}_i - U_i\tilde{\boldsymbol{\alpha}}), \\
\tilde{\mathbf{b}}_i &= E(\mathbf{b}_i|\mathbf{z}_i, \mathbf{y}_i, \tilde{\boldsymbol{\theta}}) = \tilde{B}H_{2i}^T L_{1i}^{-1}(\tilde{\mathbf{y}}_i - P_{1i}\tilde{\boldsymbol{\alpha}} - P_{2i}\tilde{\boldsymbol{\beta}} - H_{1i}\tilde{\mathbf{a}}_i), \\
L_{1i} &= \tilde{\delta}^2 I + (H_{1i}, H_{2i})\tilde{C}(H_{1i}, H_{2i})^T, \\
L_{2i} &= \tilde{\sigma}^2 I + V_i\tilde{A}V_i^T,
\end{aligned}
$$

C is defined next, and all the partial derivatives in the foregoing expressions are evaluated at $(\tilde{\boldsymbol{\alpha}}, \tilde{\boldsymbol{\beta}}, \tilde{\mathbf{a}}_i, \tilde{\mathbf{b}}_i)$.

Let $\tilde{\mathbf{s}}_i = (\tilde{\mathbf{y}}_i, \mathbf{z}_i)$, and let

$$Q_i = \begin{pmatrix} P_{1i} & P_{2i} \\ U_i & 0 \end{pmatrix}, \qquad T_i = \begin{pmatrix} H_{1i} & H_{2i} \\ V_i & 0 \end{pmatrix}, \qquad C = \begin{pmatrix} B & 0 \\ 0 & A \end{pmatrix}.$$

We obtain the following joint "working" LME model

$$
\begin{aligned}
\tilde{\mathbf{s}}_i &= Q_i\boldsymbol{\gamma} + T_i\mathbf{d}_i + \tilde{\mathbf{e}}_i, \tag{9.24} \\
\mathbf{d}_i &\sim N(0, C), \qquad \tilde{\mathbf{e}}_i \sim N(0, R_i).
\end{aligned}
$$

Let $L_i = R_i + T_i C T_i^T$ and let $\boldsymbol{\eta} = (\delta^2, \sigma^2, C)$ be the variance-covariance parameters. We can iteratively solve the LME model (9.24). At each iteration, following Richardson (1997), estimates of the parameters in model (9.24) can be obtained by solving the following estimating equations:

$$\sum_{i=1}^{n} Q_i^T L_i^{-1}(\tilde{s}_i - Q_i\gamma) = 0, \tag{9.25}$$

$$\sum_{i=1}^{n} \left[(\tilde{s}_i - Q_i\gamma)^T A_i(\tilde{s}_i - Q_i\gamma) - \operatorname{tr}\left(L_i^{-1}\frac{\partial L_i}{\partial \eta} \right) \right] = 0, \tag{9.26}$$

where $A_i = L_i^{-1}(\partial L_i/\partial \eta)L_i^{-1}$.

Let $A_i^* = L_i^{-1/2}(\partial L_i/\partial \eta)L_i^{-1/2}$. We consider the following robust versions of estimating equations (9.25) and (9.26):

$$\sum_{i=1}^{n} Q_i^T W_i(\mathbf{x}_i)L_i^{-1/2}\psi_i\left(L_i^{-1/2}(\tilde{s}_i - Q_i\gamma) \right) = 0, \tag{9.27}$$

$$\sum_{i=1}^{n} \left[\psi_i\left(L_i^{-1/2}(\tilde{s}_i - Q_i\gamma) \right)^T W_i^{-1/2} A_i^* W_i^{-1/2}\psi_i\left(L_i^{-1/2}(\tilde{s}_i - Q_i\gamma) \right) \right.$$
$$\left. -\operatorname{tr}\left(L_i^{-1}\frac{\partial L_i}{\partial \eta} \right) \right] = 0. \tag{9.28}$$

Estimating equations (9.27) and (9.28) can again be solved using the Newton-Raphson method. For example, the k-th iteration for parameter γ can be written as

$$\gamma^{(k+1)} = \gamma^{(k)} - \left[\sum_{i=1}^{n} Q_i^T W_i(\mathbf{x}_i)L_i^{-1/2}\frac{\partial \psi_i(\tilde{r}_i(\gamma^{(k)}))}{\partial \gamma} \right]^{-1}$$
$$\times \left[\sum_{i=1}^{n} Q_i^T W_i(\mathbf{x}_i)L_i^{-1/2}\psi_i(\tilde{r}_i(\gamma^{(k)})) \right],$$

where $\tilde{r}_i(\gamma) = L_i^{-1/2}(\tilde{s}_i - Q_i\gamma)$. Equations for other parameters can be written similarly.

Next, consider the asymptotic properties of the robust approximate estimates from the robust estimating equations (9.27) and (9.28). Let

$$\phi_i^*(\gamma) = E\left(Q_i^T W_i(\mathbf{x}_i)L_i^{-1/2}\psi_i(L_i^{-1/2}(\tilde{s}_i - Q_i\gamma)) \right),$$

$$\Omega^*(\gamma) = \operatorname{limit}_{n\to\infty} \frac{1}{n}\sum_{i=1}^{n} \phi_i^*(\gamma)\phi_i^*(\gamma)^T, \quad \text{and}$$

$$\Lambda^*(\gamma) = \operatorname{limit}_{n\to\infty} \frac{1}{n}\sum_{i=1}^{n} \frac{\partial \phi_i^*(\gamma)}{\partial \gamma}.$$

Under regularity conditions similar to R1* – R9*, the following theorem shows that the approximate robust estimates are consistent and asymptotically normal when both the sample size and the number of measurements per individual go to infinite, i.e., when both $n \to \infty$ and $\min_i(n_i) \to \infty$.

Theorem 9.5 *Let $\tilde{\gamma}_n$ be the approximate robust estimate of the true parameter value γ_0. Under suitable regularity conditions, we have*
(a) consistency:

$$\tilde{\gamma}_n \xrightarrow{p} \gamma_0, \qquad as \ n \to \infty, \ \min_i(n_i) \to \infty,$$

(b) asymptotic normality:

$$\sqrt{n}(\tilde{\gamma}_n - \gamma_0) \xrightarrow{d} N(0, \Lambda^*(\gamma_0)^{-1}\Omega^*(\gamma_0)\Lambda^*(\gamma_0)^{-1}),$$
$$as \ n \to \infty, \ \min_i(n_i) \to \infty.$$

The proof is given in the Appendix (Section 9.5.4).

9.5.3 Robust Inference with Non-Ignorable Missing Data

For non-ignorable missing covariates, we need to incorporate a missing data model to avoid biased inference. In the following, we focus on a non-ignorable missing data model which links the missing probability to the unobserved random effects. Let δ_{ij} be a missing covariate indicator such that δ_{ij} is 0 if z_{ij} is missing and δ_{ij} is 1 if z_{ij} is observed, and let $\boldsymbol{\delta}_i = (\delta_{i1}, \ldots, \delta_{in_i})$. Let $\mathbf{z}_{mis,i}$ be a collection of missing components in $\mathbf{z}_i$ and $\mathbf{z}_{obs,i}$ be a collection of observed components in $\mathbf{z}_i$. Let $f(\boldsymbol{\delta}_i|\mathbf{a}_i, \mathbf{b}_i, \boldsymbol{\xi})$ be a non-ignorable missing data model which links the missingness of the covariates to the random effects, where $\boldsymbol{\xi}$ are unknown parameters.

We can assume the following missing data model:

$$f(\boldsymbol{\delta}_i|\mathbf{a}_i, \mathbf{b}_i, \boldsymbol{\xi}) = f(\delta_{i1}|\mathbf{a}_i, \mathbf{b}_i, \boldsymbol{\xi}_1) \prod_{k=2}^{n_i} f(\delta_{ik}|\delta_{i,k-1}, \mathbf{a}_i, \mathbf{b}_i, \boldsymbol{\xi}_k).$$

Since each δ_{ik} is a binary variable, a series of logistic regression models may be assumed for $f(\delta_{ik}|\delta_{i,k-1}, \mathbf{a}_i, \mathbf{b}_i, \boldsymbol{\xi}_k)$, $k = 1, \ldots, n_i$. The observed data are $\{(\mathbf{y}_i, \mathbf{z}_{obs,i}, \mathbf{x}_i, \boldsymbol{\delta}_i), \ i = 1, \ldots, n\}$. An estimating equation for robust inference of γ is given by

$$\sum_{i=1}^{n} E\left\{\left[D_i^T R_i^{-\frac{1}{2}} W_i(\mathbf{x}_i)\psi_i(\mathbf{r}_i) - a(\gamma)\right] \Big| \mathbf{y}_i, \mathbf{z}_{obs,i}, \boldsymbol{\delta}_i, \boldsymbol{\theta}\right\} = 0, \quad (9.29)$$

where $D_i = D_i(\gamma, \mathbf{d}_i)$ and $\mathbf{r}_i = \mathbf{r}_i(\gamma, \mathbf{d}_i)$, and the expectation is taken with respect to the conditional distribution $f(\mathbf{z}_{mis,i}, \mathbf{a}_i, \mathbf{b}_i|\mathbf{y}_i, \mathbf{z}_{obs,i}, \boldsymbol{\delta}_i, \boldsymbol{\theta})$.

A Monte Carlo Newton Raphson method for solving equation (9.29) then involves sampling the unobserved values $(\mathbf{z}_{mis,i}, \mathbf{a}_i, \mathbf{b}_i)$ from the predictive distribution $f(\mathbf{z}_{mis,i}, \mathbf{a}_i, \mathbf{b}_i | \mathbf{y}_i, \mathbf{z}_{obs,i}, \boldsymbol{\delta}_i, \boldsymbol{\theta})$, which can again be accomplished using the Gibbs sampler along with rejection sampling methods, as in previous chapters. We can also consider computationally more efficient approximate methods, similar to that in previous sections.

There may also be missing data in the error-free covariate $\mathbf{x}_i$ and nonignorable missing data in the response $\mathbf{y}_i$. For simplicity of presentation, assume that missing data in $\mathbf{x}_i$ are ignorable. Define $\mathbf{x}_{mis,i}, \mathbf{x}_{obs,i}, \mathbf{y}_{mis,i}, \mathbf{y}_{obs,i}$ in a similar way as for $\mathbf{z}_{mis,i}$ and $\mathbf{z}_{obs,i}$. Let $\mathbf{w}_i = (\mathbf{y}_i, \mathbf{x}_i, \mathbf{z}_i)$ and $\mathbf{w}_{obs,i} = (\mathbf{y}_{obs,i}, \mathbf{z}_{obs,i}, \mathbf{x}_{obs,i})$. Let the missing response indicator be $\boldsymbol{\omega}_i = (\omega_{i1}, \cdots, \omega_{in_i})$ such that ω_{ij} is 0 if y_{ij} is missing and ω_{ij} is 1 if y_{ij} is observed. Then, an estimating equation for robust inference of $\boldsymbol{\gamma}$ is given by

$$\sum_{i=1}^{n} E\Big\{ \Big[D_i^T(\boldsymbol{\gamma}, \mathbf{d}_i) R_i^{-1/2} W_i(\mathbf{x}_{obs,i}) \psi_i(\mathbf{r}_i(\boldsymbol{\gamma}, \mathbf{d}_i))$$

$$-a(\boldsymbol{\gamma}) \Big] \Big| \mathbf{w}_{obs,i}, \boldsymbol{\delta}_i, \boldsymbol{\omega}_i, \boldsymbol{\theta} \Big\} = 0.$$

A Monte Carlo Newton Raphson method for solving the above equation involves sampling from $f(\mathbf{y}_{mis,i}, \mathbf{z}_{mis,i}, \mathbf{x}_{mis,i}, \mathbf{a}_i, \mathbf{b}_i | \mathbf{w}_{obs,i}, \boldsymbol{\delta}_i, \boldsymbol{\omega}_i, \boldsymbol{\theta})$. This can again be done using the Gibbs sampler combined with rejection sampling methods in principle, but the computation can be extremely intensive and sometimes may even be practically infeasible. In this case, a computationally much more efficient approximate method, similar to the ones in previous sections, can offer huge computational advantages and may be the only realistic approach in many cases. Therefore, approximate methods are particularly valuable for such missing data problems.

Similar asymptotic results of the robust estimates from the above estimating equations can be obtained. Specifically, let $\hat{\boldsymbol{\gamma}}_n^{**}$ be the solution of the above estimating equations. Under suitable regularity conditions, we have

$$\sqrt{n}(\hat{\boldsymbol{\gamma}}_n^{**} - \boldsymbol{\gamma}_0) \xrightarrow{d} N(0, \Lambda^{**}(\boldsymbol{\gamma}_0)^{-1} \Omega^{**}(\boldsymbol{\gamma}_0) \Lambda^{**}(\boldsymbol{\gamma}_0)^{-1}), \qquad \text{as } n \to \infty.$$

The matrices Λ^{**} and Ω^{**} can be consistently estimated by

$$\hat{\Omega}^{**}(\boldsymbol{\gamma}) = n^{-1} \sum_{i=1}^{n} E\Big(\phi^{**}(\boldsymbol{\gamma}, \mathbf{w}_i) \phi^{**}(\boldsymbol{\gamma}, \mathbf{w}_i)^T \Big| \mathbf{w}_{obs,i}, \boldsymbol{\delta}_i, \boldsymbol{\omega}_i, \boldsymbol{\theta} \Big),$$

$$\hat{\Lambda}^{**}(\boldsymbol{\gamma}) = n^{-1} \sum_{i=1}^{n} E\Big(\frac{\partial \phi^{**}(\boldsymbol{\gamma}, \mathbf{w}_i)}{\partial \boldsymbol{\gamma}} \Big| \mathbf{w}_{obs,i}, \boldsymbol{\delta}_i, \boldsymbol{\omega}_i, \boldsymbol{\theta} \Big),$$

where

$$\phi^{**}(\boldsymbol{\gamma}, \mathbf{w}_i) = E\Big(D_i^T(\boldsymbol{\gamma}, \mathbf{d}_i) R_i^{-1/2} W_i \Delta_i(\boldsymbol{\gamma}, \mathbf{d}_i) \Big| \mathbf{w}_{obs,i}, \boldsymbol{\delta}_i, \boldsymbol{\omega}_i, \boldsymbol{\theta} \Big),$$

and the conditional expectations can be approximated by their Monte Carlo estimates and the unknown parameters can be replaced by their consistent estimates. The proof is similar to Theorem 9.1, with some straightforward modifications.

9.5.4 Appendix

The following proof follows Wu (2009). Let

$$\phi_n(\gamma) = \sum_{i=1}^{n} \phi_i(\gamma) = \sum_{i=1}^{n} Q_i^T W_i L_i^{-1/2} \psi_i \left(L_i^{-1/2} (\tilde{s}_i - Q_i \gamma) \right).$$

Let $\tilde{\gamma}_n$ be the approximate robust estimate of γ, satisfying $\phi_n(\tilde{\gamma}_n) = 0$. Taking Taylor expansion of $\phi_n(\gamma)$ about the true value γ_0, and by the mean value theorem, we have

$$0 = \phi_n(\tilde{\gamma}_n) = \phi_n(\gamma_0) + \frac{\partial \phi_n(\gamma^*)}{\partial \gamma} (\tilde{\gamma}_n - \gamma_0),$$

where $||\gamma^* - \gamma_0|| \leq ||\tilde{\gamma}_n - \gamma_0||$. Thus we have

$$\sqrt{n}(\tilde{\gamma}_n - \gamma_0) = \left[-\frac{1}{n} \frac{\partial \phi_n(\gamma^*)}{\partial \gamma} \right]^{-1} \left(\frac{1}{\sqrt{n}} \phi_n(\gamma_0) \right).$$

Let

$$\phi_i^*(\gamma) = E \left(Q_i^T W_i L_i^{-1/2} \psi_i (L_i^{-1/2} (\tilde{s}_i - Q_i \gamma)) \right),$$

$$\Omega^*(\gamma) = \lim_{n\to\infty} \frac{1}{n} \sum_{i=1}^{n} \phi_i^*(\gamma) \phi_i^*(\gamma)^T,$$

$$\Lambda^*(\gamma) = \lim_{n\to\infty} \frac{1}{n} \sum_{i=1}^{n} \frac{\partial \phi_i^*(\gamma)}{\partial \gamma}.$$

Following Richardson (1997) and by Lindeberg's Central Limit Theorem, we have

$$\frac{1}{\sqrt{n}} \phi_n(\gamma_0) \xrightarrow{d} N(0, \Omega^*(\gamma_0)), \qquad \text{as } n \to \infty.$$

Let $o_p(1_{n_i,n})$ denote convergence in probability when both n_i and n go to ∞. Let

$$\tilde{d}_i(\tilde{\gamma}_n) = CT_i^T (T_i CT_i^T + R_i)^{-1} (\tilde{s}_i - Q_i \tilde{r}_n)$$

be an estimate of the random effects d_i. Following Vonesh et al. (2002), it can be shown that

$$\tilde{d}_i(\gamma) = d_i + O_p(n_i^{-1/2}), \quad \text{and} \quad \tilde{\gamma}_n = \gamma_0 + o_p(1_{\min_i(n_i),n}). \quad (9.30)$$

Therefore, $\tilde{\gamma}_n \xrightarrow{P} \gamma_0$, as $n \to \infty$, and $\min_i(n_i) \to \infty$. This proves part (a).

By Lemma 1 and condition R8, we have

$$
\begin{aligned}
\tilde{\mathbf{d}}_i(\tilde{\gamma}_n) &= \tilde{\mathbf{d}}_i(\gamma_0) + o_p(1_{\min_i(n_i),n}) \\
&= \mathbf{d}_i + O_p(\min_i(n_i)^{-1/2}) + o_p(1_{\min_i(n_i),n}) \\
&= \mathbf{d}_i + o_p(1_{\min_i(n_i),n}).
\end{aligned}
$$

Therefore, it follows by the Law of Large Numbers

$$
\frac{1}{n}\frac{\partial \phi_n(\gamma^*)}{\partial \gamma} \xrightarrow{p} \Lambda^*(\gamma_0), \qquad \text{as } n \to \infty, \ \min_i(n_i) \to \infty.
$$

Finally, by Slutsky's theorem, we have

$$
\sqrt{n}(\tilde{\gamma}_n - \gamma_0) \xrightarrow{d} N(0, \Lambda^{*-1}(\gamma_0)\Omega^*(\gamma_0)\Lambda^{*-1}(\gamma_0)),
$$
$$
\text{as } n \to \infty, \ \min_i(n_i) \to \infty.
$$

This proves Theorem 9.5.

CHAPTER 10

Generalized Estimating Equations (GEEs)

10.1 Introduction

In previous chapters, we have mostly focused on likelihood methods for estimation and inference. The likelihood methods are based on distributional assumptions for the data. For example, given the random effects, in GLMMs we assume that the responses in the models follow distributions in the exponential family, and in LME and NLME models we assume that the responses in the models follow normal distributions. *If* the distributional assumptions hold, the likelihood methods are very attractive since the MLEs are asymptotically most efficient and asymptotically normal under some regularity conditions (see Chapter 12).

In practice, however, the distributional assumptions for regression models may not hold, especially for GLMs or GLMMs. For example, in a binomial GLM or a Poisson GLM the variance is completely determined by the mean, but this may not be consistent with the observed data. If the observed variation in the data is larger than the variance determined by the distribution assumed in the model, we have an *over-dispersion* problem and the distributional assumptions for the model do not hold. Thus, in this case the likelihood method, which is based on the distributional assumptions, may not perform well.

When the observed variation in the data is inconsistent with the theoretical variation determined by the assumed model, one approach is to introduce additional dispersion parameters to account for the extra variation in the data. In this case, one can still obtain parameter estimates by solving the resulting "score" equations, but the corresponding "likelihood" is no longer a true likelihood but rather a *quasi-likelihood*, and these "score" equations are called *generalized estimating equations (GEEs)*. For longitudinal or clustered data, Liang and Zeger (1986) introduced the idea of using a working correlation matrix in the GEEs. That is, GEEs keep the same assumptions about the mean and covariance structures as in quasi-likelihood methods but introduce the working covariance matrix which may depend on fewer nuisance parameters to simplify the correlation structure.

In quasi-likelihood and GEE methods, we assume a mean structure and a variance-covariance structure based on the *observed data* without distributional assumptions, and we can still construct a "likelihood" and "score" equations for parameter estimation in forms similar to standard likelihood equations. These "likelihood" and "score" equations with working correlation matrices are quasi-likelihoods and GEEs respectively. It can be shown that the parameter estimates based on GEEs are still consistent and asymptotically normally distributed, if the mean structure is correctly specified, but the GEE estimates are not necessarily most efficient, unlike MLE when the distributional assumptions hold.

In the analysis of longitudinal data or clustered data, often we are mainly interested in the relationship between the (marginal) mean of the response and covariates, i.e., we are mainly interested in correct specification of the mean structure (or the first moment). The association between response values in the same clusters or same individuals is usually of secondary interest. Thus, the mean structure and the covariance structure can be modeled *separately*, with the primary scientific objective being modeling the mean of the response. The association between responses may be specified based on the nature of the data and based on simplicity, not necessarily based on any parametric distributions. Such models are often called *marginal models* or *GEE models* (Diggle et al. 2002). Statistical inference for marginal models is usually based on the corresponding GEEs, which have similar forms as score equations from likelihoods even though no distributional assumptions are made. Therefore, in GEE models or marginal models we only need to specify the first two moments, without distributional assumptions.

In a marginal model or GEE model, the parameters are fixed across clusters or individuals, so marginal models are appropriate for analyzing *population-average* effects. GEE models are different from *mixed effects models* in which random effects are introduced to allow for *cluster-specific* or *individual-specific* inference. GEE models are also different from *transitional models* in which Markov structures are introduced to account for the dependence within individuals. In both mixed effects models and transitional models, distributional assumptions are typically made, so likelihood methods are standard for inference. GEE models or marginal models, on the other hand, are perhaps more useful for non-normal data such as binary data or count data, since over-dispersion or under-dispersion problems are more common for these data.

In summary, GEE models or marginal models have the following advantages in data analysis:

- GEE models do not require distributional assumptions but only specifications of the first two moments;
- GEE estimates are *consistent* even if the variance-covariance structure (i.e., the second moments) is mis-specified, as long as the mean structure is correctly specified;
- GEE estimates are *asymptotically normal* under suitable conditions.

For these reasons, GEE methods are widely used in the analysis of clustered data or longitudinal data. However, GEE models also have some limitations as follows:

- GEE estimates may not be fully efficient, i.e., there may be a loss of efficiency if distributional assumptions hold or if the variance-covariance structure is mis-specified;
- GEE models do not allow for cluster-specific or individual-specific inference, so they may not be the best choices when there are large variations between clusters or individuals;
- in the presence of missing data, especially non-ignorable missing data, inference based on GEE models is often less straightforward than likelihood methods.

In the analyses of longitudinal data or clustered data, both mixed effects models and GEE models are widely used. In practice, it may be a good strategy to analyze a dataset using different approaches to gain additional insights.

In the presence of missing data, likelihood methods and multiple imputation methods described in previous chapters are typically based on assumed distributions for the missing data given the observed data: for likelihood methods, the observed-data likelihoods are obtained by integrating (or averaging) over the assumed distribution of the missing data, and for multiple imputation methods, imputations are usually generated from the conditional distributions of the missing data given the observed data. A disadvantage of these methods is that the final results may be sensitive to the distributional assumptions, especially when the missing rate is high. For a GEE model with missing data, however, we do not need to assume a distribution for the missing data. Instead, we appropriately *weight* complete observations in GEE equations using *inverse probability weighting*. So GEE methods for missing data are robust against distributional assumptions, but they may be less efficient than the likelihood or multiple imputation methods if distributional assumptions hold.

In this chapter, we focus on GEE or marginal models for non-normal responses for which generalized linear models (GLMs) are usually used. Since over-dispersion or under-dispersion problems often arise in GLMs in practice, quasi-likelihood models or GEE models are particularly useful for non-normal data.

Thus, we first consider quasi-likelihood models for cross-sectional data which motivate more general GEE methods. We then extend GEE models to longitudinal data or clustered data. In the presence of missing data, we describe the inverse probability weighting methods for GEE models. GEE approaches are very general, and the ideas can be applied to normal models such as NLME models, even if the normality assumptions in these models do not hold. There are also various extensions to standard GEE models, such as GEE2 methods and quadratic inference methods.

10.2 Marginal Models

Marginal or GEE models may be viewed as models motivated from quasi-likelihood methods. Quasi-likelihood methods are useful in situations where the assumed models are inconsistent with the observed data but likelihood-type methods can still be considered based on the first two moments. Such approaches are particularly useful for GLMs in which the theoretical variances based on the assumed models may be inconsistent with the variations in the observed data. Thus, we begin with quasi-likelihood methods in the following section.

10.2.1 Quasi-Likelihood and GEE

Quasi-likelihood models are perhaps most useful for non-normal data for which one often attempts to fit a GLM. For example, for binary data one usually chooses logistic regression models to fit the data, and for count data one usually chooses Poisson regression models for the data. A common problem for GLMs, such as a logistic regression model or a Poisson regression model, is that the variance function is often closely related to (or completely determined by) the mean function. For example, in a Poisson distribution the theoretical variance is equal to the mean. In practice, such a relationship between the mean and variance may be too restrictive, since the variation in the observed data may not agree with the theoretical variance determined by the assumed GLM. Note that this problem does not exist for normal regression models since the mean parameters and variance parameters in a normal distribution are distinct and can vary freely.

The idea of a quasi-likelihood method for a GLM is to specify a variance function *based on the data or scientific considerations or simplicity*, rather than on the parametric distribution assumed for the model. In other words, the variance function is determined *independent* of the mean function. Thus, the resulting model is no longer a GLM and the mean and variance functions may *not* correspond to any parametric distributions. However, one can still *mimic a score equation* which has a similar form as that from a true likelihood equation based on the assumed mean and variance functions. The corresponding "likelihood" is called *quasi-likelihood*, as it may not be a true likelihood from any parametric distributions. Moreover, one can show that the resulting estimates are still *consistent* and asymptotically normal, as long as the mean structure is correctly specified. The specification of the variance structure only affects the *efficiency* of the estimates. These estimating equations are also called *generalized estimating equations (GEE)*. We describe the approach for cross-sectional data as follows.

Let $\{y_1, y_2, \cdots, y_n\}$ be an independent sample from a cross-sectional study. For a GLM, the response is assumed to follow a distribution in the exponential family, with the density function given by

$$f(y_i | \theta_i, \phi, w_i) = \exp \left\{ \frac{y_i \theta_i - b(\theta_i)}{a(\phi)} + c(y_i, \phi) \right\}, \qquad (10.1)$$

where $a(\cdot)$, $b(\cdot)$ and $c(\cdot)$ are known functions, θ_i is the natural parameter corresponding to the location of the distribution, and ϕ is the dispersion parameter representing the scale of the distribution. In regression settings, we model the effects of covariates x_i on the response y_i by a link function. Let $\mu_i = E(y_i | \mathbf{x}_i)$ be the mean of the response for individual i. A GLM for cross-sectional data can be written as

$$g(\mu_i) = \boldsymbol{x}_i^T \boldsymbol{\beta}, \qquad \text{or} \qquad \mu_i = h(\mathbf{x}_i^T \boldsymbol{\beta}), \qquad i = 1, 2, \cdots, n, \qquad (10.2)$$

where $g(\cdot)$ is a monotone link function, $h(\cdot) = g^{-1}(\cdot)$, and

$$\boldsymbol{x}_i^T \boldsymbol{\beta} = \eta_i = x_{i1}\beta_1 + x_{i2}\beta_2 + \cdots + x_{ip}\beta_p$$

is the linear predictor.

The mean and variance functions for the GLM (10.2) are

$$\mu_i = \mu_i(\boldsymbol{\beta}) = h(\mathbf{x}_i^T \boldsymbol{\beta}), \qquad \sigma_i^2(\boldsymbol{\beta}) \equiv var(y_i | \mathbf{x}_i) = \phi\, v(\mu_i),$$

where $v(\mu_i)$ is the variance function uniquely determined by the specific exponential family through the relation

$$v(\mu_i) = \partial^2 b(\theta_i) / \partial \theta_i^2,$$

and θ_i is a function of μ_i.

For some GLMs, the variance functions are completely determined by the mean functions. For example, in a logistic regression model for binary response, we have

$$var(y_i) = \mu_i(1 - \mu_i),$$

where $\mu_i = P(y_i = 1)$, and for a Poisson regression model for count response, we have

$$var(y_i) = \mu_i.$$

In practice, such relationships can be too restrictive since the variance functions determined from the parametric distributions assumed for the GLMs may be inconsistent with the variation observed in the real data, leading to over-dispersion or under-dispersion problems.

Example 10.1 *Over-dispersion problems*

As an example, consider the Copenhagen Housing Conditions Survey dataset (e.g., Cox and Snell 1984). In this dataset, 1681 householders in Copenhagen were surveyed on the type of rental accommodation they occupied (Type), the degree of contact they had with other residents (Cont), their feeling of influence on apartment management (Infl), and their level of satisfaction with their housing conditions (Sat). The response variable y_i is the number of residents in each class. Since the response is a count, the following Poisson GLM is a natural choice and is fitted to the data

$$\log(y_i) = \beta_0 + \beta_1 Sat_i + \beta_2 Infl_i + \beta_3 Type_i + \beta_4 Cont_i,$$

where y_i is assumed to follow a Poisson distribution. Figure 10.1 shows the data and estimated means and variances based on the above GLM.

In a Poisson GLM, the mean and the variance should be the same, if the assumed Poisson distribution holds. However, from Figure 10.1, one can see that the estimated values of the mean and variance are clearly not the same: the variance seems much larger than the mean, indicating an over-dispersion problem. In fact, the estimated dispersion parameter is 4.85, which is much larger than 1. Therefore, one should take the over-dispersion problem into account and fit a model which adjusts the over-dispersion. A quasi-likelihood or a GEE model can be fitted to the data.

Over-dispersion or under-dispersion problems do not exist for normal regression models since in a normal regression model the variance is unrelated to the mean, which provides much flexibility in modeling real data. Thus, in practice a limitation of standard GLMs for non-normal data is the restriction on the variance function. This motivates the following quasi-likelihood models.

For a GLM, the log-likelihood for individual i is given by

$$l_i(\theta_i) = \log f(y_i|\theta_i, \phi) = \frac{y_i\theta_i - b(\theta_i)}{a(\phi)} + c(y_i, \phi).$$

Likelihood inference can be based on the score functions. For example, the *score function* for the mean parameters β is given by

$$s(\beta) = \sum_{i=1}^{n} \frac{\partial l_i(\theta_i)}{\partial \beta} = \sum_{i=1}^{n} \left[\Delta_i(\beta)\sigma_i^{-2}(\beta)(y_i - \mu_i(\beta))\right], \qquad (10.3)$$

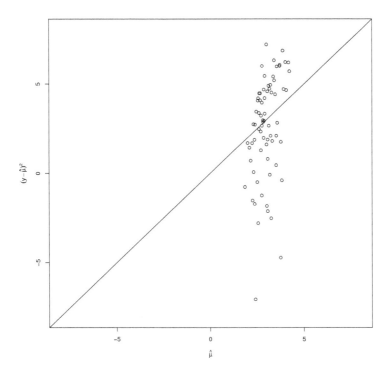

Figure 10.1 *Estimates of the mean and variance for the Copenhagen Housing Conditions Survey data.*

where

$$\Delta_i(\boldsymbol{\beta}) = \frac{\partial \mu_i(\boldsymbol{\beta})}{\partial \boldsymbol{\beta}}.$$

Then, the MLE of $\boldsymbol{\beta}$ can be obtained by solving the following *score equation*

$$s(\boldsymbol{\beta}) = \sum_{i=1}^{n} \left[\Delta_i(\boldsymbol{\beta}) \sigma_i^{-2}(\boldsymbol{\beta})(y_i - \mu_i(\boldsymbol{\beta})) \right] = 0. \qquad (10.4)$$

Score functions and score equations for other parameters can be written similarly.

In practice, the variance function may be inconsistent with the variation in the observed data, so we may consider modeling the mean and variance structures *separately*. That is, the variance function $\sigma_i(\boldsymbol{\beta})$ is *not* necessarily chosen to be the same as that determined by a parametric distribution in the exponential family. Instead, we can specify a variance function *freely* based on the data or based on scientific considerations or based on simplicity. The resulting "score equation" for estimating β still has the same form as (10.4), i.e.,

$$\sum_{i=1}^{n} \left[\Delta_i(\beta)\sigma_i^{-2}(\beta)(y_i - \mu_i(\beta)) \right] = 0, \qquad (10.5)$$

but with the variance function $\sigma_i^2(\beta)$ not necessarily equal to the theoretical variance $\phi v(\mu_i) = \phi \, \partial^2 b(\theta)/\partial\theta^2$. In this case, equation (10.5) is not a true score equation since it is not from a true likelihood of a parametric distribution. However, we can still construct a "likelihood" based on (10.5), which is called a *quasi-likelihood* (Wedderburn 1974; McCullagh 1983; McCullagh and Nelder 1989).

Estimating equation (10.5), with the second moment $\sigma_i^2(\beta)$ specified *separately* from the first moment $\mu_i(\beta)$, is called a *generalized estimating equation (GEE)*. Therefore, quasi-likelihood models and GEE models do not require distributional assumptions. They are only based on the specifications of the first and second moments. This gives us much flexibility in modeling real data. Moreover, under appropriate conditions, estimates based on the quasi-likelihood or GEE are still consistent and asymptotically normal, as shown below.

In quasi-likelihood models, if the mean function $\mu_i = h(\mathbf{x}_i^T \beta)$ is correctly specified, the estimates from GEE (10.5) will be *consistent* even if the variance function $\sigma_i^2(\beta)$ is mis-specified. Thus, we should focus more attention on correct specification of the mean structure. The variance function can be treated as a *working variance function*, which is often chosen to be close to the true variance based on the data. Usually, the closer the working variance function to the true variance, the more efficient the resulting GEE estimates. On the other hand, we should avoid a complex variance structure in order to reduce the number of nuisance parameters. In practice, choice of the variance function is typically a compromise between simplicity and efficiency.

Let

$$U(\beta) = -E \left(\frac{\partial s(\beta)}{\partial \beta} \right), \qquad V(\beta) = Cov(s(\beta)),$$

where the expectations are taken with respect to the true (but possibly incorrectly specified) data-generating probability distribution. Let $\hat{\beta}$ be a solution of GEE (10.5). Under appropriate conditions, it can be shown that

$$\hat{\beta} \xrightarrow{d} N(\beta, \, U(\beta)^{-1}V(\beta)U(\beta)^{-1}), \qquad \text{as } n \to \infty.$$

Note that, in general $U(\beta) \neq V(\beta)$, but when the variance structure is *correctly* specified, we have $U(\beta) = V(\beta)$ and so have an efficient estimate. In practice, the variance of the GEE estimate $\hat{\beta}$ can be estimated by

$$\widehat{Var}(\hat{\beta}) = U(\hat{\beta})^{-1}V(\hat{\beta})U(\hat{\beta})^{-1},$$

which is called the *sandwich estimate* of the covariance matrix of $\hat{\beta}$. The sandwich estimate adjusts the loss of efficiency due to possible misspecification of the variance function. To get efficient estimates, we should try to specify a working variance function that is as close as possible to the true variance structure.

10.2.2 Marginal Models for Longitudinal Data or Cluster Data

The ideas of quasi-likelihood methods in the previous section can be extended to longitudinal data or clustered data. In the analysis of longitudinal data, the primary scientific objective is often to model the regression relationship, i.e., the mean structure or the effects of covariates on the mean of the response, while the correlation between the responses within each cluster is usually of secondary interest. Therefore, we can model the mean structure and the variance-covariance structure *separately*, without distributional assumptions, and pay more attention to the correct specification of the mean structure or the regression relationship. This leads to marginal or GEE models for longitudinal data or clustered data, which was first proposed by Liang and Zeger (1986) and Zeger and Liang (1986). A detailed discussion of marginal models can be found in Diggle, Heagerty, Liang, and Zeger (2002). Fahrmeir and Tutz (2001) provided a overview for multivariate marginal models.

For marginal models, we only need to specify the first two moments, i.e., the mean structure and the variance-covariance structure. No distributional assumption is needed. The mean and variance-covariance structures are specified separately without any restrictions on their relationship, i.e., they do not need to be consistent with any parametric distributions, although the variance function may depend on the mean function. The parameters in marginal models are fixed or common across individuals or clusters, so marginal models are appropriate for analyzing *population average* effects and for longitudinal data with relatively small between-individual variations. Mixed effects models, on the other hand, are often more appropriate for analyzing longitudinal data with moderate or large between-individual variations, and mixed effects models allow *both* population average inference *and* subject-specific inference.

Parameter estimation for marginal models is based on a set of GEEs, as in Section 10.2.1. Much of the ideas and results in Section 10.2.1 for quasi-likelihood models carry over to marginal models for longitudinal data or clustered data with relatively minor modification. For example, for marginal models the GEE estimates are consistent even if the variance-covariance structure is mis-specified, as long as the mean structure is correctly specified. The choice of the variance-covariance structure only affects the efficiency of the estimates: the closer the variance-covariance structure to the true one, the more efficient the resulting GEE estimates. Marginal models for longitudinal data or clustered data are described in more details as follows.

Let $\mathbf{y}_i = (y_{i1}, \cdots, y_{in_i})^T$ be the response values on individual i or cluster i, and let $\mathbf{x}_i$ contains covariates, $i = 1, 2, \cdots, n$. In marginal models, the dependence of the response on the covariates and the association between responses within each cluster are modeled separately, i.e., the mean structure and the variance-covariance structure are modeled separately, without distributional assumptions. Specifically, a marginal model for longitudinal data can be defined as follows:

- the *marginal mean* of the response y_{ij} is correctly specified by

$$\mu_{ij}(\boldsymbol{\beta}) = E(y_{ij}|\mathbf{x}_i, \boldsymbol{\beta}) = h(\mathbf{x}_i^T \boldsymbol{\beta}),$$

where $h(\cdot)$ is a known function.

- the *variance-covariance structure* of $\mathbf{y}_i$ is separately specified by a *working covariance matrix*

$$Cov(\mathbf{y}_i) = \Sigma_i(\boldsymbol{\beta}, \boldsymbol{\alpha}),$$

where $\boldsymbol{\alpha}$ contains parameters for the variance-covariance structure.

The working covariance matrix $\Sigma_i(\boldsymbol{\beta}, \boldsymbol{\alpha})$ can be written as follows (Liang and Zeger 1986; Prentice 1988):

$$\Sigma_i(\boldsymbol{\beta}, \boldsymbol{\alpha}) = V_i^{1/2}(\boldsymbol{\beta}) R_i(\boldsymbol{\alpha}) V_i^{1/2}(\boldsymbol{\beta}),$$

where $V_i(\boldsymbol{\beta}) = diag(\sigma_{i1}^2, \cdots, \sigma_{in_i}^2)$, with $\sigma_{ik}^2 = var(y_{ik}|\mathbf{x}_i, \boldsymbol{\beta})$, are the variances of the responses in cluster i, and the matrix $R_i(\boldsymbol{\alpha})$ is called the *working correlation matrix* which measures the correlation between the responses in cluster i.

We should choose the working correlation matrix $R_i(\boldsymbol{\alpha})$ to be close to the true correlation structure in the data, and in the meantime choose a *simple* structure to avoid too many nuisance parameters, especially if the observed data are not rich enough to estimate all parameters well. The following are common choices of the working correlation matrix $R_i(\boldsymbol{\alpha})$:

- *independence model*:

$$R_i(\boldsymbol{\alpha}) = I_{n_i},$$

where I_{n_i} is the $n_i \times n_i$ identity matrix, for which we assume that the responses in cluster i are independent of each other;

- *equicorrelation model*:

$$(R_i(\boldsymbol{\alpha}))_{jk} = corr(y_{ij}, y_{ik}) = \alpha, \qquad \text{for } j \neq k,$$

where $corr(y_{ij}, y_{ik})$ is the correlation between y_{ij} and y_{ik}, for which we assume that the correlation between each pair of the responses in cluster i is the same;

- *stationary correlation model*:

$$(R_i(\boldsymbol{\alpha}))_{jk} = corr(y_{ij}, y_{ik}) = \alpha|t_{ij} - t_{ik}|, \qquad \text{for } j \neq k,$$

for which we assume that the correlation between two response values in cluster i is proportional to the time distance between them;

- *unstructured model*:

$$(R_i(\boldsymbol{\alpha}))_{jk} = corr(y_{ij}, y_{ik}) = \alpha_{jk}, \qquad \text{for } j \neq k,$$

which is the most general correlation structure and may be considered when the data is rich. In data analysis, one may try more than one correlation structures and check whether analysis results are sensitive to the choices of correlation structure.

For continuous responses, we use correlation as a measure of association between the responses. For binary responses, however, an alternative and probably more appropriate measure of association is the *odds ratio* (Lipsitz, Laird, and Harrington 1991; Fahrmeir and Tutz 2001). Suppose that the binary response y_{ij} takes two possible values: 0 or 1. Let $p_{ij} = P(y_{ij} = 1) = E(y_{ij}|\mathbf{x}_i)$. The *odds ratio* between y_{ij} and y_{ik} is defined as

$$\gamma_{ijk} = \frac{P(y_{ij} = 1, y_{ik} = 1)\, P(y_{ij} = 0, y_{ik} = 0)}{P(y_{ij} = 1, y_{ik} = 0)\, P(y_{ij} = 0, y_{ik} = 1)},$$
$$i = 1, 2, \cdots, n; \; j, k = 1, 2, \cdots, n_i,$$

which measures the association between two binary responses y_{ij} and y_{ik}.

To reduce the number of nuisance parameters, we can write $\gamma_{ijk} = \gamma_{ijk}(\boldsymbol{\alpha})$ so that the odds ratios depend on a common set of parameters $\boldsymbol{\alpha}$. For example, we may consider a *log-linear model*

$$\log(\gamma_{ijk}) = \mathbf{x}_i^T \boldsymbol{\alpha},$$

or we may consider the following simple structure

$$\gamma_{ijk} = \alpha,$$

for all i, j, k.

Note that the usual covariance matrix of $\mathbf{y}_i$ can be expressed as a function of the odds ratios based on the following relationship (Lipsitz, Laird, and Harrington 1991; Fahrmeir and Tutz 2001):

$$E(y_{ij} y_{ik}) = \begin{cases} \frac{1 - (p_{ij} + p_{ik})(1 - \gamma_{ijk}) - a(p_{ij}, p_{ik}, \gamma_{ijk})}{2(\gamma_{ijk} - 1)}, & \gamma_{ijk} \neq 1 \\ p_{ij} p_{ik}, & \gamma_{ijk} = 1, \end{cases}$$

where

$$a(p_{ij}, p_{ik}, \gamma_{ijk}) = \left[(1 - (p_{ij} + p_{ik})(1 - \gamma_{ijk}))^2 - 4(\gamma_{ijk} - 1)\gamma_{ijk} p_{ij} p_{ik} \right]^{1/2},$$

so there is a relationship between odds ratio and correlation.

Example 10.2 *Marginal models*

Suppose that one wishes to consider a marginal model for a regression be-
tween a longitudinal response $\mathbf{y}_i$ and covariates $\mathbf{x}_i$. If the response is a continuous variable, a marginal linear regression model with constant variance but
unstructured correlations can be specified as follows:

$$
\begin{aligned}
E(\mathbf{y}_i|\mathbf{x}_i) &= \mathbf{x}_i^T\boldsymbol{\beta}, & i &= 1, \cdots, n, \\
var(\mathbf{y}_i|\mathbf{x}_i) &= \sigma^2, \\
corr(y_{ij}, y_{ik}) &= \alpha_{jk}, & j \neq k, \quad j, k &= 1, \cdots, n_i,
\end{aligned}
$$

where $\{(\boldsymbol{\beta}, \sigma, \alpha_{jk}), \ j \neq k, \ j, k = 1, \cdots, \max_i\{n_i\}\}$ are unknown parameters. Here no distributional assumption is made for the response.

If the response is a binary variable, a marginal model with a simple association
structure can be specified as follows:

$$
\begin{aligned}
E(\mathbf{y}_i|\mathbf{x}_i) &= \frac{\exp(\mathbf{x}_i^T\boldsymbol{\beta})}{1+\exp(\mathbf{x}_i^T\boldsymbol{\beta})}, & \text{or} \quad \log\frac{p_{ij}}{1-p_{ij}} &= \mathbf{x}_i^T\boldsymbol{\beta}, \\
var(\mathbf{y}_i|\mathbf{x}_i) &= \phi\, p_{ij}(1-p_{ij}), \\
\gamma_{ijk} &= \alpha, & i = 1, \cdots, n; \quad j, k &= 1, \cdots, n_i,
\end{aligned}
$$

where $p_{ij} = E(y_{ij}|\mathbf{x}_i)$, γ_{ijk} is the odds ratio between y_{ij} and y_{ik}, and
$(\boldsymbol{\beta}, \phi, \alpha)$ are unknown parameters. This model has a similar form as a logistic
regression model but it contains a dispersion parameter ϕ and it incorporates
the correlation between the responses within each cluster via a common odds
ratio α. Here the response y_{ij} does not follow any parametric distribution since
no parametric distribution has the above mean and variance-covariance structures.

10.2.3 GEE for Marginal Models

As in Section 10.2.1, parameters in marginal models for longitudinal data or
clustered data can be estimated based on the GEE method. Specifically, given
estimates of the association parameters $\boldsymbol{\alpha}$ and the dispersion parameter ϕ, the
generalized estimating equation (GEE) for estimating the mean parameters $\boldsymbol{\beta}$
in a marginal model is given by

$$
S_{\boldsymbol{\beta}}(\boldsymbol{\beta}, \boldsymbol{\alpha}) = \sum_{i=1}^{n} \left[\Delta_i(\boldsymbol{\beta})\Sigma_i^{-1}(\boldsymbol{\beta}, \boldsymbol{\alpha})(\mathbf{y}_i - \boldsymbol{\mu}_i(\boldsymbol{\beta})) \right] = 0, \tag{10.6}
$$

where $\boldsymbol{\mu}_i(\boldsymbol{\beta}) = E(\mathbf{y}_i|\mathbf{x}_i, \boldsymbol{\beta})$, $\Delta_i(\boldsymbol{\beta}) = \partial\boldsymbol{\mu}_i(\boldsymbol{\beta})/\partial\boldsymbol{\beta}$, and $\Sigma_i(\boldsymbol{\beta}, \boldsymbol{\alpha})$ is a work-
ing covariance matrix.

Note that GEE (10.6) has the same form as a standard likelihood equation, but *no* distributional assumption is made for the response. Moreover, the use of the working covariance matrix $\Sigma_i(\boldsymbol{\beta}, \boldsymbol{\alpha})$, which often has a simple structure, avoids potential difficulties of a singular estimated covariance matrix or too many nuisance parameters. The solutions of GEE (10.6), called *GEE estimates*, have similar properties as MLEs, such as consistency and asymptotic normality, but they are usually not fully (asymptotically) efficient unless the covariance matrix is correctly specified.

GEE (10.6) can be solved by the following iterative algorithm. At iteration k, given estimates $\hat{\boldsymbol{\alpha}}^{(k)}$ and $\hat{\boldsymbol{\beta}}^{(k)}$, an updated estimate of $\boldsymbol{\beta}$ is given by

$$\hat{\boldsymbol{\beta}}^{(k+1)} = \hat{\boldsymbol{\beta}}^{(k)} + (\hat{A}^{(k)})^{-1} S_{\boldsymbol{\beta}}(\hat{\boldsymbol{\beta}}^{(k)}, \hat{\boldsymbol{\alpha}}^{(k)}), \qquad k = 0, 1, 2, \cdots,$$

where

$$\hat{A}^{(k)} = \sum_{i=1}^{n} \left[\Delta_i^T(\hat{\boldsymbol{\beta}}^{(k)}) \Sigma_i^{-1}(\hat{\boldsymbol{\beta}}^{(k)}, \hat{\boldsymbol{\alpha}}^{(k)}) \Delta_i(\hat{\boldsymbol{\beta}}^{(k)}) \right].$$

Iterating the above procedure until convergence, we obtain a GEE estimate $\hat{\boldsymbol{\beta}}$, which may not be unique.

Estimation of the association parameters $\boldsymbol{\alpha}$ (and the dispersion parameters ϕ if present) can be based on *Pearson residuals*:

$$\hat{r}_{ij}^{(k)} = v(\hat{\mu}_{ij}^{(k)})^{-1/2} (y_{ij} - \hat{\mu}_{ij}^{(k)}),$$

where $\hat{\mu}_{ij}^{(k)} = E(y_{ij}|\mathbf{x}_i, \hat{\boldsymbol{\beta}}^{(k)})$ and $v(\hat{\mu}_{ij}^{(k)}) = var(y_{ij}|\mathbf{x}_i, \hat{\boldsymbol{\beta}}^{(k)})$. For example, if Σ is unstructured, we have

$$\hat{\Sigma}^{(k)} = \sum_{i=1}^{n} (\mathbf{y}_i - \hat{\boldsymbol{\mu}}_i^{(k)})(\mathbf{y}_i - \hat{\boldsymbol{\mu}}_i^{(k)})^T \Big/ (n \hat{\phi}^{(k)}),$$

$$\hat{\phi}^{(k)} = \sum_{i=1}^{n} \sum_{j=1}^{n_i} \hat{r}_{ij}^{2(k)} \Big/ (\sum_{i=1}^{n} n_i - p).$$

Prentice (1988) suggested that estimation of the association parameters $\boldsymbol{\alpha}$ (and the dispersion parameters ϕ if present) can be based on a second GEE for $\boldsymbol{\alpha}$. Specifically, let

$$w_{ijk} = (y_{ij} - \mu_{ij})(y_{ik} - \mu_{ik}),$$

and $\mathbf{w}_i = (w_{i12}, w_{i13}, \cdots, w_{in_{i-1}n_i})^T$. Then, the second GEE for $\boldsymbol{\alpha}$ is given by

$$S_{\boldsymbol{\alpha}}(\boldsymbol{\beta}, \boldsymbol{\alpha}) = \sum_{i=1}^{n} \left(\frac{\partial v_i}{\partial \boldsymbol{\alpha}} \right) W_i^{-1}(\mathbf{w}_i - E(\mathbf{w}_i)) = 0,$$

where W_i is a working covariance matrix for $\mathbf{w}_i$.

Let $\hat{\beta}$ be the GEE estimate at convergence. Under some regularity conditions, $\hat{\beta}$ is consistent and asymptotically normal:

$$\hat{\beta} \xrightarrow{d} N(\beta, \, U^{-1}VU^{-1}), \qquad \text{as } n \to \infty,$$

where

$$U = \sum_{i=1}^{n} \Delta_i \Sigma_i^{-1} \Delta_i, \qquad V = \sum_{i=1}^{n} \Delta_i \Sigma_i^{-1} (Cov(\mathbf{y}_i)) \Sigma_i^{-1} \Delta_i.$$

This result can be used to obtain approximate standard error of $\hat{\beta}$, to construct confidence interval for β, and to conduct hypothesis testing such as a Wald-type test for β.

We note again that, if the mean structure $\mu_i(\beta)$ is correctly specified, the GEE estimate $\hat{\beta}$ is consistent and asymptotically normal even if the covariance matrix $\Sigma_i = \Sigma_i(\beta, \alpha)$ is mis-specified. The choice of the working covariance matrix $\Sigma_i(\beta, \alpha)$ will only affect the efficiency of the GEE estimates. Note also that no distributional assumptions are required for the GEE estimates and their asymptotic results.

For either continuous responses or discrete responses, the forms of the GEE are the same. When choosing the working correlation structures, however, one should be careful with discrete responses since some parameter values in the working covariance matrices may be restricted. Chaganty and Joe (2004) discussed the efficiency of GEE for binary responses and noted that the ranges for admissible correlations can be narrow. Zeng and Cook (2007) used odds ratio to measure association for multivariate longitudinal binary data, which may allow more flexibility.

Example 10.3 *Analyses of mental distress data*

For the mental distress data described in Section 1.3.1 of Chapter 1, one may consider either a mixed effects model or a GEE model for data analyses. A mixed effects model makes distributional assumptions, while a GEE model only assumes the first two moments. In the following, we analyze the data using both approaches and compare the results. For simplicity, we delete all incomplete data here. We focus on modeling the GSI scores over time for illustration.

For a GEE model, we assume the following simple mean structure

$$E(y_{ij}) = \beta_0 + \beta_1 t_{ij}, \qquad i = 1, \cdots, n; \; j = 1, \cdots, n_i, \qquad (10.7)$$

where y_{ij} is the GSI score for subject i measured at time t_{ij}. We consider three working correlation structures for the GEE model: independence, exchangeable correlation, and unstructured correlation structures. For a mixed effects

Table 10.1 *Parameter estimates based on the GEE model and the LME model*

Model	Par.	Est.	S.E.	Par.	Est.	S.E.
GEE Model I	β_0	1.01	0.041	β_1	−0.27	0.045
GEE Model II	β_0	1.02	0.041	β_1	−0.21	0.038
GEE Model III	β_0	1.02	0.041	β_1	−0.25	0.041
LME Model (10.8)	β_0	1.02	0.041	β_1	−0.22	0.038

The three GEE models are all based on model (10.7) but with different covariance structures. GEE model I: independent correlation structure, GEE model II: exchangeable correlation structure, GEE model III: unstructured correlation. Par.: parameter, Est.: estimate, S.E.: standard error.

model, we consider the following LME model

$$y_{ij} = \beta_0 + b_{0i} + (\beta_1 + b_{1i})t_{ij} + e_{ij},$$
$$e_{ij} \quad \text{i.i.d.} \quad \sim N(0, \sigma^2), \qquad (b_{0i}, b_{1i})^T \sim N(0, D), \qquad (10.8)$$

where (b_{0i}, b_{1i}) are random effects. Note that the GEE model (10.7) and the LME model (10.8) have the same marginal mean

$$E(y_{ij}) = \beta_0 + \beta_1 t_{ij}.$$

In the following data analysis, we standardize the time (in month) to avoid very large or small estimates.

Table 10.1 shows the parameter estimates and their standard errors based on the four models. One can see that the parameter estimates are quite close, although the estimates based on the LME model may be slightly more efficient (smaller standard errors), possibly due to distributional assumptions. An advantage of a mixed effects model is that it allows for individual-specific inference, while an advantage of a GEE model is that it is robust against distributional assumptions. Note that the choices of correlation structures seem to only have minor effects on the parameter estimates in this example. The GEE model with unstructured correlation is the least restrictive among the models. Overall, the four models give similar results. A significant decreasing trend of the GSI scores is observed and confirmed by this analysis, i.e., the GSI scores decrease over time.

10.3 Estimating Equations with Incomplete Data

10.3.1 Weighted GEE for Missing Data

In the presence of missing data, GEE models can be modified to incorporate the missing data. The idea is to appropriately weight the complete observations so that the resulting estimates are still consistent even if the missing data are not missing completely at random (MCAR). Note that a major advantage of GEE models is that there are no distributional assumptions. To incorporate missing data in GEE models, we again make no distributional assumptions for the missing data, but we need to assume a missing data model and then weight the complete-data estimating equations by the inverse probability of the missingness. It can then be shown that the weighted GEE models are unbiased if the missing data model is correctly specified. These weighted GEE methods were first developed in a series of papers by Robins and his colleagues (e.g., Robins et al. 1994, 1995; Robins and Rotnitzky 1995).

In the following, we briefly describe the basic idea of weighted GEE for cross-sectional data with missing values. We focus on missing covariates problems in regression models and assume that the missing data are missing at random (MAR).

Suppose that $\mathbf{x}_i$ is a vector of covariates with missing data. We write $\mathbf{x}_i = (\mathbf{x}_{mis,i}, \mathbf{x}_{obs,i})$, where $\mathbf{x}_{mis,i}$ contains the missing components of $\mathbf{x}_i$ and $\mathbf{x}_{obs,i}$ contains the observed components of $\mathbf{x}_i$, $i = 1, 2, \cdots, n$. Let r_i be a missingness indicator of observation i such that $r_i = 1$ if $\mathbf{x}_i$ is completely observed and $r_i = 0$ if $\mathbf{x}_i$ has missing components. As in previous sections, let $\mu_i = \mu_i(\beta) = E(y_i|\mathbf{x}_i, \beta)$, $\sigma_i^2(\beta) = var(y_i|\mathbf{x}_i, \beta)$, and $\Delta_i(\beta) = \partial\mu_i(\beta)/\partial\beta$. Then, the GEE for β based on the *complete cases*, i.e., cases with $r_i = 1$, can be written as

$$\sum_{i=1}^{n} r_i \left[\Delta_i(\beta)\sigma_i^{-2}(\beta)(y_i - \mu_i(\beta)) \right] = 0. \tag{10.9}$$

The above equation leads to unbiased estimates when the missing data are MCAR, but may lead to biased estimates when the missing data are MAR or MNAR (missing not at random or non-ignorably missing).

To incorporate missing data in GEE when the missingness is MAR, Robins et al. (1994) suggested the following *weighted GEE (WGEE)*

$$\sum_{i=1}^{n} \frac{r_i}{\pi_i} \left[\Delta_i(\beta)\sigma_i^{-2}(\beta)(y_i - \mu_i(\beta)) \right] = 0, \tag{10.10}$$

where

$$\pi_i = P(r_i = 1|y_i, \mathbf{x}_{obs,i}, \psi) \tag{10.11}$$

is the probability of missingness. Thus, the complete cases in weighted GEE (10.10) are weighted by the inverse probability of missingness, i.e., the complete cases have weights greater than 1.

Note that, in weighted GEE (10.10), only complete observations contribute to the estimating equations, and these complete cases are weighted by the probability of missingness π_i. We may allow the missingness probability π_i to depend on the observed data $(y_i, \mathbf{x}_{obs,i})$, i.e., we allow the missing data to be MAR. If the missing data model (10.11) is known and correctly specified, the resulting GEE estimates are consistent and asymptotically normal. If the missingness probability π_i is unknown, we can substitute π_i in (10.10) by a consistent estimate of π_i. Since r_i is a binary variable, we may consider a logistic regression model to estimate π_i.

The weighted GEE methods are only based on complete observations. That is, incompletely observed data or observations with partially missing data are not used. This may lead to some loss of information and thus some loss of efficiency. However, the weighted GEE methods do not require distributional assumptions, so they are robust. On the other hand, the likelihood methods in previous chapters use all observed data, so the resulting MLEs are usually more efficient than weighted GEE estimators if the distributional assumptions hold.

10.3.2 Weighted GEE for Measurement Errors and Missing Data

The weighted GEE method for missing data in Section 10.3.1 can be extended to models for incomplete longitudinal data or clustered data. It can also be extended to models with measurement errors in covariates. One approach is to introduce additional estimating equations to incorporate missing data mechanisms and address associations within clusters. Yi (2008) proposed a unified approach to address both measurement errors and missing data simultaneously. This approach is briefly described as follows.

Let $\mathbf{y}_i = (y_{i1}, \cdots, y_{im})^T$ be m response measurements on individual or cluster i, $i = 1, \cdots, n$. Let $\mathbf{z}_{ij}$ be a vector of error-prone covariates for individual i at measurement j, $\mathbf{z}_i = (\mathbf{z}_{i1}^T, \cdots, \mathbf{z}_{im}^T)^T$, $\mathbf{z}_i^*$ be the corresponding true but unobserved covariate values, and $\mathbf{x}_i = (\mathbf{x}_{i1}^T, \cdots, \mathbf{x}_{im}^T)^T$ be a vector of error-free covariates. Let

$$\mu_{ij} = \mu_{ij}(\boldsymbol{\beta}) = E(y_{ij}|\mathbf{z}_i^*, \mathbf{x}_i, \boldsymbol{\beta})$$

be the mean response given covariates for individual i at measurement j, with $\boldsymbol{\beta}$ being the regression parameters, $\boldsymbol{\mu}_i = (\mu_{i1}, \cdots, \mu_{im})^T$, and $\Sigma_i = \Sigma_i(\boldsymbol{\beta}) = Cov(\mathbf{y}_i|\mathbf{z}_i^*, \mathbf{x}_i, \boldsymbol{\beta})$ being the variance-covariance matrix of the response vector. We consider the following marginal model

$$g(\mu_{ij}) = \boldsymbol{\beta}_z^T \mathbf{z}_{ij}^* + \boldsymbol{\beta}_x^T \mathbf{x}_{ij}, \qquad i = 1, \cdots, n; \ j = 1, \cdots, m, \qquad (10.12)$$

where $\boldsymbol{\beta} = (\boldsymbol{\beta}_z^T, \boldsymbol{\beta}_x^T)^T$ is a vector of regression parameters and $g(\cdot)$ is a known monotone function.

To address measurement errors in covariates $\mathbf{z}_i$, we consider the following classical measurement error model

$$\mathbf{z}_{ij} \quad = \quad \mathbf{z}_{ij}^* + \epsilon_{ij}, \qquad i = 1, \cdots, n; \; j = 1, \cdots, m, \qquad (10.13)$$

$$\epsilon_{ij} \quad \sim \quad N(0, A), \qquad\qquad\qquad\qquad\qquad\qquad (10.14)$$

where A is an unknown covariance matrix. Validation data or repeated measurements of the covariates are needed to address measurement errors and to identify model parameters, or one may conduct sensitivity analysis based on background information (Yi 2008).

Suppose that the responses $\mathbf{y}_i$ contain missing data, with a MAR missing mechanism and a monotone missing data pattern. Let $r_{ij} = 1$ if y_{ij} is observed and $r_{ij} = 0$ if y_{ij} is missing, and let $\mathbf{r}_i = (r_{i1}, \cdots, r_{im})^T$. Write $\mathbf{y}_i = (\mathbf{y}_{mis,i}, \mathbf{y}_{obs,i})$, where $\mathbf{y}_{mis,i}$ contains the missing components of $\mathbf{y}_i$ and $\mathbf{y}_{obs,i}$ contains the observed components of $\mathbf{y}_i$. Let

$$\lambda_{ij} = P(r_{ij} = 1 | r_{i,j-1} = 1, \mathbf{y}_{obs,i}, \mathbf{z}_i^*, \mathbf{x}_i, \psi),$$

and let

$$\pi_{ij} = P(r_{ij} = 1 | \mathbf{y}_{obs,i}, \mathbf{z}_i^*, \mathbf{x}_i, \psi).$$

We may assume that the missingness of response y_{ij} may depend on its previous missing status and the observed data, i.e., one may consider the following logistic regression model

$$\log(\lambda_{ij}/(1 - \lambda_{ij})) = \alpha^T \mathbf{u}_i,$$

where $\mathbf{u}_i = (\mathbf{y}_{obs,i}, \mathbf{z}_i^*, \mathbf{x}_i)$ and α is a vector of regression parameters.

In the following, we let t_{m_i} be the dropout time for individual i, and let

$$f(\mathbf{r}_i, \alpha) = (1 - \lambda_{im_i}) \prod_{j=2}^{m_i-1} \lambda_{ij}.$$

Let $\Delta_i(\beta) = \partial \mu_i(\beta)/\partial\beta$, and let $W_i = diag(I(r_{ij} = 1)/\pi_{ij})$ be a diagonal weight matrix whose (j,j)-th element is $I(r_{ij} = 1)/\pi_{ij}$, where $I(\cdot)$ is the indicator function. In the absence of measurement errors in $\mathbf{z}_i$ (i.e., when $\mathbf{z}_i^* = \mathbf{z}_i$), Yi (2008) proposed to solve the following estimating equations

$$\sum_{i=1}^n \Delta_i \Sigma_i^{-1} W_i (\mathbf{y}_i - \mu_i) \quad = \quad 0, \qquad\qquad (10.15)$$

$$\sum_{i=1}^n \log(\partial f(\mathbf{r}_i, \alpha)/\partial\alpha) \quad = \quad 0, \qquad\qquad (10.16)$$

which lead to consistent estimators of the parameters. In the presence of measurement errors in $\mathbf{z}_i$, bias may arise in the above estimating equations if one still substitutes $\mathbf{z}_{ij}^*$ by $\mathbf{z}_{ij}$ in the estimating equations. To adjust the bias,

Yi (2008) proposed a simulation extrapolation (SIMEX) method, which is a simulation-based approach, and established the asymptotic normality of the SIMEX estimators.

In practice, longitudinal data may also arise in clusters. For example, in family studies if observations are repeatedly measured over time, the data are clustered longitudinal data where each family may be viewed as a cluster. Similarly, for a longitudinal school-based study or a longitudinal hospital-based study, each school or hospital may be viewed as a cluster. In clustered longitudinal studies, two types of correlation arise: the correlation between the repeated measurements over time and the correlation within each cluster. Although in marginal GEE models, the association parameters are often viewed as nuisance, in some situations interest may lies in the strength of both types of association. Yi and Cook (2002) proposed a GEE method for incomplete longitudinal binary data arising in clusters. They proposed a second set of estimating equations for the association parameters, and used odds ratio to measure the association in binary data.

10.4 Discussion

GEE models are often used for modeling non-normal data, without distributional assumptions, but the general approach is more widely applicable. In GEE models, working correlation matrices are used to incorporate the correlation within clusters, and these working correlation matrices typically contain only a small set of nuisance parameters. An advantage of using a working correlation matrix rather than an empirical unstructured correlation matrix is that a unstructured covariance matrix may contain too many nuisance parameters, especially for large clusters, and thus its estimate sometimes may not be positive definite and may be singular or nearly singular, which may cause computational difficulties.

GEE estimators are consistent even if the working correlation matrix is mis-specified, as long as the mean structure is correctly specified. GEE estimators are also efficient when the working correlation is correctly specified. However, when the working correlation is mis-specified, the GEE estimators are not optimal. In some cases, estimates of the parameters in working correlation matrices even may not exist. For binary data, Chaganty and Joe (2004) argued that usual working correlation matrices cannot in general be the true correlation matrices of the data. They showed that the parameters in correlation matrices for binary data are *bounded*, which is often ignored in current softwares, and these bounds can be quite narrow in some cases. Chaganty and Joe (2004) suggested that it may be more appropriate to view working correlation matrices as weight matrices. For binary data, odds ratios can be used to measure association within clusters (e.g., Yi and Cook 2002).

Qu, Lindsay, and Li (2000) introduced the idea of *quadratic inference functions*. They argued that their approach produces optimal estimates of regression parameters within an assumed family even if the working correlation is misspecified. Their approach may be viewed as to minimize the empirical asymptotic variances of regression parameters to obtain estimates of the parameters in working correlation matrices. The idea is to express the inverse of a working correlation matrix by a linear combination of some basis matrices. They showed that the quadratic inference function estimators are more efficient than standard GEE estimators.

For GEE methods, the focus is usually on inference for the mean parameters, with the variance-covariance (association) parameters being viewed as nuisance parameters. In some cases, however, the association parameters may also be of great interest, as one may wish to understand the strength of the association within clusters. In these cases, a second set of estimating equations may be specified in a way similar to the estimating equations for the mean parameters (e.g., Prentice 1988; Yi and Cook 2002). One then proceeds to *jointly solve* the estimating equations. This approach, however, requires specifications of higher order moments.

The idea of GEE methods can be applied to a wide variety of problems. There is an extensive literature on GEE-type methods. Consistency and asymptotic normality of GEE-type estimators can usually be established. For mixed effects models, one can also consider GEE-type methods in which one only needs to specify the first two moments without distributional assumptions for the data and then solves a set of estimating equations. However, GEE-type estimators are usually less efficient than maximum likelihood estimators if the distributional assumptions hold.

Missing data problems, especially non-ignorable missing data problems, are usually more conveniently handled by likelihood methods. Likelihood methods for missing data or measurement error problems are often conceptually straightforward, although the computation can be tedious. GEE methods for missing data or measurement error problems, on the other hand, have the advantage of avoiding distributional assumptions.

A main advantage of GEE-type estimators is the robustness against distributional assumptions, while a main advantage of likelihood-based estimators is the asymptotic optimality. Both types of estimators are asymptotically normal. In real data analyses, one may wish to consider both approaches to gain additional insights.

CHAPTER 11

Bayesian Mixed Effects Models

11.1 Introduction

For the mixed effects models considered in previous chapters, parametric models are often assumed for the response, for the random effects, for the incompletely observed covariates, and for the missing data or measurement error models. So the joint models often contain too many parameters, and this may lead to poor estimation of the main parameters or may lead to identifiability problems, especially if the observed data are not rich. Bayesian methods offer the advantage of *borrowing information* from similar studies or from experts, which are then incorporated in the current analysis in the forms of *prior* distributions for the parameters. Such prior information helps estimating parameters that may be poorly identified by the current observed data alone.

Typically computation is a main challenge for Bayesian inference. Due to the availability of modern computers and computational tools such as the Markov chain Monte Carlo (MCMC) methods, however, Bayesian inference for many problems become feasible and even straightforward. In Bayesian inference, we can still get maximum likelihood type estimates with non-informative priors. That is, when the prior distributions are non-informative, Bayesian methods are equivalent to likelihood methods. Thus, Bayesian methods are closely related to likelihood methods. In fact, they share some similar computational challenges and use similar computational tools.

For a Bayesian approach, model parameters are treated as random variables, which is a major difference with likelihood methods. The assumed distributions for the parameters are called *prior distributions*. Bayesian inference is based on the *posterior distribution*, which is the conditional distribution of unobserved quantities, such as the parameters or unobserved covariates, given the observed data. The posterior distribution summarizes all the information about the unobservables. For example, we can use the mean, or the median, or the mode of the posterior distribution as point estimators. A Bayesian analog to a confidence interval is the *credit set*, which is a region with probability $1 - \alpha$

under the posterior distribution. Choices of prior distributions are important. In fact, much of the controversy regarding Bayesian methods revolves around the prior distributions.

Computation of posterior distributions is often non-trivial and computer intensive. We can use MCMC methods, especially the Gibbs sampler, to approximate posterior distributions. The Gibbs sampler (see Chapter 12) generates a Markov chain whose stationary distribution is the target posterior distribution. In many cases, Bayesian computation becomes relatively straightforward with availability of Bayesian softwares such as *WinBUGS*, which is based on the BUGS (Bayesian inference Using Gibbs Sampling) project.

There have been extensive developments of Bayesian methods in the last few decades, mainly due to the availability of modern computers and the breakthrough in computational tools such as the MCMC methods, which make tedious Bayesian computation feasible. The basic idea of Bayesian inference is as follows:

- assume some *prior* distributions for the unknown parameters in the models;
- make inference based on the *posterior* distributions of the parameters given the data.

Computation of the posterior distributions is often challenging but it is now feasible using MCMC methods such as the Gibbs sampler method. Gelman et al. (2003) provided a comprehensive overview of modern Bayesian methods.

In the following sections, we give a brief review of the general ideas of Bayesian methods.

11.2 Bayesian Methods

11.2.1 General Concepts

In this section, we describe some general concepts and approaches for Bayesian inference. These general concepts and approaches can be easily extended to more specific models, such as mixed effects models.

Let $\mathbf{y}$ be the data following an assumed parametric distribution with probability density function $f(\mathbf{y}|\boldsymbol{\theta})$, where $\boldsymbol{\theta}$ contains unknown parameters. A *Bayesian method* assumes that the unknown parameters $\boldsymbol{\theta}$ are random variables following a distribution with probability density function $f(\boldsymbol{\theta}) = f(\boldsymbol{\theta}|\boldsymbol{\theta}_0)$, called a *prior distribution*. The parameters $\boldsymbol{\theta}_0$ in the prior distribution are called *hyperparameters* and are often assumed to be known, which can be chosen based on similar studies or expert opinion or even non-informative.

Bayesian inference for the unknown parameters θ is based on the *posterior distribution* $f(\theta|\mathbf{y})$ given the data $\mathbf{y}$. Specifically, given the prior distribution $f(\theta)$, the posterior distribution $f(\theta|\mathbf{y})$ can be obtained via the Bayes's theorem:

$$f(\theta|\mathbf{y}) = \frac{f(\mathbf{y}|\theta)f(\theta)}{f(\mathbf{y})} = \frac{f(\mathbf{y}|\theta)f(\theta)}{\int f(\mathbf{y}|\theta)f(\theta)d\theta} \propto f(\mathbf{y}|\theta)f(\theta). \quad (11.1)$$

Bayesian inference for θ is then based on the posterior distribution $f(\theta|\mathbf{y})$. For example, a Bayesian estimator of θ is the *posterior mean*:

$$\hat{\theta}_B = E(\theta|\mathbf{y}) = \int \theta f(\theta|\mathbf{y})d\theta,$$

with its precision being measured by the *posterior variance*:

$$Cov(\hat{\theta}_B) = Cov(\theta|\mathbf{y}) = \int (\theta - E(\theta|\mathbf{y}))(\theta - E(\theta|\mathbf{y}))^T f(\theta|\mathbf{y}) \, d\theta.$$

The posterior mean is an optimal estimator of θ under the quadratic loss.

When the hyper-parameters are unknown, one approach is to estimate them from the data, and the resulting Bayesian estimates are called *empirical Bayesian estimates*.

The choice of prior distribution $f(\theta)$ may affect Bayesian estimation. In other words, Bayesian inference may be influenced by a strong prior. In practice, we can try different prior distributions or different values of the hyper-parameters for *sensitivity analysis*. In the absence of any prior information, we may choose a *non-informative prior*: $f(\theta) \propto 1$. Note that the likelihood $L(\theta|\mathbf{y}) = f(\mathbf{y}|\theta)$, so we have

$$f(\theta|\mathbf{y}) \propto L(\theta|\mathbf{y})f(\theta).$$

Therefore, Bayesian methods and likelihood methods are linked. In particular, if we choose a non-informative prior distribution for $f(\theta)$, Bayesian inference is equivalent to likelihood inference.

Although the Bayesian paradigm seems conceptually straightforward, implementation of a Bayesian method is often non-trivial since the integrations involved in Bayesian computation, such as that in (11.1), are often of high dimensions and intractable, and these integrals usually do not have closed-form or analytic expressions, except in some special cases. Computational challenges are partially due to possible high dimensionality of the parameters θ. The developments of MCMC methods, such as the Gibbs sampler, make such tedious computation feasible. A detailed description of MCMC methods is given in Chapter 12.

MCMC methods are often used to generate large samples from the posterior distribution $f(\theta|\mathbf{y})$, and these samples are then used for Bayesian inference.

For example, the widely used Gibbs sampler method breaks down the dimensionality of $\boldsymbol{\theta}$ by iteratively sampling from lower dimensional distributions which are easier to sample. These MCMC methods are often combined with rejection sampling methods or importance sampling methods. Note that, however, although modern computers are increasingly fast, these MCMC methods can still be computationally very intensive and it is not always easy to check the convergence of these iterative algorithms.

To avoid numerical integrations or simulation methods, which may be computationally intensive, alternatively we can consider Bayesian estimation based on *posterior mode* rather than posterior mean (e.g., Santner and Duffy 1989). The idea is to find an estimator $\tilde{\boldsymbol{\theta}}_B$, called *posterior mode estimator*, which maximizes the posterior density $f(\boldsymbol{\theta}|\mathbf{y})$ or maximizes the log posterior likelihood:

$$l_p(\boldsymbol{\theta}|\mathbf{y}) = \log L(\boldsymbol{\theta}|\mathbf{y}) + \log f(\boldsymbol{\theta}).$$

If a non-informative prior is chosen for $\boldsymbol{\theta}$, the posterior mode estimator coincides with the MLE. Note that the posterior mode estimation is also closely related to the Laplace approximation method (Breslow and Clayton 1993).

It can be shown that, under similar regularity conditions as that for asymptotic normality of MLE, the posterior mode estimator $\tilde{\boldsymbol{\theta}}_B$ is asymptotically normal:

$$\tilde{\boldsymbol{\theta}}_B \xrightarrow{d} N(\boldsymbol{\theta}, I_p^{-1}(\boldsymbol{\theta})), \qquad \text{as } n \to \infty,$$

where

$$I_p(\boldsymbol{\theta}) = -E\left(\frac{\partial^2 l_p(\boldsymbol{\theta}|\mathbf{y})}{\partial\boldsymbol{\theta}\partial\boldsymbol{\theta}^T}\right).$$

The posterior mode $\tilde{\boldsymbol{\theta}}_B$ and the curvature $I_p^{-1}(\tilde{\boldsymbol{\theta}}_B)$ can be used to approximate the posterior mean and covariance matrix when they are difficult to compute.

11.2.2 Prior Distributions

In Bayesian inference, the choice of prior distributions is important since it may affect the final results. In choosing the prior distributions, if there is no inherent reason to prefer one prior distribution over another, a conjugate prior is sometimes chosen for simplicity. A *conjugate prior* is a (parametric) prior distribution for which the resulting posterior distribution also belongs to the same family of distributions. This is important since Bayesian inference is based on the posterior distribution. Specifically, the prior distribution $f(\boldsymbol{\theta})$ is *conjugate* to $f(\mathbf{y}|\boldsymbol{\theta})$ if the posterior distribution $f(\boldsymbol{\theta}|\mathbf{y})$ is in the same family as the prior distribution $f(\boldsymbol{\theta})$.

For example, the normal distribution (Gaussian family) is conjugate to itself, i.e., if a prior distribution is normal then the posterior distribution is also normal. In fact, all members of the exponential family have conjugate priors (Gelman et al. 2003). In regression models, we typically choose a multivariate normal distribution as a prior distribution for the mean parameters β, i.e., we typically assume that $\beta \sim N(\beta_0, \Sigma_0)$, where β_0 and Σ_0 are hyper-parameters. For a non-informative prior, we can choose $\Sigma_0^{-1} = 0$ or $\beta \sim \text{uniform}(-\infty, \infty)$. For variance-covariance matrices, we typically choose Wishart distributions as prior distributions, which are described as follows.

The Wishart distribution is a generalization of the χ^2 distribution to multiple dimensions or a generalization of the gamma distribution. It is useful for estimation of covariance matrices. Suppose that Z is an $n \times p$ matrix, with i-th row $\mathbf{z}_i \sim N_p(0, V)$ independently, where the $p \times p$ covariance matrix V is positive definite. Then, the probability distribution of

$$W = Z^T Z$$

has a *Wishart distribution* with degrees of freedom n, denoted by $W_p(V, n)$ or $W(V, n)$, and a density function given by

$$f(W) = \frac{|W|^{(n-p-1)/2}}{2^{np/2}|V|^{n/2}\Gamma_p(\frac{n}{2})} \exp\left(-\frac{1}{2}tr(V^{-1}W)\right),$$

where $W > 0$ (positive definite), and $\Gamma_p(\cdot)$ is the multivariate gamma function defined as

$$\Gamma_p(n/2) = \pi^{p(p-1)/4} \prod_{j=1}^{p} \Gamma((n+1-j)/2).$$

The Wishart distribution $W_p(V, n)$ has the mean nV and the mode $(n-p-1)V$ for $n \geq p + 1$. When $p = 1$ and $V = 1$, the Wishart distribution $W_p(V, n)$ reduces to a χ_n^2-distribution. Note that the Wishart distribution is the distribution of the MLE for the covariance matrix in a multivariate normal distribution.

In Bayesian inference, a conjugate prior for the covariance matrix of a multivariate normal distribution is the inverse Wishart distribution, defined as follows. If a $p \times p$ random matrix $A \sim W_p(V, n)$, then $B = A^{-1}$ has an *inverse Wishart distribution* (or *inverted Wishart distribution*), denoted by $W_p^{-1}(V^{-1}, n)$ or $W^{-1}(V^{-1}, n)$, with probability density function

$$f(B) = \frac{|V|^{-n/2}|B|^{-(n+p+1)/2}\exp(-tr(V^{-1}B^{-1})/2)}{2^{np/2}\Gamma_p(n/2)}.$$

The mean of $B \sim W_p^{-1}(V^{-1}, n)$ is given by

$$E(B) = V^{-1}/(n - p - 1).$$

Let $X = (\mathbf{x}_1, \cdots, \mathbf{x}_n)$, with $\mathbf{x}_i \sim N_p(0, \Sigma)$. If we assume a prior distribution

$\Sigma \sim W_p^{-1}(\Phi, m)$, then the posterior distribution is given by

$$\Sigma | X \sim W_p^{-1}(XX^T + \Phi, \ m + n).$$

When $p = 1$, the inverse Wishart distribution becomes a inverse gamma distribution.

11.3 Bayesian Mixed Effects Models

The general Bayesian approach described in Section 11.1 can be applied to mixed effects models. In the following sections, we follow Searle et al. (1992) and Davidian and Giltinan (1995), and consider Bayesian inference for mixed effects models.

As noted in Section 11.1, in Bayesian estimation analytic expressions are often unavailable, so Monte Carlo methods are often used, which can be computationally intensive. For some special cases such as linear models, however, some analytic expressions can be obtained. These analytic expressions may offer some insights about Bayesian estimators. Therefore, in the following we first focus on Bayesian LME models in which some analytic expressions are presented. We then discuss Bayesian methods for GLMM and NLME models in which estimation is typically based on MCMC methods.

11.3.1 Bayesian LME Models

We first consider Bayesian LME models in which prior distributions are assumed for the mean parameters, but the variance components are assumed to be known. Then, we consider estimation of the variance components.

Let $\mathbf{y}_i = (y_{i1}, \cdots, y_{in_i})^T$ be the responses in cluster i or individual i, $i = 1, 2, \cdots, n$. We first consider the following Bayesian LME model

$$
\begin{aligned}
\mathbf{y}_i &= X_i \boldsymbol{\beta} + Z_i \mathbf{b}_i + \mathbf{e}_i, \qquad i = 1, \cdots, n, & (11.2) \\
\mathbf{b}_i &\sim N(0, D), \qquad \mathbf{e}_i \sim N(0, R_i), \\
\boldsymbol{\beta} &\sim N(\boldsymbol{\beta}_0, \Sigma_0), & (11.3)
\end{aligned}
$$

where $X_i(n_i \times p)$ and $Z_i(n_i \times q)$ are known design matrices, $\mathbf{b}_i = (b_{i1}, \cdots, b_{iq})^T$ are random effects, $\mathbf{e}_i = (e_{i1}, \cdots, e_{in_i})^T$ are within-individual errors, and (11.3) specifies a prior distribution for the mean fixed parameters $\boldsymbol{\beta} = (\beta_1, \cdots, \beta_p)^T$. We assume that the hyper-parameters $\boldsymbol{\beta}_0$ and Σ_0 are known, and we also assume that $R_i = \sigma^2 I_{n_i}$ for simplicity, where I_{n_i} is an $n_i \times n_i$ identity matrix.

For convenience of presentation, we write the Bayesian LME model (11.2) –

(11.3) in a more compact form as follows. Let $N = \sum_{i=1}^{n} n_i$, and let

$$
\mathbf{y} = \begin{pmatrix} \mathbf{y}_1 \\ \vdots \\ \mathbf{y}_n \end{pmatrix}_{N \times 1}, \qquad
\mathbf{b} = \begin{pmatrix} \mathbf{b}_1 \\ \vdots \\ \mathbf{b}_n \end{pmatrix}_{nq \times 1}, \qquad
\mathbf{X} = \begin{pmatrix} X_1 \\ \vdots \\ X_n \end{pmatrix}_{N \times p},
$$

$D^* = diag(D, \cdots, D)$ $(nq \times nq)$, $Z = diag(Z_1, \cdots, Z_n)$ $(N \times nq)$, and $R = diag(R_1, \cdots, R_n)$ $(N \times N)$. Then, Bayesian LME model (11.2) – (11.3) can be combined as a single model:

$$
\begin{aligned}
\mathbf{y} &= X\boldsymbol{\beta} + Z\mathbf{b} + \mathbf{e}, & \text{(11.4)} \\
\mathbf{b} &\sim N(0, D^*), & \mathbf{e} \sim N(0, R), \\
\boldsymbol{\beta} &\sim N(\boldsymbol{\beta}_0, \Sigma_0). & \text{(11.5)}
\end{aligned}
$$

Note that

$$
\mathbf{y}|\mathbf{b} \sim N(X\boldsymbol{\beta} + Z\mathbf{b}, R), \qquad \mathbf{y} \sim N(X\boldsymbol{\beta}, V),
$$

where $V = R + ZD^*Z^T$.

Bayesian estimation of the mean parameter $\boldsymbol{\beta}$ can be based on the following posterior distribution

$$
f(\boldsymbol{\theta}|\mathbf{y}) = \frac{f(\mathbf{y}|\boldsymbol{\theta})f(\boldsymbol{\theta})}{f(\mathbf{y})} = \frac{\int f(\mathbf{y}|\boldsymbol{\theta}, \mathbf{b})f(\boldsymbol{\theta})f(\mathbf{b}) \, d\mathbf{b}}{\int \int f(\mathbf{y}|\boldsymbol{\theta}, \mathbf{b})f(\boldsymbol{\theta})f(\mathbf{b}) \, d\mathbf{b} \, d\boldsymbol{\theta}}.
$$

Since the distributions of $\mathbf{y}$, $\mathbf{b}$, and $\boldsymbol{\beta}$ are all multivariate normal, the joint distribution of $(\mathbf{y}, \mathbf{b}, \boldsymbol{\beta})$ is also multivariate normal. Then, based on properties of multivariate normal distributions, it can be shown that the posterior distribution of $\boldsymbol{\beta}$ is given by (Searle et al. 1992; Davidian and Giltinan 1995)

$$
\boldsymbol{\beta}|\mathbf{y} \sim N(C^{-1}(X^T V^{-1}\mathbf{y} + \Sigma_0^{-1}\boldsymbol{\beta}_0), \ C^{-1}),
$$

where $C = X^T V^{-1} X + \Sigma_0^{-1}$. Bayesian inference for $\boldsymbol{\beta}$ can then be based on this posterior distribution.

For example, a Bayesian estimate of $\boldsymbol{\beta}$ is given by

$$
\hat{\boldsymbol{\beta}}_B = E(\boldsymbol{\beta}|\mathbf{y}) = C^{-1}(X^T V^{-1}\mathbf{y} + \Sigma_0^{-1}\boldsymbol{\beta}_0),
$$

with variance-covariance given by

$$
Cov(\hat{\boldsymbol{\beta}}_B) = Cov(E(\boldsymbol{\beta}|\mathbf{y})) = C^{-1}(X^T V^{-1} X)C^{-1}.
$$

Similarly, it can be shown that the Bayesian estimate of the random effects is given by

$$
\hat{\mathbf{b}}_B = E(\mathbf{b}|\mathbf{y}) = (Z^T L Z + D^*)^{-1} Z^T L(\mathbf{y} - X\boldsymbol{\beta}_0),
$$

where $L = (R + X\Sigma_0 X^T)^{-1}$, with variance-covariance

$$
Cov(\hat{\mathbf{b}}_B) = Cov(\mathbf{b}|\mathbf{y}) = (Z^T L Z + D^*)^{-1}.
$$

If we assume a non-informative prior for β, i.e., $\Sigma_0^{-1} = 0$ or $\beta \sim$ uniform$(-\infty, \infty)$, the Bayesian estimator of β becomes the familiar MLE (with known variance components):

$$\hat{\beta}_{MLE} = (X^T V^{-1} X)^{-1} X^T V^{-1} \mathbf{y},$$

with variance

$$Cov(\hat{\beta}_{MLE}) = (X^T V^{-1} X)^{-1}.$$

In other words, the MLE is a Bayesian estimator with non-informative prior.

When the variance components D and R are unknown, we can substitute them by their usual estimates. The resulting Bayesian estimators of the mean parameters β and random effects $\mathbf{b}$ are then called *empirical Bayes estimators*.

Note that, if we assume non-informative prior for β, i.e., $\Sigma_0^{-1} = 0$ or $\beta \sim$ uniform$(-\infty, \infty)$, estimation of the variance components D and R can be based on the *marginal likelihood*, obtained by integrating out β and $\mathbf{b}$,

$$L_m(D, R|\mathbf{y}) = \int \int f(\mathbf{y}|\mathbf{b}, \beta) f(\mathbf{b}|D) f(\beta) \, d\beta \, d\mathbf{b}.$$

It can be shown that estimates of D and R based on the above marginal likelihood is equivalent to the *restricted maximum likelihood estimates (REMLs)* of D and R (Searle et al. 1992; Davidian and Giltinan 1995). In other words, the REMLs of D and R can be obtained from a Bayesian framework by assuming an non-informative prior for the mean parameters β and then integrating out β and $\mathbf{b}$ in the likelihood.

For a *full Bayesian* inference of LME models, we should also choose prior distributions for the variance components D and R. We may consider the following prior distributions for the variance-covariance matrices D and R (note $R_i = \sigma^2 I_{n_i}$):

$$D \sim W_q^{-1}(D_0^{-1}, \rho_0), \qquad \sigma^{-2} \sim G\left(\frac{\nu_0}{2}, \frac{\nu_0 \tau_0}{2}\right), \tag{11.6}$$

where $W^{-1}(\cdot)$ and $G(\cdot)$ denote the inverse Wishart distribution and the gamma distribution respectively. The joint posterior distribution for all unknown parameters and random effects is then given by

$$f(\beta, \sigma^2, D, \mathbf{b}|\mathbf{y}) = \frac{f(\mathbf{y}|\beta, \sigma, \mathbf{b}) f(\mathbf{b}|D) f(\beta, \sigma, D)}{f(\mathbf{y})},$$

where

$$f(\mathbf{y}) = \int \int \int \int f(\mathbf{y}|\beta, \sigma, \mathbf{b}) f(\mathbf{b}|D) f(\beta, \sigma, D) \, d\beta \, d\sigma \, dD \, d\mathbf{b}.$$

Thus, for example, Bayesian inference for the mean parameters β can be based on the following posterior distribution

$$f(\beta|\mathbf{y}) = \int \int \int f(\mathbf{y}|\beta, \sigma, \mathbf{b}) f(\mathbf{b}|D) f(\beta, \sigma, D) \, d\sigma \, dD \, d\mathbf{b} \Big/ f(\mathbf{y}).$$

It is a common practice to assume that the prior distributions are independent, i.e.,

$$f(\beta, \sigma, D) = f(\beta) f(\sigma) f(D),$$

which greatly simplifies the computation. Sometimes we also assume diagonal matrices for the covariance matrices Σ_0 and D_0 to reduce the number of parameters.

Consider the simpler case where $X_i = Z_i$ and $\beta_i = \beta + \mathbf{b}_i$. Let $\beta_* = \{\beta_1, \cdots, \beta_n\}$ be the individual-specific parameters. The full conditional distributions of any of the parameters given the remaining parameters can be written as (Wakefield et al. 1994):

$$[\beta|\mathbf{y}, \sigma, D, \beta_*] \sim N\left(U(nD^{-1}\bar{\beta} + \Sigma_0^{-1}\beta_0), \ U\right),$$

$$[D|\mathbf{y}, \sigma, \beta, \beta_*] \sim W^{-1}\left(\left[\sum_{i=1}^{n}(\beta_i - \beta)(\beta_i - \beta)^T + D_0\right]^{-1}, \ n + \rho_0\right),$$

$$[\sigma^{-2}|\mathbf{y}, \beta, D, \beta_*] \sim G\left(\frac{\nu_0 + N}{2}, \frac{1}{2}\left[\sum_{i=1}^{n}(\mathbf{y}_i - X_i\beta_i)^T(\ \cdot\ ) + \nu_0\tau_0\right]\right),$$

$$[\beta_i|\mathbf{y}, \beta, \sigma, D, \beta_j, j \neq i] \sim N\left(H_i(\sigma^{-2}X_i^T\mathbf{y}_i + D^{-1}\beta), \ H_i\right),$$

where

$$\bar{\beta} = \sum_{i=1}^{n}\beta_i/n, \quad H_i = \sigma^{-2}X_i^TX_i + D^{-1}, \quad U^{-1} = nD^{-1} + \sigma_0^{-1}.$$

Note that many of the foregoing analytic expressions are derived mainly based on two facts for LME models: (i) the response $\mathbf{y}_i$, the parameters β, the random effects $\mathbf{b}_i$, and the random errors $\mathbf{e}_i$ are all linked in a *linear* form in the model; and (ii) the parameters β, the random effects $\mathbf{b}_i$, and the random errors $\mathbf{e}_i$ are all assumed to be normally distributed. Many of the above analytic expressions, however, are no longer available for Bayesian GLMM or NLME models since these models are nonlinear.

11.3.2 Bayesian GLMMs

Conceptually, the Bayesian LME models in Section 11.3.1 can be extended to Bayesian GLMMs in a straightforward way. However, a Bayesian GLMM differs from a Bayesian LME model in two ways:

- there is a *nonlinear* relationship between the response and the parameters and the random effects,

- the response does not follow a normal distribution but instead a distribution in the exponential family.

These differences make some of the analytic expressions in Section 11.3.1 unavailable, so MCMC methods are typically used for inference of a Bayesian GLMM. We briefly describe Bayesian GLMMs and inference as follows, based on Zeger and Karim (1991) and Gelman et al. (2003).

Suppose that the responses $\{y_{i1}, \cdots, y_{in_i}\}$ in the i-th cluster are conditionally independent given the mean parameters β and random effects $\mathbf{b}_i$. Let $\mathbf{y}_i = (y_{i1}, \cdots, y_{in_i})^T$. A full *Bayesian generalized linear mixed model (GLMM)* can be written as

$$E(\mathbf{y}_i|\beta, \mathbf{b}_i) = h(X_i\beta + Z_i\mathbf{b}_i), \qquad i = 1, \cdots, n, \qquad (11.7)$$
$$\mathbf{b}_i \sim N(0, D), \qquad (11.8)$$
$$\beta \sim N(\beta_0, \Sigma_0), \qquad D \sim W_q^{-1}(\eta, \Psi), \qquad (11.9)$$

where $h(\cdot)$ is a known link function and X_i and Z_i are known design matrices. Let $\mathbf{b} = (\mathbf{b}_1, \cdots, \mathbf{b}_n)$. We assume that the prior distributions are independent, so

$$f(\beta, D) = f(\beta)f(D).$$

Then, the posterior distribution of all parameters can be written as

$$f(\beta, D, \mathbf{b}|\mathbf{y}) \propto \left[\prod_{i=1}^{n}\prod_{j=1}^{n_i} f(y_{ij}|\beta, \mathbf{b}_i)f(\beta)\right]\left[\prod_{i=1}^{n} f(\mathbf{b}_i|D)f(D)\right].$$

For Bayesian inference, note that the full conditionals are given by

$$f(\beta|D, \mathbf{b}, \mathbf{y}) \propto \prod_{i=1}^{n}\prod_{j=1}^{n_i} f(y_{ij}|\beta, \mathbf{b}_i)f(\beta), \qquad (11.10)$$

$$f(\mathbf{b}|\beta, D, \mathbf{y}) \propto \prod_{i=1}^{n}\prod_{j=1}^{n_i} f(y_{ij}|\beta, \mathbf{b}_i)f(\mathbf{b}_i|D), \qquad (11.11)$$

$$f(D|\beta, \mathbf{b}, \mathbf{y}) \propto \prod_{i=1}^{n} f(\mathbf{b}_i|D)f(D), \qquad (11.12)$$

where

$$[D|\beta, \mathbf{b}, \mathbf{y}] \sim W_q^{-1}(\eta + n/2, \ \Psi + \sum_{i=1}^{n} \mathbf{b}_i\mathbf{b}_i^T/2).$$

Bayesian inference can then be based on the Gibbs sampler along with rejection sampling methods (Zeger and Karim 1991; Gelman et al. 2003). A

Gibbs sampler method to generate samples from the posterior distribution $f(\boldsymbol{\beta}, D, \mathbf{b}|\mathbf{y})$ is described as follows. At k-th iteration

- sample $\boldsymbol{\beta}^{(k)}$ from $f(\boldsymbol{\beta}|D^{(k-1)}, \mathbf{b}^{(k-1)}, \mathbf{y})$;
- sample $D^{(k)}$ from $f(D|\boldsymbol{\beta}^{(k)}, \mathbf{b}^{(k-1)}, \mathbf{y})$;
- sample $\mathbf{b}^{(k)}$ from $f(\mathbf{b}|\boldsymbol{\beta}^{(k)}, D^{(k)}, \mathbf{y})$, $k = 1, 2, 3, \cdots$.

Beginning with starting values $(\boldsymbol{\beta}^{(0)}, D^{(0)}, \mathbf{b}^{(0)})$, after a warm-up period we obtain a sample of $(\boldsymbol{\beta}, D, \mathbf{b})$ from the posterior distribution $f(\boldsymbol{\beta}, D, \mathbf{b}|\mathbf{y})$. Once we generate many such samples, the posterior mean and posterior covariance can be approximated by the sample mean and sample covariance based on the simulated samples. As one can imagine, this procedure can be computationally intensive.

11.3.3 Bayesian NLME Models

Bayesian NLME models are similar to Bayesian GLMMs, but in Bayesian NLME models the responses are assumed to follow normal distributions. Bayesian NLME models differ from Bayesian LME models in that the responses in NLME models are nonlinearly related to the parameters and random effects. We briefly describe Bayesian NLME models as follows, following Davidian and Giltinan (1995) and Wakefield (1995).

Let $\mathbf{y}_i = (y_{i1}, \cdots, y_{in_i})^T$ be the responses from cluster i. We consider the following Bayesian NLME model

$$
\begin{aligned}
\mathbf{y}_i &= \mathbf{f}_i(\boldsymbol{\beta}_i) + \mathbf{e}_i, & \mathbf{e}_i &\sim N(0, \sigma^2 I_{n_i}), & &\quad(11.13)\\
\boldsymbol{\beta}_i &= \boldsymbol{\beta} + \mathbf{b}_i, & \mathbf{b}_i &\sim N(0, D), & i = 1, \cdots, n,&\\
\sigma^{-2} &\sim G(\nu_0, \tau_0), & \boldsymbol{\beta} \sim N(\boldsymbol{\beta}_0, \Sigma_0), & \quad D \sim W_q^{-1}(D_0, \rho_0), & &\quad(11.14)
\end{aligned}
$$

where $\mathbf{f}_i(\cdot)$ is a vector-valued known nonlinear function. We assume that the prior distributions are mutually independent:

$$ f(\boldsymbol{\beta}, \sigma, D) = f(\boldsymbol{\beta})f(\sigma)f(D). $$

Let $\mathbf{b} = (\mathbf{b}_1, \cdots, \mathbf{b}_n)$ be the random effects. The posterior distribution of all parameters and random effects can be written as

$$ f(\boldsymbol{\beta}, D, \sigma, \mathbf{b}|\mathbf{y}) \propto \prod_{i=1}^{n} \left[f(\mathbf{y}_i|\boldsymbol{\beta}, \sigma, \mathbf{b}_i) f(\mathbf{b}_i|D) f(D) f(\boldsymbol{\beta}) f(\sigma) \right]. $$

Let $\boldsymbol{\beta}_* = (\boldsymbol{\beta}_1, \cdots, \boldsymbol{\beta}_n)$ be the individual-specific parameters, and let $\mathbf{y} = (\mathbf{y}_1, \cdots, \mathbf{y}_n)$. Following Gelfand and Smith (1990) and Davidian and Giltinan (1995), the full conditional distributions can be shown to have the following distributions

$$[\beta|\sigma^2, D, \beta_*, \mathbf{y}] \sim N\left((nD^{-1} + \Sigma_0^{-1})^{-1}\left(D^{-1}\sum_{i=1}^{n}\beta_i + \Sigma_0^{-1}\beta_0\right),\right.$$

$$\left.(nD^{-1} + \Sigma_0^{-1})^{-1}\right),$$

$$[\sigma^{-2}|\beta, D, \beta_*, \mathbf{y}] \sim G\left(\nu_0 + \frac{1}{2}\sum_{i=1}^{n}n_i, \left[\tau_0^{-1} + \frac{1}{2}\sum_{i=1}^{n}\sum_{j=1}^{n_i}(y_{ij} - f_{ij}(\beta_i))^2\right]^{-1}\right),$$

$$[D|\sigma, \beta, \beta_*, \mathbf{y}] \sim W_q^{-1}\left(\left[D_0^{-1}\sum_{i=1}^{n}(\beta_i - \beta)(\beta_i - \beta)^T\right]^{-1}, n + \rho\right).$$

The full conditional distribution of β_i does not have a closed form expression, but we have

$$f(\beta_i|\beta, \sigma, D, \mathbf{y}) \propto \exp\left\{-\frac{\sigma^{-2}}{2}\sum_{j=1}^{n_i}(y_{ij} - f_{ij}(\beta_i))^2 - \frac{1}{2}(\beta_i - \beta)^T D^{-1}(\beta_i - \beta)\right\}.$$

Now we can use the Gibbs sampler method to generate samples from the posterior distribution $f(\beta, D, \sigma, \mathbf{b}|\mathbf{y})$ as follows. At k-th iteration

- sample $\beta^{(k)}$ from $f(\beta|\sigma^{2(k-1)}, D^{(k-1)}, \beta_*^{(k-1)}, \mathbf{y})$;
- sample $\sigma^{-2(k)}$ from $f(\sigma^{-2}|\beta^{(k)}, D^{(k-1)}, \beta_*^{(k-1)}, \mathbf{y})$;
- sample $D^{(k)}$ from $f(D|\sigma^{-2(k)}, \beta^{(k)}, \beta_*^{(k-1)}, \mathbf{y})$;
- sample $\beta_i^{(k)}$ from $f(\beta_i|\beta^{(k)}, \sigma^{(k)}, D^{(k)}, \mathbf{y})$, $k = 1, 2, 3, \cdots$.

Beginning with starting values $(\beta^{(0)}, D^{(0)}, \sigma^{(0)}, \mathbf{b}^{(0)})$, after a burn-in period we obtain a sample of $(\beta, D, \sigma, \mathbf{b})$ from the posterior distribution $f(\beta, D, \sigma, \mathbf{b}|\mathbf{y})$. Repeating this process many times, we can obtain many independent samples from the target posterior distribution. Then, we can approximate the posterior means and covariances by their corresponding sample means and sample covariances based on the simulated samples, which are the approximate Bayesian estimates of the means and covariances.

Example 11.1 *A Bayesian NLME model for HIV viral dynamics*

For illustration, we consider a Bayesian model for long-term HIV viral dynamics with time-varying drug efficacy, following Huang and Lu (2009). HIV viral dynamic models describe the interaction between cells susceptible to target cells (T), infected cells (T^*), and free virus (V). Let λ be the rate at which new T cells are created from sources within the body, d_T be the death rate of T cells, k be the infection rate of T cells infected by virus, δ be the death rate

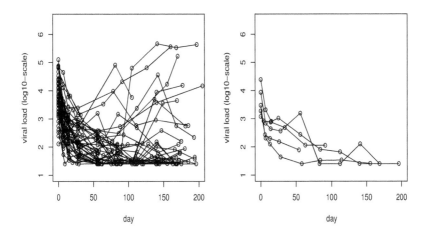

Figure 11.1 *Viral loads trajectories (in* log_{10} *scale). The open circles are observed values. Left figure: all patients. Right figure: five randomly selected patients.*

for infected cells, m be the number of new virions produced from each of the infected cells during their life time, and c be the clearance rate of free virions.

A long-term HIV viral dynamic model can be written as a solution of the following set of differential equations (Perelson and Nelson 1999, Huang and Lu 2009):

$$\frac{d}{dt}\tilde{T} = d_T[1 - \tilde{T} - (1 - \gamma(t))\tilde{T}\tilde{V}],$$
$$\frac{d}{dt}\tilde{T}^* = \delta[(1 - \gamma(t))\tilde{T}\tilde{V} - \tilde{T}^*], \qquad (11.15)$$
$$\frac{d}{dt}\tilde{V} = c(r\tilde{T}^* - \tilde{V}),$$

where $\tilde{T} = (d_T/\lambda)T$, $\tilde{T}^* = (\delta/\lambda)T^*$, $\tilde{V} = (k/d_T)V$, and $\gamma(t)$ is a time-varying parameter quantifying the antiviral drug efficacy and it can be written as a function of a parameter ϕ (Huang and Lu 2009). Model (11.15) is nonlinear and it does not have a closed-form solution.

Figure 11.1 shows the AIDS dataset used in the analysis which has a sample size of 42. We see that there are substantial variations between patients, so we should allow the parameters in model (11.15) to vary across patients by introducing random effects to the parameters. There are also random errors for within-patient viral load measurements. By incorporating between-patient ran-

Table 11.1 *Estimated posterior means and the corresponding 95% equal-tail credible intervals for the parameters*

Parameter	ϕ	c	δ	d_T	ρ	r
Posterior mean	14.21	4.73	0.36	0.016	3.70	2.61
Left credit limit	3.39	1.57	0.11	0.003	0.58	1.52
Right credit limit	37.13	8.40	0.53	0.075	7.14	4.07

dom effects and within-patient random errors in the nonlinear model (11.15), we obtain a NLME model, although the NLME model does not have an analytic expression. Furthermore, by introducing prior distributions for the model parameters, we obtain a Bayesian NLME model, as described below.

Let $\boldsymbol{\beta} = (\log \phi, \log c, \log \delta, \log d_T, \log \rho, \log r)^T$ be the fixed mean parameters in the NLME model. A Bayesian NLME model corresponding to model (11.15) can be written in the following general form

$$y_{ij} = g_{ij}(\boldsymbol{\beta}_i, t_{ij}) + e_{ij}, \qquad e_{ij} \text{ i.i.d. } \sim N(0, \sigma^2), \qquad (11.16)$$

$$\boldsymbol{\beta}_i = \boldsymbol{\beta} + \mathbf{b}_i, \qquad \mathbf{b}_i \sim N(0, D), \quad i = 1, \cdots, n; \; j = 1, \cdots, n_i,$$

$$\boldsymbol{\beta} \sim N(\boldsymbol{\beta}_0, \Gamma_0), \quad \sigma^{-2} \sim G(a_0, b_0), \quad D^{-1} \sim W(\Omega_0, \nu_0), \qquad (11.17)$$

where y_{ij} is the $\log_{10}$-transformation of viral load for patient i at time t_{ij}, $G(a_0, b_0)$ is a gamma distribution, $W(\Omega_0, \nu_0)$ is a Wishart distribution, $g_{ij}(\cdot)$ is a nonlinear function without analytic expression, which is a solution of model (11.15), and $(\boldsymbol{\beta}_0, \Gamma_0, a_0, b_0, \Omega_0, \nu_0)$ are known hyper-parameters.

Bayesian inference of the parameters is based on their posterior distributions given the observed data. Here the posterior distributions are quite intractable, but one can generate many samples from the posterior distribution using the Gibbs sampler and then obtain summary statistics from these samples to conduct approximate Bayesian inference.

Huang and Lu (2009) analyzed the data based on the above Bayesian NLME model. The values of the hyper-parameters were chosen as follows: $a = 4.5$, $b = 9.0$, $\nu_0 = 8.0$, $\boldsymbol{\beta}_0 = (4.0, 1.1, -1.0, -2.5, 1.4, 0.28)^T$, $\Gamma_0 = diag(1000, 1000, 1000, 1000, 1000, 1000)$, $\Omega_0 = diag(2.0, 2.0, 2.0, 2.0, 2.0, 2.0)$. In the Gibbs sampler, they took 30,000 iterations as a burn-in period, and then took every 5th MCMC sample which led to 24,000 samples from the targeted posterior distribution. Estimated posterior means and the corresponding 95% equal-tail credible intervals for the parameters were then obtained from these samples. Table 11.1 shows the results. Huang and Lu (2009) compared these Bayesian estimates with estimates in previous studies using non-Bayesian methods and discussed advantages of the Bayesian method.

11.4 Bayesian Mixed Models with Missing Data

11.4.1 Bayesian Models with Missing Data

The Bayesian methods described in the previous sections can be extended to models with missing data. In the following, we focus on Bayesian regression models for cross-sectional data with missing covariates and a non-ignorable missing data mechanism, following Ibrahim, Chen, and Lipsitz (2002). When covariates have missing data, we need to assume a model for the covariates in Bayesian inference, We assume prior distributions for all model parameters, including prior distributions for the parameters in the covariate model. Bayesian estimation is then based on the posterior distribution of the parameters given the observed data. We briefly describe the approach as follows.

Let $f(y_i|\mathbf{x}_i, \boldsymbol{\beta})$ be the density function of a regression model for cross-sectional data, such as a GLM, where $\mathbf{x}_i$ contains covariates. In the presence of missing data in the covariates $\mathbf{x}_i$, we assume a model for the covariates, with density $f(\mathbf{x}_i|\boldsymbol{\alpha})$. For simplicity, we assume $\mathbf{x}_i \sim N(\boldsymbol{\alpha}_1, \boldsymbol{\alpha}_2)$ with $\boldsymbol{\alpha} = (\boldsymbol{\alpha}_1, \boldsymbol{\alpha}_2)$. Write $\mathbf{x}_i = (\mathbf{x}_{mis,i}, \mathbf{x}_{obs,i})$, where $\mathbf{x}_{mis,i}$ contains the missing components and $\mathbf{x}_{obs,i}$ contains the observed components of $\mathbf{x}_i$. Let $(\mathbf{y}, \mathbf{x}_{obs}) = \{(y_i, \mathbf{x}_{obs,i}), i = 1, \cdots, n\}$ be the observed data. The joint posterior density for all the parameters $(\boldsymbol{\beta}, \boldsymbol{\alpha})$ can be written as

$$f(\boldsymbol{\beta}, \boldsymbol{\alpha}|\mathbf{y}, \mathbf{x}_{obs}) \propto \left\{ \prod_{i=1}^{n} \int f(y_i|\mathbf{x}_i, \boldsymbol{\beta}) f(\mathbf{x}_{mis,i}|\mathbf{x}_{obs,i}, \boldsymbol{\alpha}) \, d\mathbf{x}_{mis,i} \right\}$$
$$\times f(\boldsymbol{\beta}, \boldsymbol{\alpha}),$$

where $f(\boldsymbol{\beta}, \boldsymbol{\alpha})$ is the joint density of the prior distributions for $\boldsymbol{\beta}$ and $\boldsymbol{\alpha}$. Bayesian inference for the parameters $(\boldsymbol{\beta}, \boldsymbol{\alpha})$ can then be based on the posterior distribution $f(\boldsymbol{\beta}, \boldsymbol{\alpha}|\mathbf{y}, \mathbf{x}_{obs})$. In the following, we consider the Gibbs sampler method to generate many samples from the posterior distribution and then carry out approximate Bayesian inference based on the simulated samples.

Sampling from the posterior distribution $f(\boldsymbol{\beta}, \boldsymbol{\alpha}|\mathbf{y}, \mathbf{x}_{obs})$ can be accomplished using the Gibbs sampler by iteratively sampling from the following full conditionals:

- sample $\boldsymbol{\beta}$ from $f(\boldsymbol{\beta}|\mathbf{y}, \mathbf{x}, \boldsymbol{\alpha}_1, \boldsymbol{\alpha}_2)$;
- sample $\boldsymbol{\alpha}_1$ from $f(\boldsymbol{\alpha}_1|\mathbf{y}, \mathbf{x}, \boldsymbol{\beta}, \boldsymbol{\alpha}_2)$;
- sample $\boldsymbol{\alpha}_2$ from $f(\boldsymbol{\alpha}_2|\mathbf{y}, \mathbf{x}, \boldsymbol{\beta}, \boldsymbol{\alpha}_1)$;
- sample $\mathbf{x}_{mis,i}$ from $f(\mathbf{x}_{mis,i}|\mathbf{y}, \mathbf{x}_{obs,i}, \boldsymbol{\beta}, \boldsymbol{\alpha}_1, \boldsymbol{\alpha}_2)$, $\quad i = 1, \cdots, n$.

After a burn-in period, we obtain a sample from $f(\boldsymbol{\beta}, \boldsymbol{\alpha}|\mathbf{y}, \mathbf{x}_{obs})$. Repeating the procedure many times leads to many independent samples from the posterior distribution. Then, approximate Bayesian inference can be based on these samples.

For example, suppose that the following prior distributions are assumed:

$$\boldsymbol{\beta} \sim N(\boldsymbol{\beta}_0, \Sigma_0), \quad \boldsymbol{\alpha}_1 \sim N(\boldsymbol{\alpha}_{10}, R_0), \quad \text{and} \quad \boldsymbol{\alpha}_2 \sim W^{-1}(\nu_0, G_0),$$

where $(\boldsymbol{\beta}_0, \Sigma_0, \boldsymbol{\alpha}_{10}, R_0, \nu_0, G_0)$ are known hyper-parameters. To sample from the above full conditionals, note that

$$f(\boldsymbol{\beta}|\mathbf{y}, \mathbf{x}, \boldsymbol{\alpha}) \propto \prod_{i=1}^{n} f(y_i|\mathbf{x}_i, \boldsymbol{\beta}) f(\boldsymbol{\beta}),$$

$$f(\mathbf{x}_{mis,i}|\mathbf{y}, \mathbf{x}_{obs,i}, \boldsymbol{\beta}, \boldsymbol{\alpha}) \propto \prod_{i=1}^{n} f(y_i|\mathbf{x}_i, \boldsymbol{\beta}) f(\mathbf{x}_{mis,i}|\mathbf{x}_{obs,i}, \boldsymbol{\alpha}),$$

where the density functions on the right-hand sides are known. Thus, rejection sampling methods may be used to sample from the full conditionals. Note that sampling from $f(\boldsymbol{\alpha}_1|\mathbf{y}, \mathbf{x}, \boldsymbol{\beta}, \boldsymbol{\alpha}_2)$ and $f(\boldsymbol{\alpha}_2|\mathbf{y}, \mathbf{x}, \boldsymbol{\beta}, \boldsymbol{\alpha}_1)$ is straightforward. Therefore, the Gibbs sampler combined with rejection sampling methods can be used to generate samples from the posterior distribution $f(\boldsymbol{\beta}, \boldsymbol{\alpha}|\mathbf{y}, \mathbf{x}_{obs})$. Bayesian inference then proceeds in the usual way.

11.4.2 Bayesian Mixed Models with Missing Data

Bayesian methods for mixed effects models with missing data are similar to that in Section 11.4.1. The only modification is to introduce prior distributions for the parameters in the random effects distributions. We briefly describe the method as follows, assuming the same setting as that in Section 11.4.1.

Let $f(y_i|\mathbf{x}_i, \mathbf{b}_i, \boldsymbol{\beta})$ be the conditional density function for a mixed effects model given the random effects $\mathbf{b}_i$, such as a LME, or GLMM, or NLME model. We assume the random effects $\mathbf{b}_i \sim N(0, D)$, with a prior distribution $D \sim W_q^{-1}(D_0, \rho_0)$. The joint posterior density for the parameters $(\boldsymbol{\beta}, D, \boldsymbol{\alpha})$ is given by

$$f(\boldsymbol{\beta}, D, \boldsymbol{\alpha}|\mathbf{y}, \mathbf{x}_{obs}) \propto \left\{ \prod_{i=1}^{n} \int \int f(y_i|\mathbf{x}_i, \mathbf{b}_i, \boldsymbol{\beta}) f(\mathbf{b}_i|D) \right.$$

$$\left. \times f(\mathbf{x}_{mis,i}|\mathbf{x}_{obs,i}, \boldsymbol{\alpha}) \, d\mathbf{b}_i \, d\mathbf{x}_{mis,i} \right\} f(\boldsymbol{\beta}, D, \boldsymbol{\alpha}),$$

where $f(\boldsymbol{\beta}, D, \boldsymbol{\alpha})$ is the joint prior distribution. We assume that the prior distributions are independent:

$$f(\boldsymbol{\beta}, D, \boldsymbol{\alpha}) = f(\boldsymbol{\beta}) f(D) f(\boldsymbol{\alpha}).$$

As in Section 11.4.1, we can generate many samples from the posterior distribution $f(\beta, D, \alpha | \mathbf{y}, \mathbf{x}_{obs})$ using the Gibbs sampler method, and then perform approximate Bayesian inference based on the generated samples.

Sampling from the posterior distribution $f(\beta, D, \alpha | \mathbf{y}, \mathbf{x}_{obs})$ can be accomplished using the Gibbs sampler by iteratively sampling from the following full conditionals:

- sample β from $f(\beta | \mathbf{y}, \mathbf{x}, \mathbf{b}, \alpha_1, \alpha_2)$;
- sample b from $f(\mathbf{b} | \mathbf{y}, \mathbf{x}, \beta, \alpha_1, \alpha_2)$;
- sample α_1 from $f(\alpha_1 | \mathbf{y}, \mathbf{x}, \mathbf{b}, \beta, \alpha_2)$;
- sample α_2 from $f(\alpha_2 | \mathbf{y}, \mathbf{x}, \mathbf{b}, \beta, \alpha_1)$;
- sample $\mathbf{x}_{mis,i}$ from $f(\mathbf{x}_{mis,i} | \mathbf{y}, \mathbf{x}_{obs,i}, \mathbf{b}, \beta, \alpha_1, \alpha_2)$, $i = 1, \cdots, n$.

After a burn-in period, one obtains a sample of (β, D, α) from the posterior distribution $f(\beta, D, \alpha | \mathbf{y}, \mathbf{x}_{obs})$. Repeating the procedure many times, one obtains many samples from the posterior distribution and then proceeds with approximate Bayesian inference.

11.5 Bayesian Models with Covariate Measurement Errors

11.5.1 Bayesian Regression Models with Covariate Measurement Errors

Consider a regression model for cross-sectional data, with response y_i and covariates $\mathbf{z}_i$ and $\mathbf{x}_i$, $i = 1, 2, \cdots, n$. Suppose that the observed covariates $\mathbf{z}_i$ are measured with errors, with the corresponding unobserved true values being $\mathbf{z}_i^*$, and suppose that covariates $\mathbf{x}_i$ are accurately observed without measurement errors. Let $\mathbf{y} = (y_1, \cdots, y_n)^T$, $\mathbf{z} = (\mathbf{z}_1^T, \cdots, \mathbf{z}_n^T)^T$, and define $\mathbf{z}^*$ and $\mathbf{x}$ similarly. We may write the regression model as $f(y_i | \mathbf{z}_i^*, \mathbf{x}_i, \theta)$, where the response y_i depends on the unobserved true covariate values $\mathbf{z}_i^*$ rather than the observed but mis-measured covariate values $\mathbf{z}_i$ and θ contains all unknown parameters.

A Bayesian approach typically treats the unobserved true covariates $\mathbf{z}_i^*$ as missing data and essentially imputes them many times by sampling from the conditional distribution of $\mathbf{z}_i^*$ given all other variables and observed data. Specifically, by treating the unobserved true covariates $\mathbf{z}_i^*$ as missing data, we can write the "complete data" likelihood for individual i as follows

$$f(y_i, \mathbf{z}_i, \mathbf{z}_i^* | \mathbf{x}_i, \theta) = f(y_i | \mathbf{z}_i^*, \mathbf{x}_i, \theta) f(\mathbf{z}_i | \mathbf{z}_i^*, \mathbf{x}_i, \theta) f(\mathbf{z}_i^* | \mathbf{x}_i, \theta),$$

where θ contains all unknown parameters. In measurement error literature, model $f(y_i | \mathbf{z}_i^*, \mathbf{x}_i, \theta)$ is sometimes called *outcome model*, which can be a linear, or nonlinear, or generalized linear model. Model $f(\mathbf{z}_i | \mathbf{z}_i^*, \mathbf{x}_i, \theta)$ is sometimes called *measurement model*. For example, a classical measurement error model can be written as

$$\mathbf{z}_i = \mathbf{z}_i^* + \boldsymbol{\epsilon}_i, \qquad E(\boldsymbol{\epsilon}_i) = 0.$$

Model $f(\mathbf{z}_i^*|\mathbf{x}_i, \boldsymbol{\theta})$ is sometimes called *exposure model*. The term exposure model comes from epidemiology, where $\mathbf{z}_i^*$ is often the true exposure to a toxicant. To address measurement errors, replicates or validation data are typically required for the mis-measured covariates $\mathbf{z}_i$.

For Bayesian inference, we consider a prior distribution $f(\boldsymbol{\theta})$ for the parameters in $\boldsymbol{\theta}$, and consider the "likelihood"

$$f(\boldsymbol{\theta}) \prod_{i=1}^{n} f(y_i, \mathbf{z}_i, \mathbf{z}_i^*|\mathbf{x}_i, \boldsymbol{\theta}).$$

Bayesian inference is then based on the following posterior distribution of $\boldsymbol{\theta}$

$$f(\boldsymbol{\theta}|\mathbf{y}, \mathbf{z}, \mathbf{x}) = \frac{f(\boldsymbol{\theta}) \int f(\mathbf{y}, \mathbf{z}, \mathbf{z}^*|\mathbf{x}, \boldsymbol{\theta}) \, d\mathbf{z}^*}{\int \int f(\boldsymbol{\theta}) f(\mathbf{y}, \mathbf{z}, \mathbf{z}^*|\mathbf{x}, \boldsymbol{\theta}) \, d\mathbf{z}^* d\boldsymbol{\theta}}. \tag{11.18}$$

Note that the integral in the denominator of (11.18) is typically very complicated, since the parameter vector $\boldsymbol{\theta}$ is typically of a high dimension and consists of components with different types, such as mean parameters and variance-covariance parameters. Thus, a major difficulty in Bayesian inference is the computational challenge. However, we may again use the Gibbs sampler to generate many samples from the posterior distribution $f(\boldsymbol{\theta}|\mathbf{y}, \mathbf{z}, \mathbf{x})$ and conduct approximate Bayesian inference based on the simulated samples. This avoids highly intractable integrals and is thus a solution to the dilemma.

Starting with initial values of $\boldsymbol{\theta}$ and $\mathbf{z}_i^*$, the Gibbs sampler consists of the following steps:

- Generate a sample of the unobserved $\mathbf{z}_i^*$ from its posterior distribution given the observed data and current estimate of $\boldsymbol{\theta}$, i.e.,

$$\mathbf{z}_i^* \sim f(\mathbf{z}_i^*|y_i, \mathbf{z}_i, \mathbf{x}_i, \boldsymbol{\theta}) \propto f(y_i|\mathbf{z}_i^*, \mathbf{x}_i, \boldsymbol{\theta}) f(\mathbf{z}_i|\mathbf{z}_i^*, \mathbf{x}_i, \boldsymbol{\theta}) f(\mathbf{z}_i^*|\mathbf{x}_i, \boldsymbol{\theta}).$$

- Generate a sample of $\boldsymbol{\theta}$ from its posterior distribution given the observed data and current generated values of $\mathbf{z}_i^*$, i.e.,

$$\boldsymbol{\theta} \sim f(\boldsymbol{\theta}|\mathbf{y}, \mathbf{z}, \mathbf{z}^*, \mathbf{x}) \propto f(\boldsymbol{\theta}) f(\mathbf{y}|\mathbf{z}^*, \mathbf{x}, \boldsymbol{\theta}) f(\mathbf{z}|\mathbf{z}^*, \mathbf{x}, \boldsymbol{\theta}) f(\mathbf{z}^*|\mathbf{x}, \boldsymbol{\theta}),$$

which can be done one element of $\boldsymbol{\theta}$ at a time, holding other elements fixed.

Sampling in the above two steps can be done using the rejection sampling methods since the density functions on the right-hand sides of the above expressions are known. Repeating the two steps many times for a burn-in period, we obtain a desired sample from the posterior distribution $f(\boldsymbol{\theta}|\mathbf{y}, \mathbf{z}, \mathbf{x})$.

In some cases, it may be more desirable to use the Metropolis-Hastings algorithm, which is a very versatile and flexible tool and includes the Gibbs sampler as a special case. See Chapter 12 for a more detailed description of the Metropolis-Hastings algorithm. Carroll et al. (2006) provided a detailed description of a Metropolis-Hastings algorithm for measurement error problems and other approaches for Bayesian inference.

11.5.2 Bayesian Mixed Models with Covariate Measurement Errors

The Bayesian method for regression models with covariate measurement errors for cross-sectional data in the previous section can be extended to mixed effects models for longitudinal data or clustered data with covariate measurement errors. Note that a mixed effects model may be viewed as an extension of the corresponding regression model for cross-sectional data by introducing random effects, which are used to incorporate the correlation between the repeated measurements within each individual or cluster and the variations between individuals or clusters. For time-dependent covariates with measurement errors, the repeated measurements within each individual or cluster may be viewed as "replicates" and thus may be used to partially address measurement errors. For a mixed effects model, we can treat the random effects as additional unobserved error-free "covariates" and then proceed for Bayesian inference in a way similar to that in the previous section. The idea is briefly described as follows.

The conditional density function for a mixed effects model for longitudinal or clustered data with error-prone covariates can be written as $f(\mathbf{y}_i|\mathbf{z}_i^*, \mathbf{x}_i, \mathbf{b}_i, \boldsymbol{\theta})$, given random effects $\mathbf{b}_i$, where $\mathbf{y}_i = (y_{i1}, y_{i2}, \cdots, y_{in_i})^T$, $\mathbf{z}_i^*$ are unobserved true covariates (either time dependent or time independent), $\mathbf{x}_i$ are covariates without measurement errors, and $\boldsymbol{\theta}$ contains all unknown parameters, $i = 1, 2, \cdots, n$. Let $\mathbf{b} = (\mathbf{b}_1, \cdots, \mathbf{b}_n)$. We can treat the unknown parameters $\boldsymbol{\theta}$, the unobserved true covariates $\mathbf{z}_i^*$, and the unobserved random effects $\mathbf{b}_i$ all as "missing data". Then, the "complete data likelihood" can be written as

$$f(\boldsymbol{\theta})f(\mathbf{y}|\mathbf{z}^*, \mathbf{x}, \mathbf{b}, \boldsymbol{\theta})f(\mathbf{b}|\boldsymbol{\theta})f(\mathbf{z}|\mathbf{z}^*, \mathbf{x}, \boldsymbol{\theta})f(\mathbf{z}^*|\mathbf{x}, \boldsymbol{\theta}),$$

where $f(\boldsymbol{\theta})$ is the prior distribution for the parameters $\boldsymbol{\theta}$.

Bayesian inference is based on the following posterior distribution of $\boldsymbol{\theta}$ given the observed data $(\mathbf{y}, \mathbf{z}, \mathbf{x})$:

$$f(\boldsymbol{\theta}|\mathbf{y}, \mathbf{z}, \mathbf{x}) = \frac{f(\boldsymbol{\theta}) \int \int f(\mathbf{y}, \mathbf{z}, \mathbf{z}^*, \mathbf{b}|\mathbf{x}, \boldsymbol{\theta}) \, d\mathbf{z}^* d\mathbf{b}}{\int \int \int f(\boldsymbol{\theta}) f(\mathbf{y}, \mathbf{z}, \mathbf{z}^*, \mathbf{b}|\mathbf{x}, \boldsymbol{\theta}) \, d\mathbf{z}^* d\boldsymbol{\theta} d\mathbf{b}}. \tag{11.19}$$

The above posterior distribution is quite intractable and generally does not have an analytic expression. Numerical integration methods such as the Gauss-Hermite quadrature generally do not work well, due to the high dimensional integrations. Thus, the main challenge for Bayesian inference is computation. We can again use the Gibbs sampler method, which generates many samples from the posterior distribution by iteratively sampling from lower dimensional conditional distributions.

Starting with initial values of the unobserved quantities $(\boldsymbol{\theta}, \mathbf{z}_i^*, \mathbf{b}_i)$, the Gibbs sampler iterates between the following steps:

- Generate a sample of the unobserved covariate $\mathbf{z}_i^*$ from its posterior distribution given the observed data and the last generated random effects $\mathbf{b}_i$ and parameters $\boldsymbol{\theta}$, i.e.,

$$\mathbf{z}_i^* \sim f(\mathbf{z}_i^*|y_i, \mathbf{z}_i, \mathbf{x}_i, \mathbf{b}_i, \boldsymbol{\theta})$$
$$\propto f(y_i|\mathbf{z}_i^*, \mathbf{x}_i, \mathbf{b}_i, \boldsymbol{\theta})f(\mathbf{z}_i|\mathbf{z}_i^*, \mathbf{x}_i, \boldsymbol{\theta})f(\mathbf{z}_i^*|\mathbf{x}_i, \boldsymbol{\theta}), \quad i = 1, \cdots, n.$$

- Generate a sample of the random effects $\mathbf{b}_i$ from its posterior distribution given the observed data and the last generated values of $\mathbf{z}_i^*$ and $\boldsymbol{\theta}$, i.e.,

$$\mathbf{b}_i \sim f(\mathbf{b}_i|y_i, \mathbf{z}_i, \mathbf{x}_i, \mathbf{z}_i^*, \boldsymbol{\theta}) \propto f(y_i|\mathbf{z}_i^*, \mathbf{x}_i, \mathbf{b}_i, \boldsymbol{\theta})f(\mathbf{b}_i|\boldsymbol{\theta}). \quad i = 1, \cdots, n.$$

- Generate a sample of the parameter $\boldsymbol{\theta}$ from its posterior distribution given the observed data and last generated values of $\mathbf{z}_i^*$ and $\mathbf{b}_i$, i.e.,

$$\boldsymbol{\theta} \sim f(\boldsymbol{\theta}|\mathbf{y}, \mathbf{z}, \mathbf{z}^*, \mathbf{x}, \mathbf{b})$$
$$\propto f(\boldsymbol{\theta})f(\mathbf{y}|\mathbf{z}^*, \mathbf{x}, \mathbf{b}, \boldsymbol{\theta})f(\mathbf{b}|\boldsymbol{\theta})f(\mathbf{z}|\mathbf{z}^*, \mathbf{x}, \boldsymbol{\theta})f(\mathbf{z}^*|\mathbf{x}, \boldsymbol{\theta}).$$

This step is usually done one element of $\boldsymbol{\theta}$ at a time, holding other elements fixed, since the components of $\boldsymbol{\theta}$ are of different types, such as mean and variance-covariance parameters.

Iterating the above three steps for a burn-in period, the resulting sequence is a Markov chain which will converge to its stationary distribution, the target posterior distribution, So one obtains a sample from the target posterior distribution. Sampling from the above full conditionals may be done using rejection sampling methods, such as the adaptive rejection sampling method, since the full conditionals are proportional to a product of known density functions.

11.6 Bayesian Joint Models of Longitudinal and Survival Data

In joint modeling of longitudinal data and survival data, the joint models often contain many parameters, which may lead to identifiability problems or may be poorly estimated. Moreover, the computation associated with joint model inference is typically very challenging. A Bayesian approach for joint models is thus very appealing since it can use prior information or borrow information from similar studies to better estimate the parameters. Moreover, a Bayesian approach may avoid many complicated approximations required by a frequentist approach. In a Bayesian approach, when the prior distributions are non-informative, the Bayesian estimates are equivalent to maximum likelihood estimates. In this section, we follow Brown and Ibrahim (2003) and describe a Bayesian approach for joint modeling of longitudinal data and survival data.

We focus on a joint model arising from a survival model with measurement errors in a time-dependent covariate. For individual i, let s_i be the true survival time, which may not be observed due to censoring, and let c_i be the censoring time. We assume non-informative right censoring. The observed survival times are

$$t_i = \min(s_i, c_i), \qquad i = 1, 2, \cdots, n. \tag{11.20}$$

Let

$$\delta_i = I(s_i \leq c_i) = \begin{cases} 1 & \text{an event is observed for individual } i \\ 0 & \text{the event time is censored for individual } i \end{cases}$$

be the censoring indicator. Then, the observed data can be written as

$$\{(t_i, \delta_i), \ i = 1, 2, \cdots, n\}. \tag{11.21}$$

Let $z_i(t)$ be a time-dependent covariate with measurement error, whose unobserved true value is $z_i^*(t)$.

For the survival model, we consider the following Cox proportional hazards model

$$\lambda(t_i) = \lambda_0(t_i) \exp(\beta_1 z_i^*(t_i) + \mathbf{x}_i^T \boldsymbol{\beta}_2), \tag{11.22}$$

where $\lambda_0(t)$ is a baseline hazard function, and $\mathbf{x}_i$ are baseline covariates without measurement errors. As in Brown and Ibrahim (2003), we may assume that the baseline hazard function is piecewise constant as follows:

$$\lambda(u) = \lambda_j, \qquad \text{for } u_{j-1} \leq u < u_j, \qquad j = 1, \cdots, J,$$

where the u_j's define the intervals for the baseline hazard function and can be selected based on the quantiles of the observed event times.

For the covariate process, we consider the following classical measurement error model

$$z_{ij} \quad = \quad g(t_{ij}, \boldsymbol{\alpha}) + \epsilon_{ij} = z_{ij}^* + \epsilon_{ij}, \tag{11.23}$$

$$\epsilon_{ij} \quad \text{i.i.d.} \quad \sim N(0, \sigma^2), \qquad i = 1, \cdots, n; \ j = 1, \cdots, m_i, \tag{11.24}$$

where $z_{ij} = z_i(t_{ij})$ and $g(t, \boldsymbol{\alpha})$ is a parametric or nonparametric smooth function of time t with unknown parameters $\boldsymbol{\alpha}$, such as a polynomial function of time.

The joint likelihood function for individual i can then be written as

$$L(\boldsymbol{\theta}) = \lambda_0(s_i)^{\delta_i} \exp\left[\delta_i(\beta_1 z_i^*(t_i) + \mathbf{x}_i^T\boldsymbol{\beta}_2) - \exp(\mathbf{x}_i^T\boldsymbol{\beta}_2)\sum_{j=1}^{J}\Lambda_{ij}(\boldsymbol{\theta})\right]$$

$$\times \frac{1}{(2\pi\sigma^2)^{m_i/2}} \exp\left[-\frac{1}{2\sigma^2}\sum_{j=1}^{m_i}(z_{ij} - g(t_{ij}, \boldsymbol{\alpha}))^2\right],$$

where

$$\Lambda_{ij}(\boldsymbol{\theta}) = I(s_i > u_{j-1})\lambda_j \int_{u_{j-1}}^{\min(u_j, s_i)} \exp(\beta_1 z_i^*(u))du,$$

$I(s_i > u_{j-1})$ is the usual indicator function, and $\boldsymbol{\theta}$ contains all unknown parameters.

For Bayesian inference, we assume prior distributions for the parameters and then conduct estimation and inference based on the posterior distribution of the parameters given the data. For non-informative proper priors, we may assume the following conjugate prior for the hazard

$$\begin{aligned} \lambda_j &\sim \Gamma(a_j, b_j), \qquad j = 1, \cdots, J, \\ \sigma^2 &\sim IG(a, b), \end{aligned}$$

where $\Gamma(a, b)$ is the gamma distribution with shape parameter a and scale parameter b, and $IG(a, b)$ is the inverse gamma distribution. For the regression parameters $\boldsymbol{\beta}$, we can consider a normal prior

$$\boldsymbol{\beta} \sim N(\boldsymbol{\beta}_0, \Sigma_0).$$

For the parameters $\boldsymbol{\alpha}$ in the covariate model, Brown and Ibrahim (2003) considered a unspecified nonparametric function and then place a Dirichlet process prior on this distribution, which allows a more flexible modeling of the covariate process. The Gibbs sampler can then be used to obtain Bayesian estimates from the posterior distribution.

The above Bayesian approach can be extended in different ways. For example, we may consider a parametric, or a semiparametric, or a nonparametric mixed effects model to model the covariate process, which incorporates the correlation between the repeated covariate measurements and the variations between individuals or clusters. We may also consider a frailty models for the survival models by introducing random effects in the survival models. In the presence of random effects, one only needs to add another step of sampling the random effects in the Gibbs sampler, so the extension is straightforward. Other survival models such as accelerated failure time models or parametric survival models may also be considered.

CHAPTER 12

Appendix: Background Materials

In the following sections, we provide some background materials for the standard methods that have been repeatedly used throughout the book, including the likelihood methods and MCMC methods. We will focus on essential ideas and results, without going too much details. More detailed discussions of these topics can be found in many books, which are listed in the corresponding sections.

12.1 Likelihood Methods

Likelihood methods are widely used in statistical inference, due to general applicability of likelihood methods and attractive asymptotic properties of MLEs such as asymptotic most efficiency and asymptotic normality. Moreover, the *likelihood principle* says that likelihood functions contain all of the information in the data about unknown parameters in the assumed models. Maximum likelihood estimation is often viewed as the "gold standard" of estimation procedures. Likelihood functions also play an integral role in Bayesian inference. In the following, we provide a brief overview of likelihood methods.

For a likelihood method, once the likelihood for the *observed data* is specified based on the assumed distributions, the MLEs of unknown parameters in the assumed distributions can be obtained by maximizing the likelihood using standard optimization procedures or the EM algorithms. The resulting MLEs will be asymptotically consistent, most efficient (in the sense of attaining the Cramer-Rao lower bound for the variances of the MLEs), and normally distributed, if some common regularity conditions hold. In other words, when the sample size is large, the MLE is approximately optimal if the assumed distributions and some regularity conditions hold. In many problems, the sample sizes do not have to be very large in order for the MLEs to perform well, and the regularity conditions are often satisfied. Violations of the regularity conditions may arise, for example, when the parameters are on the boundary of the parameter space. Therefore, likelihood methods are conceptually straightforward. In practice, difficulties often lie in computation since the observed-data likelihoods can be highly intractable for some complex problems.

The asymptotic normality of MLEs can be used for (approximate) inference in practice where the sample size is finite. For example, we may use the asymptotic normal distributions of MLEs to construct approximate confidence intervals for the unknown parameters and to perform hypothesis testing such as Wald-type tests, the likelihood ratio test, and the score test (these tests will be described below). Likelihood methods are very general and can be used in almost any situations where probability distributions are assumed. Potential drawbacks of likelihood methods are that MLEs are often sensitive to outliers, the assumed distributions may not hold, and MLEs may be biased for finite samples (but the bias should decrease as the sample size increases). Restricted maximum likelihood estimates (REML) are often used to correct some of the biases in MLEs for variance components.

Let $y_1, y_2, \cdots, y_n$ be a sample of independent and identically distributed (i.i.d.) observations drawn from a distribution with a probability density function (for continuous variables) or a probability mass function (for discrete variables) $f(y; \boldsymbol{\theta})$, where $\boldsymbol{\theta} = (\theta_1, \cdots, \theta_p)^T$ are unknown parameters. Note that the results below also apply when the observations are independent but not identically distributed, e.g., in regression settings where the mean and variance of y_i may depend on covariates $\mathbf{x}_i$. Let $\mathbf{y} = (y_1, y_2, \cdots, y_n)^T$. The *likelihood function* for the observed data $\mathbf{y}$ is defined as

$$L(\boldsymbol{\theta}) = L(\boldsymbol{\theta}|\mathbf{y}) = \prod_{i=1}^{n} f(y_i; \boldsymbol{\theta}),$$

which may be roughly interpreted as the probability of observing the data $\mathbf{y}$ under the assumed distribution for the data.

The *maximum likelihood estimate (MLE)* of $\boldsymbol{\theta}$, denoted by $\hat{\boldsymbol{\theta}}$, is the value of $\boldsymbol{\theta}$ which maximizes the likelihood $L(\boldsymbol{\theta})$, i.e., the MLE is the value of the parameter which makes the observed data most likely to occur. The corresponding *log-likelihood* is given by

$$l(\boldsymbol{\theta}) \equiv \log L(\boldsymbol{\theta}) = \sum_{i=1}^{n} \log f(y_i; \boldsymbol{\theta}).$$

Since the log-likelihood $l(\boldsymbol{\theta})$ is a monotone function of the likelihood $L(\boldsymbol{\theta})$, maximization of the likelihood $L(\boldsymbol{\theta})$ is equivalent to maximization of the log-likelihood $l(\boldsymbol{\theta})$, but the log-likelihood is easier to handle since a summation is mathematically more manageable than a product. Thus, likelihood inference is often based on the log-likelihood.

The MLE $\hat{\boldsymbol{\theta}}$ satisfies the following estimating equation (likelihood equation)

$$\frac{\partial l(\boldsymbol{\theta})}{\partial \boldsymbol{\theta}} = \sum_{i=1}^{n} \frac{\partial \log f(y_i; \boldsymbol{\theta})}{\partial \boldsymbol{\theta}} = 0.$$

Note that the MLE may not be unique or may not even exist, but if the MLE exists it should satisfy the above estimating equation. Note also that, if the likelihood function has multiple modes, a solution to the above estimating equation may be a local maximum/minimum, depending on the choice of starting values. So the choice of starting values is important for complex likelihood functions. In practice, a simple approach is to try different starting values and check if the solutions differ. The vector

$$\mathbf{s}(\boldsymbol{\theta}) = \sum_{i=1}^{n} \frac{\partial \log f(y_i; \boldsymbol{\theta})}{\partial \boldsymbol{\theta}} = \left(\sum_{i=1}^{n} \frac{\partial \log f(y_i; \boldsymbol{\theta})}{\partial \theta_1}, \cdots, \sum_{i=1}^{n} \frac{\partial \log f(y_i; \boldsymbol{\theta})}{\partial \theta_p} \right)^T$$

is called the *Fisher efficient score* or the *score*. It can be shown that $E(\mathbf{s}(\boldsymbol{\theta})) = 0$.

The *Fisher's information function (matrix)* is defined by

$$I(\boldsymbol{\theta}) = -E\left(\frac{\partial^2 l(\boldsymbol{\theta})}{\partial \boldsymbol{\theta}^2} \right) = (I(\boldsymbol{\theta})_{jk})_{p \times p}, \quad \text{with} \quad I(\boldsymbol{\theta})_{jk} = -E\left(\frac{\partial^2 l(\boldsymbol{\theta})}{\partial \theta_j \partial \theta_k} \right).$$

The information matrix $I(\boldsymbol{\theta})$ quantifies the expected amount of information in the data about the unknown parameters $\boldsymbol{\theta}$. Note that the second derivatives $\partial^2 l(\boldsymbol{\theta})/\partial \theta_j^2$ describe the curvature of the likelihood in the neighborhood of θ_j, so the greater the value of $-\partial^2 l(\boldsymbol{\theta})/\partial \theta_j^2$, the sharper is the peak of the likelihood function and thus the greater is the information about θ_j. The Fisher's information matrix can also be expressed as

$$I(\boldsymbol{\theta}) = E\left[\left(\frac{\partial l(\boldsymbol{\theta})}{\partial \boldsymbol{\theta}} \right) \left(\frac{\partial l(\boldsymbol{\theta})}{\partial \boldsymbol{\theta}} \right)^T \right],$$

which only involves the first derivatives so sometimes may be easier to evaluate. The matrix

$$H(\boldsymbol{\theta}) = \frac{\partial^2 l(\boldsymbol{\theta})}{\partial \boldsymbol{\theta}^2} = \frac{\partial \mathbf{s}(\boldsymbol{\theta})}{\partial \boldsymbol{\theta}},$$

is called the *Hessian matrix*. The *observed information* is defined as

$$i(\boldsymbol{\theta}) = -H(\boldsymbol{\theta}) = -\frac{\partial^2 l(\boldsymbol{\theta})}{\partial \boldsymbol{\theta}^2},$$

which sometimes can be used to approximate $I(\boldsymbol{\theta})$ since $I(\boldsymbol{\theta}) = E(i(\boldsymbol{\theta}))$.

Under some regularity conditions, the MLE is consistent, asymptotically efficient, and asymptotically normally distributed. These regularity conditions can be stated as follows:

R1. The parameter space Θ of $\boldsymbol{\theta}$ is an open subset of the whole space R^p.

R2. The set $A = \{y : f(y; \boldsymbol{\theta}) > 0\}$ does not depend on $\boldsymbol{\theta}$.

R3. The function $f(y; \boldsymbol{\theta})$ is three times continuously differentiable with respect to $\boldsymbol{\theta}$ for all y.

R4. The following equations hold

$$E(\partial l(y; \boldsymbol{\theta})/\partial \boldsymbol{\theta}) = 0, \qquad Cov(\partial l(y; \boldsymbol{\theta})/\partial \boldsymbol{\theta}) = I(\boldsymbol{\theta}), \qquad \text{for all } \boldsymbol{\theta}.$$

R5. The expectations of all the derivatives of $f(y; \boldsymbol{\theta})$ with respect to $\boldsymbol{\theta}$ exist and are finite.

The above regularity conditions are satisfied for a wide variety of models and are relatively easy to verify. Note that there are variations of these conditions, and weaker conditions are available.

Under the regularity conditions R1 – R5, the MLE $\hat{\boldsymbol{\theta}}$ of $\boldsymbol{\theta}$ has the following large-sample properties:

- The MLE $\hat{\boldsymbol{\theta}}$ is *consistent*, i.e.,

$$\hat{\boldsymbol{\theta}} \xrightarrow{p} \boldsymbol{\theta}, \qquad \text{as } n \to \infty;$$

- The MLE $\hat{\boldsymbol{\theta}}$ is *asymptotically efficient*, i.e., the asymptotic variance of $\hat{\boldsymbol{\theta}}$ attains the *Cramer-Rao lower bound*, which is $I^{-1}(\boldsymbol{\theta})$;

- The MLE $\hat{\boldsymbol{\theta}}$ is *asymptotically normal*, i.e.,

$$\hat{\boldsymbol{\theta}} \xrightarrow{d} N(\boldsymbol{\theta}, I^{-1}(\boldsymbol{\theta})), \qquad \text{as } n \to \infty.$$

Thus, the MLE is asymptotically *optimal*. Note that, however, the MLE is not necessary unbiased for finite samples. In some cases, the bias of MLE may be substantial. On the other hand, the MLE is asymptotically unbiased, i.e., its bias tends to zero as the sample size increases. Due to the above attractive asymptotic properties of MLEs, likelihood methods are widely used in statistical inference.

Based on the asymptotic normality of the MLE $\hat{\boldsymbol{\theta}}$, in practice when the sample size is finite, an approximate level $1 - \alpha$ confident interval for θ_j, the j-th component of $\boldsymbol{\theta}$, is given by

$$\hat{\theta}_j \pm z_{\alpha/2} \cdot s.e.(\hat{\theta}_j),$$

where $z_{\alpha/2}$ is the $1 - \alpha/2$ percentile of the standard normal distribution $N(0, 1)$ and $s.e.(\hat{\theta}_j) = I^{-1/2}(\hat{\boldsymbol{\theta}})_{jj}$ is the approximate standard error of the MLE $\hat{\theta}_j$. For hypothesis testing, the following three likelihood-based large-sample tests are widely used: the Wald test, the likelihood ratio test (LRT), and the efficient score test. These three tests are briefly described as follows.

Consider testing the hypotheses

$$H_0 : \boldsymbol{\theta} = \boldsymbol{\theta}_0 \qquad \text{versus} \qquad H_1 : \boldsymbol{\theta} \neq \boldsymbol{\theta}_0.$$

The following three tests are based on asymptotic results and widely used in practice:

- *Wald-type test.* The Wald-type test statistic for testing H_0 versus H_1 is given by

$$T_W = (\hat{\boldsymbol{\theta}} - \boldsymbol{\theta}_0)^T \hat{\Sigma}^{-1} (\hat{\boldsymbol{\theta}} - \boldsymbol{\theta}_0),$$

where $\hat{\Sigma} = I(\hat{\boldsymbol{\theta}})^{-1}$ is an estimate of the covariance matrix of $\hat{\boldsymbol{\theta}}$. The test statistic $T_W \sim \chi_p^2$ asymptotically under H_0, where p is the dimension of parameter $\boldsymbol{\theta}$. To test an individual component of $\boldsymbol{\theta}$, say $H_{0j} : \theta_j = \theta_{j0}$ versus $H_1 : \theta_j \neq \theta_{j0}$, we may consider individual Wald-type test statistic

$$T_W^{(j)} = \frac{(\hat{\theta}_j - \theta_0)^2}{\widehat{var}(\hat{\theta}_j)}$$

where $\widehat{var}(\hat{\theta}_j) = (I(\hat{\boldsymbol{\theta}})^{-1})_{jj}$. The test statistic $T_W^{(j)} \sim \chi_1^2$ asymptotically under H_{0j}.

- *Likelihood ratio test (LRT).* Let $\hat{\boldsymbol{\theta}}$ be the MLE of $\boldsymbol{\theta}$, and let $L(\boldsymbol{\theta}_0)$ and $L(\hat{\boldsymbol{\theta}})$ be the likelihood functions evaluated at $\boldsymbol{\theta}_0$ and $\hat{\boldsymbol{\theta}}$ respectively. The LRT test statistic for testing H_0 versus H_1 is given by

$$T_L = -2 \log \left(\frac{L(\boldsymbol{\theta}_0)}{L(\hat{\boldsymbol{\theta}})} \right) = 2 \log L(\hat{\boldsymbol{\theta}}) - 2 \log L(\boldsymbol{\theta}_0).$$

The test statistic $T_L \sim \chi_p^2$ asymptotically under H_0.

- *Score test.* The score test statistic for testing H_0 versus H_1 is given by

$$T_S = \mathbf{s}(\boldsymbol{\theta}_0)^T I(\boldsymbol{\theta}_0)^{-1} \mathbf{s}(\boldsymbol{\theta}_0),$$

where $\mathbf{s}(\boldsymbol{\theta}_0)$ is the score function at $\boldsymbol{\theta}_0$. The test statistic $T_S \sim \chi_p^2$ asymptotically under H_0.

The above three tests are asymptotically equivalent, but they may differ with finite samples. The LRT is equivalent to the deviance test which is widely used in GLMs. The Wald test requires the least computational effort. The score test does not require computing the MLE since the test statistic is evaluated under the null hypothesis.

Note that the above asymptotic results do not hold for order-restricted tests or constrained tests, such as the one-sided test $H_0 : \boldsymbol{\theta} = 0$ versus $H_1 : \boldsymbol{\theta} > 0$. In this case, the above three tests can be constructed in a similar way, but their asymptotic distributions are no longer χ^2-distributions but are mixtures of χ^2-distributions. See Section 2.7.1 (page 89) for a more detailed discussion.

12.2 The Gibbs Sampler and MCMC Methods

In likelihood inference of mixed effects models with incomplete data, Monte Carlo EM algorithms are often used in which the E-step requires sampling from multi-dimensional and intractable distributions. Similarly, in Bayesian inference the target posterior distributions are often highly complicated, and approximate Bayesian inference is usually based on large samples drawn from the target posterior distributions. In both cases, one needs to generate large numbers of samples from highly complicated and multi-dimensional distributions. *Markov chain Monte Carlo (MCMC)* methods are great tools for such tasks and they have become very popular.

MCMC methods are algorithms for generating samples from intractable distributions. The key idea of MCMC methods is to construct *Markov chains* that have the desired distributions as their stationary distributions. After a large number of steps, called a *burn-in period*, the Markov chain will converge to its stationary distribution, and thus the last state of the chain can be used as a sample from the desired distribution. A key characteristic of a Markov chain is that the current state depends on the previous one, so there may be many ways to construct a Markov chain which converges to the same target distribution.

MCMC methods have revolutionized Bayesian inference since they have made highly complicated Bayesian computations feasible. These MCMC methods are also very useful tools in likelihood inference since many likelihood computations encount similar problems as in Bayesian inference. The most useful MCMC method is probably the *Gibbs sampler*, which is briefly described below. Detailed discussions of MCMC methods can be found in Gilks et al. (1996), Gelman et al. (2003), and Robert and Casella (2004).

The Gibbs sampler

Gibbs sampling or the *Gibbs sampler* is an example of MCMC methods, which is perhaps the most widely used MCMC method. It was devised by Geman and Geman (1984). The Gibbs sampler is typically used to obtain random samples from a multi-dimensional probability distribution, which is either intractable or is not known explicitly. The desired samples can be obtained by sequentially sampling from lower-dimensional conditional distributions which are easier to sample from. These samples then comprise a Markov chain, whose stationary distribution is the target distribution. The Gibbs sampler is widely used because it is often easier to sample from the *lower-dimensional* conditional distributions than the original distribution. We describe the details as follows.

Suppose that we wish to generate samples from the probability distribution $f(\mathbf{u}|\boldsymbol{\theta})$, where $\mathbf{u} = (\mathbf{u}_1^T, \mathbf{u}_2^T, \cdots, \mathbf{u}_q^T)^T$ is a random vector, with each component $\mathbf{u}_j$ being possibly also a random vector. Suppose also that $f(\mathbf{u}|\boldsymbol{\theta})$ is highly intractable or even not known explicitly, so it is difficult to generate samples from $f(\mathbf{u}|\boldsymbol{\theta})$ directly. Note that the components $\mathbf{u}_j$'s are typically *unobserved* quantities which may have different dimensions or different types. For example, for missing covariates $\mathbf{x}_i$ in a mixed effects model $f(\mathbf{y}_i|\mathbf{x}_i, \mathbf{b}_i, \boldsymbol{\theta})$, with $\mathbf{b}_i$ being the random effects, we may want to generate samples from the intractable distribution $f(\mathbf{x}_{mis,i}, \mathbf{b}_i|\mathbf{x}_{obs,i}, \mathbf{y}_i, \boldsymbol{\theta})$ in the EM algorithm for likelihood estimation. In this case, we can choose

$$q = 2, \quad \mathbf{u}_1 = \mathbf{x}_{mis,i}, \quad \mathbf{u}_2 = \mathbf{b}_i.$$

As another example, let $\mathbf{y} \sim N(\boldsymbol{\mu}, \Sigma)$. Suppose that we want to simulate from the posterior distribution $f(\mu, \Sigma|\mathbf{y})$ in Bayesian inference. In this case, we can choose

$$q = 2, \quad \mathbf{u}_1 = \boldsymbol{\mu}, \quad \mathbf{u}_2 = \Sigma.$$

In the following, we describe the Gibbs sampler method to generate samples from $f(\mathbf{u}|\boldsymbol{\theta})$, assuming $\boldsymbol{\theta}$ is known for simplicity. Let

$$\mathbf{u}_{-j} = (\mathbf{u}_1^T, \cdots, \mathbf{u}_{j-1}^T, \mathbf{u}_{j+1}^T, \cdots, \mathbf{u}_q^T)^T, \qquad j = 1, 2, \cdots, q,$$

be the sub-vector of $\mathbf{u}$ without component $\mathbf{u}_j$. It is often easier to generate samples from the lower-dimensional conditional distributions $f(\mathbf{u}_j|\mathbf{u}_{-j}, \boldsymbol{\theta}), j = 1, 2, \cdots, q$, which are called *full conditionals*. The Gibbs sampler proceeds as follows: beginning with starting values $(\mathbf{u}_1^{(0)}, \cdots, \mathbf{u}_q^{(0)})$, at step k,

- sample $\mathbf{u}_1^{(k)}$ from $f(\mathbf{u}_1|\mathbf{u}_2^{(k-1)}, \mathbf{u}_3^{(k-1)}, \cdots, \mathbf{u}_q^{(k-1)}, \boldsymbol{\theta})$;
- sample $\mathbf{u}_2^{(k)}$ from $f(\mathbf{u}_2|\mathbf{u}_1^{(k)}, \mathbf{u}_3^{(k-1)}, \cdots, \mathbf{u}_q^{(k-1)}, \boldsymbol{\theta})$;
- $\cdots$;
- sample $\mathbf{u}_q^{(k)}$ from $f(\mathbf{u}_q|\mathbf{u}_1^{(k)}, \cdots, \mathbf{u}_{q-1}^{(k)}, \boldsymbol{\theta})$, $\quad k = 1, 2, \cdots$.

The sequence $\{(\mathbf{u}_1^{(k)}, \cdots, \mathbf{u}_q^{(k)}), \ k = 1, 2, 3, \cdots\}$ then comprises a Markov chain with stationary distribution $f(\mathbf{u}|\boldsymbol{\theta})$. Therefore, when k is large enough (after a burn-in period), we can view $\mathbf{u}^{(k)} = (\mathbf{u}_1^{(k)}, \cdots, \mathbf{u}_q^{(k)})^T$ as a sample generated from the target distribution $f(\mathbf{u}|\boldsymbol{\theta})$.

Repeating the above process m times, or taking an independent sample of size m after burn-in, we obtain a sample of size m from the intractable distribution $f(\mathbf{u}|\boldsymbol{\theta})$. When m is large, we can approximate the mean and variance of the distribution $f(\mathbf{u}|\boldsymbol{\theta})$ by the sample mean and sample variance respectively, or we can approximate the density curve $f(\mathbf{u}|\boldsymbol{\theta})$ by the empirical density function based on the simulated samples.

The key idea of the Gibbs sampler is to sequentially sample from lower dimensional conditional distributions in order to generate samples from the original higher dimensional and intractable distribution. Usually it is easier to sample from these lower dimensional conditional distributions (full conditionals) than the original distribution. Sometimes, however, sampling from the lower-dimensional conditional distributions may not be easy either. In this case, we can use rejection sampling methods to sample from these full conditionals. That is, we can combine the Gibbs sampler with rejection sampling methods or other sampling methods.

Note that the Gibbs sampler or other MCMC methods can only *approximate* the target distributions. The accuracy of the approximation improves as the number of steps (burn-in period) increases. It may not be easy to determine the burn-in period for the Markov chain to converge to the stationary distribution within acceptable random errors. Determining the convergence criteria is an important issue. See Gelman et al. (2003) for a more detailed discussion.

WinBUGS is a statistical software that is widely used to do Gibbs sampling. It is based on the BUGS project (Bayesian inference Using Gibbs Sampling). For details of the WinBUGS software, see webpage:
http://www.mrc-bsu.cam.ac.uk/bugs/winbugs/contents.shtml.

Example 12.1 Consider a simple example of generating samples from the bivariate normal distribution $\mathbf{u} = (u_1, u_2)^T \sim N(\boldsymbol{\mu}, \Sigma)$ where $\boldsymbol{\mu} = (\mu_1, \mu_2)^T$ and Σ is a 2×2 covariance matrix with diagonal elements being 1 and off-diagonal elements being ρ ($|\rho| < 1$). Assume that $\boldsymbol{\mu}$ and Σ are known. Here it is not difficult to sample from $N(\boldsymbol{\mu}, \Sigma)$, but we consider a Gibbs sampler for illustration purpose.

Consider the Gibbs sampler to sample from the target distribution $N(\boldsymbol{\mu}, \Sigma)$. The full conditionals are

$$u_1 | u_2, \boldsymbol{\mu}, \rho \sim N(\mu_1 + \rho(u_2 - \mu_2),\ 1 - \rho^2),$$
$$u_2 | u_1, \boldsymbol{\mu}, \rho \sim N(\mu_2 + \rho(u_1 - \mu_1),\ 1 - \rho^2).$$

The Gibbs sampler proceeds as follows: beginning with starting value $(u_1^{(0)}, u_2^{(0)})$, at k-th step,

- generate $u_1^{(k)}$ from $N(\mu_1 + \rho(u_2^{(k-1)} - \mu_2),\ 1 - \rho^2)$;
- generate $u_2^{(k)}$ from $N(\mu_1 + \rho(u_1^{(k)} - \mu_2),\ 1 - \rho^2)$, $\quad k = 1, 2, \cdots$.

Then, the sequence $\{(u_1^{(k)}, u_2^{(k)}), k = 0, 1, 2, \cdots\}$ forms a Markov chain with stationary distribution $N(\boldsymbol{\mu}, \Sigma)$. Thus, when k is large (say 200), we may consider $(u_1^{(k)}, u_2^{(k)})$ as a sample from $N(\boldsymbol{\mu}, \Sigma)$.

The Metropolis-Hastings Algorithm

The Gibbs sampling algorithm is a special case of the *Metropolis-Hastings algorithm*. That is, the Metropolis-Hastings algorithm is a more general method for creating a Markov chain that can be used to generate samples from an intractable probability distributions, and it is often faster and easier to use than the Gibbs sampler but it is less generally applicable. The algorithm can be used to draw samples from a distribution without knowing the normalization factor, i.e., from $g(\mathbf{u}) \propto f(\mathbf{u})$, where $f(\mathbf{u})$ is the target distribution of interest but the normalization factor may be very difficult to compute (e.g., in Bayesian inference).

Suppose that we wish to simulate a sample from an intractable distribution $f(\mathbf{u})$. Let $p(\mathbf{u}|\mathbf{v})$ be a proposal density. Beginning with a starting value $\mathbf{u}^{(0)}$, at step k we generate a value $\mathbf{u}^*$ from $p(\mathbf{u}|\mathbf{v})$. Then the Metropolis-Hastings algorithm proceeds as follows: simulate a value a from the uniform distribution $U(0,1)$ on $(0,1)$, then accept $\mathbf{u}^*$ as the next value $\mathbf{u}^{(k+1)}$ (i.e., choose $\mathbf{u}^{(k+1)} = \mathbf{u}^*$) *if*

$$a < \frac{f(\mathbf{u}^*)\,p(\mathbf{u}^{(k)}|\mathbf{u}^*)}{f(\mathbf{u}^{(k)})\,p(\mathbf{u}^*|\mathbf{u}^{(k)})}.$$

Otherwise, the proposal is not accepted and the current value is retained (i.e., $\mathbf{u}^{(k+1)} = \mathbf{u}^{(k)}$). The sequence $\{\mathbf{u}^{(k)}, k = 0, 1, 2, \cdots\}$ then forms a Markov chain with stationary distribution $f(\mathbf{u})$. After a burn-in period, we can view $\mathbf{u}^{(k)}$ as a sample from $f(\mathbf{u})$.

The algorithm works best if the proposal density $p(\mathbf{u}|\mathbf{v})$ is close to $f(\mathbf{u})$, but often this is difficult to do. Gibbs sampling is a special case of the Metropolis-Hastings sampling where the proposal is always accepted (see, e.g., Gelman et al. 2004, page 328).

12.3 Rejection Sampling and Importance Sampling Methods

In likelihood or Bayesian inference, we often need to evaluate intractable integrals which do not have analytic or closed-form expressions. Monte Carlo methods are widely used to approximate these integrals. In this section, we briefly describe two popular Monte Carlo methods: rejection sampling methods and importance sampling methods.

We first consider the following integral

$$I = \int g(x)f(x)dx,$$

where $g(\cdot)$ is a continuous function and $f(\cdot)$ is a probability density function. Suppose that I does not have an analytic expression. A Monte Carlo method can be used as follows. If we can generate an i.i.d. sample $x_1, x_2, \cdots, x_m$ from the density $f(x)$, we can then approximate the integral I by

$$I \approx \hat{I} = \frac{1}{m} \sum_{j=1}^{m} g(x_j).$$

The accuracy of this approximation increases as the number m increases. Thus, the problem is how to sample from the distribution $f(x)$. Unfortunately, in many problems the density function $f(x)$ is highly complicated, especially when x is a vector, so it may not be straightforward to generate these samples. The rejection sampling methods and the importance sampling methods are two classes of general and widely used methods to generate samples from intractable distributions.

Rejection Sampling Methods

Suppose that we wish to generate a sample from a complicated density function $f(x)$. Since $f(x)$ is complicated, it may be hard to sample directly from $f(x)$. Suppose, however, that we know how to sample from the distribution with density $h(x)$, and that there is a known constant c such that

$$f(x) \leq c\, h(x), \qquad \text{for all } x.$$

Then, a rejection sampling method may proceed as follows:

- generate a value x^* from the distribution $h(x)$;
- generate a value u from the uniform distribution on (0,1);
- accept x^* if

$$u < \frac{f(x^*)}{c\, h(x^*)},$$

otherwise reject x^*.

Repeating this procedure and retaining only accepted values of x^*, we obtain a sample $x_1^*, x_2^*, \cdots$, from the target distribution $f(x)$.

The efficiency of the rejection sampling method depends strongly on how well the function $h^*(x) = c\, h(x)$, often called the *envelope function*, approximates the target function $f(x)$. If the ratio $f(x)/h^*(x)$ is small, the probability of acceptance will be small, so the algorithm will spend most of time rejecting x^* values. A multivariate version of the rejection sampling method is similar. See Evans and Swartz (2000) and Robert and Casella (2004) for more details.

It is often not easy to find a good envelope function. The following adaptive rejection sampling method is widely applicable and very popular.

The Adaptive Rejection Sampling Method

The *adaptive rejection sampling method* (Gilks and Wild 1992) is a very useful rejection sampling method when the target density function $f(x)$ is *log-concave*, i.e., function $\log f(x)$ is concave. It is particularly useful in Gibbs sampling where the full conditional distributions may be intractable but are known to be log-concave. In its original version, the adaptive rejection sampling method constructs the envelope functions based on a set of *tangents* to the function $\log f(x)$. In later versions, the method updates the envelope functions to correspond more closely to the target function $\log f(x)$ whenever a point is rejected, and thus improves the efficiency of the method. More flexible algorithms, which relax the log-concavity requirement, have also been proposed (Evans and Swartz 2000).

Importance Sampling Methods

Let $g(x)$ be a continuous function and $f(x)$ be a probability density function. Consider again the integral

$$I = \int g(x)f(x)dx = E(g(x)).$$

Suppose that it is hard to directly sample from $f(x)$, but we know how to sample from a distribution $h(x)$, which does not need to be an envelope function as in a rejection sampling method. Since

$$I = \int g(x)f(x)dx = \int \frac{g(x)f(x)}{h(x)}h(x)dx,$$

if $x_1, x_2, \cdots, x_m$ is a sample generated from the distribution $h(x)$, we then have

$$I \approx \hat{I} = \frac{1}{m}\sum_{i=1}^{m}\frac{g(x_i)f(x_i)}{h(x_i)} = \frac{1}{m}\sum_{i=1}^{m}w_i g(x_i),$$

where $w_i = f(x_i)/h(x_i)$, $i = 1, \cdots, m$, are called importance weights and $h(x)$ is called the *importance function*. Note that here we do not require the condition $f(x) \le ch(x)$, and, unlike rejection sampling methods, here we use all the generated x_j's. The efficiency of the importance sampling method depends on the choice of the importance function. The closer the function $h(x)$ approximates $f(x)$, the better the importance sampling method. Often, the function $h(x)$ is chosen to have larger tails than $f(x)$. Evans and Swartz (2000) and Robert and Casella (2004) provided more detailed discussions.

12.4 Numerical Integration and the Gauss-Hermite Quadrature Method

To evaluate intractable integrals, numerical integration methods are alternatives to Monte Carlo methods. Numerical integration methods approximate an integral by a weighted sum, with suitably chosen points and weights. These methods include the Simpson's rule and quadrature methods. In the following, we briefly describe the popular Gauss-Hermite quadrature method. Evans and Swartz (2000) provided a detailed discussion of various approaches.

Consider the following integral

$$I = \int g(x)f(x)dx,$$

where $g(x)$ is a continuous function and $f(x)$ is a normal density function. We first consider the $N(0, 1)$ density and let $f(x) = \exp(-x^2)$. The *Gauss-Hermite quadrature* method approximates the integral by

$$I = \int \exp(-x^2)g(x)dx \approx \sum_{i=1}^{k} w_i(x_i)g(x_i),$$

where the node x_i is the i-th root of the Hermite polynomial $H_k(x)$ with degree of k, $w_i(x_i)$ is the weight which depends on $H_{k-1}(x_i)$:

$$w_i(x_i) = \frac{2^{k-1}k!\sqrt{\pi}}{k^2(H_{k-1}(x_i))^2},$$

and the *Hermite polynomials* are orthogonal polynomials defined by

$$H_k(x) = (-1)^k e^{x^2/2} \frac{d^k \, e^{-x^2/2}}{dx^k}.$$

The above approximation can be arbitrarily accurate when the number k of nodes increases. When $g(x)$ is a polynomial of degree up to $2k - 1$, the approximation is exact. Note that the first several Hermite polynomials are:

$$\begin{aligned} H_0(x) &= 1, & H_1(x) &= x, & H_2(x) &= x^2 - 1, \\ H_3(x) &= x^3 - 3x, & H_4(x) &= x^4 - 6x^2 + 3, & \cdots. \end{aligned}$$

If $f(x)$ is the density function of a general normal distribution $N(\mu, \sigma^2)$, we may consider transformation $x = \sqrt{2}\sigma z + \mu$, and we then have

$$I = \int g(x)f(x)dx \approx \sum_{i=1}^{k} w_i^*(x_i)g(\sqrt{2}\sigma z_i + \mu),$$

where $w_i^*(x_i) = \pi^{-1/2}w_i(x_i)$.

If $\mathbf{x} = (x_1, \cdots, x_m)^T$ is a m-dimensional vector, we have

$$I = \int_{R^m} g(\mathbf{x})f(\mathbf{x})d\mathbf{x} \approx \sum_{i_1=1}^{k_1} w_{i_1}^{(1)} \cdots \sum_{i_m=1}^{k_m} w_{i_m}^{(m)} g(x_{i_1}^{(1)}, \cdots, (x_{i_m}^{(m)})$$

where $x_{i_j}^{(j)}$ is the i_j-th root of the Hermite polynomial with degree k_j and $w_{i_j}^{(j)}$ is the corresponding weight. Note that the number of nodes increases exponentially with the number m of dimensions, so the method can be very inefficient for high-dimensional integrals. In practice, one may use the method for integrals with dimensions up to 5 or 6. See Evans and Swartz (2000) and Fahrmeir and Tutz (2001) for more detailed discussions.

12.5 Optimization Methods and the Newton-Raphson Algorithm

In estimation problems, especially in likelihood methods, one often needs to find a maxima or minima of a function, say $L(\boldsymbol{\theta})$. This problem is often equivalent to finding a root of the function $g(\boldsymbol{\theta}) \equiv \partial \log L(\boldsymbol{\theta})/\partial\boldsymbol{\theta} = \partial l(\boldsymbol{\theta})/\partial\boldsymbol{\theta}$, or solving equation

$$g(\boldsymbol{\theta}) = 0.$$

There are different general optimization procedures to accomplish this. The most widely used one is perhaps the *Newton-Raphson method* or the *Newton's method*, which is briefly described below.

The Newton-Raphson method is an iterative algorithm based on a Taylor series expansion. In the univariate case $\boldsymbol{\theta} = \theta$, the Newton-Raphson method iteratively solves the following equation

$$\theta_k = \theta_{k-1} - \frac{g(\theta_{k-1})}{g'(\theta_{k-1})}, \qquad k = 1, 2, \cdots,$$

where $g'(\theta)$ is the derivative of $g(\theta)$. Beginning with an initial value θ_0, the algorithm will usually converge to a possibly local maxima or minima of $L(\theta)$. In the multi-dimensional case, the algorithm can be written as

$$\boldsymbol{\theta}_k = \boldsymbol{\theta}_{k-1} - \left(\frac{\partial g(\boldsymbol{\theta})}{\partial\boldsymbol{\theta}}\right)^{-1}\Big|_{\boldsymbol{\theta}=\boldsymbol{\theta}_{k-1}} \cdot g(\boldsymbol{\theta}_{k-1}), \qquad k = 1, 2, \cdots,$$

where

$$\frac{\partial g(\boldsymbol{\theta})}{\partial\boldsymbol{\theta}} = \left(\frac{\partial g(\boldsymbol{\theta})}{\partial\theta_1}, \cdots, \frac{\partial g(\boldsymbol{\theta})}{\partial\theta_p}\right) = \left(\frac{\partial^2 l(\boldsymbol{\theta})}{\partial\theta_i \partial\theta_j}\right)_{p\times p}.$$

Note that the choice of the initial value $\boldsymbol{\theta}_0$ is important. The closer $\boldsymbol{\theta}_0$ to the root of $g(\boldsymbol{\theta})$, the better. However, since we often do not know the root of $g(\boldsymbol{\theta})$, a guessed value $\boldsymbol{\theta}_0$ is usually chosen. If the function $l(\boldsymbol{\theta})$ has multi-mode (so $l(\boldsymbol{\theta})$ has two or more roots), the Newton-Raphson algorithm only converges to a local maxima. To find a global maxima, one should try several different initial values. Note that the Newton-Raphson algorithm may not converge in some cases, such as the initial value being too far from the true zero or the derivatives being zeros or near zeros (so the tangent line is nearly horizontal). In practice, one may want to put an upper limit on the number of iterations. For more detailed discussions of the Newton-Raphson methods and other optimization methods, see, e.g., Deuflhard (2004) and Press et al. (2007).

12.6 Bootstrap Methods

Bootstrap methods are widely used general approaches for statistical inference. They are computer-intensive resampling methods, and are very useful for some difficult problems, such as variance estimations for intractable estimators and statistical inference when parametric assumptions are in doubt or when parametric inference is highly complicated. For example, it is usually difficult to compute the variances of a sample median or a sample percentile or a sample correlation coefficient. In these cases, it is straightforward to use the bootstrap method to compute estimates of standard errors and confidence intervals of these estimators. Bootstrap methods are often easy to implement, though may be computationally intensive, and can be applied to a wide variety of problems. Therefore, bootstrap has become a very popular statistical tool in modern statistics.

The idea of a bootstrap method is usually to approximate a distribution by the empirical distribution of the *observed data*, implemented by repeatedly resampling from the observed dataset *with replacement* (with the same sample size as the observed dataset). For example, suppose that $(x_1, x_2, \cdots, x_n)$ is an observed dataset, and suppose that one wishes to estimate the variance of the sample median. A simple bootstrap method proceeds as follows. We can sample from this observed dataset with replacement. The resulting sample, denoted by $(x_1^*, x_2^*, \cdots, x_n^*)$, is called a *bootstrap sample.* Then, we compute the sample median of this bootstrap sample. Repeating this process B times (B is often large, say $B = 1000$), we obtain B median estimates from the B bootstrap samples. We then compute the sample variance of these B median estimates and obtain a bootstrap estimate of the variance of the sample median from the original dataset. The *sampling distribution* of these B estimates is an approximation to the "true" distribution of the sample median from the original dataset.

As another example, we know that the MLE of a parameter is asymptotically normally distributed. In practice, this asymptotic distribution is often used to construct approximate confidence intervals and hypothesis testing where the sample size is in fact finite. Since the sample size is finite in practice, we may want to know how close the distribution of the MLE is to normality, so that we can judge how reliable the approximate confidence intervals and testing results are. We can use a bootstrap method to check this, as illustrated as follow.

Suppose that we fit a mixed effects model, such as an NLME model, to a longitudinal dataset (with sample size n) using the likelihood method, and we wish to check if the resulting MLEs of the parameters are approximately normal. A simple bootstrap method can be performed as follows:

- sample from the original dataset with replacement and obtain a bootstrap sample;
- fit the mixed effects model to the bootstrap sample using the likelihood method and obtain MLEs of the parameters;

- Repeating the procedure B times, one obtains B sets of parameter estimates (MLEs).

The sampling distribution of the B estimates of a parameter is an approximation to the "true" sampling distribution of the MLE of this parameter based on the original dataset. One can then, for example, obtain an approximate confidence interval from the bootstrap samples by taking the α and $1 - \alpha$ (say, $\alpha = 0.05$) quantiles of the B estimates. A bootstrap estimate of the standard error of the parameter estimate is the sample standard error of the B estimates.

For a parametric bootstrap method, one would fit a parametric model and obtain bootstrap samples from the fitted parametric model. The estimates are again computed from the bootstrap samples.

For more detailed discussions of Bootstrap methods, see Efron and Tibshirani (1993) and Davison and Hinkley (2006).

12.7 Matrix Algebra and Vector Differential Calculus

For statistical models with more than one parameters, it is often more convenient to write the models in vector and matrix forms. For example, a multiple linear regression model can be written in the compact form: $\mathbf{y} \sim N(X\boldsymbol{\beta}, \sigma^2 I)$, where $\mathbf{y}$ and $\boldsymbol{\beta}$ are vectors and X and I are matrices. The least square can be written as $L(\boldsymbol{\beta}) = (\mathbf{y} - X\boldsymbol{\beta})^T(\mathbf{y} - X\boldsymbol{\beta})$, and the least square estimate of $\boldsymbol{\beta}$, obtained by solving $\partial L(\boldsymbol{\beta})/\partial \boldsymbol{\beta} = 0$, is given by $\hat{\boldsymbol{\beta}} = (X^T X)^{-1} X^T \mathbf{y}$. Thus, the compact forms are quite convenient, and some basic matrix algebra and vector differential calculus are important. In this section, we present some useful results, which may be helpful in understanding some of the results and derivations presented in the book.

Let $f(\mathbf{x})$ be a scalar function of a vector variable $\mathbf{x} = (x_1, \cdots, x_p)^T$. Let

$$\frac{\partial f(\mathbf{x})}{\partial \mathbf{x}} = \left(\frac{\partial f(\mathbf{x})}{\partial x_1}, \cdots, \frac{\partial f(\mathbf{x})}{\partial x_p} \right)^T,$$

and let

$$\frac{\partial^2 f(\mathbf{x})}{\partial \mathbf{x}^2} = \frac{\partial^2 f(\mathbf{x})}{\partial \mathbf{x} \partial \mathbf{x}^T} = \left(\frac{\partial^2 f(\mathbf{x})}{\partial x_i \partial x_j} \right)_{p \times p}$$

be the $p \times p$ matrix with (i, j)-th element being $\partial^2 f(\mathbf{x})/\partial x_i \partial x_j$. Note that, if $df(\mathbf{x})/d\mathbf{x} = \mathbf{a}$, then $df(\mathbf{x}) = \mathbf{a} d\mathbf{x}$. Let $det(A)$ be the determinant of matrix A and $tr(A)$ be the trace of A, and let I denote the identity matrix.

Let A and B are matrices. We first present the following rules for matrix algebra, which are often useful,

$$det(AB) \quad = \quad det(A)det(B),$$

$$
\begin{aligned}
det(A^{-1}) &= det(A)^{-1}, \\
det(I + AB^T) &= det(I + B^T A), \\
det(\exp(A)) &= \exp(tr(A)), \\
tr(AB) &= tr(BA), \\
tr(A^T B) &= vec(A)^T vec(B),
\end{aligned}
$$

where $vec(A)$ is the vectorization of matrix A.

Let $\mathbf{x}$ and $\mathbf{y}$ are vector functions and X and Y be matrix functions. Let A and B be constant matrices. The following are some useful rules for differentials

$$
\begin{aligned}
d(tr(X)) &= tr(dX), \\
d(XY) &= (dX)Y + X(dY), \\
dX^{-1} &= -X^{-1}(dX)X^{-1}, \\
d\,det(A) &= det(A)tr(A^{-1}dA).
\end{aligned}
$$

The following are some specific results which are useful in many cases:

$$
\frac{\partial(A^T\mathbf{x})}{\partial \mathbf{x}} = A, \qquad \frac{\partial^2(A^T\mathbf{x})}{\partial \mathbf{x}^2} = 0,
$$

$$
\frac{\partial(\mathbf{x}^T A\mathbf{y})}{\partial \mathbf{x}} = A\mathbf{y}, \qquad \frac{\partial^2(\mathbf{x}^T A\mathbf{y})}{\partial \mathbf{x}^2} = 0,
$$

$$
\frac{\partial(\mathbf{x}^T\mathbf{x})}{\partial \mathbf{x}} = 2\mathbf{x}, \qquad \frac{\partial^2(\mathbf{x}^T\mathbf{x})}{\partial \mathbf{x}^2} = 2I,
$$

$$
\frac{\partial(\mathbf{x}^T A\mathbf{x})}{\partial \mathbf{x}} = 2A\mathbf{x}, \qquad \frac{\partial^2(\mathbf{x}^T A\mathbf{x})}{\partial \mathbf{x}^2} = 2A, \qquad \text{if } A \text{ is symmetric}
$$

$$
\frac{\partial \log(det(A(\mathbf{x})))}{\partial \mathbf{x}} = tr\left(A^{-1}(\mathbf{x})\frac{\partial A(\mathbf{x})}{\partial \mathbf{x}}\right)
$$

where $A(\mathbf{x})$ means that matrix A is a function of variable $\mathbf{x}$.

Let $f(A)$ be a scalar function of a matrix $A = (a_{ij})_{p \times p}$, and let $\partial f(A)/\partial A$ be the matrix with (i,j)-th element being $\partial f(A)/\partial a_{ij}$. We have

$$
\frac{\partial(\mathbf{x}^T A\mathbf{y})}{\partial A} = \mathbf{x}\mathbf{y}^T,
$$

$$
\frac{\partial(tr(A))}{\partial A} = I,
$$

$$
\frac{\partial(\mathbf{x}^T A\mathbf{x})}{\partial A} = 2\mathbf{x}\mathbf{x}^T - diag(\mathbf{x}\mathbf{x}^T), \qquad \text{if } A \text{ is symmetric},
$$

$$
\frac{\partial|A|}{\partial A} = |A|(2A^{-1} - diag(A^{-1})), \qquad \text{if } A \text{ is symmetric},
$$

$$
\frac{\partial tr(AB)}{\partial A} = B + B^T - diag(B), \qquad \text{if } A \text{ is symmetric},
$$

where $diag(A)$ is the diagonal matrix of A.

The above results are only partial lists of results for matrix algebra and vector differential calculus. For more results and details, see Magnus and Neudecker (1988) and Wand (2002).

Example 12.2 *Maximum likelihood estimation in linear models*

Consider the linear regression model

$$\mathbf{y} \sim N(X\beta, \sigma^2 I).$$

The log-likelihood for estimating β is given by

$$l(\beta) = -\frac{1}{2\sigma^2}(\mathbf{y} - X\beta)^T(\mathbf{y} - X\beta).$$

For likelihood estimation, we have

$$\frac{\partial l(\beta)}{\partial \beta} = \frac{1}{\sigma^2}(\mathbf{y} - X\beta)^T X = 0.$$

Thus, the MLE of β is given by

$$\hat{\beta} = (X^T X)^{-1} X^T \mathbf{y},$$

which is the same as the least-square estimate. To obtain the variance-covariance matrix of $\hat{\beta}$, note that

$$\frac{\partial^2 l(\beta)}{\partial \beta^2} = -\frac{1}{\sigma^2} X^T X.$$

Thus, the Fisher information matrix is given by

$$I(\beta) = \frac{1}{\sigma^2}(X^T X).$$

The asymptotic variance-covariance matrix of $\hat{\beta}$ is

$$Var(\hat{\beta}) = I^{-1}(\beta).$$

References

Albert, P.S. and Shih, J.H. (2009). On estimating the relationship between longitudinal measurements and time-to-event data using a simple two-stage procedure. *Biometrics*, in press.

Amemiya, T. (1985). *Advanced Econometrics*, Oxford:Blackwell.

Andersen, P.K., Borgan, O., Gill, R.D., and Keiding, N. (1993). *Statistical Models Based on Counting Processes*. Berlin: Springer-Verlag.

Andersson, S.A. and Perlman, M.D. (1993). Lattice models for conditional independence in a multivariate normal distribution. *Annals of Statistics* **21**, 1318-1358.

Bates, D.M. and Watts, D.G. (2007). *Nonlinear Regression Analysis and Its Applications*. New York: John Wiley.

Beal, S.L., Boeckmann, A.J., and Sheiner, L.B. (1992), *NONMEM User's Guide VI*. San Francisco: University of California.

Beal, S.L. and Sheiner, L.B. (1982). Estimating population kinetics. *CRC Critical Reviews in Biomedical Engineering* **8**, 195-222.

Berkson, J. (1950). Are there two regressions? *Journal of the American Statistical Association* **45**, 164-180.

Bernaards, C.A., Belin, T.R., and Schafer, J.L. (2007). Robustness of a multivariate normal approximation for imputation of incomplete binary data. *Statistics in Medicine* **26**, 1368-1382.

Booth, J.G. and Hobert, J.P. (1999). Maximizing generalized linear mixed model likelihoods with an automated Monte Carlo EM algorithm. *Journal of the Royal Statistical Society, Ser. B* **61**, 265-285.

Breslaw, J.A. (1994). Random sampling from a truncated multivariate normal distribution. *Applied Mathematics Letters* **7**, 1-6.

Breslow, N.E. and Clayton, D.G. (1993). Approximate inference in generalized linear mixed models. *Journal of the American Statistical Association* **88**, 9-25.

Breslow, N.E. and Lin, X. (1995). Bias correction in generalized linear mixed models with a single component of dispersion. *Biometrika* **82**, 81-91.

Brown, E.R. and Ibrahim, J.G. (2003). A Bayesian semiparametric joint hierarchical model for longitudinal and survival data. *Biometrics* **59**, 221-228.

Brunner, J. and Austin, P.C. (2009). Inflation of type I error rate in multiple regression when independent variables are measured with error. *The Canadian Journal of Statistics* **37**, 33-46.

Buonaccorsi, J., Demidenko, E., and Tosteson, T. (2000). Estimation in longitudinal random effects models with measurement error. *Statistica Sinica* **10**, 885-903.

Caffo, B.S., Jank, W., and Jones, G.L. (2005). Ascent-based Monte Carlo expectation–maximization. *Journal of the Royal Statistical Society: Series B* **67**, 235-251.

Cantoni, E. and Ronchetti, E. (2001). Robust inference for generalized linear models. *Journal of the American Statistical Association* **96**, 1022-1030.

Carpenter, J.R., Kenward, M.G., Evans, S., and White, I. (2004). Last observation carry-forward and last observation analysis. *Statistics in Medicine* **23**, 3241-3244.

Carpenter, J.R., Kenward, M.G., and Vansteelandt, S. (2006). A comparison of multiple imputation and doubly robust estimation for analyses with missing data. *Journal of the Royal Statistical Society, Series A* **169**, 571-584.

Carroll, R.J., Delaigle, A., and Hall, P. (2007). Non-parametric regression estimation from data contaminated by a mixture of Berkson and classical errors. *Journal of the Royal Statistical Society, Series B* **69**, 859-878.

Carroll, R.J., Midthune, D., Freedman, L.S., and Kipnis, V. (2006). Seemingly unrelated measurement error models, with application to nutritional epidemiology. *Biometrics* **62**, 75-84.

Carroll, R.J., Ruppert, D., Stefanski, L.A., and Crainiceanu, C. (2006). *Measurement Error in Nonlinear Models: A Modern Perspective*, 2nd edition, London: Chapman and Hall.

Chaganty, N.R. and Joe, H. (2004). Efficiency of the generalised estimating equations for binary response. *Journal of the Royal Statistical Society Series B* **66**, 851-860.

Chan, K.S. and Ledolter, J. (1995). Monte Carlo EM estimation for time series models involving counts. *Journal of the American Statistical Association* **90**, 242-252.

Chen, H.Y. and Little, R.J. (1999). Proportional hazards regression with missing covariates. *Journal of the American Statistical Association* **94**, 896-908.

Chen, J., Zhang, D., and Davidian, M. (2002). A Monte-Carlo EM algorithm for generalized linear mixed models with flexible random effects distributions. *Biostatistics* **3**, 347-360.

Clayton, D.G. and Cuzick, J. (1985). Multivariate generalizations of the proportional hazards model (with discussion). *Journal of the Royal Statistical Society Series A* **148**, 82-117.

Collett, D. (2003). *Modelling Survival Data in Medical Research*, 2nd Edition. Boca Raton, FL: Chapman & Hall/CRC.

Cook, R.J. and Lawless, J.F. (2007). *The Statistical Analysis of Recurrent Events*. New York: Springer.

Cook, R.J., Yi, G.Y., Lee, K.A., and Gladman, D.D. (2004). A conditional Markov model for clustered progressive multistate processes under incomplete observation. *Biometrics* **60**, 436-443.

Cook, R.J., Zeng, L., and Yi, G.Y. (2004). Marginal analysis of incomplete longitudinal binary data: a cautionary note on LOCF imputation. *Biometrics* **60**, 820-828.

Copt, S. and Victoria-Feser, M.P. (2006). High-breakdown inference for mixed linear models. *Journal of the American Statistical Association* **101**, 292-300.

Corbeil, R.R. and Searle, S.R. (1976). A comparison of variance component estimators. *Biometrics* **32**, 779-791.

Cox, D.R. (1972). Regression models and life tables (with discussion). *Journal of Royal Statistical Society, Series B* **34**, 187-200.

Cox, D.R. (1975). Partial likelihood. *Biometrika* **62**, 269-276.

Cox, D.R. and Snell, E.J. (1984). *Applied Statistics, Principles and Examples*. London: Chapman & Hall.

Dafni, U.G. and Tsiatis, A.A. (1998). Evaluating surrogate markers of clinical outcome when measured with error. *Biometrics* **54**, 1445-1462.

Davidian, M. and Gallant, A.R. (1993). The nonlinear mixed effects model with a smooth random effects density. *Biometrika* **80**, 475-488.

Davidian, M. and Giltinan, D.M. (1995). *Nonlinear Models for Repeated Measurements Data*. London: Chapman & Hall.

Davidian, M. and Giltinan, D.M. (2003). Nonlinear models for repeated measurement data: an overview and update. *Journal of Agricultural, Biological, and Environmental Statistics* **8**, 387-419.

Davison, A.C. and Hinkley, D. (2006). *Bootstrap Methods and Their Applications*, 8th edition. Cambridge, UK: Cambridge Series in Statistical and Probabilistic Mathematics.

de Boor, C. (1978). *A Practical Guide to Splines*. New York: Springer-Verlag.

Dean, C.B. (1998). Overdispersion. *Encyclopedia of Biostatistics*, P. Armitage and T. Colton (eds.), London: Wiley, 467-472.

Dean, C.B. and Nielsen, J.D. (2007). Generalized linear mixed models: A review and some extensions. *Lifetime Data Analysis* **13**, 497-512.

DeGruttola, V. and Tu, X.M. (1994). Modeling progression of CD4-lymphocyte count and its relationship to survival time. *Biometrics* **50**, 1003-1014.

Demidenko, E. (2004). *Mixed Models: Theory and Applications*. New York: Wiley.

Dempster, A.P., Laird, N.M., and Rubin, D.B. (1977). Maximum likelihood estimation from incomplete data via the EM algorithm. *Journal of the Royal Statistical Society, Ser. B* **39**, 1-38.

Deuflhard, P. (2004). *Newton Methods for Nonlinear Problems, Affine Invariance and Adaptive Algorithms*, Springer Series in Computational Mathematics, Vol. 35. Berlin: Springer.

Diggle, P., Heagerty, P., Liang, K.Y., and Zeger, S. (2002) *Analysis of Longitudinal Data*, 2nd edition. Oxford, England: Oxford University Press.

Diggle, P. and Kenward, M.G. (1994). Informative drop-out in longitudinal data analysis (with Discussion). *Applied Statistics* **43**, 49-93.

Ding, A.A. and Wu, H. (1999), Relationships between antiviral treatment effects and biphasic viral decay rates in modeling HIV dynamics. *Mathematical Biosciences* **160**, 63-82.

Ding, A.A. and Wu, H. (2001). Assessing antiviral potency of anti-HIV therapies in vivo by comparing viral decay rates in viral dynamic models. *Biostatistics* **2(1)**, 13-29.

Domowitz, I. and White, H. (1982). Misspecified models with dependent observations. *Journal of Econometrics* **20**, 35-58.

Draper, N.R. and Smith, H. (1998). *Applied Regression Analysis*, 3rd edition. New York: Wiley-Interscience.

Efron, B. and Tibshirani, R.J. (1993). *An Introduction to the Bootstrap*. New York: Chapman & Hall/CRC.

Eubank, R.L. (1988). *Spline Smoothing and Nonparametric Regression*. New York: Marcel Dekker.

Evans, M. and Swartz, T.B. (1998). Random variable generation using concavity properties of transformed densities. *Journal of Computational and Graphical Statistics* **7**, 514-528.

Evans, M. and Swartz, T.B. (2000). *Approximating Integrals via Monte Carlo and Deterministic Methods*. Oxford, England: Oxford University Press.

Fahrmeir, L. and Tutz, G. (2001). *Multivariate Statistical Modelling Based on Generalized Linear Models*, 2nd edition. New York: Springer.

Fan, J. and Gijbels, I. (1996). *Local Polynomial Modelling and Its Applications*. London: Chapman & Hall.

Faraway, J.J. (2004). *Linear Models with R*. Boca Raton, FL: Chapman & Hall/CRC.

Faraway, J.J. (2005). *Extending the Linear Model with R: Generalized Linear, Mixed Effects and Nonparametric Regression Models*. Boca Raton, FL: Chapman & Hall/CRC.

Faucett, C.L., Schenker, N., and Taylor, J.M.G. (2002). Survival analysis using auxiliary variables via multiple imputation, with application to AIDS clinical trials data. *Biometrics* **58**, 37-47.

Fay, R.E. (1996). Alternative paradigms for the analysis of imputed survey data. *Journal of the American Statistical Association* **91**, 490-498.

Fitzgerald, A.P., DeGruttola, V.G., and Vaida, F. (2002). Modeling HIV viral rebound using non-linear mixed effects models. *Statistics in Medicine* **21**, 2093-2108.

Fitzmaurice, G., Davidian, M., Molenberghs, G., and Verbeke, G. (2008). *Longitudinal Data Analysis*. Boca Raton, FL: Chapman & Hall/CRC .

Fitzmaurice, G.M., Laird, N.M., and Zahner, G.E.P. (1996). Multivariate logistic models for incomplete binary responses. *Journal of the American Statistical Association* **91**, 99-108.

Fleming, T.R. and Harrington, D.P. (1991). *Counting Processes and Survival Analysis*. New York: Wiley.

Fort, G. and Moulines, E. (2003). Convergence of the Monte-Carlo EM for curved exponential families. *Annals of Statistics* **31**, 1220-1259.

Follmann, D. and Wu, M. (1995). An approximate generalized linear model with random effects for informative missing data. *Biometrics* **51**, 151-168.

Fuller, W.A. (1987). *Measurement Error Models*. New York: John Wiley & Sons.

Gelfand, A.E., Sahu, S.K., and Carlin, B.P. (1996). Efficient parameterizations for generalized linear models (with Discussion). *Bayesian Statistics* **5**, 165-180.

Gelfand, A.E. and Smith, A.F.M. (1990). Sampling-based approaches to calculating marginal densities. *Journal of the American Statistical Association* **85**, 398-409.

Gelman, A., Carlin, J.B., Stern, H.S., and Rubin, D.B. (2003). *Bayesian Data Analysis*, 2nd edition. London: Chapman & Hall/CRC.

Geman, S. and Geman, D. (1984). Stochastic relaxation, Gibbs distributions, and the Bayesian restoration of images. *IEEE Transactions on Pattern Analysis and Machine Intelligence* **6**, 721-741.

Geweke, J. (1996). *Handbook of Computational Economics*, Ch. 15. Amsterdam: North-Holland.

Gibaldi, M. and Perrier, D. (1982). *Pharmacokinetics*, 2nd edition. New York: Marcel Dekker.

Gilks, W.R., Richardson, S., and Spiegelhalter, D.J. (1996). *Markov Chain Monte Carlo in Practice*. London: Chapman & Hall/CRC.

Gilks, W.R. and Wild, P. (1992). Adaptive rejection sampling for Gibbs sampling. *Applied Statistics* **41**, 337-348.

Goldstein, H. (1987). *Multilevel Models in Education and Social Research*. Oxford: Oxford University Press.

Gong, G. and Sammaniego, F.J. (1981). Pseudo maximum likelihood estimation: theory and applications. *Annals of Statistics* **9**, 861-869.

Green, P.J. and Silverman, B.W. (1994). *Nonparametric Regression and Generalized Linear Models*. London: Chapman & Hall.

Guo, X. and Carlin, B.P. (2004). Separate and joint modeling of longitudinal and event time data using standard computer packages. *The American Statistician* **58**, 1–9.

Gustafson, P. (2004). *Measurement Error and Misclassification in Statistics and Epidemiology*. Boca Raton, FL: Chapman & Hall/CRC.

Gustafson, P. (2007). Measurement error modeling with an approximate instrumental variable. *Journal of the Royal Statistical Society, Series B* **69**, 797-815.

Gutfreund, H. (1995). *Kinetics for the Life Sciences, Receptors, Transmitters, and Catalysts*. Cambridge, UK: Cambridge University Press.

Hampel, F.R., Ronchetti, E.M., Rousseeuw, P.J., and Stahel, W.A. (1986). *Robust Statistics*. New York: Wiley.

Hardle, W., Liang, H., and Gao, J. (2000). *Partially Linear Models*. Heidelberg: Springer.

Hartford, A. and Davidian, M. (2000). Consequences of misspecifying assumptions in nonlinear mixed effects models. *Computational Statistics and Data Analysis* **34**, 139-164.

Harville, D.A. (1977). Maximum likelihood approaches to variance component estimation and to related problems. *Journal of the American Statistical Association* **72**, 320-340.

Hastie, T. and Tibshirani, R. (1990). *Generalized Additive Models*. London: Chapman & Hall.

Heagerty, P. (1999). Marginally specified logistic-normal models for longitudinal binary data. *Biometrics* **55**, 688-698.

Hedeker, D. and Gibbons, R.D. (2006). *Longitudinal Data Analysis*, New York: Wiley-Interscience.

Henderson, R., Diggle, P.J., and Dobson, A. (2002). Joint modeling of longitudinal measurements and event time data. *Biostatistics* **1**, 465-480.

Henry, W.K., Tebas, P., and Lane, H.C. (2006). Explaining, predicting, and treating HIV-associated CD4 cell loss, after 25 years still a puzzle. *Journal of American Medical Association* **296**, 1523-1525.

Herring, A.H. and Ibrahim, J.G. (2001). Likelihood-based methods for missing covariates in the Cox proportional hazards model, *Journal of the American Statistical Association* **96**, 292-302.

Herring, A.H., Ibrahim, J.G., and Lipsitz, S.R. (2002). Frailty models with missing covariates. *Biometrics* **58**, 98-109.

Higgins, D.M., Davidian, M., and Giltinan, D.M. (1997). A two-step approach to measurement error in time-dependent covariates in nonlinear mixed-effects models, with application to IGF-I pharmacokinetics. *Journal of the American Statistical association* **92**, 436-448.

Ho, D.D., Neumann, A.U., Perelson, A.S., Chen, W., Leonard, J.M., and Markowitz, M. (1995). Rapid turnover of plasma virions and CD4 lymphocytes in HIV-1 infection. *Nature* **373**, 123-126.

Hogan, J.W. and Laird, N.M. (1997). Mixture models for the joint distribution of repeated measures and event times. *Statistics in Medicine* **16**, 239-258.

Horton, N.J. and Kleinman, K.P. (2007). Much ado about nothing: a comparison of missing data methods and software to fit incomplete data regression models. *The American Statistician* **61**, 79-90.

Hougaard, P. (1999). Fundamentals of survival data. *Biometrics* **55**, 13-22.

Hu, P., Tsiatis, A.A., and Davidian, M. (1998). Estimating the parameters in the Cox model when covariate variables are measured with error. *Biometrics* **54**, 1407-1419.

Huang, Y. and Lu, T. (2008). Modeling long-term longitudinal HIV dynamics with application to an AIDS clinical study. *The Annals of Applied Statistics* **2**, 1384-1408.

Huber, P.J. (1981). *Robust Statistics*. New York: Wiley.

Hughes, J.P. (1999). Mixed effects models with censored data with applications to HIV RNA levels. *Biometrics* **55**, 625-629.

Hughes, M.D. (2000). Analysis and design issues for studies using censored biomarker measurements with an example of viral load measurements in HIV clinical trials. *Statistics in Medicine* **19**, 3171-3191.

Ibrahim J.G. (1990). Incomplete data in generalized linear models. *Journal of the American Statistical Association* **85**, 765-769.

Ibrahim, J.G., Chen, M.H., and Lipsitz, S.R. (2001). Missing responses in generalized linear mixed models when the missing data mechanism is nonignorable. *Biometrika* **88**, 551-564.

Ibrahim, J.G., Lipsitz, S.R., and Chen, M.H. (1999). Missing covariates in generalized linear models when the missing data mechanism is nonignorable. *Journal of the Royal Statistical Society, Ser. B.* **61**, 173-190.

Ibrahim, J.G., Lipsitz, S.R., and Chen, M.H. (2002). Bayesian methods for generalized linear models with missing covariates. *The Canadian Journal of Statistics* **30**, 55-78.

Ibrahim, J.G., Chen, M.H., Lipsitz, S.R., and Herring, A.H. (2005). Missing data methods for generalized linear models: a comparative review. *Journal of the American Statistical Association* **100**, 332-346.

Jennrich, R.I. and Schluchter, M.D. (1986). Unbalanced repeated measures models with structural covariance matrix. *Biometrics* **42**, 805-820.

Jiang, J. (2007). *Linear and Generalized Linear Mixed Models and Their Applications*. New York: Springer.

Jiang, J., Rao, J.S., Gu, Z., and Nguyen, T. (2008). Fence methods for mixed model selection. *Annals of Statistics* **36**, 1669-1692.

Joe, H. (2008). Accuracy of Laplace approximation for discrete response mixed models. *Computational Statistics & Data Analysis* **52**, 5066-5074.

Johnson, N.L. and Kotz, S. (1972). *Distributions in Statistics: Continuous Multivariate Distributions*. New York: Wiley.

Kalbfleisch, J.D. and Prentice, R. L. (2002). *The Statistical Analysis of Failure Time Data,* 2nd edition, New York: John Wiley.

Ke, C. and Wang, Y. (2001). Semiparametric nonlinear mixed-effects models and their applications (with Discussions). *Journal of the American Statistical Association* **96**, 1272-1298.

Klein, J.P. (1992). Semiparametric estimation of random effects using the Cox model based on the EM algorithm. *Biometrics* **48**, 795-806.

Klein, J.P., Pelz, C., and Zhang, M. (1999). Modelling random effects for censored data by a multivariate normal regression model. *Biometrics* **55**, 497-506.

Ko, H.J. and Davidian, M. (2000). Correcting for measurement error in individual-level covariates in nonlinear mixed effects models. *Biometrics* **56**, 368-375.

Kotz, S. and Nadarajah, S. (2004). *Multivariate T Distributions and Their Applications*. Cambridge, UK: Cambridge University Press.

Lai, T.L. and Shih, M.C. (2003a). Nonparametric estimation in nonlinear mixed effects models. *Biometrika* **90**, 1-13.

Lai, T.L. and Shih, M.C. (2003b). A hybrid estimator in nonlinear and generalized linear mixed effects models. *Biometrika* **90**, 859-879.

Lai, T.L., Shih, M.C., and Wong, S.P.S. (2006). Flexible modeling via a hybrid estimation scheme in generalized mixed models for longitudinal data. *Biometrics* **62**, 159-167.

Laird, N.M., Lange, N., and Stram, D. (1987). Maximum likelihood computations with repeated measures: Application of the EM algorithm. *Journal of the American Statistical Association* **82**, 97-105.

Laird, N.M. and Ware, J.H. (1982). Random-effects models for longitudinal data. *Biometrics* **38**, 963-974.

Lange, K.L., Little, R.J.A., and Taylor, J.M.G. (1989). Robust statistical modeling using the t distribution. *Journal of the American Statistical Association* **84**, 881-896.

Lauritzen, S. (1996). *Graphical Models*. Oxford: Oxford University Press.

Lawless, J.F. (2003). *Statistical Models and Methods for Lifetime Data*, 2nd edition. Hoboken, NJ: Wiley.

Lederman, M.M., Connick, E., Landay, A., Kuritzkes, D.R., Spritzler, J., St. Clair, M., Kotzin, B.L., Fox, L., Chiozzi, M.H., Leonard, J.M., Rousseau, F., Wade, M., Roe, J.D., Martinez, A., and Kessler, H. (1998). Immunologic responses associated with 12 weeks of combination antiretroviral therapy consisting of zidovudine, lamivudine and ritonavir: results of AIDS Clinical Trials Group Protocol 315. *The Journal of Infectious Diseases* **178**, 70-79.

Lee, Y., Nelder, J.A., and Pawitan, Y. (2006). *Generalized Linear Models with Random Effects: Unified Analysis via H-likelihood*. London: Chapman & Hall/CRC.

Levine, R.A. and Casella, G. (2001). Implementations of the Monte Carlo EM algorithm. *Journal of Computational and Graphical Statistics* **10**, 422-439.

Li, H. and Thompson, E. (1997). Semiparametric estimation of major gene and family-specific random effects for age of onset. *Biometrics* **53**, 282-293.

Li, K.H., Raghunathan T.E., and Rubin D.B. (1991). Large-sample significance levels from multiply imputed data using moment-based statistics and an F reference distribution. *Journal of the American Statistical Association* **86**, 1065-1073.

Li, Y. and Lin, X. (2003) Functional inference in frailty measurement error models for clustered survival data using the SIMEX approach. *Journal of the American Statistical Association* **98**, 191-204.

Liang, H., Wu, H., and Carroll, R. (2003). The relationship between virologic and immunologic responses in AIDS clinical research using mixed-effects varying-coefficient models with measurement errors. *Biostatistics* **4**, 297-312.

Liang, K.Y., Self, S.G., Bandeen-Roche, K.J., and Zeger, S.L. (1995). Some recent developments for regression analysis of multivariate failure time data. *Lifetime Data Analysis* **1**, 403-415.

Liang, K.Y. and Zeger, S.L. (1986). Longitudinal data analysis using generalized linear models. *Biometrika* **73**, 13–22.

Lin, X. and Breslow, N.E. (1996). Bias correction in generalized linear mixed models with multiple components of dispersion. *Journal of the American Statistical Association* **91**, 1007-1016.

Lin, X. and Carroll, R.J. (2000). Nonparametric function estimation for clustered data when the predictor is measured without/with error. *Journal of the American Statistical Association* **95**, 520-534.

Lin, X. and Zhang, D. (1999). Inference in generalized additive mixed models by using smoothing splines. *Journal of the Royal Statistical Society, Series B* **61**, 381-400.

Lindsey, J.K. (2001). *Nonlinear Models in Medical Statistics*. Oxford: Oxford University Press.

Lindstrom, M.J. and Bates, D.M. (1988). Newton-Raphson and EM algorithms for linear mixed-effects models for repeated-measures data. *Journal of the American Statistical Association* **83**, 1014-1022.

Lindstrom, M.J. and Bates, D.M. (1990). Nonlinear mixed effects models for repeated measures data. *Biometrics* **46**, 673-687.

Lipsitz, S., Laird, N., and Harrington, D. (1991). Generalized estimation equations for correlated binary data: Using the odds ratio as a measure of association in unbalanced mixed models with nested random effects. *Biometrika* **78**, 153-160.

Littell, R.C., Milliken, G.A., Stroup, W.W., Wolfinger, R.D., and Schabenberger, O. (2006). *SAS for Mixed Models*, 2nd edition. Cary, NC: SAS Publishing.

Littell, R.C., Stroup, W.W., and Freund, R.J. (2002). *SAS for Linear Models*, 4th edition. Cary, NC: SAS Publishing.

Little, R.J.A. (1992). Regression with missing X's: a review. *Journal of the American Statistical Association* **87**, 1227-1237.

Little, R.J.A. (1993). Pattern-mixture models for multivariate incomplete data. *Journal of the American Statistical Association* **88**, 125-134.

Little, R.J.A. (1994). A class of pattern-mixture models for normal incomplete data. *Biometrika* **81**, 471-483.

Little, R.J.A. (1995). Modeling the drop-out mechanism in repeated measures studies. *Journal of the American Statistical Association* **90**, 1112-1121.

Little, R.J.A. and Rubin, D.B. (2002). *Statistical Analysis with Missing Data*, 2nd edition. New York: Wiley.

Liu, C., Rubin, D.B., and Wu, Y.N. (1998). Parameter expansion for EM acceleration: The PX-EM algorithm. *Biometrika* **85**, 755-770.

Liu, Q. and Pierce, D.A. (1994). A note on Gauss-Hermite quadrature. *Biometrika* **81**, 624-629.

Liu, W. and Wu, L. (2007). Simultaneous inference for semiparametric nonlinear mixed-effects models with covariate measurement errors and missing responses. *Biometrics* **63**, 342-350.

Liu, W. and Wu, L. (2009). Two-step and likelihood methods for semiparametric nonlinear mixed effects models with measurement errors and missing covariates. *Technical report*, Department of Mathematics and Statistics, York University.

Louis, T.A. (1982). Finding the observed information matrix when using the EM algorithm. *Journal of the Royal Statistical Society, Series B* **44**, 226-233.

Magnus, J.R. and Neudecker, H. (1988). *Matrix Differential Calculus with Applications in Statistics and Econometrics*. Chichester, UK: Wiley.

Mallows, C.L. (1975). On some topics in robustness. Unpublished memorandum, Bell Telephone Laboratories, Murray Hill, NJ.

Maronna, R., Martin, D., and Yohai, V., (2006). *Robust Statistics – Theory and Methods*. New York: Wiley.

McCullagh, P. (1983). Quasi-likelihood functions. *Annals of Statistics* **11**, 59-67.

McCullagh, P. and Nelder, J.A. (1989). *Generalized Linear Models*, 2nd edition. New York: Chapman & Hall.

McCulloch, C.E. (1994). Maximum likelihood variance components estimation for binary data. *Journal of the American Statistical Association* **89**, 330-335.

McCulloch, C.E. (1997). Maximum likelihood algorithms for generalized linear mixed models. *Journal of the American Statistical Association* **92**, 162-170.

McCulloch, C.E., Searle, S.R., and Neuhaus, J.M. (2008). *Generalized, Linear, and Mixed Models*, 2nd edition. New York: Wiley.

McLachlan, G.J. and Krishnan, T. (1997). *The EM-Algorithm and Extension*. New York: Wiley.

Meng, X.L. and Rubin, D.B. (1991). Using EM to obtain asymptotic variance covariance matrices: The SEM algorithm. *Journal of the American Statistical Association* **86**, 899-909.

Meng X.L. and Rubin D.B. (1992). Performing likelihood ratio tests with multiply-imputed data sets. *Biometrika* **79**, 103-111.

Meng, X.L. and Rubin, D.B. (1993). Maximum likelihood estimation via the ECM algorithm: a general framework. *Biometrika* **80**, 267-278.

Meng, X.L. and van Dyk, D.A. (1997). The EM algorithm — an old folk-song sung to a fast new tune (with discussion). *Journal of the Royal Statistical Society, Series B* **59**, 511-567.

Meng, X.L. and van Dyk, D.A. (1998). Fast EM implementations for mixed-effects models. *Journal of the Royal Statistical Society, Series B* **60**, 559-578.

Miller, J.J. (1977). Asymptotic properties of maximum likelihood estimates in the mixed model of the analysis of variance. *The Annals of Statistics* **5**, 746-762.

Molenberghs, G. and Kenward, M.G. (2007). *Missing Data in Clinical Studies*. Chichester, UK: Wiley.

Murphy, S.A. (1995). Asymptotic theory for the frailty model. *Annals of Statistics* **23**, 182-198.

Murphy, S., Das Gupta, A., Cain, K.C., Johnson, L.C. Lohan, J., Wu, L., and Mekwa, J. (1999). Changes in parents' mental distress after the violent death of an adolescent or young adult child: a longitudinal prospective analysis. *Death Studies* **23**, 129-159.

Murphy, S.A., Johnson, L.C., Wu, L., Fan, J.J., Lohan, J. (2003). Bereaved parents' outcomes 4 to 60 months after their children's deaths by accident, suicide, or homicide: a comparative study demonstrating differences. *Death Studies* **27**, 39-61.

Nathoo, F.S. and Dean, C.B. (2008). Spatial multi-state transitional models for longitudinal event data. *Biometrics* **64**, 271-279.

Nielsen, G.G., Gill, R.D., Andersen, P.K., and Sorensen, T.I.A. (1992). A counting process approach to maximum likelihood estimation in frailty models. *Scandinavian Journal of Statistics* **19**, 25-43.

Noh, M. and Lee, Y. (2007). Robust modeling for inference from generalized linear model classes. *Journal of the American Statistical Association* **102**, 1059-1072.

Oakes, D. (1989). Bivariate survival models induced by frailties. *Journal of the American Statistical Association* **84**, 487-493.

Ogden, R.T. and Tarpey, T. (2006). Estimation in regression models with externally estimated parameters. *Biostatistics* **7**, 115-129.

Paxton, W.B., Coombs, R.W., McElrath, M.J. Keefer, M.C., Hughes, J., Sinangil, F., Chernoff, D., Demeter, L., Williams, B., and Corey, L. (1997). Longitudinal analysis of quantitative virologic measures in human immunodeficiency virus – infected subjects with $\geq$ 400 CD4 lymphocytes: implications for applying measurements to individual patients. *Journal of Infectious Diseases* **175**, 247-254.

Perelson, A.S., Essunger, P., Cao, Y., Vesanen, M., Hurley, A., Saksela, K., Markowitz, M., and Ho, D.D. (1997). Decay characteristics of HIV-1-infected compartments during combination therapy. *Nature* **387**, 188-191.

Perelson, A.S. and Nelson, P.W. (1999). Mathematical analysis of HIV-1 dynamics *in vivo*. *SIAM Review* **41**, 3-44.

Perelson, A.S., Neumann, A.U., Markowitz, M., Leonard, J.M., and Ho, D.D. (1996). HIV-1 dynamics *in vivo*: virion clearance rate, infected cell lifespan, and viral generation time. *Science* **271**, 1582-1586.

Perlman M.D. (1969). One-sided testing problems in multivariate analysis. *Annals of Mathematical Statistics* **40**, 549-567.

Perlman, M.D. and Wu, L. (1999a). The emperor's new tests (with Discussion). *Statistical Science* **14**, 355-369.

Perlman, M.D. and Wu, L. (1999b). Lattice conditional independence models for contingency tables with missing data. *Journal of Statistical Planning and Inference* **79**, 259-287.

Perlman, M.D. and Wu, L. (2002). A class of conditional tests for a multivariate one-sided alternative. *Journal of Statistical Planning and Inference* **107**, 155-171.

Pettitt, A.N. (1986). Censored observations, repeated measures and mixed effects models: an approach using the EM algorithm and normal errors. *Biometrika* **73**, 635-643.

Pierce, D.A. and Kellerer, A.M. (2004). Adjusting for covariate errors with nonparametric assessment of the true covariate distribution. *Biometrika* **91**, 863-876.

Pinheiro, J.C. and Bates, D.M. (1995). Approximations to the log-likelihood function in the nonlinear mixed-effects model. *Journal of Computational and Graphical Statistics* **4**, 12-35.

Pinheiro, J.C. and Bates, D.M. (2002). *Mixed-Effects Models in S and S-PLUS*. New York: Springer.

Pinheiro, J.C., Liu, C., and Wu, Y.N. (2001). Efficient algorithms for robust estimation in linear mixed effects models using the multivariate t-distribution. *Journal of Computational and Graphical Statistics* **10**, 249-276.

Preisser, J.S. and Qaqish, B.F. (1999). Robust regression to clustered data with application to binary responses. *Biometrics* **55**, 574-579.

Prentice, R.L. (1982). Covariate measurement errors and parameter estimates in a failure time regression model. *Biometrika* **69**, 331-342.

Prentice, R.L. (1988). Correlated binary regression with covariates specific to each binary observation. *Biometrics* **44**, 1033-1084.

Press, W.H., Flannery, B.P., Teukolsky, S.A., and Vetterling, W.T. (2007). *Numerical Recipes: The Art of Scientific Computing*, Cambridge, UK: Cambridge University Press.

Qu, A., Lindsay, B.G., and Li, B. (2000). Improving generalized estimating equations using quadratic inference functions. *Biometrika* **87**, 823-836.

Raghunathan, T.E., Lepkowski, J.M., van Hoewyk, J., and Solenberger, P. (2001). A multivariate technique for multiply imputed missing values using a sequence of regression models. *Survey Methodology* **27**, 85-95.

Ramsay, J.O., Hooker, G., Campbell, D., and Cao, J. (2007), Parameter estimation for differential equations: a generalized smoothing approach (with Discussion), *Journal of the Royal Statistical Society, Series B* **69**,741-796.

Ramsay, J.O. and Silverman, B.W. (2005). *Functional Data Analysis*, 2nd edition. New York: Springer.

Rao, J.N.K. and Shao, J. (1992). Jackknife variance estimation with survey data under hot deck imputation. *Biometrika* **79**, 811-822.

Raudenbush, S.W., Yang, M.L., and Yosef, M. (2000). Maximum likelihood for generalized linear models with nested random effects via high-order, multivariate Laplace approximation. *Journal of Computational and Graphical Statistics* **9**, 141-157.

Rice, J.A. and Wu, C.O. (2001). Nonparametric mixed-effects models for unequally sampled noisy curves. *Biometrics* **57**, 253-259.

Richardson, A.M. (1997). Bounded influence estimation in the mixed linear model. *Journal of the American Statistical Association* **92**, 154-161.

Robert, C.P. and Casella, G. (2004). *Monte Carlo Statistical Methods*, 2nd edition. New York: Springer-Verlag.

Robins, J.M. and Rotnitzky, A. (1995). Semiparametric efficiency in multivariate regression models with missing data. *Journal of the American Statistical Association* **90**, 122-129.

Robins, J.M., Rotnitzky, A., and Zhao, L.P. (1994). Estimation of regression coefficients when some regressors are not always observed. *Journal of the American Statistical Association* **89**, 846-866.

Robins, J.M., Rotnitzky, A., and Zhao, L.P. (1995). Analysis of semiparametric regression models for repeated outcomes in the presence of missing data. *Journal of the American Statistical Association* **90**, 106-121.

Rodriguez, B., Sethi, A.K., Cheruvu, V.K., Mackay, W., Bosch, R.J., Kitahata, M., Boswell, S.L., Mathews, W.C., Bangsberg, D.R., Martin, J., Whalen, C.C., Sieg, S., Yadavalli, S., Deeks, S.G., and Lederman, M.M. (2006). Predictive value of plasma HIV RNA level on rate of CD4 T-cell decline in untreated HIV infection. *Journal of American Medical Association* **296**, 1498-1525.

Rousseeuw, P.J. and Leroy, A.M. (1987). *Robust Regression and Outlier Detection*. New York: Wiley.

Rubin, D.B. (1976). Inference and missing data. *Biometrika* **63**, 581-592.

Rubin, D.B. (1987). *Multiple Imputation for Nonresponse in Sample Surveys*. New York: Wiley.

Rubin, D.B. (1996). Multiple imputation after 18+ years. *Journal of the American Statistical Association* **91**, 473-489.

Ruppert, D., Wand, M.P., and Carroll, R.J. (2003). *Semiparametric Regression*. Cambridge, UK: Cambridge University Press.

Sakamoto, Y., Ishiguro, M., and Kitagawa, G. (1986). *Akaike Information Criterion Statistics*, Dordrecht: Reidel.

Santner, T.J. and Duffy, D.E. (1989). *The Statistical Analysis of Discrete Data*. New York: Springer-Verlag.

Schwarz, G. (1978). Estimating the dimension of a model. *Annals of Statistics* **6**, 461-464.

Schafer, J.L. (1997). *Analysis of Incomplete Multivariate Data*. London: Chapman & Hall.

Schafer, J.L. and Schenker, N. (2002). Inference with imputed conditional means. *Journal of American Medical Association* **95**, 144-154.

Schafer, J.L. and Yucel, R.M. (2002). Computational strategies for multivariate linear mixed models with missing values. *Journal of Computational and Graphical Statistics* **11**, 421-442.

Searle, S.R., Casella, G., and McCulloch, C.E. (1992). *Variance Components*. New York: Wiley.

Seber, G.A.F. and Wild, C.J. (2003). *Nonlinear Regression*. New York: John Wiley.

Shah, A., Laird, N., and Schoenfeld, D. (1997). A random-effects model for multiple characteristics with possibly missing data. *Journal of the American Statistical Association* **92**, 775-779.

Shao, J. (2002). Replication methods for variance estimation in complex surveys with imputed data, Chapter 20, in *Survey Nonresponse* (R.M. Groves, D.A. Dillman, J.L. Eltinge, and R.J.A. Littel, eds.), New York: Wiley.

Shi, M., Weiss, R.E., and Taylor, J.M.G. (1996). An analysis of pediatric aids cd4 counts using flexible random curves. *Applied Statistics* **45**, 151-164.

Silvapulle, M.J. (1997a). A curious example involving the likelihood ratio test against one-sided hypotheses. *American Statistician* **51**, 178-180.

Silvapulle, M.J. (1997b). On order restricted inference in some mixed linear models. *Statistics and Probability Letters* **36**, 23-27.

Silvapulle, M.J. and Sen, P.K. (2004). *Constrained Statistical Inference: Inequality, Order, and Shape Restrictions*. New York: Wiley.

Sinha, S.K. (2004). Robust analysis of generalized linear mixed models. *Journal of the American Statistical Association* **99**, 451-460.

Solomon, P.J. and Cox, D.R. (1992). Nonlinear component of variance models. *Biometrika* **79**, 1-11.

Song, P.X.K., Zhang, P., and Qu, A. (2007). Maximum likelihood inference in robust linear mixed-effects models using multivariate t-distributions. *Statistica Sinica* **17**, 929-943.

Song, X. and Wang, C.Y. (2008), Semiparametric approaches for joint modeling of longitudinal and survival data with time-varying coefficients. *Biometrics* **64**, 557-566.

Stram, D.O. and Lee, J.W. (1994). Variance components testing in the longitudinal mixed effects model. *Biometrics* **50**, 1171-1177.

Stubbendick, A.L. and Ibrahim, J.G. (2003). Maximum likelihood methods for nonignorable missing responses and covariates in random effects models. *Biometrics* **59**, 1140-1150.

Tamhane, A.C. and Logan, B.R. (2002). Accurate critical constants for the one-sided approximate likelihood ratio test of a normal mean vector when the covariance matrix is estimated. *Biometrics* **58**, 650-656.

Tamhane, A.C. and Logan, B.R. (2004). A superiority-equivalence approach to one-sided tests on multiple endpoints in clinical trials. *Biometrika* **91**, 715-727.

Tanner, M.A. and Wong, W.H. (1987). The calculation of posterior distributions by data augmentation. *Journal of the American Statistical Association* **82**, 528-540.

Taylor, J.M.G., Cumberland, W.G., and Sy, J.P. (1994). A stochastic model for analysis of longitudinal data. *Journal of the American Statistical Association* **89**, 727-776.

Tchetgen, E.J. and Coull, B.A. (2006). A diagnostic test for the mixing distribution in a generalized linear mixed model. *Biometrika* **93**, 1003-1010.

Ten Have, T.R., Pulkstenis, E., Kunselman, A., and Landis, J.R. (1998). Mixed effects logistics regression models for longitudinal binary response data with informative dropout. *Biometrics* **54**, 367-383.

Tibshirani, R. and Hastie, T. (1987). Local likelihood estimation. *Journal of the American Statistical Association* **82**, 559-567.

Tierney, L. and Kadane, J.B. (1986). Accurate approximations for posterior moments and densities. *Journal of the American Statistical Association* **81**, 82-86.

Tseng, Y.K., Hsieh, F., and Wang, J.L. (2005). Joint modelling of accelerated failure time and longitudinal data. *Biometrika* **92**, 587-603.

Tsiatis, A.A. and Davidian, M. (2001). A semiparametric estimator for the proportional hazards model with longitudinal covariates measured with error. *Biometrika* **88**, 447-458.

Tsiatis, A.A. and Davidian, M. (2004). An overview of joint modeling of longitudinal and time-to-event data. *Statistica Sinica* **14**, 793-818.

Tsiatis, A.A., DeGruttola, V., and Wulfsohn, M.S. (1995). Modeling the relationship of survival to longitudinal data measured with error. Applications to survival and CD4 counts in patients with AIDS. *Journal of the American Statistical Association* **90**, 27-37.

van Buuren, S. (2006). Multiple imputation online. Available at the following website: *http://www.multiple-imputation.com*.

van Buuren, S., Brand, J.P.L., Groothius-Oudshoorn, C.G.M., and Rubin, D.B. (2006). Fully conditional specification in multivariate imputation. *Journal of Statistical Computation and Simulation* **76**, 1049-1064.

van Eeden, C. (2006). *Restricted-Parameter-Space Estimation Problems - Admissibility and Minimaxity Properties*. New York: Springer.

van Dyk, D.A. (2000). Fitting mixed-effects models using efficient EM-type algorithms. *The Journal of Computational and Graphical Statistics* **9**, 78-98.

Venables, W.N. and Ripley, B.D. (2003). *Modern Applied Statistics with S*, 4th edition. New York: Springer.

Verbeke, G. and Molenberghs, G. (2001). *Linear Mixed Models for Longitudinal Data*. New York: Springer.

von Hippel, P. (2004). Biases in SPSS 12.0 missing value analysis. *The American Statistician* **58**, 160-164.

Vonesh, E.F. (1996). A note on the use of Laplace's approximation for non-linear mixed-effects models. *Biometrika* **83**, 447-452.

Vonesh, E.F. and Chinchilli, V.M. (1997). *Linear and Nonlinear Models for the Analysis of Repeated Measurements*. New York: Marcel Dekker.

Vonesh, E.F., Wang, H., Nie, L., and Majumdar, D. (2002). Conditional second-order generalized estimating equations for generalized linear and nonlinear mixed-effects models. *Journal of the American Statistical Association* **97**, 271-283.

Waagepetersen, R. (2006). A simulation-based goodness-of-fit test for random effects in generalized linear mixed models. *Scandinavian Journal of Statistics* **33**, 721-731.

Wakefield, J.C. (1995). The Bayesian analysis of population pharmacokinetic models. *Journal of the American Statistical Association* **91**, 61-76.

Wakefield, J.C., Smith, A.F.M., Racine-Poon, A., and Gelfand, A.E. (1994). Bayesian analysis of linear and non-linear population models using the Gibbs sampler. *Applied Statistics* **43**, 201-221.

Walker, S.G. (1996). An EM algorithm for nonlinear random effects models. *Biometrics* **52**, 934-944.

Wand, M.P. (2002). Vector differential calculus in statistics. *The American Statistician* **56**, 1-8.

Wang, C.Y., Wang, N., and Wang, S. (2000). Regression analysis when co-variates are regression parameters of a random effects model for observed longitudinal measurements. *Biometrics* **56**, 487-495.

Wang, L. (2004). Estimation of nonlinear models with Berkson measurement errors. *The Annals of Statistics* **32**, 2559-2579.

Wang, N. (2003) Marginal nonparametric kernel regression accounting for within-subject correlation. *Biometrika* **90**, 43-52.

Wang, N., Carroll, R.J., and Lin, X. (2005). Efficient semiparametric marginal estimation for longitudinal/clustered data. *Journal of the American Statistical Association* **100**, 147-157.

Wang, N., Lin, X., Gutierrez, R.G., and Carroll, R.J. (1998). Generalized linear mixed measurement error models. *Journal of the American Statistical Association* **93**, 249-261.

Wang, W. and Heckman, N. (2009). Identifiability in linear mixed models. *Technical report*, Department of Statistics, University of British Columbia.

Wang, Y. and Taylor, J.M.G. (2001). Jointly modeling longitudinal and event time data with application to acquired immunodeficiency syndrome. *Journal of the American Statistical Association* **96**, 895-905.

Wedderburn, R.W.M. (1974). Quasilikelihood functions, generalized linear models, and the Gauss-Newton method. *Biometrika* **61**, 439-447.

Wei, G.C. and Tanner, M.A. (1990). A Monte-Carlo implementation of the EM algorithm and the poor man's data augmentation algorithm. *Journal of the American Statistical Association* **85**, 699-704.

Weisberg, S. (2005). *Applied Linear Regression*, 3rd edition. Hoboken, NJ: Wiley-Interscience.

Wolfinger, R. (1993). Laplace's approximation for nonlinear mixed models. *Biometrika* **80**, 791-795.

Wolfinger, R. and Lin, X. (1997). Two Taylor-series approximation methods for nonlinear mixed models. *Computational Statistics and Data Analysis* **25**, 465-490.

Wu, C.F.J. (1983). On the convergence properties of the EM algorithm. *Annals of Statistics* **11**, 95-103.

Wu, H. (2005). Statistical methods for HIV dynamic studies in AIDS clinical trials. *Statistical Methods in Medical Research* **14**, 171-192.

Wu, H. and Ding, A. (1999). Population HIV-1 dynamics in vivo: applicable models and inferential tools for virological data from AIDS clinical trials. *Biometrics* **55**, 410-418.

Wu, H. and Wu, L. (2002). Identification of significant host factors for HIV dynamics models by nonlinear mixed-effect models. *Statistics in Medicine* **21**, 753-771.

Wu, H. and Zhang, J. (2002). The study of long-term HIV dynamics using semi-parametric non-linear mixed-effects models. *Statistics in Medicine* **21**, 3655-3675.

Wu, H. and Zhang, J. (2006). *Nonparametric Regression Methods for Longitudinal Data Analysis*. New York: Wiley-Interscience.

Wu, L. (2002). A joint model for nonlinear mixed-effects models with censoring and covariates measured with error, with application to AIDS studies. *Journal of the American Statistical Association* **97**, 955-964.

Wu, L. (2004a). Exact and approximate inferences for nonlinear mixed-effects models with missing covariates. *Journal of the American Statistical Association* **99**, 700-709.

Wu, L. (2004b). A unified approach for linear mixed effects models with censored response and measurement errors and missing data in covariates. *Statistics in Medicine* **23**, 1715-1731.

Wu, L. (2007). HIV viral dynamic models with dropouts and missing covariates. *Statistics in Medicine* **26**, 3342-3357.

Wu, L. (2009). Approximate bounded influence estimation for incompletely observed longitudinal data. Technical report, Department of Statistics, University of British Columbia.

Wu, L. and Gilbert, P. (2002). Flexible weighted log-rank tests optimal for detecting early and/or late survival differences. *Biometrics* **85**, 97-104.

Wu, L., Hu, J., and Wu, H. (2008). Joint inference for nonlinear mixed effects models and time-to-event at the presence of missing data, *Biostatistics* **9**, 308-320.

Wu, L., Liu, W., and Hu, J. (2009). Joint inference on HIV viral dynamics and immune suppression in presence of measurement errors. *Biometrics*, in press.

Wu, L., Liu, W., and Liu, J. (2009). A longitudinal study of children's aggressive behaviors and their relationships with potential predictors. *Canadian Journal of Statistics*, in press.

Wu, L. and Wu, H. (2002). Missing time-dependent covariates in HIV viral dynamic models. *Journal of the Royal Statistical Society, Ser. C (Applied Statistics)* **51**, 297-318.

Wu, M.C. and Bailey, K.R. (1989). Estimation and comparison of changes in the presence of informative right censoring: conditional linear model. *Biometrics* **45**, 939-955.

Wu, M.C. and Carroll, R.J. (1988). Estimation and comparison of changes in the presence of informative right censoring by modeling the censoring process. *Biometrics* **44**, 175-188.

Wu, M.C. and Follmann, D.A. (1999). Use of summary measurements to adjust for informative missingness repeated measures data with random effects. *Biometrics* **55**, 75-84.

Wulfsohn, M.S. and Tsiatis, A.A. (1997). A joint model for survival and longitudinal data measured with error. *Biometrics* **53**, 330-339.

Xu, J. and Zeger, S.L. (2001). Joint analysis of longitudinal data comprising repeated measures and times to events. *Applied Statistics* **50**, 375-387.

Ye, W., Lin, X., and Taylor, J.M.G. (2008). Semiparametric modeling of longitudinal measurements and time-to-event data – a two-stage regression calibration approach. *Biometrics* **64**, 1238-1246.

Yeap, B.Y. and Davidian, M. (2001). Robust two-stage estimation in hierarchical nonlinear models. *Biometrics* **57**, 266-272.

Yi, G.Y. (2008). A simulation-based marginal method for longitudinal data with dropout and mismeasured covariates. *Biostatistics* **9**, 501-512.

Yi, G.Y. and Cook, R.J. (2002). Marginal methods for incomplete longitudinal data arising in clusters. *Journal of the American Statistical Association* **97**, 1071-1080.

Yi, G.Y., Liu, W., and Wu, L. (2009). Bias analysis and simultaneous inference for longitudinal data with covariate measurement error and missing responses under generalized linear mixed models. Technical report, Department of Statistics, University of Waterloo.

Zhang, D., Lin, X., Raz, J., and Sowers, M. (1998). Semiparametric stochastic mixed models for longitudinal data. *Journal of the American Statistical Association* **93**, 710-719.

Zeger, S.L. and Karim, M.R. (1991). Generalized linear models with random effects: a Gibbs sampling approach. *Journal of the American Statistical Association* **86**, 79-95.

Zeger, S.L. and Liang, K.Y. (1986). Longitudinal data analysis for discrete and continuous outcomes. *Biometrics* **42**, 121-130.

Zeng, D. and Cai, J. (2005). Asymptotic results for maximum likelihood estimators in joint analysis of repeated measurements and survival time. *The Annals of Statistics* **33**, 2132-2163.

Zeng, L. and Cook, R.J. (2007). Transition models for multivariate longitudinal binary data, *Journal of the American Statistical Association* **102**, 211-223.

Zidek, J.V., Le, N.D., Wong, H., and Burnett, R.T. (1998). Including structural measurement errors in the nonlinear regression analysis of clustered data. *The Canadian Journal of Statistics* **26**, 537-548.

Index

Abstract

In the analysis of longitudinal data or clustered data, mixed effects models have been widely used. In practice, data are often complex in the sense that there may be missing values, dropouts, measurement errors, censoring, and outliers. These problems are especially common in longitudinal studies. Statistical analyses ignoring any of these problems or the use of naive methods in data analysis may lead to biased or misleading results. This book provides an overview of commonly used mixed effects models and discusses methods to appropriately handle the common complications in practice. The book focuses on likelihood methods, but other methods are also discussed, including generalized estimating equations (GEEs) methods and Bayesian methods. Specifically, the book

- covers the following mixed effects models: linear mixed effects (LME) models, generalized linear mixed models (GLMMs), nonlinear mixed effects (NLME) models, frailty models, and semiparametric and nonparametric mixed effects models;
- describes common approaches to handle missing data, dropouts, measurement errors, censoring, and outliers, including EM algorithms, multiple imputations, and M-estimators;
- reviews joint modeling of longitudinal data and survival data;
- discusses different approaches for analysis of longitudinal data or clustered data.

The book should be accessible to graduate students or senior undergraduate students in statistics, practitioners, researchers in statistics and biostatistics, and researchers in other fields with particular interest in mixed effects models. The reader is assumed to have a basic knowledge of statistical methods and some familiarity with regression models. The mathematics level is mostly intermediate, at the level of first-year college calculus and linear algebra.

Lang Wu is an associate professor in the Department of Statistics at the University of British Columbia, Vancouver, Canada. He received a Ph.D. in statistics from the University of Washington at Seattle.

Lightning Source UK Ltd.
Milton Keynes UK
UKHW020821010921
389819UK00008B/96